METHODS IN MOLECULAR BIOLOGY

Series Editor
John M. Walker
School of Life and Medical Sciences
University of Hertfordshire
Hatfield, Hertfordshire, AL10 9AB, UK

For further volumes:
http://www.springer.com/series/7651

Neisseria gonorrhoeae

Methods and Protocols

Edited by

Myron Christodoulides

Molecular Microbiology Group, Academic Unit of Clinical and Experimental Sciences, Faculty of Medicine, University of Southampton, Southampton, UK

Humana Press

Editor
Myron Christodoulides
Molecular Microbiology Group
Academic Unit of Clinical and Experimental Sciences
Faculty of Medicine
University of Southampton
Southampton, UK

ISSN 1064-3745 ISSN 1940-6029 (electronic)
Methods in Molecular Biology
ISBN 978-1-4939-9495-3 ISBN 978-1-4939-9496-0 (eBook)
https://doi.org/10.1007/978-1-4939-9496-0

Cover Caption: *Neisseria gonorrhoeae* infection remodels microvilli in human 3-dimensional (3-D) endometrial cell model. Scanning electron microscopy image of 3-D human cells infected with *N. gonorrhoeae* MS11 at a multiplicity of infection of 10 for 4h. Gonococci were pseudo-colored red using Photoshop C5.1 software (Adobe). Image courtesy of Paweł Łaniewski and Melissa M. Herbst-Kralovetz.

This Humana Press imprint is published by the registered company Springer Science+Business Media, LLC part of Springer Nature
The registered company address is: 233 Spring Street, New York, NY 10013, U.S.A.

Preface

Introduction

"Gonorrhoea is as old as the world" (George Luys) [1]. *Neisseria gonorrhoeae* (gonococcus) is an obligate human pathogen that has coexisted with humanity and causes the sexually transmitted disease gonorrhea. Gonococci infect principally the mucosal epithelium of the genitourinary tract and can also infect anorectal and pharyngeal mucosal surfaces [2]. In men, infection of the urethra causes urethritis and painful discharge, and untreated infection may cause epididymitis leading to infertility. In women, localized infection of the ectocervix and endocervix leads to a mucopurulent cervicitis: however, infection is frequently asymptomatic, and in ~10–25% of untreated women, gonococci can ascend into the upper reproductive tract (URT). The host response to ascending gonococcal infection is pelvic inflammatory disease syndrome, which is an umbrella term for severe inflammatory conditions affecting all parts of the URT, e.g., endometritis, pelvic peritonitis (tubal, ovarian), and fallopian tube salpingitis [3]. Disseminated infection is rare but can present as arthritis, perihepatitis, meningitis, or endocarditis. Long-term and permanent sequelae resulting from untreated infection include chronic pelvic pain, tubal damage, ectopic pregnancy, and infertility. Chorioamnionitis, septic abortion, preterm delivery, and premature rupture of membranes can occur in infected pregnant women. Gonococci can also infect neonates by vertical transmission during birth and cause *ophthalmia neonatorum* (neonatal conjunctivitis) that leads to irreparable corneal tissue destruction and blindness [2]. In addition, there is a strong association between maternal gonorrhea with premature delivery and low neonatal birth weight [4]. More rarely observed in neonates are abscesses, meningitis, and sepsis.

The WHO estimates that ~78 million people are infected with gonorrhea worldwide, which is probably an underestimate (http://www.who.int/en/news-room/fact-sheets/detail/sexually-transmitted-infections-(stis)) [5]. Antibiotics have been tremendously successful for treating gonorrhea, but treatment is now severely compromised by the emergence of gonococci resistant to "last-resort" antibiotics [6]. Gonococci are now colloquially referred to as a "superbug" and listed by the WHO as a high-priority pathogen for research and development of new antimicrobials and vaccines. Despite the good intentions of global initiatives and funding for developing antimicrobials, the current pipeline for new anti-gonococcal treatments is dismal. Potentially untreatable gonorrhea has arrived, and prevention through vaccination is now a priority [7], as highlighted by the summary and recommendations of a NIAID workshop on gonorrhea vaccines held in 2015 [8].

A Brief History of Gonorrhea

Sexually transmitted diseases (STDs), known also as venereal diseases (VD), as derived from Venus, the Roman goddess of love, are known to have existed from antiquity and are recorded in many of humanity's earliest surviving texts. The compelling descriptions offered by these texts suggest that sexually transmitted infections such as syphilis, gonorrhea/chlamydia, herpes, and genital warts have been a constant scourge to human health.

However, in the absence of microbiological proof, the provenance of many of these diseases could be attributable to fungal, protozoan, and ectoparasitic infection.

There is sufficient circumstantial evidence that gonorrhea existed as an STD in early human cultures. As a starting reference point, the George Ebers papyrus, which was found in a temple at the city of Abydos in upper Egypt, is believed to date from 1555 BC but most certainly reflects circumstances dating back to 3000 BC (http://digi.ub.uniheidelberg.de/diglit/ebers1875bd1, http://digi.ub.uni-heidelberg.de/diglit/ebers1875bd2) [9]. Gynecology is a major theme of this Egyptian text, and mention is made of pelvic (inflammatory) disease (e.g., salpingitis, pelvic adhesions), leukorrhea (mucus discharge from the vagina), and "burning micturition," all clinical signs suggestive of gonorrhea infection [10]. Unfortunately, there are no Egyptian records of painful urination accompanying urethral discharge in men, although orchitis, a reference to epididymitis and suggestive of gonorrhea, is mentioned [10]. However, urethral gonorrhea is a recognized disease in the Bible, wherein it is commonly described as the "issue" and explicit measures are provided for disease control. In Leviticus 15:2–12, there is a description of male discharge that can render him "unclean," and in Leviticus 22:4, "any descendent. . .who has a male discharge may not eat any sacred offerings." Recommendations for disease control included social exclusion (quarantine) for 7 days after cessation of the discharge and disinfection of the patient and belongings [11]. In the Book of Numbers (fourth Book of the Law of Moses/Torah or Pentateuch), which chronicles the exodus of Moses and the Israelites from Egypt and their 40-year wandering, mention is made in Chapter 5:2 of the command to "cast (the unclean) out of the camp. . .whosoever hath an issue of seed." There is also the record in Chapter 31: 35 of the army returning from war with the Midianites (who had drawn the Israelites "into sin"), with "32,000 of the female sex, that had not known men," and explicit instructions (31:19–20) to "stay without the camp seven days." Moses' troops were considered to have returned with the "issue," and the chapter recites how Moses dealt with infection by quarantine and disinfection of individuals and their garments and possessions [10].

By the fourth century BC, Hippocrates (460–375 BC) had noted that the introduction of sexual mores and practices from Asia and Egypt into Greece was accompanied by STDs [12]. He described a disease suggestive of a chronic suppurative renal infection or a sexually transmitted urethritis, complicated by renal involvement. Hippocrates termed this "strangury" ("Their urine was copious, thick, varied, mixed with pus, and passed with pain"). Interestingly, gonorrhea in women was not recognized easily. Weatherhead presents a summary of the history of venereal disease, including gonorrhea [13], and cites passages by Herodotus (484–425 BC) that report a "female disease" that is surmised to be gonorrhea. As the Roman Empire superseded Greek dominion, many Greek physicians immigrated to Rome to practice medicine. The Roman Gaius Plinius Secundus, (23–79 AD, Pliny the Elder) in his *Naturalis Historia* describes a *profluvia geniturae viris*, which is generally thought to describe a male urethral discharge. The Greco-Roman gynecologist Soranus of Ephesus (practicing between 98 and 138 AD) in his remarkable book titled *Gynecology* provides instructions for "treating the eyes (of newborns) by an injection of olive oil; for it is good thus to wash off the thickest moisture in them; if it is not done, in most cases the nurslings become dim-sighted" (Book 2; viii)—a possible treatment for gonococcal conjunctivitis (http://hdl.handle.net/2027/fulcrum.n870zr06z). Weatherhead also cites Actuarius' description of a flux and a "profluvium seminis sine voluptate" (i.e., seminal discharge without pleasure) [13]. But it is to Aelius Galen (Claudius Galen, Galen of Pergamon, 129–216 AD), arguably the most important Greek physician in Rome, to whom we owe the first mention of the word "gonorrhea." However, Galen mistook the discharge of urethral pus from a male as semen, hence from the Greek words *gonos* (semen) and *rhoia* (flow).

Galen's ideas on medicine prevailed into the Middle Ages and were preserved along with the major works of the Greco-Roman period in Byzantine medicine. Gonorrhea and other venereal diseases were rife in the Middle Ages and are mentioned in both Christian and Arabic texts [14]. The surgeon John Arderne (English, 1307–1392), in 1376, described a condition called *la chaude-pisse* (literally "hot piss"), i.e., a burning sensation during micturition in both men and women, and the term has been colloquially adopted to describe gonorrhea. Other colloquialisms include "the clap," which may originate from "le clapier," French for "house of prostitutes." It is interesting to note that until the sixteenth century, syphilis and gonorrhea were considered different diseases. In 1496, the humanist and physician Joseph Grünpeck (German, 1473–1532) gave clear descriptions of mixed infections of syphilis and gonorrhea, or colloquially known as the "French disease" [15]. VD has been a common subject in (nonmedical) literature from Greco-Roman times to the modern day [16, 17]. Gonorrhea does not escape even the attentions of Shakespeare (1564–1616); it is probably no accident that in *King Lear* (written in 1603–1606) the evil daughter is named Goneril (the name itself a reference back to Geoffrey of Monmouth's character name Gonorilla in *The History of the Kings of Britain*, written in 1136). Shakespeare intended Goneril to be like a "disease full of venom," and the imagery of STDs, putrescence, and pathogenic bodies is very intense in this particular play.

Later, distinguished scientists/clinicians, such as Ambroise Pare (French, 1510–1590), Thomas Sydenham (English, 1624–1689), and John Hunter (Scottish, 1728–1793), all thought that syphilis and gonorrhea were one and the same [18–20]. However, the surgeon Benjamin Bell (Scottish, 1749–1806) published in 1793 *A Treatise on Gonorrhoea Virulenta and Lues Venerea* in which he contended that the signs and symptoms of syphilis and gonorrhea were indeed distinct and the diseases separate [21], a conclusion supported by the physician Philippe Ricord (French, 1800–1889) in 1838. In 1815, the venereologist Francois Swediaur (Austrian, 1748–1824) coined the terms blennorrhagia and blennorrhea to describe acute and chronic gonorrhea [22]. Interestingly, Ricord recommended the rapid application of silver nitrate to preserve sight when "gonorrheal matter" was found in the conjunctiva of the eyes, thereby laying down the first recorded use of this compound for the treatment of gonococcal ocular infection [23]. Both Bell and Ricord also argued that mercury was effective for treating syphilis, but not gonorrhea. For the interested reader, Benedek provides an excellent review of research on ocular inflammation associated with gonorrhea and the studies in the eighteenth to early twentieth century on the experimental induction of gonorrheal ophthalmia in humans (including the ethically unacceptable practice of application of gonorrheal cultures to the eyes of sick children), following the observations that animals were refractory to gonococcal infection [24]. It is noteworthy that the ophthalmologist Joseph F. Piringer (Austrian, 1800–1879) inoculated the eyes of blind people with blennorrheal (gonorrheal) pus, some 38 years before the discovery of the gonococcus, as an attempt to cure those patients with eyes damaged by trachoma [24, 25]. In 1872, the gynecologist Emil Jacob Noeggerath (American, 1827–1895) described latent gonorrhea in women and subsequently on its influence especially on female fertility [26, 27].

Albert Ludwig Siegmund Neisser (German, 1855–1916) is credited with the discovery of the gonococcus and is arguably the first microbiologist to attribute a chronic human disease to a microorganism. In his seminal paper of 1879, Neisser used Robert Koch's (German, 1843–1910) methyl violet staining technique to study smears from 35 men and

9 women with purulent urethritis and 2 patients with acute ophthalmia. Neisser noted that "If gonorrheal pus is spread out in a layer, allowed to dry, stained by methyl violet, a number of masses of micrococci are seen. They have a characteristic, typical form. These characteristic micrococci appear to be a constant mark of all gonorrheal affections" [28]. The organism was named *Neisseria gonorrhoeae* to respect Neisser [19, 29].

Progress in gonococcal biology was rapid over the following decade. Following Neisser's first identification and isolation, Frédéric Weiss (French) isolated *Neisseria gonorrhoeae* in 1880 [30], and Leo Leistikow (German, 1847–1917) followed suit in 1882 [31, 32]. In 1880–1881, the surgeon Alexander Ogston (Scottish, 1844–1929) examined a variety of lesions associated with gonorrhea, soft chancre, sycosis (hair follicle inflammation), sputa from pulmonary tuberculosis, and discharges from wounds and ulcers [33, 34]. He reported the presence of micrococci within gonorrhea lesions, which, reportedly, he was able to culture in fresh eggs. The surgeon and pioneering microbiologist George Sternberg (American, 1838–1915) also confirmed that no microorganism other than the gonococcus (micrococcus) was detectable in gonorrhea secretions [35].

Neisser was unable to satisfy Koch's postulates with the organism he had isolated on a meat extract—gelatin medium, but his identification of the pathogen led to decades of experimental gonococcal infection studies in humans that, in retrospect, are central to the development of clinical research ethics [24]. Koch's postulates were subsequently satisfied by the experiments of the physicians Arpad Bokai (Hungarian, 1856–1919) and Max Bockhart (German, 1883–1921) and the gynecologists Ernst von Bumm (German, 1858–1925) and Ernst Wertheim (Austrian, 1864–1920). In 1880, Bokai confirmed Neisser's microscopy observations and also inoculated six students urethrally with gonococcal culture fluid and reported that three of the patients developed acute gonorrhea [36]. In 1883, Bockhart inoculated the urethra of a man and reported the development of classical gonococcal urethritis after 3 days [37]. In 1885, von Bumm grew axenic cultures of Neisser's gonococcus, *Neisseria gonorrhoeae*, and proved by inoculations of humans that it causes gonorrhea (urethritis) [38, 39]. This experimental urethritis was reproduced by Wertheim in 1891 [40]. In 1890, Wertheim had already demonstrated the existence of gonococci in fallopian tube tissue, and in 1892, he established the hypothesis that gonococci could ascend the female reproductive tract [41]. In 1895, he further demonstrated the presence of gonococci in acute cystitis (bladder gonorrhea) [42] and, a year later, the importance of latent uterine gonorrhea [43]. In 1893, Steinschneider, working in Neisser's laboratory, was able to induce gonorrhea in a colleague following urethral instillation of bacteria [44]. In the same year, the dermato-venereologist Ernst Finger (Austrian, 1856–1939) and his colleagues asked the question as to whether previous gonorrhea infection conferred immunity against reinfection. In this pioneering experiment, he instilled gonococci into the urethra of six men who had a history of gonorrhea but were currently "healthy." Each subject subsequently developed gonorrhea, leading him to conclude that "the gonorrheal process is capable of re-infection and super-infection" [45]. Other inoculation experiments by the surgeon Edward Martin (American, 1859–1938) in 1982 [46], the pediatrician Henry Heiman (American) in 1895 [47], and Jundell and Ahman (Swedish) in 1897 [48] served to reinforce the now obvious conclusion that gonococci experimentally instilled into the urethra can induce gonorrhea. By the turn of the twentieth century, nothing further could be learned about gonococcal pathogenesis from these human inoculation experiments, and they were essentially abandoned. Ethical criticism of some of these studies was probably also a major factor in their discontinuation, given that some were particularly unpalatable, e.g., Heiman's instillation of gonococci into the urethra of intellectually disabled children aged 4 and 16.

Gonorrhea was now accepted as an STD caused by a singular pathogen, *N. gonorrhoeae*, and attention turned to the pathogen's involvement in extragenital infections and to pathological studies. Gonococcal conjunctivitis was well known, and there were anecdotal reports of gonococci associated with arthritis, which was eventually proved in 1894 by the clinician Guido Bordoni-Uffreduzzi (Italian, 1859–1943), who was able to culture gram-negative cocci from pus obtained from the ankle of a woman with gonorrhea. Incontrovertible proof that this organism was a gonococcus was provided by his subsequent inoculation of the urethra of a healthy male volunteer, with no history of STDs, with the cultured organism, which resulted in typical gonorrhea [49]. By the end of this decade, the neurosurgeon and pathologist Harvey Williams Cushing (American, 1869–1939) had reported also that a vaginal gonococcus infection could eventually lead to acute diffuse peritonitis, demonstrating further the ability of the gonococcus to disseminate [50]. An extensive historical literature is also available for pathological findings from studying patient cadavers and biopsy materials from diseased organs [25].

Isolation and culture of the gonococcus now provided an opportunity in the twentieth century for fundamental studies to understand the ultrastructure of the pathogen, its virulence factors, the immunobiology of gonococcal interactions with the host, the nature of the host innate, and adaptive immune responses and to develop effective treatments and potential prophylactics. But it was not until 1938 that the first drug to reliably cure gonorrhea was introduced, sulfanilamide; however, gonococcal resistance rapidly emerged to this drug, a portent of the ensuing and enduring arms race of new antimicrobial chemotherapies and gonococcal adaptability. In addition, vaccines for gonorrhea remain an elusive goal.

Book Synopsis

The first chapter of *Neisseria gonorrhoeae: Methods and Protocols* is a review of the biology of the gonococcus, and this is followed by a brief description of an algorithm for examining infection in women. Antimicrobial resistance (AMR) in gonococci is a major global health problem, and a review examines comprehensively the development of AMR in this pathogen and the strategies for treatment going forward. This links with a chapter that provides a protocol for whole-genome sequencing to predict antimicrobial resistance in gonococci. A seminal chapter follows on colony phenotyping of gonococci, which is an essential method that underpins many laboratory studies. Methods chapters are also provided for producing major gonococcal antigens to high purity, i.e., peptidoglycan, pilus, and lipooligosaccharide (LOS), and outer membranes (OM). High-purity antigens are useful for structural and biochemical studies, for vaccine studies, and for examining their roles in pathogen interactions with the host.

A couple of chapters follow that provide protocols for genetic transformation of gonococci and strategies for global RNA sequencing, in order to study the gonococcal regulatory responses to specific host environments. With no vaccines currently available for *Neisseria gonorrhoeae*, much effort has gone toward identifying potential vaccine candidates. In this book, we have chapters that provide protocols for using high-throughput processes, e.g., bioinformatics workflows for gonococcal proteomics and phenotypic microarray screening of gonococci in chemically defined liquid medium, to evaluate potential vaccine/drug targets and the assessment of hypothetical protein function. There is a companion chapter describing how to grow gonococci in metal-depleted conditions, which leads to the expression of virulence-associated factors that may be important vaccine targets. [Other vaccine strategies, such as immuno-proteomics, antigen identification starting from the genome (reverse vaccinology), and DNA vaccination, can be found in a previous volume on the sister organism *N. meningitidis*, and the interested reader is directed there for methods that are transferable to the gonococcus [51].] In anticipation that new gonococcal vaccines probably will contain recombinant protein antigens, a chapter is provided that comprehensively guides the reader through all the steps required for protein production. One potential measure of vaccine efficacy will be the demonstration of an induced bactericidal response, and a following chapter is provided that explains how to do gonococcal human serum bactericidal assays (hSBA).

Finally, a large proportion of the book is devoted to methods and protocols for studying the biological interactions of the gonococcus and the host. A great deal is still unknown about how gonococci cause disease, and methods and protocols are provided for cell culture models (with primary human macrophages, neutrophils, epithelial cells in three dimensions), ex vivo tissue (human fallopian tubes, bovine eye tissue), in vivo murine models, and finally a guide to the experimental human challenge model.

Acknowledgments

Many of the techniques described herein should have broad appeal not only to the experienced *Neisseria* researcher but also to new researchers seeking to work with this pathogen. Many of the methods may appear daunting and/or require specialist training, but I do hope that they offer an opportunity for inquisitive researchers to engage and collaborate. The book could not have been possible without the contributions of many: principally, I would like to express my gratitude to all authors, all of whom contributed their articles with enthusiasm and showed patience with my editing; to the staff at Humana Press for commissioning this volume; and to the series editor, John Walker, who provided prompt guidance and advice.

The First International Pathogenic Neisseria Conference, held in San Francisco, CA, USA, on 18th–20th January, 1978, was devoted to the immunobiology of *Neisseria gonorrhoeae* [52]. Yet, 40 years on, gonorrhea continues to prove as an intractable disease, compounded by the emergence of antibiotic-resistant strains. However, the continual development and application of new methods and protocols to understand the biology of this pathogen, such as those described in this book, should provide a basis for eventual control of the disease and a reduction in global case numbers.

Southampton, UK *Myron Christodoulides*

References

1. Luys G (1921) Traite de la blennorragie et de ses complications. Librairie Octave Doin, Paris
2. Mayor MT, Roett MA, Uduhiri KA (2012) Diagnosis and management of gonococcal infections. Am Fam Physician 86(10):931–938
3. Mitchell C, Prabhu M (2013) Pelvic inflammatory disease: current concepts in pathogenesis, diagnosis and treatment. Infect Dis Clin North Am 27(4):793–809. https://doi.org/10.1016/j.idc.2013.08.004
4. Donders GGG, Desmyter J, Dewet DH et al (1993) The association of gonorrhea and syphilis with premature birth and low-birth-weight. Genitourin Med 69:98–101
5. Rice PA, Shafer WM, Ram S et al (2017) Neisseria gonorrhoeae: drug resistance, mouse models, and vaccine development. Annu Rev Microbiol 71:665–686. https://doi.org/10.1146/annurev-micro-090816-093530
6. Unemo M, Del Rio C, Shafer WM (2016) Antimicrobial resistance expressed by Neisseria gonorrhoeae: a major global public health problem in the 21st century. Microbiol Spectr 4(3). https://doi.org/10.1128/microbiolspec.EI1110-0009-2015
7. Jerse AE, Deal CD (2013) Vaccine research for gonococcal infections: where are we? Sex Transm Infect 89 Suppl 4:iv63–68. https://doi.org/10.1136/sextrans-2013-051225
8. Wetzler LM, Feavers IM, Gray-Owen SD et al (2016) Summary and recommendations from the National Institute of Allergy and Infectious Diseases (NIAID) workshop "Gonorrhea Vaccines: the Way Forward". Clin Vacc Immunol 23(8):656–663. https://doi.org/10.1128/CVI.00230-16
9. Ghalioungui P (1987) The Ebers Papyrus: a new English translation, commentaries and glossaries. Academy of Scientific Research and Technology, Cairo.
10. Morton RS (1995) Sexual attitudes, preferences and infections in Ancient Egypt. Genitourin Med 71 (3):180–186
11. Willcox RR (1949) Venereal disease in the Bible. Br J Vener Dis 25(1):28–33
12. Morton RS (1991) Sexual attitudes, preferences and infections in Ancient Greece: has antiquity anything useful for us today? Genitourin Med 67(1):59–66
13. Weatherhead G (1861) The history of the early and present state of venereal disease. S. Highley, London
14. Gruber F, Lipozencic J, Kehler T (2015) History of venereal diseases from antiquity to the renaissance. Acta Dermatovenerol Croat 23(1):1–11
15. Grünpeck J (1496) Tractatus de Pestilentiali Scorra Sive Mala de Franzos [Treatise on the pestilence of Scorra or The French Disease]. Caspar Hochfelder, Nuremberg
16. Rolleston JD (1934) Venereal disease in literature. Br J Vener Dis 10(3):147–174
17. Anon (1934) Venereal disease in literature: discussion. Br J Vener Dis 10(3):175–182
18. Steen JS Chronology of biology and medicine, part 2B. http://ntweb.deltastate.edu/jtift/scbm/scbm-2b.html. Accessed 29 Aug 2018
19. Ligon BL (2005) Albert Ludwig Sigesmund Neisser: discoverer of the cause of gonorrhoea. Semin Pediatr Infect Dis 16(4):336–341. https://doi.org/10.1053/j.spid.2005.07.001.
20. Thomas C (1961) The light of many minds: royal faculty of physicians and surgeons of Glasgow. Glasgow Lock Hospital Trust Prize Essay. Brit J Ven Dis 37:33–58.
21. Bell B (1793) A treatise on gonorrhoea virulenta, and lues venerea. Watson and Mudie, Edinburgh
22. Swediaur F (1815) A complete treatise on the symptoms, effects, nature and treatment of syphilis. T. Dobson, Philadelphia
23. Ricord P (1838) Traité pratique des maladies vénériennes [A practical treatise on venereal diseases]. De Just Rouvier and E. Le Bouvier, Paris.
24. Benedek TG (2005) Gonorrhea and the beginnings of clinical research ethics. Perspect Biol Med 48 (1):54–73. https://doi.org/10.1353/pbm.2005.0003
25. Harkness AH (1948) The pathology of gonorrhoea. Brit J Ven Dis 24:137–147
26. Noeggerath EJ (1872) Die Latente Gonorrhoe im Weiblichen Geschlecht [The latent gonorrhea in females]. M. Cohen and Sonn, Bonn

27. Noeggerath EJ (1876) Latent gonorrhea especially with regard to its influence on fertility in women. In: American Gynecological Society, 1st meeting, Boston, MA
28. Neisser AL (1879) Über eine der Gonorrhoe eigentümliche Micrococcusform. Centralbl Med Wissensch 17:497–500
29. Oriel JD (1989) Eminent venereologists. 1. Albert Neisser. Genitourin Med 65(4):229–234
30. Weiss F (1880) Le Microbe du Pus Blennorrhagique [The microbe of Blenorrheal Pus]. Nancy
31. Leistikow L (1882) Über bacterien bei den venerischen krankheiten [On bacteria in the venereal diseases]. Charite-Annalen 7:750–772
32. Leistikow L (1882) Resultate seiner Untersuchungen uber die Tripperbacterien. Berliner Klinische Wochenschrift 19:500
33. Ogston A (1880) Ueber Abscesse. Arch Klin Chir 225:588–600
34. Ogston A (1881) Report upon micro-organisms in surgical disease. BMJ 1:369–375
35. Sternberg GM (1893) The micrococcus of gonorrhoeal pus: infective virulence is not due to the presence of this parasitic micro-organism. Med News 42:62–70 and 96–99
36. Bokai A (1880) Uber das contagium der acuten blennorhoe. Allge medizin Central-Zeit 49:900–903
37. Bockhart M (1883) Beitrag zur aetiologie und pathologie des Harnrohrentrippers. Vrtlschr fur Derm und Syph 10:3–18
38. von Bumm E (1885) Der mikro-organismus der gonorrhöischen schleimhaut-erkrankungen, gonococcus Neisser [The microorganism of gonorrhea mucosal disease, "Gonococcus Neisser"]. Dtsch Med Wochenschr 11:508–509
39. von Bumm E (1885) Der Mikro-Organismus der Gonorrhoischen Schleimhaut-Erkrankungen 'Gonococcus Neisser'. Nach Untersuchungen Beim Weibe und an der Conjunctiva der Neugeborenen Dargestellt [The microorganism of gonorrhea mucosal disease "Gonococcus Neisser". after investigations in the female and the conjunctiva of the newborn shown]. Bergmann, Wiesbaden
40. Wertheim E (1891) Reinzuchtung des Gonococcous Neisser mittels des Plattenverfahrens. Dtsch Med Wochenschr 17:1351–1352
41. Wertheim E (1892) Die aszendierende gonorrhoe beim weibe. Bakteriologische und klinische studien zur biologie des gonococcus neisser. Archiv für Gynakologie 42:1–86
42. Wertheim E (1895) Über blasen-gonorrhöe [Gonorrhea of the bladder]. Zeitschrift für Geburtshilfe und Gynäkologie 35:1–10
43. Wertheim E (1896) Über uterus-gonorrhöe [About gonorrhea of the uterus]. Verhandlungen der Deutschen Gesellschaft für Gynakologie 6:199–223
44. Steinschneider (1893) Uber die culture der gonokokken. Berl Klin Wochenschr 30:696–699 and 728–731
45. Finger E, Ghon A, Schlagenhaufer F (1894) Beitrage zur biologie des gonococcus und zur pathologischen anatomie der gonorrhoischen processes. Arch Dermatol Syph 28:277–344
46. Martin E (1892) Vulvo-vaginitis in children. J Cutan G-U Med 10:415–427
47. Heiman H (1895) Clinical and bacteriological study of the gonococcus (Neisser) as found in the male urethra and in the vulvo-vaginal tract of children. Med Record 47:769–778
48. Jundell I, Ahman C (1897) Uber die reinzuchtung des gonococcus Neisser. Arch Dermatol Syph 38:59–68
49. Bordoni-Uffreduzzi G (1894) Uber die localisationen des gonococcus im innern des organismus (durch den gonococcus hervorgerufene pleuritis und arthritis). Dtsch Med Wochenschr 20:484–485
50. Cushing HW (1899) Acute diffuse gonococcus peritonitis. Johns Hopkins Hosp Bull 10:75
51. Christodoulides M (ed) (2012) Neisseria meningitidis: advanced methods and protocols. Methods in molecular biology, vol 799. Humana, New York
52. Brooks GF, Gotschlich EC, Holmes KK, Sawyer WD, Young FE (eds) (1978) Immunobiology of Neisseria gonorrhoeae. Proceedings of a conference held in San Francisco, California, 18th–20th Jan. 1978. ASM, Washington, DC

Contents

Contributors

A. SAID ÁLAMOS-MUSRE • *Laboratory of Medical Microbiology and Pathogenesis, Faculty of Medicine, Universidad Andres Bello, Concepción, Región del Bío-Bío, Chile*
HANNIA LILIANA ALMONACID-MENDOZA • *Molecular Microbiology Group, Academic Unit of Clinical and Experimental Sciences, Faculty of Medicine, University of Southampton, Southampton, UK*
TUBA ALTINDAL • *Molecular Biology and Biochemistry, Simon Fraser University, Burnaby, BC, Canada*
BENJAMIN I. BAARDA • *Department of Pharmaceutical Sciences, College of Pharmacy, Oregon State University, Corvallis, OR, USA*
MELANIE M. CALLAGHAN • *Department of Medical Microbiology and Immunology, University of Wisconsin-Madison, Madison, WI, USA*
MYRON CHRISTODOULIDES • *Molecular Microbiology Group, Academic Unit of Clinical and Experimental Sciences, Faculty of Medicine, University of Southampton, Southampton, UK*
COLIN P. CHURCHWARD • *School of Life Sciences, Pharmacy, and Chemistry, Kingston University, London, UK; National Heart and Lung Institute, Imperial College London, London, UK*
CYNTHIA NAU CORNELISSEN • *Department of Microbiology and Immunology, School of Medicine, Virginia Commonwealth University, Richmond, VA, USA; Institute for Biomedical Sciences, Center for Translational Immunology, Georgia State University, Atlanta, GA, USA*
LISA CRAIG • *Molecular Biology and Biochemistry, Simon Fraser University, Burnaby, BC, Canada*
ALISON K. CRISS • *Department of Microbiology, Immunology, and Cancer Biology, University of Virginia, Charlottesville, VA, USA*
JOSEPH P. DILLARD • *Department of Medical Microbiology and Immunology, University of Wisconsin-Madison, Madison, WI, USA*
JOSEPH A. DUNCAN • *Department of Medicine, University of North Carolina at Chapel Hill, Chapel Hill, NC, USA*
FADI E. EL-RAMI • *Department of Pharmaceutical Sciences, College of Pharmacy, Oregon State University, Corvallis, OR, USA*
ALEJANDRO ESCOBAR • *Laboratorio Biología celular y molecular, Instituto de Ciencias Odontológicas, Facultad de Odontología, Universidad de Chile, Santiago, Región Metropolitana, Chile*
DAVID W. EYRE • *Big Data Institute, University of Oxford, Oxford, UK; Oxford University Hospitals NHS Foundation Trust, Oxford, UK*
IAN P. FRANCIS • *Section of Infectious Diseases, Department of Medicine, Boston Medical Center, Boston, MA, USA*
JIMENA GATICA • *Instituto de Investigación en Ciencias Odontológicas, Facultad de Odontología, Universidad de Chile, Santiago, Chile*
CAROLINE A. GENCO • *Department of Immunology, Tufts University School of Medicine, Boston, MA, USA*
DANIEL GOLPARIAN • *National Reference Laboratory for STIs, WHO Collaborating Centre for Gonorrhoea and Other STIs, Örebro University Hospital, Örebro, Sweden*

J. McLeod Griffiss • *Department of Laboratory Medicine, University of California, San Francisco, CA, USA*

Sunita Gulati • *Division of Infectious Diseases and Immunology, Department of Medicine, University of Massachusetts Medical School, Worcester, MA, USA*

John E. Heckels • *Molecular Microbiology Group, Academic Unit of Clinical and Experimental Sciences, Faculty of Medicine, University of Southampton, Southampton, UK*

Melissa M. Herbst-Kralovetz • *Department of Basic Medical Sciences, The University of Arizona College of Medicine-Phoenix, Phoenix, AZ, USA; Department of Obstetrics and Gynecology, The University of Arizona College of Medicine-Phoenix, Phoenix, AZ, USA*

Marcia M. Hobbs • *Department of Medicine, University of North Carolina at Chapel Hill, Chapel Hill, NC, USA*

María Victoria Humbert • *Molecular Microbiology Group, Academic Unit of Clinical and Experimental Sciences, Faculty of Medicine, University of Southampton, Southampton, UK*

Ann E. Jerse • *Uniformed Services, University of the Health Sciences, Bethesda, MD, USA*

Paweł Łaniewski • *Department of Basic Medical Sciences, The University of Arizona College of Medicine-Phoenix, Phoenix, AZ, USA*

Xiuping Lui • *Department of Microbiology, Boston University School of Medicine, Boston, MA, USA*

Mancheong Ma • *Department of Immunobiology, BIO5 Institute, University of Arizona, Tucson, AZ, USA*

Ryan McClure • *Biological Sciences Division, Pacific Northwest National Laboratory, Richland, WA, USA*

Stephanie E. McLaughlin • *Department of Epidemiology, John Hopkins Bloomberg School of Public Health, Baltimore, MD, USA*

Stephanie A. Ragland • *Department of Microbiology, Immunology, and Cancer Biology, University of Virginia, Charlottesville, VA, USA*

Sanjay Ram • *Division of Infectious Diseases and Immunology, Department of Medicine, University of Massachusetts Medical School, Worcester, MA, USA*

Erica L. Raterman • *Uniformed Services, University of the Health Sciences, Bethesda, MD, USA*

Katherine Rhodes • *Department of Immunobiology, BIO5 Institute, University of Arizona, Tucson, AZ, USA*

Peter A. Rice • *Division of Infectious Diseases and Immunology, Department of Medicine, University of Massachusetts Medical School, Worcester, MA, USA*

Paula I. Rodas • *Laboratory of Medical Microbiology and Pathogenesis, Faculty of Medicine, Universidad Andres Bello, Concepción, Región del Bío-Bío, Chile*

Ryan E. Schaub • *Department of Medical Microbiology and Immunology, University of Wisconsin-Madison, Madison, WI, USA*

Jutamas Shaughnessy • *Division of Infectious Diseases and Immunology, Department of Medicine, University of Massachusetts Medical School, Worcester, MA, USA*

Aleksandra E. Sikora • *Department of Pharmaceutical Sciences, College of Pharmacy, Oregon State University, Corvallis, OR, USA*

Lori A. S. Snyder • *School of Life Sciences, Pharmacy, and Chemistry, Kingston University, London, UK*

Magdalene So • *Department of Immunobiology, BIO5 Institute, University of Arizona, Tucson, AZ, USA*

CECILIA V. TAPIA • *Laboratorio de Especialidad, Clínica Dávila, Santiago, Región Metropolitana, Chile*
MAGNUS UNEMO • *National Reference Laboratory for STIs, WHO Collaborating Centre for Gonorrhoea and Other STIs, Örebro University Hospital, Örebro, Sweden*
LEE M. WETZLER • *Section of Infectious Diseases, Department of Medicine, Boston Medical Center, Boston, MA, USA; Department of Microbiology, Boston University School of Medicine, Boston, MA, USA*

Chapter 1

Biology of the Gonococcus: Disease and Pathogenesis

Jutamas Shaughnessy, Sanjay Ram, and Peter A. Rice

Abstract

Neisseria gonorrhoeae infection is a major public health problem worldwide. The increasing incidence of gonorrhea coupled with global spread of multidrug-resistant isolates of gonococci has ushered in an era of potentially untreatable infection. Gonococcal disease elicits limited immunity, and individuals are susceptible to repeated infections. In this chapter, we describe gonococcal disease and epidemiology and the structure and function of major surface components involved in pathogenesis. We also discuss the mechanisms that gonococci use to evade host immune responses and the immune responses following immunization with selected bacterial components that may overcome evasion. Understanding the biology of the gonococcus may aid in preventing the spread of gonorrhea and also facilitate the development of gonococcal vaccines and treatments.

Key words Gonococcus, *N. gonorrhoeae*, Pathogenesis, Porin, Opa, Lipooligosaccharide

1 Disease and Epidemiology

Neisseria gonorrhoeae (gonococcus) is an obligate human pathogen and the etiologic agent of gonorrhea. Gonococcal transmission occurs from an infected individual by direct contact between the mucosal membranes of the urogenital tract, anal canal, oropharynx, and occasionally the eye (conjunctivitis), usually during sexual activity that may result in urethritis in men and cervicitis in women. Most men who develop symptoms (usually urethritis) seek medical attention and are no longer infectious after treatment. However, asymptomatic men constitute about two-thirds of all infected men and are the principal source of spread of infection. Greater than half of infected women ("uncomplicated cervicitis") do not complain of symptoms but often manifest physical signs of inflammation upon direct (speculum) examination of the cervix. Untreated cervical infection with *N. gonorrhoeae* with or without *Chlamydia trachomatis*, which frequently coinfects with *N. gonorrhoeae*, may ascend upward into the uterus and fallopian tubes and can cause pelvic inflammatory disease (PID

Myron Christodoulides (ed.), *Neisseria gonorrhoeae: Methods and Protocols*, Methods in Molecular Biology, vol. 1997,
https://doi.org/10.1007/978-1-4939-9496-0_1, © Springer Science+Business Media, LLC, part of Springer Nature 2019

encompassing endometritis and/or salpingitis), which may result in tubal factor infertility (TFI), ectopic pregnancy, and chronic pelvic pain. Rarely, ectopic pregnancy that occurs in a fallopian tube scarred from a prior infection leads to severe hemorrhage. Rectal or pharyngeal gonorrhea is often asymptomatic. Occasionally, *N. gonorrhoeae* enter the bloodstream, disseminate, and cause skin and/or joint/tendon infection, and more rarely endocarditis or meningitis [1].

Infected mothers may transmit gonococcal infection to their newborns during delivery that can result in eye infection (*ophthalmia neonatorum*) and/or rarely disseminated infection. The absence of routine prenatal screening for gonococcal infection can result in a fivefold increased risk of preterm delivery [2] and 30–45% increased risk of *ophthalmia neonatorum*, which leads to scarring and blindness if untreated [3]. Concomitant infection with HIV and gonorrhea can enhance the rate of HIV transmission fivefold [4–6].

Gonorrhea is a major global public health problem and the World Health Organization (WHO) estimated that in 2012, there were 78.3 million new cases in adults (15–49 years of age) worldwide [7]. In the USA, the Centers for Disease Control and Prevention (CDC) report gonorrhea as the second most common notifiable disease; *Chlamydia trachomatis* is first and often coinfects with *N. gonorrhoeae* [8–13]. In 2017, ~556,000 cases of gonorrhea were reported in the USA; the total burden is estimated to be ~800,000 cases/year nationally. During 2016–2017, the rate of reported gonorrhea cases increased by 18.6% in the U.S., and has increased 75.2% since a historic low in 2009 [14]. The US national rate of gonorrhea in 2017 was 172 new cases per 100,000 population, representing a 1-year increase of 18% and was the highest rate reported among industrialized countries [14] . The true number of cases is much higher, particularly from resource-poor settings where the lack of laboratory testing/sensitive diagnostics and incomplete case reporting contribute to underrecognition and underreporting.

Gonorrhea is on the rise worldwide, and strains resistant to many antibiotics are emerging. *N. gonorrhoeae* has developed resistance to every antimicrobial it has encountered. Isolates that are highly resistant to ceftriaxone have emerged in Asia [15], Europe [16], and Australia [17], and azithromycin-resistant isolates have been identified in Asia, Hawaii, Canada, and Europe [18–23]. These alarming developments have ushered in an era of potentially untreatable gonococcal infection. A recent US estimate predicts that the increased number of gonococcal infections and gonorrhea-associated HIV cases due to emerging ceftriaxone resistance over a 10-year period will cost 378.2 million USD [24]. *See* Chapter 3 by Unemo and colleagues for an in-depth analysis of antimicrobial resistance in *N. gonorrhoeae* and treatment strategies for gonorrhea.

Appropriate treatments and preventive measures, including effective vaccines, are urgently needed. To guide research in this direction, a human experimental model in men [25] and a cervicovaginal colonization model in mice [26] have been developed to study the mechanisms by which *N. gonorrhoeae* evades host-derived antimicrobial factors and to identify protective and immunosuppressive pathways.

2 Gonococcal Pathogenesis

N. gonorrhoeae colonizes and invades the epithelium of the lower and upper genitourinary tract that can incite an inflammatory response, which results in localized influx of polymorphonuclear leukocytes (PMNs) primarily, and macrophages. Studies, performed ex vivo, in primary urethral epithelial cell and fallopian tube organ culture systems have proposed four phases of infection [27, 28]. In the first phase, gonococci attach and intimately associate with microvilli of epithelial cells. This is followed by focal polarization of actin, extension of microvilli (second phase) and uptake of organisms into vacuoles at the apical surface of epithelial cells. In the third phase, gonococci replicate in the vacuoles and can transcytose to the basolateral surface of epithelial cells. In the fourth phase, organisms rupture through the basolateral surface into the subepithelial space, inducing an inflammatory process and infecting other epithelial cells. Bacterial surface components that are involved in attachment and invasion include pili (assembled from pilin protein), opacity associated proteins (Opa proteins), lipooligosaccharide (LOS), and porin (Por) [29–32]. *N. gonorrhoeae* also uses several mechanisms to survive host defenses; in particular, they resist complement-mediated killing by serum, which may also result in impaired opsonization and ingestion of organisms by professional phagocytes.

The gonococcal cell wall is similar to other Gram-negative bacteria and consists of an inner cytoplasmic membrane, a middle layer containing peptidoglycan and the periplasmic space and an outer membrane (OM), the latter comprising a complex mixture of phospholipids, LOS and proteins, organized as an asymmetric lipid bilayer (Fig. 1). Much of gonococcal pathogenesis is driven by structures present in the OM and the most widely examined of these are discussed below.

2.1 Pili

Pili are filamentous structures composed of the major pilus subunit (pilin, PilE) with an adherence-associated protein (PilC) located at the pilus tip and also within the bacterial membrane [33–35]. Gonococcal pili are grouped as Type IVa pili based on the deduced amino acid sequence of pilin genes and their assembly

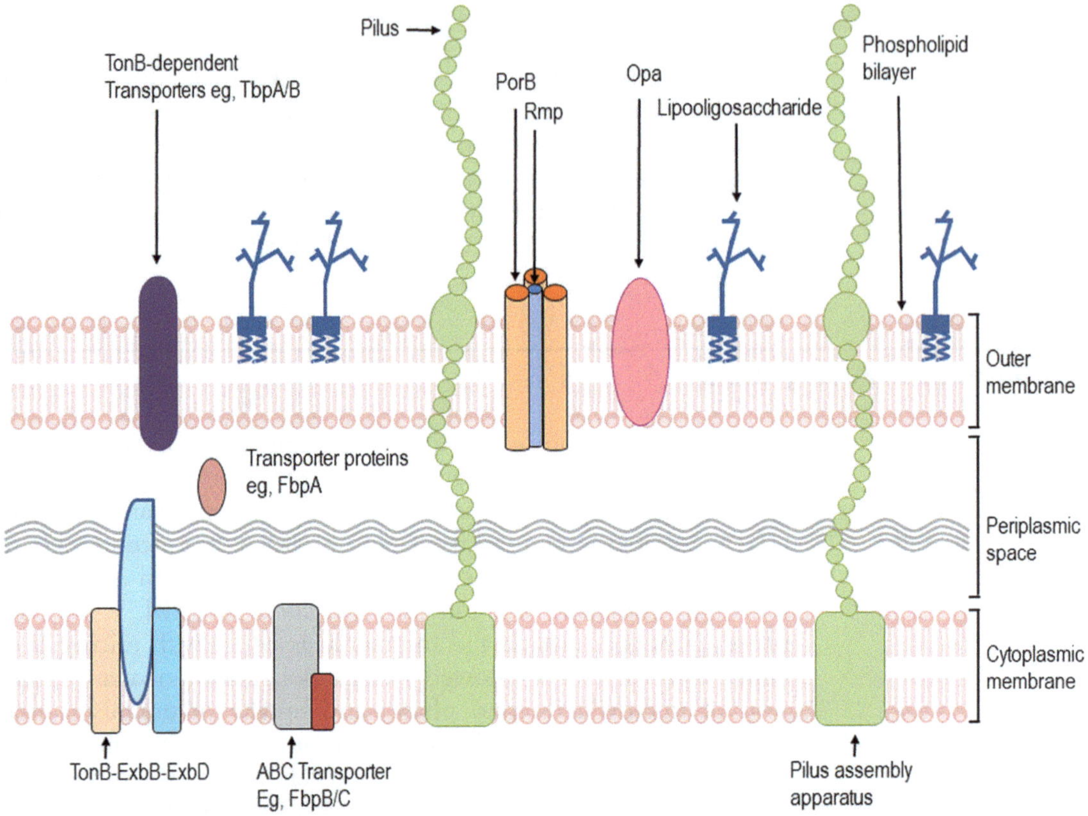

Fig. 1 Schematic representation of the surface structure of *Neisseria gonorrhoeae* (modified from [236]. These surface structures include type IV pili, porin, lipooligosaccharide (LOS), opacity-associated protein (Opa) and iron and zinc regulated proteins (shown as TonB dependent transporters (TdTs) on the OM surface). Iron transporter systems include TdTs, TonB-ExbB-ExbD complexes that provide energy to transport iron across the OM, transporter proteins (e.g., ferric binding protein A (FbpA)), which binds to iron in the periplasmic space, and the ABC transporter (e.g., FbpB/FbpC complexes), which transports iron across the cytoplasmic membrane

mechanisms [36]. Pilus subunits vary in M_r from about 17 to 21 kDa in different gonococcal strains. Pili function as adhesins for a variety of epithelial and endothelial cells. They also are involved in biofilm formation, DNA uptake and a flagella-independent form of movement known as twitching motility. A single gonococcal type IV pilus can exert retractile forces in the range of 100 pN [37, 38]. Piliated strains adhere better to cells derived from human mucosal surfaces and are more virulent than nonpiliated variants in organ culture models and in the human experimental model. In a fallopian tube explant model, pili mediate gonococcal attachment to nonciliated columnar epithelial cells. This event initiates "nonprofessional" phagocytosis and transport of gonococci through these cells to intercellular spaces near the basement membrane or directly into subepithelial tissue.

Piliated gonococci are naturally competent for DNA transformation at a frequency of 3–4 logs higher than their nonpiliated counterparts [39], which permits horizontal transfer of genetic material between different gonococcal lineages and perhaps to other competent bacteria in vivo [40–42]. Distinct colony morphological variants of *N. gonorrhoeae* have been linked to virulence in the human experimental model [43]. Types 1 and 2 gonococci have pili and are virulent, and they form small, round and convex colonies on translucent agar; type 2 colonies are larger. Type 3 and type 4 colonies, which comprise nonpiliated gonococci, are larger, minimally convex, and flat-edged. Primary isolates from infected humans with gonorrhea are piliated and form mostly type 1 and 2 colonies, which revert to types 3 and 4 upon repeated transfer on nonselective media. A standardized intraurethral inoculum of piliated gonococci is infectious in the human experimental model and the infectious dose can be as low as 5×10^2 bacteria [44]. Nonpiliated types 3 and 4 gonococci are less virulent and have failed to cause disease in the human experimental model [43, 45]. Types 1 and 2 (piliated) organisms more readily resist phagocytosis by PMNs, compared to types 3 and 4 [46, 47]. Gonococcal pili undergo phase and antigenic variation; they switch between the piliated and nonpiliated state at a frequency of approximately 10^{-4} per colony-forming units (CFU) per generation in vitro [48, 49]. Piliated gonococci change over to nonpiliated during transcellular passage through the human endometrial cell line HEC-1-B [50]. In vitro, selection for nonpiliated variants may result from signals derived from host cells that induce loss of pili or from a growth advantage of non-piliated over piliated gonococci [50].

The major subunit of pilus—PilE, pilin protein—undergoes antigenic variation in vitro and in vivo. For example, the identical strain with different pilus types has been isolated from different sites of infection in a single person [51]. During an epidemic, a single gonococcal strain expressing one pilus type can give rise to isolates whose pili have undergone antigenic variation to express different pili [52]. This antigenic variation occurs predominantly by nonreciprocal recombination events between the expressed *pilE* gene and one of the silent *pilS* loci [53]. PilE can be expressed on the gonococcal surface or secreted as a soluble 16 kDa truncated pilin (S-pilin) in culture supernatants [54]. S-pilin is produced by cleavage of a 39-amino acid peptide from the N-terminus of mature pilin protein. S-pilin-producing gonococci can appear nonpiliated by colony morphology (i.e., types 3 or 4) but can revert to a piliated state at high frequency. PilE consists of an approximately 160-amino acid-long protein possessing a highly conserved N-terminal portion and an extensive variable region in the C-terminus [52, 55]. It can be divided into three regions [52]. The first 53 amino acids at the N-terminus are always conserved. The middle portion (amino acid residues 54–114) is semivariable and the C-terminus (amino acid residues 114–160) is hypervariable. The crystallographic structure of pilin protein shows it to be O-linked to a disaccharide (*N*-acetyl glucosamine*α*1, 3-galactose) at position 63 (Ser) [56]. A pilus assembly model suggests that pilin forms a core of coiled helices, leaving disaccharide and hypervariable regions exposed to solvent [56]. This may explain the immunodominance of the hypervariable region compared to the conserved region and the low immunological cross-reactivity (5–50%) of pili from different strains [52, 57].

An adherence-associated protein, PilC, has been shown to function as an adhesin at the tip of the pilus [35]. PilC is a 110 kDa protein and is encoded by two variant genes (*pilC1* and *pilC2*) in *N. gonorrhoeae* strain MS11 [33]. The expression of PilC is controlled by frequent frameshift phase mutations within a polyguanine (G) tract located in the signal peptide-coding region [33]. PilC has been proposed to play a central role in pilus-mediated epithelial cell adherence [29]. Purified PilC2 inhibits the adherence to culture cells of both PilC1- and PilC2-expressing gonococci [35]. Mutations in both *pilC* genes generate variants that form nonpiliated colony morphological types, produce the majority of S-pilin, and acquire new *pilE* sequences. This has led to an additional proposed role for PilC in pilus assembly [33, 34].

The role of PilC in pilus-mediated epithelial cell adherence is complicated by observed antigenic variation of PilE that can change the adherence properties of gonococci to epithelial cells [58, 59]. PilE may modulate the binding of gonococcal pili to epithelial cells, perhaps by recognizing a secondary receptor or by influencing the recognition of the receptor of PilC [60]. Human membrane cofactor protein CD46 and complement receptor 3 (CR3) have been identified as putative pilus receptors [61–63]. The precise component of the gonococcal pilus that binds to CD46 has not been identified.

Antibodies directed against pili in vaginal secretions inhibit attachment of gonococci to epithelial cells [64]. However, studies in the human experimental model [65] have demonstrated that pili are not always necessary for colonization of the male urethra; for example, a *pilE* mutant of strain FA1090 unable to express pilin was able to colonize most inoculated subjects and was capable of causing mild urethritis in a small proportion of subjects.

Anti-pilus antibodies (IgG, IgA, and IgM) in serum also inhibit the attachment of gonococci to epithelial cells [66]. A pilin vaccine demonstrated an immune response in serum and semen [44, 67] and showed some protection against infection with homologous organisms in the human experimental model [66], but was not protective against heterologous infection in a large clinical trial [67].

2.2 Porin (Por, Protein I, PI)

Porin (Por) comprises 60% of gonococcal OM protein [68] and ranges in M_r from 32 to 39 kDa [69]. It is noncovalently associated with reduction modifiable protein (Rmp, *see* below) [70, 71]. Porin functions as a hydrophilic channel that allows the passage of small nutrients and waste products (with sizes up to 1.4 kDa) across the OM [72, 73]. Pores are formed by a trimer of porin monomers organized in antiparallel β-strands that fold into cylinder-like molecules [74, 75]. Mature gonococcal porin contains 16 membrane-spanning regions with eight surface-exposed regions (loops) that vary in length [76]. Variation is largely restricted to the surface-exposed regions of porin protein [77, 78].

Por is encoded by a single copy of the *por* gene [68]. There are two major variants of porin protein in *N. gonorrhoeae*, PorB.1A and PorB.1B; these are further classified into serovars based on reactivity to a panel of Por-specific monoclonal antibodies [76, 79]. PorB.1A is often associated with strains that cause disseminated gonococcal infection (DGI). DGI strains usually resist complement (C)-dependent killing by normal human serum (NHS) even in the absence of sialylation of gonococcal lipooligosaccharide (LOS) (*see* Subheading 2.7) and exhibit stable serum resistance. PorB.1A itself also resists proteolytic digestion by trypsin and α-chymotrypsin because it lacks a portion of loop 5 that contains the cleavage sites for these enzymes. The M_r of PorB.1A is less than PorB.1B because of the deletion in loop 5 [69, 80, 81]. PorB.1B strains usually cause local gonococcal infection and usually are susceptible to killing by NHS ("serum sensitive") [80, 82]. Por is involved in the sensitivity of gonococci to NHS because certain Por motifs bind C regulatory proteins including factor H (FH) and C4b binding protein (C4BP) that downregulate the alternative and classical pathways of C, respectively, which results in "serum resistance" of gonococci. FH preferentially binds to loop 5 of PorB.1A strains [83]. Also, C4BP can bind to loop 1 of PorB.1A or loops 5 and 7 of PorB.1B resulting in "serum resistance" of these strains as well [84].

PorB molecules do not undergo high-frequency variation: however, clinical isolates do exhibit antigenic differences in surface-exposed loops, presumably as a result of low-frequency mutations or recombination events in the *por* gene [79, 85].

Gonococcal PorB can also modulate host immune responses. PorB proteosomes activated host Toll-Like Receptor 2 (TLR2) signaling and act as immune-stimulating adjuvants [86–88]. However, recombinant PorB that is properly folded (e.g., in native OM vesicles) suppresses dendritic cell-induced T cell proliferation and may offset the stimulating effect of TLR2 [89].

Humans with complicated gonococcal infection including pelvic inflamatory disease (PID) and DGI develop antibodies against Por. Epidemiologic and clinical studies suggest that an immune response directed against PorB evoked by infection may confer partial, serovar-specific immunity against gonococcal cervicitis [90] and gonococcal salpingitis [91], although protective serovar-specific immunity has not been shown in men [92]. Monoclonal antibodies (mAbs) that bind PorB surface-exposed loops show C-mediated bactericidal activity in vitro [81, 93–95]. Human IgG anti-PorB antibodies also possess complement (C)-dependent bactericidal activity [96].

In 1985, a candidate OM vaccine enriched with PorB was tested for efficacy in the human experimental model. Sixty-three male volunteers were immunized either with the vaccine ($n = 35$) or given a placebo ($n = 28$). There were no significant differences in infection rates between vaccinees and placebo recipients; anti-PorB concentrations rose twofold in vaccinees (*see* Subheading 2.4 for additional vaccine discussion).

2.3 Opacity-Associated Protein (Opa, Protein II, PII)

Opa proteins are a family of heat-modifiable OM proteins with subunit molecular masses ranging between 24 and 30 kDa [97, 98]. Opa mediates self-adherence of gonococci and also promotes adherence to and invasion of eukaryotic cells [98, 99]. Opa proteins adhere to the lacto-*N*-neotetraose epitope of LOS (galactoseβ1-4-*N*-acetyl-glucosamineβ1-3-galactoseβ1-4-glucose) on opposing bacteria, thereby facilitating self-adherence [100]. Opa expression can influence colony opacity and color; for example, strains that express Opa (Opa^{+}) often (but not always) appear as opaque colonies on clear media in contrast to transparent colonies that are formed by strains that lack Opa (Opa^{-}) [101].

Opa expression exhibits both phase and antigenic variation [102], and gonococci harbor a full repertoire of up to 11 distinct *opa* genes [103]. The sequences of all 11 Opa proteins are ~70% identical, however, conserved regions are not surface exposed [102–104]. Opa proteins are made up of eight antiparallel strands that form barrel structures in the OM, which are linked by four surface-exposed loops [105]. Loops 2 and 3 show sequence variation and have been termed hypervariable HV1 and HV2 loops, respectively [102, 103]. Gonococcal strains may express none or several Opa proteins simultaneously [106]. Changes in Opa expression patterns occur at approximately 1 per 10^3 cell divisions resulting from phase shifts in the reading frames of pentanucleotide coding repeat units (CTCTT) that are located in the *opa* leader sequence [102, 107, 108]. Expression of different Opa proteins results in antigenic variation.

Several studies have suggested an important role of Opa in gonococcal pathogenesis. Phenotypically Opa^{+} gonococci often are associated with localized urogenital, cervical and/or rectal infections [109]; Opa^{-} gonococci are generally recovered in DGI [110]. Opa expression is also under hormonal influence; for example, Opa^{-} gonococci predominate during menses [99]. In the human experimental model, inoculated Opa^{-} strains (FA1090 and MS11) convert to mostly Opa^+ on isolation, suggesting that Opa^+ phenotypes are selected for in vivo [111, 112].

Opa proteins play a major role in the adherence of *N. gonorrhoeae* to host cells. Two classes of Opa protein receptors have been identified on eukaryotic cells: heparan sulfate proteoglycan (HSPG) and the CD66 family (carcinoembryonic antigens). Epithelial cells with HSPG bind to a specific Opa protein (Opa_{50} or Opa_{30}) [113, 114], while the CD66 family present on epithelial cells, lymphocytes and PMNs interact with multiple Opa proteins [115–117]. Opa_{50} also binds to the extracellular matrix protein vitronectin and stimulates bacterial uptake in vitro by HeLa cells in an integrin-dependent manner [118]. These results suggest that distinct Opa expression patterns demonstrate tissue tropism. Neisserial Opa proteins have been shown to bind *c*arcino*e*mbryonic *a*ntigen related *c*ell *a*dhesion *m*olecules (CEACAMs): CEACAM1, CEACAM3, CEACAM5, and CEACAM6. Binding of CEACAM by Opa requires a conformation interaction between both HV1 and HV2 regions of Opa. Binding of Opa to CEACAM1, which is expressed by primary CD4+ T lymphocytes, suppresses the activation and proliferation of these lymphocytes [119]. Nonopsonic uptake of gonococci by human PMNs involves Opa-CEACAM3 interactions [120, 121]. Sialylation of LOS (*see* Subheading 2.7) inhibits adherence of Opa-positive gonococci to neutrophils and stimulation of the oxidative burst [122]. CEACAM5-binding gonococci prevent exfoliation of the epithelium and may interfere with bacterial clearance [123]. Select Opa proteins can engage CEACAM3, which is expressed on neutrophils and results in nonopsonic phagocytosis (i.e., phagocytosis independent of antibody and complement) and killing of bacteria. Antibodies directed against the HV2 region of a gonococcal Opa protein inhibits interaction with neutrophils [124].

Immunization of rabbits with whole gonococci elicits anti-Opa antibodies that recognize only homologous Opa proteins [104]. Anti-Opa antibodies, elicited by isolated Opa proteins, cross-react with all Opa variants in Western blot, but only recognize homologous Opa proteins in whole-organism immunoprecipitation experiments, indicating that the hypervariable sequences are surface exposed. Although mAbs against particular Opa may decrease adherence of gonococci to host cells, antibodies to Opa are specific and do not cross-react with different Opa types [125, 126]. This diminishes the possibility of using Opa as a vaccine candidate. Anti-Opa antibody can be detected in the convalescent sera and vaginal secretions of patients with gonorrhea [109]. Modeling of the immune status of subjects exposed to *N. gonorrhoeae* reported that antibodies against several OM components, including Opa and PorB, considered individually, predicted protection from gonococcal infection [90, 127].

2.4 Reduction Modifiable Protein (Rmp, Protein III, PIII)

Rmp is a highly conserved, surface-exposed OM protein that is present in both pathogenic *Neisseria* spp. (*N. meningitidis* and *N. gonorrhoeae*) [128]. Rmp does not exhibit phase or antigenic variation. Rmp migrates on SDS-PAGE, with an estimated M_r of 30 kDa in its oxidized state and more slowly in its reduced state [129]. The M_r of Rmp (unheated protein) identified from the amino acid sequence is 23 kDa [130]. Rmp is noncovalently associated with Por in the OM [131, 132]. The gonococcal *rmp* gene has been cloned and amino acid sequence analysis shows that the carboxyl terminus is homologous with enterobacterial OmpA proteins [130]. A model for surface organization of Rmp proposes that the amino terminus of the protein (upstream of the OmpA shared region) forms a 16-amino acid cysteine disulfide loop that is unique to *Neisseria* species and is also exposed on the gonococcal surface [130].

Anti-Rmp antibodies directed against an upstream 16-amino acid disulfide loop are elicited after natural gonococcal infection. These antibodies block the bactericidal action of anti-LOS and anti-Por antibodies present in immune serum [133]. Natural antibodies to the carboxyl terminal portion of enterobacterial OmpA proteins cross-react with Rmp and also block the bactericidal action of NHS and human immune serum against gonococci [134]. Antibodies to Rmp also enhance susceptibility to repeat gonococcal infection [135]; female commercial sex workers in Nairobi, Kenya, who possessed serum anti-Rmp antibodies were 3.4-fold (adjusted odds ratio) more likely than women without anti-Rmp antibodies to encounter future bouts of gonorrhea from their male sex partners.

Further characterization of a candidate vaccine, enriched for Por used in a 1985 trial (*see* Subheading 2.2) demonstrated that the candidate also contained Rmp, which elicited a sevenfold increase in Rmp antibody concentration in vaccinees (a 3½-fold increase compared to the response against PorB) [136]. These Rmp antibodies may have undermined an effective bactericidal response, directed against PorB and lipooligosaccharide (LOS) which was shown otherwise to be present in these vaccine recipients [136–138]. These results have been corroborated experimentally in a mouse cervicovaginal colonization model where anti-Rmp antibodies were shown to reverse LOS antibody-mediated clearance of gonococcal infection [139].

Two mechanisms of blocking by anti-Rmp antibodies have been proposed. The first involves diversion of effective C component fixation on the surface of the organism away from the bactericidal antibody recognition site(s) or other critical locations on the organism where the full function of complement may not be effective. This may result in effete (or unproductive) fixation of C [140]. The second mechanism proposes that anti-Rmp antibodies might sterically inhibit the binding of bactericidal antibodies [141–143].

2.5 H.8 OM Protein (Lip)

H.8 protein or Lip (for lipoprotein) is a conserved surface-exposed lipoprotein found in both *N. gonorrhoeae* and *N. meningitidis* and also in two closely related commensals, *N. lactamica* and *N. cinerea* [144–147]. The estimated M_r of H.8 protein can vary between strains, but appears constant within a single strain [145, 148]. SDS-PAGE analysis of H.8 protein shows diffuse bands ranging from 18 to 30 kDa that stain poorly with either Coomassie Blue or silver [82, 149].

The genes that encode gonococcal H.8 proteins have been cloned and reveal two distinct genes that encode for proteins that are recognized by monoclonal H.8-specific mAb [145]: lipid-modified azurin (Laz) and H.8 OM protein (Lip) [150]. DNA sequence analysis suggests that Lip and Laz are glycoproteins, with lipoprotein signal peptide processing sites that are cleaved by signal peptidase II, resulting in an N-terminal cysteine residue modified with glycerol and fatty acid [151]. H.8 protein sequences are composed primarily of repeats of alanine–alanine–glutamic acid–alanine–proline (AAEAP) which is responsible for binding by the H.8-specific mAb [152]; for example, in strain FA1090, the predicted H.8 protein is 71 amino acids in length, and it is composed entirely of 13 AAEAP repeats [152]. Aberrant migration of H.8 protein on SDS-PAGE may result from its lipid modification or its unusual amino-acid composition and repeating structure [153]. The unusual amino-acid composition of H.8 protein is consistent with its lack of absorbance at a wavelength of 280 nm [154, 155]. It is not known whether AAEAP repeats in H.8 protein serves a structural role within the OM.

H.8 protein is present predominantly in the pathogenic *Neisseria* species leading to speculation it may be involved in pathogenesis. H.8 protein is conserved and immunogenic; antibody to H.8 is present in the convalescent sera of patients with localized gonococcal infection and DGI [156–158]; preexisting anti-H.8 antibody in humans, however, does not protect against uncomplicated gonococcal infections [159]. In addition, human IgG directed against pentapeptide repeat motifs of *N. meningitidis* Lip/H.8 and Laz lipoproteins blocks killing of serogroup B *N. meningitidis* by human sera that are otherwise bactericidal. Anti-H.8 IgG also reduces bactericidal efficacy of Factor H binding protein (FHbp) antibodies, elicited by serogroup B meningococcal vaccine [160].

2.6 Iron- and Zinc-Regulated Proteins

Iron, an essential nutrient for most bacteria [161] is a cofactor for enzyme function. Gonococci take up iron via TonB-dependent Transporters (TdT) directly from human transferrin, lactoferrin, hemoglobin, and by siderophores produced by non-gonococci (xenosiderophores) [162–165].

TdTs comprise large outer membrane (OM) β-barrel proteins that span the gonococcal OM [166]. The C-terminal loops extend into the extracellular space and the N-termini, that contain the "ton box," extend through the periplasmic space and interact with TonB, located in the inner (cytoplasmic) membrane [166]. TonB is complexed with biopolymer transport proteins ExbB and ExbD, and uses the cytoplasmic membrane's proton motive force to transport iron or zinc across the bacterial OM through the barrel of TdTs [166]. Ferric iron that is released into periplasmic space is scavenged by ferric binding protein (FbpA) [167]. Holo-FbpA then associates with the ABC transporter, FbpBC, enabling iron transport from the periplasm into the cytoplasm using energy derived from ATP hydrolysis [168]. Some TdTs have a lipoprotein that acts as a coreceptor and is anchored to the extracellular side of the OM, which mediates ligand binding to transporters (e.g., transferrin binding protein B (TbpB), lactoferrin binding protein B (LbpB), and hemoglobin binding protein A (HpuA)) [166, 169].

Well characterized gonococcal TdTs include TbpA/TbpB [165], LbpA/LbpB [164], HpuA/HpuB [170], ferric enterobactin transporter FetA (which binds to siderophores made by non-gonococci) [162], and calprotectin binding protein CbpA (which facilitates acquisition of zinc from host-derived protein calprotectin) [171].

The uptake of iron from transferrin involves two receptor proteins, TbpA and TbpB [172, 173]. An integral OM protein, TbpA, is highly conserved among pathogenic *Neisseria* [174] and is required for iron transport across the OM. The surface-exposed lipoprotein, TbpB, is not essential for iron transport, but facilitates iron uptake nonetheless [173, 175]. The lactoferrin and hemoglobin–iron acquisition systems resemble the transferrin iron acquisition system [166, 176, 177]. LbpA and HpuB serve as TonB-dependent transporters; LbpB and HpuA as the external lipoproteins [166, 176, 177]. An engineered gonococcal mutant of strain FA1090, that lacks both TbpA and TbpB (FA1090 is also naturally deficient in lactoferrin binding proteins), is unable to cause infection in the human experimental model [178], suggesting that both transferrin and lactoferrin iron acquisition are important to initiate infection in humans. A follow up study in the human experimental model using a FA1090 mutant strain that expressed LbpA/LbpB, but was deleted of the TbpA/TbpB genes, resulted in urethritis in male volunteers [179], suggesting that Lbp proteins are sufficient to restore infectivity.

All gonococcal strains can use human transferrin as a source of iron [165], but only 50–70% of clinical isolates take up lactoferrin-bound iron [164, 180]. Although the amino acid sequences of TbpA and TbpB vary among strains, these proteins do not undergo high-frequency phase and/or antigenic variation [181, 182].

Anti-TbpA and TbpB antibodies are detected in serum (at low levels), but not in genital secretions from naturally infected women and men [183]. Mice immunized intranasally with TbpA or TbpB conjugated to the B subunit of cholera toxin generate IgG-bactericidal antibodies in sera and vaginal secretions [184]. Similarly, intranasal immunization of mice with segments of TbpA or TbpB, fused to the A2 domain of cholera toxin, generate IgG antibody responses in sera and vaginal secretions [185].

Zinc (Zn) is also an essential nutrient that provides a structural or catalytic component of certain host and bacterial proteins [186, 187]. To restrict Zn availability at infection sites, humans deploy Zn-sequestering proteins, including calprotectin [188]. Calprotectin is a heterooligomeric protein complex that constitutes nearly half of the cytosolic protein content of neutrophils [189]. It is released into neutrophil extracellular traps (NETs) and possesses antimicrobial properties [190] because of its Zn-chelating ability [191]. Approximately 80% of gonococci express calprotectin binding protein, CbpA [192], a TonB dependent transporter, which overcomes Zn sequestration in Zn-restricted environments [171]. CbpA enhances gonococcal survival within NETs, and suggests that *N. gonorrhoeae* binds Zn-calprotectin and can obtain Zn from this metal-sequestering protein when entangled by NETs in this environment.

2.7 Lipooligosaccharide (LOS)

2.7.1 LOS Structure

LOS molecules are complex glycolipids found in the OMs of Gram-negative bacteria including non-Enterobacteriaceae such as *Neisseria*, *Haemophilus*, and *Bordetella* species [193–195]. LOS lacks the saccharide repeating units present in the lipopolysaccharides (LPSs) present in many enteric bacteria (hence the term oligo- instead of poly-). Gonococci can produce up to six antigenically and structurally distinct LOSs simultaneously, demonstrated by differences in migration when analyzed by SDS-PAGE (apparent subunit M_r ranges between 3200 and 7100) [195].

Gonococcal LOS consists of three oligosaccharide (OS) chains (Fig. 2). The OS chains branch from two heptose residues attached to lipid A via two 2-keto-3-deoxy-mannooctulosonic acid (KDO) molecules. One OS chain elongates from the first heptose (Hep_1); two chains extend from the second heptose (Hep_2). Glycan extensions (outward schematically in Fig. 2) from Hep_1 and Hep_2 are called the Hep_1 and Hep_2 chains, respectively. Hep_2 is also substituted (downward in Fig. 2) with *N*-acetyl glucosamine (GlcNac) or galactose (Gal) [196]. The number of branches and the length of OSs in each branch vary among gonococcal strains. The genes involved in LOS biosynthesis have been identified and characterized. The *rfaC* gene that encodes heptosyl transferase is required for the addition of Hep_1 to KDO [197]. The *rfaF* gene product, also a (different) heptosyl transferase adds Hep_2 onto Hep_1 and is essential for further elongation of the Hep_1 chain [198, 199]. The synthesis of OS chains is modulated by a series of LOS glycosyl

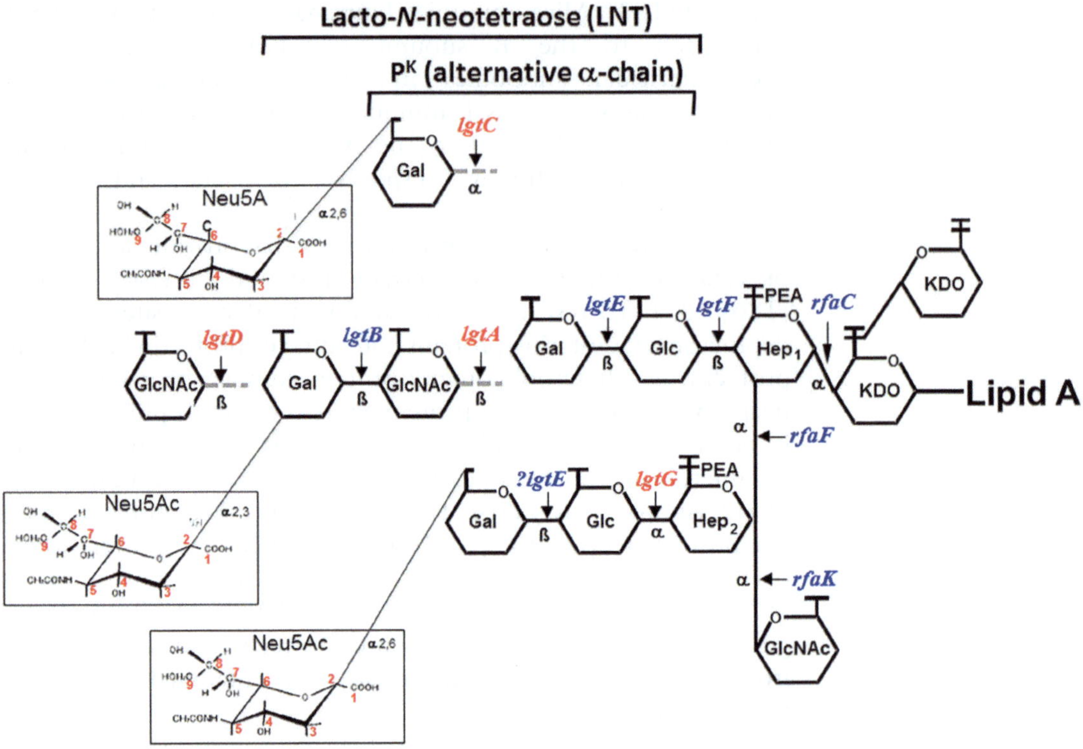

Fig. 2 General structure of gonococcal lipooligosaccharide (LOS). Gonococcal LOS consists of three oligosaccharide (OS) chains. The OS chains branch from two heptose residues attached to lipid A via two 2-keto-3-deoxy-mannooctulosonic acid (KDO) molecules. One OS chain elongates from the first heptose (Hep_1); two chains extend from the second heptose (Hep_2). Lacto-*N*-neotetraose structure (Galβ1 → 4GlcNAcβ1 → 3Galβ1 → 4Glcβ1) and P^K (Galα1 → 3Galβ1 → 4Glcβ1) extend from Hep_1. Phase variable genes involved in LOS biosynthesis are shown in red; nonvariable genes in blue. Neu5Ac (sialic acid) is shown in the boxes. Sialylation of LOS occurs via α 2–6 or α 2–3 linkage to galactose (Gal) residues. LOS branching is terminated ("capped") either by Neu5Ac (sialic acid) or otherwise extend(s) outward by adding hexose(s), linked by the broken grey lines

transferases (products of *lgt* genes). *lgtF* to *lgtD* genes are required for the stepwise addition of each sugar to extend the Hep_1-chain outward as shown in Fig. 2 [200, 201]. The *lgtF* gene product adds the first glucose (Glc) to Hep_1 [201]. Four genes (*lgtE*, -*A*, -*B*, and -*D*) encode the glycosyl transferases that add the next four hexoses onto the Hep_1-chain; *lgtC* encodes the α-galactosyl transferase responsible for adding a terminal galactose (Gal) to the sugars (also called the P^K-like LOS) extended from Hep_1, in place of continued extension.

The *lgtG* gene product adds the first glucose (Glc) onto the Hep_2 chain [202]. Hep_2 itself may be extended terminally by a single Glc or lactose (Gal-Glc) or lactose with additional sugars [203–206]. Lactose substitution of Hep_1 and Hep_2 together forms the "2C7 epitope" [206].

Phase variation of LOS results in changes in size of the LOS structures that are expressed predominantly and occur both in vitro and in vivo. LOS undergoes phase variation at a frequency of 10^{-2}–10^{-3} when grown in culture [207, 208]. The *lgtA*, *lgtC*, and *lgtD* genes (Fig. 2) contain homopolymeric tracts of guanine [poly (G)], and the *lgtG* gene, a cytosine [poly (C)] tract [200, 202, 209–211]. Slipped strand mispairing during DNA replication can result in alteration in coding sequences, which leads to premature termination of the corresponding genes and loss of function of the enzyme proteins they encode. This results in truncated LOS structures expressed by gonococci. Despite phase variation of the *lgtG* gene, the "2C7 epitope" that it controls, is expressed in at least 95% of gonococci that infect humans; its absence severely attenuates gonococcal infection in the mouse cervicovaginal colonization model [212].

Gonococci "cap" LOS molecules that terminate in the lacto-*N*-neotetraose structure (four sugars extending outward from Hep_1 (Fig. 2)) with sialic acid, using the organism's own endogenous sialyltransferase and appropriate sialic acid substrate(s) that are otherwise available in vivo because gonococci do not make sialic acid substrate (unlike most serogroups of *N. meningitidis*). In vivo, substrate(s) is present in the mammalian genital tract; in vitro (exogenous) cytidine monophospho-*N*-acetylneuraminic acid (CMP-NANA) serves as a suitable substrate. The terminal Gal of the P^K-like structure from Hep1 can also be sialylated. Recently, an additional acceptor site for sialic acid has been identified on Hep_2 lactose, when Hep_1 and Hep_2 both bear only lactose extensions [204].

2.7.2 LOS Sialylation and Complement Resistance

Complement activation on microbial surfaces occurs through three pathways: classical, lectin, and alternative. LOS sialic acid inhibits the classical and alternative pathways.

An intact classical pathway is essential for complement-dependent killing of wild-type *N. gonorrhoeae* [213]. Killing of unsialylated "serum-sensitive" gonococci by nonimmune normal human sera (NHS) is mediated principally by IgM directed against LOS [134, 214–217]; this killing is reduced by LOS sialylation [218, 219] without necessarily decreasing IgM binding [220]. Nonetheless, binding of NHS-IgG to the bacterial surface is diminished by sialylation [221]. Furthermore, sialylation also reduces binding of some [222] but not all [218, 223] antibodies directed against a separate target, PorB (but not Opa proteins) suggesting a more global role such as interfering with the ability of antibody to bind and/or to engage C1q (the first step in the initiation of the classical pathway) [224].

Sialylation of the lacto-*N*-neotetraose (LNnT) structure of gonococcal LOS enhances human factor H (FH) binding [225] to *N. gonorrhoeae*. FH (down)regulates the alternative pathway of complement by providing cofactor activity for factor I-mediated C3b cleavage and decay accelerating activity against the alternative pathway C3-convertase, C3bBb. C3 serves as the central component in the complement pathway overall, located at the junction of the two pathways. Sialylated gonococci that bind FH escape killing by complement [226, 227]. Sialylation of gonococcal LNnT LOS, but not sialylation of the P^K-like LOS, [228] or LOS in which Hep_2 extension terminate(s) with lactose [204] enhances FH binding. Nevertheless, P^K and lactose sialylated gonococci are also more resistant to killing by NHS than their unsialylated counterparts (although to a lesser degree than LNnT sialylated organisms) by mechanisms not fully understood.

2.7.3 LOS Sialylation In Vivo

Sialylation of LNnT LOS occurs in vivo in the male urethra as evidenced by electron microscopic studies [229]. LNnT sialylation has also been shown in the human experimental model using a variant of *N. gonorrhoeae* that in vitro expressed predominantly (unsialylated) lactose (Galβ1-4Glc) from Hep_1; gonococci recovered by culture from infected men expressed the sialylatable LNnT LOS species [230, 231] (presumed to be sialylated at the time of recovery but not tested). Because LNnT (unsialylated) on gonococci binds to asialoglycoprotein receptor sites in the male genital tract [232], "pre"-sialylated gonococci infect less well (1 of 5 [20%] of subjects compared with 5 of 6 [86%] infected with LNnT [unsialylated] bacteria) [233]. Use of either LNnT (unsialylated) or lactose gonococcal mutants in the model resulted in documented infection in 5/5 vs. 2/5 subjects, respectively [230]. Sialyltransferase-deficient mutants are attenuated in the mouse cervicovaginal colonization model [234, 235]. Collectively, these data support an important role for LOS sialylation in pathogenesis, thereby making it an attractive target for immunotherapeutic strategies to combat multidrug resistance.

3 Conclusions

The gonococcal research community is confronted with a modern-day gonococcus, which subverts biochemical and immune pathways, is rapidly becoming resistant to currently available antimicrobial agents and against which there is no vaccine. *Neisseria gonorrhoeae* possesses structures that are involved in numerous pathogenic mechanisms and in evasion of host defenses. These include structures that have been reviewed in this chapter and others that are addressed subsequently in this book. A more complete understanding of pathogenesis, in particular the evasion mechanisms used by the organism, holds a hope of defining weaknesses in the organism that can be exploited therapeutically.

Continued detailing of the genetics of antimicrobial resistance will inform the development of therapeutic agents to circumvent the numerous pathways that often contribute in an additive fashion to antimicrobial resistance. Vaccine development will concentrate on defining broadly representative antigenic targets that also serve as virulence factors to be neutralized and simultaneously are critical for organism survival. Such multifaceted requirements may be the only way to "corner" a skillful organism that employs a multitude of mechanisms to survive selective pressures.

References

1. Rice PA (2005) Gonococcal arthritis (disseminated gonococcal infection). Infect Dis Clin N Am 19(4):853–861. https://doi.org/10.1016/j.idc.2005.07.003
2. Christian P, Khatry SK, LeClerq SC et al (2005) Prevalence and risk factors of chlamydia and gonorrhea among rural Nepali women. Sex Transm Infect 81(3):254–258. https://doi.org/10.1136/sti.2004.011817
3. Laga M, Meheus A, Piot P (1989) Epidemiology and control of gonococcal ophthalmia neonatorum. Bull World Health Org 67 (5):471–477
4. Laga M, Manoka A, Kivuvu M et al (1993) Non-ulcerative sexually transmitted diseases as risk factors for HIV-1 transmission in women: results from a cohort study [see comments]. AIDS 7(1):95–102
5. Cohen MS (1998) Sexually transmitted diseases enhance HIV transmission: no longer a hypothesis. Lancet 351(Suppl 3):5–7
6. Cohen MS, Hoffman IF, Royce RA et al (1997) Reduction of concentration of HIV-1 in semen after treatment of urethritis: implications for prevention of sexual transmission of HIV-1. AIDSCAP Malawi Research Group. Lancet 349(9069):1868–1873
7. Newman L, Rowley J, Vander Hoorn S et al (2015) Global estimates of the prevalence and incidence of four curable sexually transmitted infections in 2012 based on systematic review and global reporting. PLoS One 10(12): e0143304. https://doi.org/10.1371/journal.pone.0143304
8. Dubbink JH, de Waaij DJ, Bos M et al (2016) Microbiological characteristics of *Chlamydia trachomatis* and *Neisseria gonorrhoeae* infections in South African women. J Clin Microbiol 54(1):200–203. https://doi.org/10.1128/JCM.02848-15
9. Ginocchio CC, Chapin K, Smith JS et al (2012) Prevalence of *Trichomonas vaginalis* and coinfection with *Chlamydia trachomatis* and *Neisseria gonorrhoeae* in the United States as determined by the Aptima *Trichomonas vaginalis* nucleic acid amplification assay. J Clin Microbiol 50(8):2601–2608. https://doi.org/10.1128/JCM.00748-12
10. Guy R, Ward J, Wand H et al (2015) Coinfection with *Chlamydia trachomatis, Neisseria gonorrhoeae* and *Trichomonas vaginalis*: a cross-sectional analysis of positivity and risk factors in remote Australian Aboriginal communities. Sex Transm Infect 91(3):201–206. https://doi.org/10.1136/sextrans-2014-051535
11. Lim RB, Wong ML, Cook AR et al (2015) Determinants of chlamydia, gonorrhea, and coinfection in heterosexual adolescents attending the national public sexually transmitted infection clinic in Singapore. Sex Transm Dis 42(8):450–456. https://doi.org/10.1097/OLQ.0000000000000316
12. Marangoni A, Foschi C, Nardini P et al (2012) *Chlamydia trachomatis* serovar distribution and other sexually transmitted coinfections in subjects attending an STD outpatients clinic in Italy. New Microbiol 35 (2):215–219
13. Papadogeorgakis H, Pittaras TE, Papaparaskevas J et al (2010) *Chlamydia trachomatis* serovar distribution and *Neisseria gonorrhoeae* coinfection in male patients with urethritis in Greece. J Clin Microbiol 48(6):2231–2234. https://doi.org/10.1128/JCM.00586-10
14. Centers for Disease Control and Prevention (CDC) (2016) 2016 Sexually transmitted diseases surveillance. CDC, Atlanta, GA
15. Ohnishi M, Golparian D, Shimuta K et al (2011) Is *Neisseria gonorrhoeae* initiating a future era of untreatable gonorrhea?: detailed characterization of the first strain with high-level resistance to ceftriaxone. Antimicrob Agents Chemother 55(7):3538–3545. https://doi.org/10.1128/aac.00325-11

16. Camara J, Serra J, Ayats J et al (2012) Molecular characterization of two high-level ceftriaxone-resistant *Neisseria gonorrhoeae* isolates detected in Catalonia, Spain. J Antimicrob Chemother 67(8):1858–1860. https://doi.org/10.1093/jac/dks162
17. Lahra MM, Ryder N, Whiley DM (2014) A new multidrug-resistant strain of *Neisseria gonorrhoeae* in Australia. N Engl J Med 371 (19):1850–1851. https://doi.org/10.1056/NEJMc1408109
18. Cole MJ, Spiteri G, Jacobsson S et al (2017) Overall low extended-spectrum cephalosporin resistance but high azithromycin resistance in *Neisseria gonorrhoeae* in 24 European countries, 2015. BMC Infect Dis 17(1):617. https://doi.org/10.1186/s12879-017-2707-z
19. Katz AR, Komeya AY, Kirkcaldy RD et al (2017) Cluster of *Neisseria gonorrhoeae* isolates with high-level azithromycin resistance and decreased ceftriaxone susceptibility, Hawaii, 2016. Clin Infect Dis 65 (6):918–923. https://doi.org/10.1093/cid/cix485
20. Liang JY, Cao WL, Li XD et al (2016) Azithromycin-resistant *Neisseria gonorrhoeae* isolates in Guangzhou, China (2009–2013): coevolution with decreased susceptibilities to ceftriaxone and genetic characteristics. BMC Infect Dis 16:152. https://doi.org/10.1186/s12879-016-1469-3
21. Vidovic S, Thakur SD, Horsman GB et al (2012) Longitudinal analysis of the evolution and dissemination of *Neisseria gonorrhoeae* strains (Saskatchewan, Canada, 2005 to 2008) reveals three major circulating strains and convergent evolution of ciprofloxacin and azithromycin resistance. J Clin Microbiol 50 (12):3823–3830. https://doi.org/10.1128/JCM.01402-12
22. Xue J, Ni C, Zhou H et al (2015) Occurrence of high-level azithromycin-resistant *Neisseria gonorrhoeae* isolates in China. J Antimicrob Chemother 70(12):3404–3405. https://doi.org/10.1093/jac/dkv266
23. Thakur SD, Levett PN, Horsman GB et al (2018) High levels of susceptibility to new and older antibiotics in *Neisseria gonorrhoeae* isolates from Saskatchewan (2003–15): time to consider point-of-care or molecular testing for precision treatment? J Antimicrob Chemother 73(1):118–125. https://doi.org/10.1093/jac/dkx333
24. Chesson HW, Kirkcaldy RD, Gift TL et al (2018) An illustration of the potential health and economic benefits of combating antibiotic-resistant gonorrhea. Sex Transm Dis 45(4):250–253. https://doi.org/10.1097/OLQ.0000000000000725
25. Hobbs MM, Sparling PF, Cohen MS et al (2011) Experimental gonococcal infection in male volunteers: cumulative experience with *Neisseria gonorrhoeae* strains FA1090 and MS11mkC. Front Microbiol 2:123. https://doi.org/10.3389/fmicb.2011.00123
26. Jerse AE (1999) Experimental gonococcal genital tract infection and opacity protein expression in estradiol-treated mice. Infect Immun 67(11):5699–5708
27. Harvey HA, Ketterer MR, Preston A et al (1997) Ultrastructural analysis of primary human urethral epithelial cell cultures infected with *Neisseria gonorrhoeae*. Infect Immun 65:2420–2427
28. McGee ZA, Stephens DS, Hoffman LH et al (1983) Mechanisms of mucosal invasion by pathogenic *Neisseria*. Rev Infect Dis 5(Suppl 4):S708–S714
29. Rudel T, van Putten JP, Gibbs CP et al (1992) Interaction of two variable proteins (PilE and PilC) required for pilus-mediated adherence of *Neisseria gonorrhoeae* to human epithelial cells. Mol Microbiol 6:3439–3450
30. van Putten JP (1993) Phase variation of lipooligosaccharide directs interconversion of invasive and immuno-resistant phenotypes of *Neisseria gonorrhoeae*. EMBO J 12:4043–4051
31. van Putten JP, Duensing TD, Carlson J (1998) Gonococcal invasion of epithelial cells driven by P.IA, a bacterial ion channel with GTP binding properties. J Exp Med 188(5):941–952
32. Weel JF, Hopman CT, van Putten JP (1991) In situ expression and localization of *Neisseria gonorrhoeae* opacity proteins in infected epithelial cells: apparent role of Opa proteins in cellular invasion. J Exp Med 173 (6):1395–1405
33. Jonsson AB, Nyberg G, Normark S (1991) Phase variation of gonococcal pili by frameshift mutation in *pilC*, a novel gene for pilus assembly. EMBO J 10:477–488
34. Jonsson AB, Pfeifer J, Normark S (1992) *Neisseria gonorrhoeae* PilC expression provides a selective mechanism for structural diversity of pili. Proc Natl Acad Sci U S A 89:3204
35. Rudel T, Scheuerpflug I, Meyer TF (1995) Neisseria PilC protein identified as type-4 pilus tip-located adhesin. Nature 373:357–359
36. Giltner CL, Nguyen Y, Burrows LL (2012) Type IV pilin proteins: versatile molecular

modules. Microbiol Mol Biol Rev 76 (4):740–772. https://doi.org/10.1128/MMBR.00035-12
37. Maier B, Potter L, So M et al (2002) Single pilus motor forces exceed 100 pN. Proc Natl Acad Sci U S A 99(25):16012–16017. https://doi.org/10.1073/pnas.242523299
38. Merz AJ, So M, Sheetz MP (2000) Pilus retraction powers bacterial twitching motility. Nature 407(6800):98–102. https://doi.org/10.1038/35024105
39. Sparling PF (1966) Genetic transformation of *Neisseria gonorrhoeae* to streptomycin resistance. J Bacteriol 92:1364–1371
40. Dubnau D (1999) DNA uptake in bacteria. Annu Rev Microbiol 53:217–244. https://doi.org/10.1146/annurev.micro.53.1.217
41. Graves JF, Biswas GD, Sparling PF (1982) Sequence-specific DNA uptake in transformation of *Neisseria gonorrhoeae*. J Bacteriol 152 (3):1071–1077
42. Lorenz MG, Wackernagel W (1994) Bacterial gene transfer by natural genetic transformation in the environment. Microbiol Rev 58 (3):563–602
43. Kellogg D Jr, Peacock W, Deacon W et al (1963) *Neisseria gonorrhoeae*. I Virulence genetically linked to clonal variation. J Bacteriol 85:1274–1279
44. Brinton CCJ, Wood SW, Brown A et al (1982) The development of a Neisserial pilus vaccine for gonorrhea and meningococcal meningitis. In: Weinstein L, Fields BN (eds) Seminars in infectious disease IV. Thieme-Stratton, New York, NY
45. Kellogg D Jr, Cohen IR, Norins LC et al (1968) *Neisseria gonorrhoeae*. II. Colonial variation and pathogenicity during 35 months in vitro. J Bacteriol 96(3):596–605
46. Ofek I, Beachey EH, Bisno AL (1974) Resistance of *Neisseria gonorrhoeae* to phagocytosis: relation to colonial morphology and surface pili. J Infect Dis 129:310–316
47. Thonthai C, Sawyer WD (1973) Studies on the virulence of *Neisseria gonorrhoeae*. I. Relation of colonial morphology and resistance to phagocytosis by polymorphonuclear lymphocytes. Infect Immun 7:373–379
48. Koomey M, Gotschlich EC, Robbins K et al (1987) Effects of *recA* mutations on pilus antigenic variation and phase transitions in *Neisseria gonorrhoeae*. Genetics 117:391–398
49. Zhang QY, DeRyckere D, Lauer P et al (1992) Gene conversion in *Neisseria gonorrhoeae*: evidence for its role in pilus antigenic variation. Proc Natl Acad Sci U S A 89:5366–5370
50. Ilver D, KAllstrom H, Normark S et al (1998) Transcellular passage of *Neisseria gonorrhoeae* involves pilus phase variation. Infect Immun 66:469–473
51. Gotschlich E (1984) Development of a gonorrhoea vaccine: prospects, strategies and tactics. Bull WHO 62:671–680
52. Hagblom P, Segal E, Billyard E et al (1985) Intragenic recombination leads to pilus antigenic variation in *Neisseria gonorrhoeae*. Nature 315(6015):156–158
53. Haas R, Meyer TF (1986) The repertoire of silent pilus genes in *Neisseria gonorrhoeae*: evidence for gene conversion. Cell 44:107–115
54. Haas R, Schwarz H, Meyer TF (1987) Release of soluble pilin antigen coupled with gene conversion in *Neisseria gonorrhoeae*. Proc Natl Acad Sci U S A 84(24):9079–9083
55. Schoolnik GK, Fernandez R, Tai JY et al (1984) Gonococcal pili; primary structure and receptor binding domain. J Exp Med 159:1351–1370
56. Parge HE, Forest KT, Hickey MJ et al (1995) Structure of the fiber-forming protein pilin at 2.6 A resolution. Nature 378:32–38
57. Buchanan TM (1975) Antigenic heterogeniety of gonococcal pili. J Exp Med 141:1470–1475
58. Lambden PR, Robertson JN, Watt PJ (1980) Biological properties of two distinct pilus types produced by isogenic variants of *Neisseria gonorrhoeae* P9. J Bacteriol 141:393–396
59. Virji M, Saunders JR, Sims G et al (1993) Pilus-facilitated adherence of *Neisseria meningitidis* to human epithelial and endothelial cells: modulation of adherence phenotype occurs concurrently with changes in primary amino acid sequence and the glycosylation status of pilin. Mol Microbiol 10:1013–1028
60. Scheuerpflug I, Rudel T, Ryll R et al (1999) Roles of PilC and PilE proteins in pilus-mediated adherence of *Neisseria gonorrhoeae* and *Neisseria menigitidis* to human erythrocytes and endothelial and epithelial cells. Infect Immun 67:834–843
61. Blom AM, Rytkonen A, Vasquez P et al (2001) A novel interaction between type IV pili of *Neisseria gonorrhoeae* and the human complement regulator C4B-binding protein. J Immunol 166(11):6764–6770
62. Edwards JL, Brown EJ, Uk-Nham S et al (2002) A co-operative interaction between *Neisseria gonorrhoeae* and complement receptor 3 mediates infection of primary cervical epithelial cells. Cell Microbiol 4(9):571–584

63. Kallstrom H, Liszewski MK, Atkinson JP et al (1997) Membrane cofactor protein (MCP or CD 46) is a cellular pilus receptor for pathogenic *Neisseria*. Mol Microbiol 25:639–647
64. Tramont EC, Ciak J, Boslego J et al (1980) Antigenic specificity of antibodies in vaginal secretions during infection with *Neisseria gonorrhoeae*. J Infect Dis 142(1):23–31
65. Cohen MS, Cannon JG (1999) Human experimentation with *Neisseria gonorrhoeae*: progress and goals. J Infect Dis 179(Suppl 2):S375–S379
66. Tramont EC, Sadoff JC, Boslego JW et al (1981) Gonococcal pilus vaccine. J Clin Invest 68:881–888
67. Boslego JW, Tramont EC, Chung RC et al (1991) Efficacy trial of a parenteral gonococcal pilus vaccine in men. Vaccine 9 (3):154–162
68. Johnston KH, Holmes KK, Gotschlich EC (1976) The serological classification of *Neisseria gonorrhoeae*. I Isolation of the outer membrane complex responsible for serotypic specificity. J Exp Med 143:741–758
69. Blake MS, Gotschlich EC (1982) Purification and partial characterization of the major outer membrane protein of *Neisseria gonorrhoeae*. Infect Immun 36:277–283
70. Johnson AP, Taylor-Robinson D, McGee ZA (1977) Species specificity of attachment and damage to oviduct mucosa by *Neisseria gonorrhoeae*. Infect Immun 18:833–839
71. Joiner KA, Warren KA, Brown EJ et al (1983) Studies on the mechanism of bacterial resistance to complement-mediated killing. IV. C5b-9 forms high molecular weight complexes with bacterial outer membrane constituents on serum-resistant but not on serum-sensitive *Neisseria gonorrhoeae*. J Immunol 131:1443–1451
72. Benz R, Ghuysen JM, Hakenbeck R (1995) Uptake of solutes through bacterial outer membranes. In: Ghuysen JM, Hakenbeck R (eds) Bacterial cell wall. Elsevier, Amsterdam, pp 397–423
73. Nikaido H (1994) Porins and specific diffusion channels in bacterial outer membranes. J Biol Chem 269:3905–3908
74. Weiss MS, Abele U, Weckesser J et al (1991) Molecular architecture and electrostatic properties of a bacterial porin. Science 254:1627–1630
75. Weiss MS, Schulz GE (1992) Structure of porin refined at 1.8 Å resolution. J Mol Biol 227:493–509
76. Tam MR, Buchanan TM, Sandstrom EG et al (1982) Serological classification of *Neisseria gonorrhoeae* with monoclonal antibodies. Infect Immun 36:1042–1053
77. Cooke SJ, de la Paz JH, La Poh C et al (1997) Variation within serovars of *Neisseria gonorrhoeae* detected by structural analysis of outer membrane protein PIB and by pulsed-field gel electrophoresis. Microbiology 143:1415–1422
78. Mee BJ, Thomas H, Cooke SJ et al (1993) Structural comparison and epitope analysis of outer-membrane protein PIA from strains of *Neisseria gonorrhoeae* with differing serovar specificities. J Gen Microbiol 139:2613–2620
79. Knapp JS, Tam MR, Nowinski RC et al (1984) Serological classification of *Neisseria gonorrhoeae* with use of monoclonal antibodies to gonococcal outer membrane protein I. J Infect Dis 150:44–48
80. Blake MS, Gotschlich EC, Swanson JL (1981) Effect of proteolytic enzymes on the outer membrane proteins of *Neisseria gonorrhoeae*. Infect Immun 34:212–222
81. Elkins C, Carbonetti N, Varela V et al (1992) Antibodies to N-terminal peptides of gonococcal porin are bactericidal when gonococcal lipooligasaccharide is not sialylated. Mol Microbiol 6:2617–2628
82. Schoolnik GK, Mietzner TA (1992) Vaccines against gonococcal infection. In: Woodrow GC, Levine MM (eds) New generation vaccines. Marcel Dekker Inc., New York, NY, pp 565–597
83. Ram S, McQuillen DP, Gulati S et al (1998) Binding of complement factor H to loop 5 of porin protein 1A: a molecular mechanism of serum resistance of nonsialylated *Neisseria gonorrhoeae*. J Exp Med 188(4):671–680
84. Ram S, Cullinane M, Blom AM et al (2001) Binding of C4b-binding protein to porin: a molecular mechanism of serum resistance of *Neisseria gonorrhoeae*. J Exp Med 193 (3):281–295
85. Fudyk TC, Maclean IW, Simonsen JN et al (1999) Genetic diversity and mosaicism at the *por* locus of *Neisseria gonorrhoeae*. J Bacteriol 181:5591–5599
86. Massari P, Visintin A, Gunawardana J et al (2006) Meningococcal porin PorB binds to TLR2 and requires TLR1 for signaling. J Immunol 176(4):2373–2380
87. Singleton TE, Massari P, Wetzler LM (2005) Neisserial porin-induced dendritic cell activation is MyD88 and TLR2 dependent. J Immunol 174(6):3545–3550
88. Toussi DN, Carraway M, Wetzler LM et al (2012) The amino acid sequence of *Neisseria lactamica* PorB surface-exposed loops

influences Toll-like receptor 2-dependent cell activation. Infect Immun 80(10):3417–3428. https://doi.org/10.1128/IAI.00683-12

89. Zhu W, Tomberg J, Knilans KJ et al (2018) Properly folded and functional PorB from *Neisseria gonorrhoeae* inhibits dendritic cell stimulation of CD4(+) T cell proliferation. J Biol Chem 293(28):11218–11229. https://doi.org/10.1074/jbc.RA117.001209
90. Plummer FA, Simonsen JN, Chubb H et al (1989) Epidemiologic evidence for the development of serovar-specific immunity after gonococcal infection. J Clin Invest 83 (5):1472–1476. https://doi.org/10.1172/jci114040
91. Buchanan TM, Eschenbach DA, Knapp JS et al (1980) Gonococcal salpingitis is less likely to recur with *Neisseria gonorrhoeae* of the same principal outer membrane protein antigenic type. Am J Obstet Gynecol 138 (7 Pt 2):978–980
92. Fox KK, Thomas JC, Weiner DH et al (1999) Longitudinal evaluation of serovar-specific immunity to *Neisseria gonorrhoeae*. Am J Epidemiol 149(4):353–358
93. Kohl P, Olsen D, Buchanan T (1989) Monoclonal antibodies to protein I for serotyping of *Neisseria gonorrhoeae*: correlation of serotype with bactericidal activity. Zbl Bakt Hyg A 270:517–526
94. Virji M, Fletcher J, Zak K et al (1987) The potential protective effect of monoclonal antibodies to gonococcal outer membrane protein IA. J Gen Microbiol 133:2639–2646
95. Virji M, Zak K, Heckels JE (1986) Monoclonal antibodies to gonococcal outer membrane protein IB: use in investigations of the potential protective effect of antibodies directed agains conserved and type-specific epitopes. J Gen Microbiol 132:1621–1629
96. Hook EW 3rd, Olsen DA, Buchanan TM (1984) Analysis of the antigen specificity of the human serum immunoglobulin G immune response to complicated gonococcal infection. Infect Immun 43(2):706–709
97. Frasch CE, Mocca LF (1978) Heat modifiable outer membrane proteins of *Neisseria meningitidis* and their organization within the membrane. J Bacteriol 136:1127–1134
98. Lambden P, Heckels J, James L et al (1979) Variations in surface protein composition associated with virulence properties in opacity types of *Neisseria gonorrhoeae*. J Gen Microbiol 114:305–312
99. James JF, Swanson J (1978) Studies on gonococcus infection. XII. Occurrence of color/opacity colonial variants in clinical cultures. Infect Immun 19:332–340
100. Blake MS, Blake CM, Apicella MA et al (1995) Gonococcal opacity: lectin-like interactions between Opa proteins and lipooligosaccharide. Infect Immun 63(4):1434–1439
101. Swanson J (1982) Colony opacity and protein II compositions of gonococci. Infect Immun 37:359–368
102. Stern A, Brown M, Nickel P et al (1986) Opacity genes in *Neisseria gonorrhoeae*: control of phase and antigenic variation. Cell 47:61–71
103. Bhat KC, Gibbs CP, Barrera O et al (1991) The opacity proteins of *Neisseria gonorrhoeae* strain MS11 are encoded by a family of 11 complete genes. Mol Microbiol 5:1889–1901
104. Swanson J, Barrera O (1983) Immunological characteristics of gonococcal outer membrane protein II assessed by immunoprecipitation, immunoblotting, and coagglutination. J Exp Med 157:1405–1420
105. de Jonge MI, Bos MP, Hamstra HJ et al (2002) Conformational analysis of opacity proteins from Neisseria meningitidis. Eur J Biochem 269(21):5215–5223
106. Swanson J (1978) Studies on gonococcus infection. XIV. Cell wall protein differences among color/opacity variants of gonococci. Infect Immun 21:292–302
107. Mayer LW (1982) Rates of *in vitro* changes of gonococci opacity phenotypes. Infect Immun 37:481–485
108. Swanson J (1981) Surface-exposed protein antigens of the gonococcal outer membrane. Infect Immun 34:804–816
109. Zak K, Diaz J, Jackson D et al (1984) Antigenic variation during infection with *Neisseria gonorrhoeae*: detection of antibodies to surface proteins in sera of patients with gonorrhea. J Infect Dis 149:166–174
110. Draper DL, James JF, Brooks GF et al (1980) Comparison of virulence markers of peritoneal and fallopian tube isolates with endocervical *Neisseria gonorrhoeae* isolates from woman with acute salpingitis. Infect Immun 27:882–888
111. Jerse AE, Cohen MS, Drown PM et al (1995) Multiple gonococcal opacity proteins are expressed during experimental urethral infection in the male. J Exp Med 179:911–920
112. Swanson J, Barrera O, Sola J et al (1988) Expression of outer membrane protein II by gonococci in experimental gonorrhea. J Exp Med 168:2121–2129

113. Kupsch EM, Knepper B, Kuroki T et al (1993) Variable opacity (Opa) outer membrane proteins account for the cell tropisms displayed by *Neisseria gonorrhoeae* for human leukocytes and epithelial cells. EMBO J 12:641–650

114. Makino S, van Putten JP, Meyer TF (1991) Phase variation of the opacity outer membrane protein controls invasion by *Neisseria gonorrhoeae* into human epithelial cells. EMBO J 10(6):1307–1315

115. Chen T, Grunert F, Medina-Marino A et al (1997) Several carciniembryonic antigens (CD66) serve as receptors for gonococcal opacity proteins. J Exp Med 185:1557–1564

116. Gray-Owen SD, Lorenzen DR, Haude A et al (1997) Differential Opa specificities for CD66 receptors influence tissue interactions and cellular response to *Neisseria gonorrhoeae*. Mol Microbiol 26:971–980

117. Virji M, Makepeace K, Ferguson DJ et al (1996) Carcinoembryonic antigens (CD66) on epithelial cells and neutrophils are receptors for Opa proteins of pathogenic neisseriae. Mol Microbiol 29:941–950

118. Gomez-Duarte OG, Dehio M, Guzman CA et al (1997) Binding of vitronectin to opa-expressing *Neisseria gonorrhoeae* mediates invasion of HeLa cells. Infect Immun 65(9):3857–3866

119. Boulton IC, Gray-Owen SD (2002) Neisserial binding to CEACAM1 arrests the activation and proliferation of CD4+ T lymphocytes. Nat Immunol 3(3):229–236. https://doi.org/10.1038/ni769

120. Sarantis H, Gray-Owen SD (2007) The specific innate immune receptor CEACAM3 triggers neutrophil bactericidal activities via a Syk kinase-dependent pathway. Cell Microbiol 9 (9):2167–2180. https://doi.org/10.1111/j.1462-5822.2007.00947.x. CMI947 [pii]

121. Schmitter T, Agerer F, Peterson L et al (2004) Granulocyte CEACAM3 is a phagocytic receptor of the innate immune system that mediates recognition and elimination of human-specific pathogens. J Exp Med 199 (1):35–46. https://doi.org/10.1084/jem.20030204

122. Rest RF, Frangipane JV (1992) Growth of *Neisseria gonorrhoeae* in CMP-N-acetylneuraminic acid inhibits nonopsonic (opacity-associated outer membrane protein-mediated) interactions with human neutrophils. Infect Immun 60(3):989–997

123. Muenzner P, Rohde M, Kneitz S et al (2005) CEACAM engagement by human pathogens enhances cell adhesion and counteracts bacteria-induced detachment of epithelial cells. J Cell Biol 170(5):825–836. https://doi.org/10.1083/jcb.200412151

124. Naids FL, Belisle B, Lee N et al (1991) Interactions of *Neisseria gonorrhoeae* with human neutrophils: studies with purified PII (Opa) outer membrane proteins and synthetic Opa peptides. Infect Immun 59(12):4628–4635

125. Elkins C, Rest RF (1990) Monoclonal antibodies to outer membrane protein PII block interactions of *Neisseria gonorrhoeae* with human neutrophils. Infect Immun 58:1078–1084

126. Sugasawara RJ, Cannon JG, Black WJ et al (1983) Inhibition of *Neisseria gonorrhoeae* attachment to HeLa cells with monoclonal antibody directed against a protein II. Infect Immun 42:980–985

127. Plummer FA, Chubb H, Simonsen JN et al (1994) Antibodies to opacity proteins (Opa) correlate with a reduced risk of gonococcal salpingitis. J Clin Invest 93(4):1748–1755. https://doi.org/10.1172/jci117159

128. Judd RC (1982) ^{125}I-peptide mapping of protein III isolated from four strains of *Neisseria gonorrhoeae*. Infect Immun 37:622–631

129. McDade RL, Johnston KH (1980) Characterization of serologically dominant outer membrane proteins of *Neisseria gonorrhoeae*. J Bacteriol 141:1183–1191

130. Gotschlich EC, Seiff M, Blake MS (1987) The DNA sequence of the structural gene of gonococcal protein III and the flanking region containing a repetitive sequence: homology of protein III with enterobacterial OmpA proteins. J Exp Med 165:471–482

131. Newhall W, Sawyer V, Haak R (1980) Cross-linking analysis of the outer membrane proteins of *Neisseria gonorrhoeae*. Infect Immun 28:785–791

132. Swanson J, Mayer LW, Tam MR (1982) Antigenicity of *Neisseria gonorrhoeae* outer membrane protein(s) III detected by immunoprecipitation and Western blot transfer with a monoclonal antibody. Infect Immun 38:668–672

133. Rice PA, McQuillen DP, Gulati S et al (1994) Serum resistance of *Neisseria gonorrhoeae*. Does it thwart the inflammatory response and facilitate the transmission of infection? Ann N Y Acad Sci 730:7–14

134. Rice PA (1989) Molecular basis for serum resistance in *Neisseria gonorrhoeae*. Clin Microbiol Rev 2S:S112–S117

135. Plummer FA, Chubb H, Simonsen JN et al (1993) Antibody to Rmp (outer membrane protein 3) increases susceptibility to

gonococcal infection. J Clin Invest 91 (1):339–343. https://doi.org/10.1172/jci116190

136. Rice PA, Gulati S, McQuillen DP et al (1996) Is there protective immunity to gonococcal disease? Proceedings of the 10th international pathogenic neisseria conference (IPNC 1996), Baltimore, MD, pp. 3–8

137. Rice PA, Tam MR, Blake MS (1985) Immunoglobulin G antibodies in normal human serum directed against protein III block killing; of serum-resistant *Neisseria gonorrhoeae* by immune human serum. In: Schoolnik G, Brooks G, Falco C et al (eds) The pathogenic Neisseriae: proceedings of the fourth international symposium. American Society for Microbiology, Washington, DC, pp 427–430

138. Rice PA, Vayo HE, Tam MR et al (1986) Immunoglobulin G antibodies directed against protein III block killing of serum-resistant *Neisseria gonorrhoeae* by immune serum. J Exp Med 164(5):1735–1748

139. Gulati S, Mu X, Zheng B et al (2015) Antibody to reduction modifiable protein increases the bacterial burden and the duration of gonococcal infection in a mouse model. J Infect Dis 212(2):311–315. https://doi.org/10.1093/infdis/jiv024

140. Joiner KA, Scales R, Warren KA et al (1985) Mechanism of action of blocking immunoglobulin G for *Neisseria gonorrhoeae*. J Clin Invest 76:1765–1772

141. Guttman RM, Waisbren BA (1975) Bacterial blocking activity of specific IgG in chronic *Pseudomonas aeruginosa* infection. Clin Exp Immunol 19:121

142. Taylor PW (1972) An antibactericidal factor in the serum of two patients with infections of the upper urinary tract. Clin Sci 43:23–30

143. Waisbren BA, Brown I (1966) A factor in the serum of patients with persisting infection that inhibits the bactericidal activity of normal serum against the organism that is causing the infection. J Immunol 97:431

144. Black WJ, Cannon JG (1985) Cloning of the gene for the common pathogenic Neisseria H.8 antigen of *Neisseria gonorrhoeae*. Infect Immun 47:322–325

145. Cannon JG, Black WJ, Nachamkin I et al (1984) Monoclonal antibody that recognizes an outer membrane antigen common to pathogenic *Neisseria* species but not to most nonpathogenic *Neisseria* species. Infect Immun 43:994–999

146. Hitchcock PJ, Brown TM (1983) Morphological heterogeneity among *Salmonella* lipopolysaccharide chemotypes in silver stained polyacrylamide gels. J Bacteriol 154:269–277

147. Zollinger WD, Ray JS, Moran EE et al (1985) Identification by monoclonal antibody of an antigen common to the pathogenic *Neisseria* species. In: SG K, Brooks GF, Falkow S et al (eds) The pathogenic neisseria. ASM, Washington, DC, pp 579–584

148. Hitchcock PJ, Hayes SF, Mayer LW et al (1985) Analysis of gonococcal H.8 antigen: surface location, inter- and intrastrain electrophoretic heterogeneity and unusual two-dimensional electrophoretic characteristics. J Exp Med 162(6):2017–2034

149. Wood JP, Spinola SM, Strobel SM et al (1989) Conserved lipoprotein H.8 of pathogenic *Neisseria* consists entirely of pentapeptide repeats. Mol Microbiol 3:43–50

150. Gotschlich EC, Blake MS, Koomey JM et al (1986) Cloning of the structural genes of three H8 antigens and of protein III of *Neisseria gonorrhoeae*. J Exp Med 164:868–881

151. Mizushima S (1984) Post-translational modification and processing of outer membrane prolipoproteins in *Escherichia coli*. Mol Cell Biochem 60(1):5–15

152. Kawula TH, Spinola SM, Klapper DG et al (1987) Localization of a conserved epitope and an azurin-like domain in the H.8 protein of pathogenic *Neisseria*. Mol Microbiol 1:179–185

153. Baehr W, Gotschlich EC, Hitchcock PJ (1989) The virulence-associated gonococcal H.8 gene encodes 14 tandemly repeated pentapeptides. Mol Microbiol 3:49–55

154. Bhattacharjee AK, Moran EE, Ray JS et al (1988) Purification and characterization of H.8 antigen from group B *Neisseria meningitidis*. Infect Immun 56:773–778

155. Strittmatter W, Hitchcock PJ (1986) Isolation and preliminary biochemical characterization of the gonococcal H.8 antigen. J Exp Med 164:2038–2048

156. Black JR, Black WJ, Cannon JG (1985) Neisserial antigen H.8 is immunogenic in patients with disseminated gonococcal and meningococcal infection. J Infect Dis 151:650–657

157. Black JR, Thompson MK, Cannon JG et al (1988) Serum immune response to common pathogenic *Neisseria* antigen H.8 in patients with uncomplicated gonococcal infection and pelvic inflammatory disease. In: Poolman JT, Zanen HC, Meyer TF et al (eds) Gonococci and meningococci. Kluwer, Dordrecht, pp 493–497

158. Lammel CJ, Sweet RL, Rice PA et al (1985) Antibody-antigen specificity in the immune response to infection with *Neisseria gonorrhoeae*. J Infect Dis 152(5):990–1001
159. Brooks GF, Lammel CJ (1989) Humoral immune response to gonococcal infections. Clin Microbiol Rev 2S:S5–S10
160. Ray TD, Lewis LA, Gulati S et al (2011) Novel blocking human IgG directed against the pentapeptide repeat motifs of *Neisseria meningitidis* Lip/H.8 and Laz lipoproteins. J Immunol 186(8):4881–4894. https://doi.org/10.4049/jimmunol.1003623
161. Griffiths E (1987) Iron and infection: molecular, physiological and clinical aspects. In: Bullen JJ, Griffiths E (eds) The iron uptake systems of pathogenic bacteria. John Wiley & Sons, Ltd., Chichester, pp 69–137
162. Carson SD, Klebba PE, Newton SM et al (1999) Ferric enterobactin binding and utilization by *Neisseria gonorrhoeae*. J Bacteriol 181(9):2895–2901
163. Dyer DW, West EP, Sparling PF (1987) Effects of serum carrier proteins on the growth of pathogenic neisseriae with heme-bound iron. Infect Immun 55(9):2171–2175
164. Mickelsen PA, Blackman E, Sparling PF (1982) Ability of *Neisseria gonorrhoeae*, *Neisseria meningitidis*, and commensal *Neisseria* species to obtain iron from lactoferrin. Infect Immun 35:915–920
165. Mickelsen PA, Sparling PF (1981) Ability of *Neisseria gonorrhoeae*, *Neisseria meningitidis*, and commensal *Neisseria* species to obtain iron from transferrin and iron compounds. Infect Immun 33:555–564
166. Noinaj N, Guillier M, Barnard TJ et al (2010) TonB-dependent transporters: regulation, structure, and function. Annu Rev Microbiol 64:43–60. https://doi.org/10.1146/annurev.micro.112408.134247
167. Chen CY, Berish SA, Morse SA et al (1993) The ferric iron-binding protein of pathogenic *Neisseria* spp. functions as a periplasmic transport protein in iron acquisition from human transferrin. Mol Microbiol 10(2):311–318. https://doi.org/10.1111/j.1365-2958.1993.tb01957.x
168. Krewulak KD, Vogel HJ (2008) Structural biology of bacterial iron uptake. Biochim Biophys Acta 1778(9):1781–1804. https://doi.org/10.1016/j.bbamem.2007.07.026
169. Cornelissen CN, Hollander A (2011) TonB-dependent transporters expressed by *Neisseria gonorrhoeae*. Front Microbiol 2:117. https://doi.org/10.3389/fmicb.2011.00117
170. Lee BC (1992) Isolation of an outer membrane hemin-binding protein of *Haemophilus influenzae* type b. Infect Immun 60 (3):810–816
171. Jean S, Juneau RA, Criss AK et al (2016) *Neisseria gonorrhoeae* evades calprotectin-mediated nutritional immunity and survives neutrophil extracellular traps by production of TdfH. Infect Immun 84(10):2982–2994. https://doi.org/10.1128/IAI.00319-16
172. Cornelissen CN, Biswas GD, Sparling PF (1993) Expression of gonococcal transferrin-binding protein I causes *Escherichia coli* to bind human transferrin. J Bacteriol 175:2448–2450
173. Cornelissen CN, Biswas GD, Tsai J et al (1992) Gonococcal transferrin-binding protein I is required for transferrin utilization and is homologous to TonB-dependent outer membrane receptors. J Bacteriol 174:5788–5797
174. Cornelissen CN, Anderson JE, Boulton IC et al (2000) Antigenic and sequence diversity in gonococcal transferrin-binding protein A (TbpA). Infect Immun 68:4725–4735
175. Anderson JE, Sparling PF, Cornelissen CN (1994) Gonococcal transferrin-binding protein 2 facilitates but is not essential for transferrin utilization. J Bacteriol 176:3162–3170
176. Brooks CL, Arutyunova E, Lemieux MJ (2014) The structure of lactoferrin-binding protein B from *Neisseria meningitidis* suggests roles in iron acquisition and neutralization of host defences. Acta Crystallogr F Struct Biol Commun 70(Pt 10):1312–1317. https://doi.org/10.1107/S2053230X14019372
177. Stojiljkovic I, Hwa V, de Saint Martin L et al (1995) The *Neisseria meningitidis* haemoglobin receptor: its role in iron utilization and virulence. Mol Microbiol 15(3):531–541
178. Cornelissen CN, Kelley M, Hobbs MM et al (1998) The transferrin receptor expressed by gonococcal strain FA 1090 is required for the experimental infection of human male volunteers. Mol Microbiol 27:611–616
179. Anderson JE, Hobbs MM, Biswas GD et al (2003) Opposing selective forces for expression of the gonococcal lactoferrin receptor. Mol Microbiol 48(5):1325–1337
180. Weisenfeld HC, Mietzner TA, Sweet RL (1994) Lactoferrin and hemoglobin iron utilization and the transmission of *Neisseria gonorrhoeae*, abstr. B-160. In: ASM (ed) Abstracts of the 94th general meeting of the American Society for Microbiology

1994. American Society for Microbiology, Washington, DC
181. Cornelissen CN, Anderson JE, Sparling PF (1997) Characterization of the diversity and the transferrin-binding domain of gonococcal transferrin-binding protein 2. Infect Immun 65:822–828
182. Meyer TF, Gibbs CP, Haas R (1990) Variation and control of protein expression in *Neisseria*. Annu Rev Microbiol 44:451–477
183. Price GA, Hobbs MM, Cornelissen CN (2004) Immunogenicity of gonococcal transferrin binding proteins during natural infections. Infect Immun 72(1):277–283
184. Price GA, Russell MW, Cornelissen CN (2005) Intranasal administration of recombinant *Neisseria gonorrhoeae* transferrin binding proteins A and B conjugated to the cholera toxin B subunit induces systemic and vaginal antibodies in mice. Infect Immun 73 (7):3945–3953. https://doi.org/10.1128/IAI.73.7.3945-3953.2005
185. Price GA, Masri HP, Hollander AM et al (2007) Gonococcal transferrin binding protein chimeras induce bactericidal and growth inhibitory antibodies in mice. Vaccine 25 (41):7247–7260. https://doi.org/10.1016/j.vaccine.2007.07.038
186. Maret W (2013) Zinc biochemistry: from a single zinc enzyme to a key element of life. Adv Nutr 4(1):82–91. https://doi.org/10.3945/an.112.003038
187. Maret W (2013) Zinc and the zinc proteome. Met Ions Life Sci 12:479–501. https://doi.org/10.1007/978-94-007-5561-1_14
188. Striz I, Trebichavsky I (2004) Calprotectin – a pleiotropic molecule in acute and chronic inflammation. Physiol Res 53(3):245–253
189. Edgeworth J, Gorman M, Bennett R et al (1991) Identification of p8,14 as a highly abundant heterodimeric calcium binding protein complex of myeloid cells. J Biol Chem 266(12):7706–7713
190. Urban CF, Ermert D, Schmid M et al (2009) Neutrophil extracellular traps contain calprotectin, a cytosolic protein complex involved in host defense against *Candida albicans*. PLoS Pathog 5(10):e1000639. https://doi.org/10.1371/journal.ppat.1000639
191. Corbin BD, Seeley EH, Raab A et al (2008) Metal chelation and inhibition of bacterial growth in tissue abscesses. Science 319 (5865):962–965. https://doi.org/10.1126/science.1152449
192. Turner PC, Thomas CE, Stojiljkovic I et al (2001) Neisserial TonB-dependent outer-membrane proteins: detection, regulation and distribution of three putative candidates identified from the genome sequences. Microbiology 147(Pt 5):1277–1290. https://doi.org/10.1099/00221287-147-5-1277
193. Campagnari AA, Spinola SM, Lesse AJ et al (1990) Lipooligosaccharides epitopes shared among gram negative non-enteric mucosal pathogens. Microb Pathog 8:353
194. Inzana TJ, Seifert WEJ, Williams RP (1985) Composition and antigenic activity of the oligosaccharide moiety of *Haemophilus influenzae* type b lipooligosaccharide. Infect Immun 48:324–330
195. Schneider H, Hale TL, Zollinger WD et al (1984) Heterogeneity of molecular size and antigenic expression within lipooligosaccharides of individual strains of *Neisseria gonorrhoeae* and *Neisseria meningitidis*. Infect Immun 45:544–549
196. Griffiss J, Brandt B, Saunders N et al (2000) Structural relationships and sialylation among meningococcal L1, L8 and L3, 7 lipooligosaccharide serotypes. J Biol Chem 275:9716–9724
197. Zhou D, Stephens DS, Gibson BW et al (1994) Lipooligosaccharide biosynthesis in pathogenic *Neisseria*. Cloning, identification, and characterization of the phosphoglucomutase gene. J Biol Chem 269 (15):11162–11169
198. Gibson BW, Melaugh W, Phillips NJ et al (1993) Investigation of the structural heterogeneity of lipooligosaccharides from pathogenic *Haemophilus* and *Neisseria* species and of R-type lipopolysaccharides from *Salmonella typhimurium* by electrospray mass spectrometry. J Bacteriol 175(9):2702–2712
199. Petricoin EF III, Danaher RJ, Stein DC (1991) Analysis of the *lsi* region involved in lipooligosaccharide biosynthesis in *Neisseria gonorrhoeae*. J Bacteriol 173:7896–7902
200. Gotschlich EC (1994) Genetic locus for the biosynthesis of the variable portion of *Neisseria gonorrhoeae* lipooligosaccharide. J Exp Med 180(6):2181–2190
201. Kahler CM, Carlson RW, Rahman MM et al (1996) Two glucosyltransferase genes, *lgtF* and *rfaK*, constitue the lipooligosaccharide *ice* (inner core extension) biosynthesis operon of *Neisseria meningitidis*. J Bacteriol 178:6677–6684
202. Banerjee A, Wang R, Uljon SN et al (1998) Identification of the gene (lgtG) encoding the lipooligosaccharide beta chain synthesizing glucosyl transferase from *Neisseria*

gonorrhoeae. Proc Natl Acad Sci U S A 95 (18):10872–10877

203. Gibson BW, Webb JW, Yamasaki R et al (1989) Structure and heterogeneity of the oligosaccharides from the lipopolysaccharides of a pyocin-resistant *Neisseria gonorrhoeae*. Proc Natl Acad Sci U S A 86:17–21
204. Ram S, Gulati S, Lewis LA et al (2018) A novel sialylation site on *Neisseria gonorrhoeae* lipooligosaccharide links heptose II lactose expression with pathogenicity. Infect Immun 86(8). https://doi.org/10.1128/IAI.00285-18
205. Yamasaki R, Kerwood DE, Schneider H et al (1994) The structure of lipooligosaccharide produced by *Neisseria gonorrhoeae*, strain 15253, isolated from a patient with disseminated infection: evidence for a new glycosylation pathway of gonococcal lipooligosaccharide. J Biol Chem 269:30345–30351
206. Yamasaki R, Koshino H, Kurono S et al (1999) Structural and immunochemical characterization of a *Neisseria gonorrhoeae* epitope defined by a monoclonal antibody 2C7; the antibody recognizes a conserved epitope on specific lipo-oligosaccharides in spite of the presence of human carbohydrate epitopes. J Biol Chem 51:36550–36558
207. Apicella MA, Shero M, Jarvis GA et al (1987) Phenotypic variation in epitope expression of the *Neisseria gonorrhoeae* lipooligosaccharide. Infect Immun 55(8):1755–1761
208. Schneider H, Hammack CA, Apicella MA et al (1988) Instability of expression of lipooligosaccharides and their epitopes in *Neisseria gonorrhoeae*. Infect Immun 56(4):942–946
209. Danaher RJ, Levin JC, Arking D et al (1995) Genetic basis of *Neisseria gonorrhoeae* lipooligosaccharide antigenic variation. J Bacteriol 177:7275–7279
210. Jennings M, Hood D, Peak R et al (1995) Molecular analysis of a locus for the biosynthesis and phase-variable expression of the lacto-N-neotetraose terminal lipopolysaccharide structure in *Neisseria meningitidis*. Mol Microbiol 18:729–740
211. Yang QL, Gotschlich EC (1996) Variation of gonococcal lipoligosaccharide structure is due to alteration in poly-G tracts in *lgt* genes encoding glycosyl transferases. J Exp Med 183:323–327
212. Gulati S, Zheng B, Reed GW et al (2013) Immunization against a saccharide epitope accelerates clearance of experimental gonococcal infection. PLoS Pathog 9(8): e1003559. https://doi.org/10.1371/journal.ppat.1003559
213. Ingwer I, Petersen BH, Brooks G (1978) Serum bactericidal action and activation of the classic and alternate complement pathways by *Neisseria gonorrhoeae*. J Lab Clin Med 92:211–220
214. Griffiss JM, Jarvis GA, O'Brien JP et al (1991) Lysis of *Neisseria gonorrhoeae* initiated by binding of normal human IgM to a hexosamine-containing lipooligosaccharide epitope(s) is augmented by strain-specific, properdin-binding-dependent alternative complement pathway activation. J Immunol 147:298–305
215. Jarvis GA (1995) Recognition and control of neisserial infection by antibody and complement. Trends Microbiol 3(5):198–201
216. Schneider H, Griffiss JM, Williams GD et al (1982) Immunological basis of serum resistance of *Neisseria gonorrhoeae*. J Gen Microbiol 128(Pt 1):13–22
217. Ward ME, Glynn AA (1972) Human antibody response to lipopolysaccharides from *Neisseria gonorrhoeae*. J Clin Pathol 25 (1):56–59
218. de la Paz H, Cooke SJ, Heckels JE (1995) Effect of sialylation of lipopolysaccharide of *Neisseria gonorrhoeae* on recognition and complement-mediated killing by monoclonal antibodies directed against different outer-membrane antigens. Microbiology 141 (Pt 4):913–920
219. Parsons NJ, Andrade JR, Patel PV et al (1989) Sialylation of lipopolysaccharide and loss of absorption of bactericidal antibody during conversion of gonococci to serum resistance by cytidine 5′-monophospho-N-acetyl neuraminic acid. Microb Pathog 7(1):63–72
220. Jarvis GA (1994) Analysis of C3 deposition and degradation on *Neisseria meningitidis* and *Neisseria gonorrhoeae*. Infect Immun 62 (5):1755–1760
221. Gulati S, Schoenhofen IC, Whitfield DM et al (2015) Utilizing CMP-sialic acid analogs to unravel *Neisseria gonorrhoeae* lipooligosaccharide-mediated complement resistance and design novel therapeutics. PLoS Pathog 11(12):e1005290. https://doi.org/10.1371/journal.ppat.1005290
222. Elkins C, Carbonetti NH, Varela VA et al (1992) Antibodies to N-terminal peptides of gonococcal porin are bactericidal when gonococcal lipopolysaccharide is not sialylated. Mol Microbiol 6(18):2617–2628
223. Wetzler LM, Barry K, Blake MS et al (1992) Gonococcal lipooligosaccharide sialylation

prevents complement-dependent killing by immune sera. Infect Immun 60(1):39–43

224. Zaleski A, Densen P (1996) Sialylation of LOS inhibits gonococcal killing primarily through an effect on classical pathway activation. In: Zollinger WD, Frasch CE, Deal CD (eds) Abstracts of the tenth international pathogenic neisseria conference. National Institutes of Health, Baltimore, MD, p 114
225. Ram S, Sharma AK, Simpson SD et al (1998) A novel sialic acid binding site on factor H mediates serum resistance of sialylated *Neisseria gonorrhoeae.* J Exp Med 187 (5):743–752
226. Blom AM, Hallstrom T, Riesbeck K (2009) Complement evasion strategies of pathogens-acquisition of inhibitors and beyond. Mol Immunol 46(14):2808–2817
227. Ram S, Shaughnessy J, DeOliveira RB et al (2016) Utilizing complement evasion strategies to design complement-based antibacterial immunotherapeutics: lessons from the pathogenic *Neisseriae.* Immunobiology 221 (10):1110–1123. https://doi.org/10.1016/j.imbio.2016.05.016
228. Gulati S, Cox A, Lewis LA et al (2005) Enhanced factor H binding to sialylated Gonococci is restricted to the sialylated lacto-N-neotetraose lipooligosaccharide species: implications for serum resistance and evidence for a bifunctional lipooligosaccharide sialyltransferase in Gonococci. Infect Immun 73(11):7390–7397
229. Apicella MA, Mandrell RE, Shero M et al (1990) Modification by sialic acid of *Neisseria gonorrhoeae* lipooligosaccharide epitope expression in human urethral exudates: an immunoelectron microscopic analysis. J Infect Dis 162:506–512
230. Schneider H, Cross AS, Kuschner RA et al (1995) Experimental human gonococcal urethritis: 250 *Neisseria gonorrhoeae* MS11mkC are infective. J Infect Dis 172(1):180–185
231. Schneider H, Griffiss JM, Boslego JW et al (1991) Expression of paragloboside-like lipooligosaccharides may be a necessary component of gonococcal pathogenesis in men. J Exp Med 174:1601–1605
232. Harvey HA, Porat N, Campbell CA et al (2000) Gonococcal lipooligosaccharide is a ligand for the asialoglycoprotein receptor on human sperm. Mol Microbiol 36:1059–1070
233. Schneider H, Schmidt KA, Skillman DR et al (1996) Sialylation lessens the infectivity of *Neisseria gonorrhoeae* MS11mkC. J Infect Dis 173(6):1422–1427
234. Lewis LA, Gulati S, Burrowes E et al (2015) Alpha-2,3-sialyltransferase expression level impacts the kinetics of lipooligosaccharide sialylation, complement resistance, and the ability of *Neisseria gonorrhoeae* to colonize the murine genital tract. MBio 6(1). https://doi.org/10.1128/mBio.02465-14
235. Wu H, Jerse AE (2006) Alpha-2,3-sialyltransferase enhances *Neisseria gonorrhoeae* survival during experimental murine genital tract infection. Infect Immun 74(7):4094–4103. https://doi.org/10.1128/iai.00433-06
236. Sadarangani M, Pollard AJ (2010) Serogroup B meningococcal vaccines-an unfinished story. Lancet Infect Dis 10(2):112–124. https://doi.org/10.1016/S1473-3099(09)70324-X

Chapter 2

How to Identify Exposed Women Who Are Infected with *Neisseria gonorrhoeae*

Stephanie E. McLaughlin and J. McLeod Griffiss

Abstract

Treatment trials of antibiotics for *Neisseria gonorrhoeae* infections frequently enroll primarily men with urethritis, as the diagnosis of acute gonococcal infection in men with urethritis is easily made by Gram stain of the urethral exudate, followed by confirmatory culture or nucleic acid amplification tests (NAATs). Enrolling women in treatment trials is of great importance, but *N. gonorrhoeae* cervical infections cause nonspecific symptoms. This makes it difficult to conduct interventional trials, as large numbers of women with nonspecific symptoms need to be screened for infection. Gram stain of cervical secretions has a strikingly low sensitivity, and culture and/or NAAT results are not available at the time of screening. This necessitates recall and delayed treatment of infected women who may not return and who may spread the infection during the interval. In this chapter we present an algorithm, derived from a comparison of women who did, or did not, become infected during exposure, which identifies those women who are highly likely to be infected before culture and/or NAAT results are available. The algorithm provides an efficient way to conduct interventional trials in women without the problem of recall and delayed treatment.

Key words *Neisseria gonorrhoeae*, Cervical infections, Cervical discharge, Vaginal pH, Cervical inflammation

1 Introduction

Treatment trials of antibiotics for *Neisseria gonorrhoeae* infections usually enroll primarily men with urethritis, as the diagnosis of acute gonococcal infection in men with urethritis is easily made by Gram stain of the urethral exudate, followed by confirmatory culture or nucleic acid amplification tests (NAATs) [1]. This allows enrollment during the initial clinic visit with subsequent censoring of the very few subjects whose infections will not be confirmed.

As *N. gonorrhoeae* acquires resistance to the currently recommended antimicrobials, enrolling women in treatment trials of newer agents is of great importance. However, *N. gonorrhoeae* cervical infections are asymptomatic in many women, cause nonspecific symptoms in others, and cannot be presumptively

Myron Christodoulides (ed.), *Neisseria gonorrhoeae: Methods and Protocols*, Methods in Molecular Biology, vol. 1997,
https://doi.org/10.1007/978-1-4939-9496-0_2,

diagnosed with Gram stain of cervical secretions, as these have a strikingly low sensitivity (<0.1) [2]. This makes it difficult to conduct intervention trials. Asymptomatic women are unlikely to be screened, and large numbers of women with nonspecific symptoms will need to be screened for infection, but not enrolled until culture and/or NAAT results are available. This necessitates recall and delayed treatment of infected women who may not return to clinic and who may spread the infection during the interval.

One way to improve the efficiency of enrollment of infected women would be to recruit women who present to clinic reporting gonococcal exposure [1], but only 60–65% of women are infected with *N. gonorrhoeae* during a single exposure [3–5]. An algorithm that identifies women who are highly likely to have been infected during exposure, before culture and/or NAAT results are available, would allow for more efficient enrolment of women in interventional trials without the problem of recall and delayed treatment. We developed such an algorithm by comparing the medical and sexual histories, symptoms and physical findings of women who resisted infection during exposure with the same information from women who were infected during exposure.

1.1 Predictors of Cervical Infection During Exposure: Comparing Women Who Resisted Infection with Those Who Did Not

Here we present the development of the algorithm and how to use it to efficiently recruit women into treatment trials. The algorithm was developed by enrolling women who presented to one of the Baltimore City Health Department Sexually Transmitted Infection (STI) clinics with a "partner notification card" indicating sexual contact with a man who had been diagnosed with gonorrhea. We then compared sexual histories, symptoms and physical findings between those who were infected and those who resisted infection.

2 Materials

1. Subjects: women who have had sexual contact with an infected man.
2. A detailed history: a questionnaire that records sexual history—past prior STIs (including prior gonococcal infections), type of sexual contact involved (oral or anal sex in addition to vaginal intercourse), and the time between exposure and reporting to the clinic.
3. A cervicovaginal physical examination by a clinician experienced in the diagnosis of STIs.
4. Amsel criteria for bacterial vaginosis (BV) [6]: (a) A thin (i.e., of low viscosity) and homogeneous-appearing vaginal discharge. (b) Vaginal pH >4.5. (c) The release of a fishy amine odor on addition of 10% (v/v) potassium hydroxide (KOH) to a drop of vaginal discharge (whiff test). (d) Clue cells on saline wet mount of vaginal discharge.

Table 1
Cervicovaginal findings and risk of gonococcal infection during exposure

	Infected		Uninfected		
	N	% of all infected	*N*	% of all uninfected	*p*[a]
Vaginal discharge observed by clinician	15	39.5	10	43.5	0.79
Cervical discharge observed by clinician	16	42.1	4	17.4	0.05
Inflammation (>5 WBC/hpf)	24	80	15	78.9	0.11
Positive "whiff" test	14	36.8	6	26.1	0.42

[a]Fisher's exact test

3 Methods

1. Recruit and enroll women who present to clinic having been identified as a partner of a man with acute gonococcal infection.
2. Record the presence of vaginal discharge on pelvic exam and whether it meets the Amsel criteria (*see* **Note 1**).
3. Measure vaginal pH by indicator strip (range: 4.0–7.0) (*see* **Note 2**).
4. Record the presence of cervical discharge on examination, and the presence of white blood cells (WBC) in cervical secretions (Table 1) (*see* **Note 3**).
5. Grade the degree of inflammation, as judged by the presence and number of leukocytes seen by the clinician on wet mount of the cervix (*see* **Note 4**).
6. Collect cervical swabs for gram stain (*see* **Note 5**), culture and NAATs (*see* **Note 6**), and wet mounts for KOH whiff test (*see* **Note 7**).
7. Scrutinize the sexual histories and symptoms recorded on the questionnaire (Table 2) (*see* **Note 8**) with particular attention to a history of oral sex and anal sex practices (*see* **Note 9**).
8. Calculate the interval between exposure and reporting to clinic (*see* **Note 10**).
9. Select female sexual contacts of men diagnosed with gonorrhea that presented within 30 days of sexual exposure and have a vaginal pH $\geq$5.0 but $\leq$6.5 for enrolment (*see* **Note 11**) and follow one of the two models in the algorithm (Fig. 1). The algorithm was constructed using the method detailed in **Note 12**.
10. Algorithm Model 1, which relies only on vaginal pH (Fig. 1—Model 1, Tables 3 and 4), enables enrolment of a larger number of infected women, but results in a lower percentage of

Table 2
Sexual history and symptoms and risk of gonococcal infection during exposure

	N	Infected		Uninfected		
All subjects	61	38	62.30%	23	37.70%	
History and symptoms		*N*	% of infected	*N*	% of uninfected	*p*[a]
Past gonococcal infection	25	15	39.5	10	42.5	0.79
Oral sex[b]	28	15	40.5	13	56.5	0.29
Mean days between contact and exam	61	24.3		25.76		
Median days between contact and exam		11 (0–120		8 (0–107)		
Any symptom	34	21	55.3	13	56.5	0.07
Vaginal discharge reported by subject	27	14	36.8	13	56.5	0.11
Abdominal pain	18	12	31.6	6	26.1	0.78
Regular douching	20	12	31.6	8	34.8	0.78
Hormonal contraception	15	8	21.1	7	30.4	0.54

[a]Fisher's exact test
[b]All reported vaginal intercourse, as well

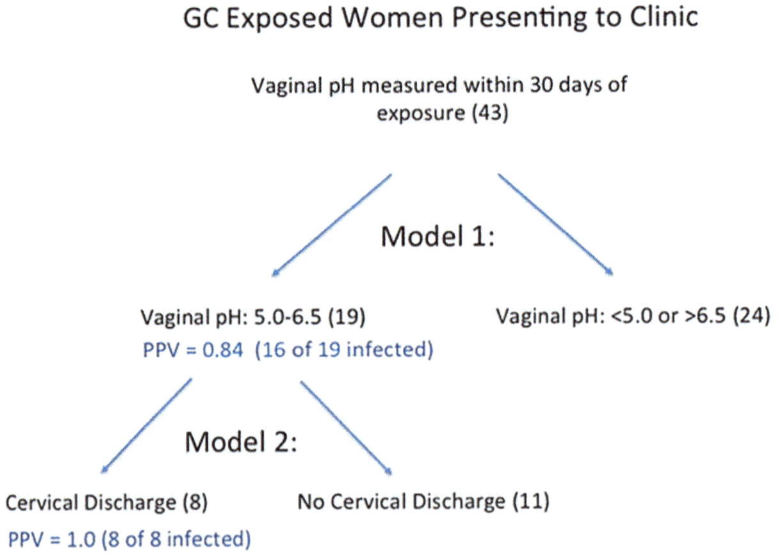

Fig. 1 Algorithm Models 1 and 2 begin selection by assessing vaginal pH of women exposed within 30 days of presentation to clinic. Model 2 continues by selecting women with a high likelihood of being infected by *Neisseria gonorrhoeae* during exposure based on vaginal pH and the presence or absence of cervical discharge

gonococcal-positive women enrolled. For example, if a study needs to enroll ~80 *N. gonorrhoeae* infected women, Model 1 could be used: 226 women who present reporting exposure to a man with gonorrhea could be screened for vaginal pH, and the 100 women with a vaginal pH of 5.0–6.5 could be enrolled. This would result in around 84 positive women enrolled (16% uninfected women and 84% of infected women enrolled).

11. Alternately, if a study needs to enroll a small number of women, Model 2 could be used; Model 2 adds cervical discharge to vaginal pH (Fig. 1—Model 2). Fewer women will be enrolled, but virtually all of them will be infected. For example, if a study needs to enroll only ~16 infected women, Model 2 could be used: 86 women who presented reporting exposure to an infected man could be screened for both vaginal pH and cervical discharge sequentially; the 16 women with vaginal pH of 5.0–6.5 and cervical discharge would be enrolled, and this would result in 16 (100%) gonococcus-infected women being enrolled.

4 Notes

1. The BV discharge has a milk-like consistency and is distinctly not floccular, granular, curd-like, stringy, or clumped. Vaginal discharge of any sort was reported *more* frequently by uninfected women but was not always consistent with BV [7–9].
2. A vaginal pH >4.5 is one of the four Amsel criteria for bacterial vaginosis [6].
3. Only cervical discharge, but not vaginal discharge, was significantly more common among infected women than those who were uninfected ($p = 0.04$, Fisher's exact test).
4. Grade 4 inflammation was seen only in infected women ($N = 10$). Scale: 1 = 0–5 neutrophils/high-powered field (hpf; ×100 magnification), 2 = 6–30 neutrophils/hpf, 3 = 31–100 neutrophils/hpf, and 4 = >100 neutrophils/hpf.
5. Gram-negative diplococci were found in cervical secretions of only four women, including one uninfected woman. Sensitivity of gram stain in this study was only 0.079, but specificity was high (0.957).
6. Negative cultures will need to be confirmed by NAAT [1]. In our study, 38 of the 61 enrolled women were infected (62%), as determined by a positive culture (35) or a subsequent positive NAAT (3) after a negative culture.
7. A positive "whiff" test—the release of a fishy amine odor on addition of 10 % (v/v) KOH to a drop of fluid on a slide—is one of the four Amsel criteria for BV. Infected women were more likely to have a positive whiff test (37%) than uninfected

women (26%), but the difference was not significant (Table 1). Women who have BV by Amsel criteria should be treated.

8. Neither sexual practices nor a history of prior gonococcal infections nor any symptoms, including vaginal discharge and abdominal pain, differed between infected and uninfected exposed women (Table 2). It is unlikely that enrolling more women would reveal any significant differences. Note, in particular, that there was no difference in symptoms between infected and uninfected exposed women, so symptoms cannot be used to identify infected women. Platt et al. reported that infected women were more likely to have abnormal adnexal findings on exam [4], but abdominal pain was infrequently experienced by women in our study and did not discriminate between infected and uninfected women.
9. Oral sex is commonly acknowledged by exposed women (Table 2) and likely is more common than is acknowledged, as it may be considered part of normal foreplay. Questions of sexual practices need to be asked in the street vernacular. All women who acknowledge performing fellatio should provide pharyngeal cultures, as they may have pharyngeal, but not cervical infections. These same considerations apply to anal sex, asked in the street vernacular, and to anal cultures.
10. Because vaginal pH varies over time, the longer the interval, the less reliable is the pH as a marker of risk (Table 3).
11. Eighty-four percent of women who have a pH ≥ 5 and ≤ 6.5, measured within 30 days (~1 menstrual cycle) of exposure will be infected. Those who have a pH ≥ 5 and ≤ 6.5, measured within 5 days, will certainly be infected (Table 4).
12. Building the algorithm (generating a "predictive factors model"): STATA 14 statistical software used for logistical regression. Logistical regression was used to investigate the relationship between possible predictive factors (e.g., pH) and gonococcal infection and any relationship between a diagnosis of gonococcal infection and associated symptoms (e.g., cervical discharge) while accounting for possible confounders

Table 3
Risk as a function of vaginal pH range

pH range	*N*	Culture positive	Percent
4.0–4.9	23	11	47.8
5.0–5.7	13	11	84.6
5.8–6.5	16	11	68.8
>6.5	4	2	50

Table 4
Effect of interval between exposure and pH measurement

		pH 5.0–6.5		pH < 5		
Interval between exposure and measurement	*N*	*N*	Infected	*N*	Infected	*p*[a]
All subjects	56	29	22 (76%)	23	11 (48%)	0.047
Within 30 days	43	19	16 (84%)	20	9 (45%)	0.019
Within 6 days	21	12	11 (92%)	9	4 (44%)	0.046
Within 5 days	17	9	9 (100%)	8	3 (38%)	0.009

[a]Fisher's exact test

(e.g., age). The "predictive factor model" was built using backward variable selection with variable inflation factor (vif) analysis to exclude possible collinear variables. Covariates included were taken from a detailed history of women presenting to clinic as gonococcal contacts with partner notification cards. This included age at sexual debut, the number of lifetime sexual partners, prior STIs, including gonococcal infections, the number of gonococcal infections, gonococcal infections in the past 6 months, whether the subject contact involved oral or anal sex in addition to vaginal intercourse, the time between exposure and reporting to the clinic, the date of last menstrual period (LMP) and the use, and type, of hormonal contraception if any. A history of regular douching was recorded, along with a history of vaginal discharge and its duration, and a history of abdominal pain. There was no difference in symptoms between infected and uninfected exposed women, so symptoms cannot be used to identify infected women (Table 2). Menstrual cycle phase, and bacterial vaginosis by Amsel Criteria [6] were excluded from the "predictive factors model" due to colinearity with pH.

References

1. Hook EW 3rd, Golden M, Jamieson BD et al (2015) A phase 2 trial of oral solithromycin 1200 mg or 1000 mg as single-dose oral therapy for uncomplicated gonorrhea. Clin Infect Dis 61 (7):1043–1048. https://doi.org/10.1093/cid/civ478
2. McLaughlin SE, Ghanem KG, Zenilman JM, Griffiss JM (2019) Risk of Gonococcal Infection During Vaginal Exposure is Associated With High Vaginal pH and Active Menstruation. Sex Transm Dis. 46(2):86–90
3. Lycke E, Lowhagen GB, Hallhagen G et al (1980) The risk of transmission of genital *Chlamydia trachomatis* infection is less than that of genital *Neisseria gonorrhoeae* infection. Sex Transm Dis 7(1):6–10
4. Platt R, Rice PA, McCormack WM (1983) Risk of acquiring gonorrhea and prevalence of abnormal adnexal findings among women recently exposed to gonorrhea. JAMA 250 (23):3205–3209

5. Lin JS, Donegan SP, Heeren TC et al (1998) Transmission of *Chlamydia trachomatis* and *Neisseria gonorrhoeae* among men with urethritis and their female sex partners. J Infect Dis 178 (6):1707–1712
6. Amsel R, Totten PA, Spiegel CA et al (1983) Nonspecific vaginitis. Diagnostic criteria and microbial and epidemiologic associations. Am J Med 74(1):14–22
7. Wiesenfeld HC, Hillier SL, Krohn MA et al (2003) Bacterial vaginosis is a strong predictor of *Neisseria gonorrhoeae* and *Chlamydia trachomatis* infection. Clin Infect Dis 36(5):663–668. https://doi.org/10.1086/367658
8. Ness RB, Kip KE, Soper DE et al (2005) Bacterial vaginosis (BV) and the risk of incident gonococcal or chlamydial genital infection in a predominantly black population. Sex Transm Dis 32(7):413–417
9. Das S, Allan S (2006) Higher vaginal pH is associated with *Neisseria gonorrhoeae* and *Chlamydia trachomatis* infection in a predominantly white population. Sex Transm Dis 33 (8):527–528. https://doi.org/10.1097/01.olq.0000230430.60398.e4

Chapter 3

Antimicrobial Resistance in *Neisseria gonorrhoeae* and Treatment of Gonorrhea

Magnus Unemo, Daniel Golparian, and David W. Eyre

Abstract

Gonorrhea and antimicrobial resistance (AMR) in *Neisseria gonorrhoeae* are major public health concerns globally. Dual antimicrobial therapy (mainly ceftriaxone 250–500 mg × 1 plus azithromycin 1–2 g × 1) is currently recommended in many countries. These dual therapies have high cure rates, have likely been involved in decreasing the level of cephalosporin resistance internationally, and inhibit the spread of AMR gonococcal strains. However, ceftriaxone-resistant strains are currently spreading internationally, predominately associated with travel to Asia. Furthermore, the first global treatment failure with recommended dual therapy was reported in 2016 and the first isolates with combined ceftriaxone resistance and high-level azithromycin resistance were reported in 2018 in the UK and Australia. New antimicrobials for treatment of gonorrhea are essential and, of the few antimicrobials in clinical development, zoliflodacin particularly appears promising. Holistic actions are imperative. These include an enhanced advocacy; prevention, early diagnosis, contact tracing, treatment, test-of-cure, and additional measures for effective management of anogenital and pharyngeal gonorrhea; antimicrobial stewardship; surveillance of infection, AMR and treatment failures; and intensified research, for example, regarding rapid molecular point-of-care detection of gonococci and AMR, novel AMR determinants, new antimicrobials, and an effective gonococcal vaccine, which is the only sustainable solution for management and control of gonorrhea.

Key words *Neisseria gonorrhoeae*, Antimicrobial resistance, Current treatment, Dual antimicrobial therapy, Ceftriaxone, Azithromycin, Future treatment

1 Introduction

Gonorrhea and antimicrobial resistance in the aetiological agent *Neisseria gonorrhoeae*, which significantly compromises the effectiveness of treatment, are major public health concerns worldwide. The World Health Organization (WHO) estimated 78 million new global cases among adults (15–49 years of age) in 2012. The highest prevalence was in the WHO Western Pacific Region (35.2 million), followed by the WHO South-East Asian Region (11.4 million), WHO African Region (11.4 million), WHO American Region (11.0 million), WHO European Region (4.7 million), and WHO Eastern Mediterranean Region (4.5 million) [1]. No

Myron Christodoulides (ed.), *Neisseria gonorrhoeae: Methods and Protocols*, Methods in Molecular Biology, vol. 1997, https://doi.org/10.1007/978-1-4939-9496-0_3, © Springer Science+Business Media, LLC, part of Springer Nature 2019

effective gonococcal vaccine is available. Effective, accessible and inexpensive antimicrobial treatment is imperative for management and control of gonorrhea, that is, in combination with adequate prevention, laboratory diagnostics, contact notification and treatment, and epidemiological surveillance. It is a grave concern that resistance to all antimicrobials currently or previously recommended for gonorrhea treatment has been developed by *N. gonorrhoeae*, which facilitates further spread of infection and emergence of severe complications and sequelae [2–7]. Dual antimicrobial therapy (mainly ceftriaxone 250–500 mg × 1 intramuscularly plus azithromycin 1–2 g × 1 orally) is now recommended for empirical first-line therapy of gonorrhea in many settings [7–13].

2 Antimicrobial Resistance in *N. gonorrhoeae*

N. gonorrhoeae was initially susceptible to many classes of antimicrobials, but since the mid-1930s, when the first antimicrobials (sulfonamides) started to be used for treatment of gonorrhea, *N. gonorrhoeae* has continually presented an extreme capacity to develop resistance to all antimicrobials introduced for treatment. *N. gonorrhoeae* has developed or acquired (through horizontal gene transfer) examples of all major known physiological antimicrobial resistance (AMR) mechanisms (e.g., inactivation/degradation of antimicrobials, alteration of antimicrobial target(s), enhanced antimicrobial efflux (by overexpressing, for example, the MtrCDE efflux pumps), and decreased antimicrobial influx (e.g., through the PorB porin)) [2, 4, 5, 14]. These AMR mechanisms have been recently reviewed elsewhere [2, 15], but the AMR determinants for antimicrobials currently recommended and/or used for treatment of gonorrhea are also summarized in Table 1. The resistance to several antimicrobials such as fluoroquinolones and extended-spectrum cephalosporins (ESCs; e.g., ceftriaxone and cefixime) has been hypothesized to initially have developed in the WHO Western Pacific Region (frequently Japan) followed by international spread [2, 5, 16, 17]. The reasons for this are complex and most likely include an overuse and misuse of antimicrobials (including inappropriate selection of antimicrobials and suboptimal dosing, and quality of antimicrobials) for many infections, including gonorrhea. This has resulted in AMR in numerous bacterial species (including non-gonococcal *Neisseria* species), which have in-turn shared their AMR determinants with *N. gonorrhoeae*, as well as AMR developing directly within *N. gonorrhoeae*. This has been combined with high incidences of gonorrhea, a lack of effective disease-control measures, a suboptimal or complete lack of gonorrhea and/or *N. gonorrhoeae* AMR and treatment failure surveillance, limited considerations of pharmacokinetics/pharmacodynamics, as well as

Table 1
Antimicrobial resistance determinants in *Neisseria gonorrhoeae* for antimicrobials used for treatment of gonorrhea [2]

Antimicrobial	Resistance determinants/mechanisms
Ceftriaxone, cefixime	– **Mosaic *penA* alleles**: encode mosaic PBP2s with decreased PBP2 acylation rate. Mosaic PBP2s amino acid substitutions confirmed to contribute to resistance are A311V, I312M, V316T, V316P, T483S, A501P, A501V, N512Y, and G545S – ***penA* SNPs**: A501V and A501T in nonmosaic *penA* alleles can increase the MICs of ESCs. Also G542S, P551S, and P551L have been statistically associated with elevated MICs of ESCs; however, their effects on resistance have not been proven with, for example, site-directed *penA* mutants into isogenic backgrounds – **Mosaic *mtr locus or mtrR* mutations**, in promoter (mainly a single nucleotide (A) deletion in the 13-bp inverted repeat sequence) or coding sequence (most common being a G45D amino acid substitution), that cause an overexpression and enhanced efflux of the MtrCDE efflux pump – ***porB1b* SNPs**: for example, G120K and G120D/A121D in loop 3 of PorB1b that decrease influx (*penB* resistance determinant). The *penB* phenotype appears only expressed in gonococcal strains that express also the *mtrR* resistance determinant – **"Factor X"**: unknown nontransformable penicillin and ESC resistance determinant
Azithromycin	– **SNPs in the 23S rRNA gene** (in 1–4 of the four alleles) that encodes 23S rRNA (peptidyltransferase loop of domain V) with decreased affinity to the 50S ribosomal target for azithromycin. The SNPs C2611T and A2059G cause low-level and high-level resistance, respectively; however, the number of mutated alleles is correlated with the MICs of azithromycin – ***mtrR* mutations**: see above – ***erm* genes (*ermB*, *ermC*, and *ermF*)**: acquired from other bacterial species and encode rRNA methylases that can methylate nucleotides in the 23S rRNA azithromycin target – **MacAB efflux pump**: overexpression can elevate the MICs of azithromycin – ***mef*-encoded efflux pump**: acquired from other bacterial species and export macrolides out of the bacterial cell and elevate the MICs of macrolides
Spectinomycin	– **16S rRNA SNP**: C1192U in the spectinomycin-binding region of helix 34 that decreases affinity to ribosomal spectinomycin target – ***rpsE* mutations** (encoding the 30S ribosomal protein S5): resulting in amino acid alterations such as T24P, deletion of V25, and K26E, which disrupt the spectinomycin binding to ribosomal target
Ciprofloxacin, ofloxacin	– ***gyrA* SNPs**: for example, S91F, D95N, and D95G in the QRDR that decrease the fluoroquinolone binding to the GyrA subcomponent of DNA gyrase – ***parC* SNPs**: for example, D86N, S88P, and E91K in the QRDR that decrease the fluoroquinolone binding to the ParC subcomponent of topoisomerase IV

PBP2 penicillin-binding protein 2 (PBP2), *MIC* minimum inhibitory concentration, *ESCs* extended-spectrum cephalosporins, *SNP* single nucleotide polymorphism, *QRDR* quinolone resistance determining region

incomplete clinical efficacy of antimicrobials for urogenital and especially extragenital gonorrhea [2, 5, 16, 17].

An enhanced understanding of the dynamics, drivers and fitness of the AMR emergence and spread of AMR *N. gonorrhoeae* strains, which can provide an improved rationale for antimicrobial management and stewardship, is essential. Whole-genome sequencing (WGS) techniques are exceedingly valuable to elucidate the emergence, spread and evolution of AMR and local, national and international spread of AMR *N. gonorrhoeae* strains [18–28]. WGS has been shown to provide a higher and more accurate resolution of AMR *N. gonorrhoeae* strains for epidemiological purposes, public health, as well as can predict AMR with relatively high accuracy [19, 22]. For further information regarding prediction of AMR, *see* Chapter 4 by Eyre et al.

The level of resistance in *N. gonorrhoeae* is presently high internationally to all antimicrobials previously used for treatment, such as sulfonamides, penicillins, tetracyclines, fluoroquinolones and early generation macrolides and cephalosporins [2–6, 14, 16, 29]. However, quality-assured AMR surveillance data is incomplete or totally lacking from large parts of the world. The WHO has since 2009 substantially strengthened the WHO Global Gonococcal Antimicrobial Surveillance Programme (WHO Global GASP; [29]), which works in close collaboration with other international and national quality-assured GASPs, including Euro-GASP [30, 31], US Gonococcal Isolate Surveillance Project (GISP) [32, 33], Canadian GASP [34], Australian Gonococcal Surveillance Programme (AGSP) [35], and UK Gonococcal Resistance to Antimicrobials Surveillance Programme (UK GRASP) [36]. In the WHO Global GASP, the number of countries reporting gonococcal AMR data for any antimicrobial increased from 56 in 2009 to 77 in 2014. However, particularly in the WHO African Region, WHO Eastern Mediterranean Region, former Soviet Union East-European countries of WHO European region, Central America, and the Caribbean, quality-assured gonococcal AMR data remains exceedingly rare [29]. However, WHO initiatives to strengthen the AMR surveillance in many countries in these regions are ongoing.

Despite the suboptimal international gonococcal AMR surveillance, resistance to ceftriaxone, the last option for empiric first-line monotherapy, has been detected in many countries and mainly all regions internationally [29]. Rare failures to treat pharyngeal gonorrhea with ceftriaxone (250–1000 mg × 1) were also rather early verified in several countries (e.g., in Japan, Australia, Slovenia, and Sweden) [5, 16]. Nearly 10 years ago, the first extensively drug-resistant (XDR) gonococcal strain (H041), displaying high-level resistance to ceftriaxone (Minimum Inhibitory Concentration (MIC) = 2–4 mg/L) due to a mosaic version of the lethal target penicillin-binding protein 2 (PBP2) and resistance to most previously used antimicrobials, was isolated in Japan [37]. This was followed by the identification of a new XDR strain (F89; ceftriaxone MIC 1–2 mg/L) in both France [38] and Spain [39]. No additional infections caused by these superbugs have been described and it has now been confirmed that the resistance-determining mosaic PBP2 in these superbugs results in a significantly decreased biological fitness in vitro and in a 17-β-estradiol-treated female BALB/c mouse model [40]. Worryingly, in the female mouse model it was also indicated that strains containing the resistance-determining mosaic PBP2 of H041 can easily develop compensatory mutations that restore the fitness (e.g., single nucleotide polymorphisms (SNPs) that increase the carbon and energy metabolism) [40].

Some additional ceftriaxone-resistant *N. gonorrhoeae* strains isolated during the recent decade have also been studied in detail. These include strains A8806 (Australia 2013, MIC = 0.5 mg/L) [41], GU140106 (Japan 2014, MIC = 0.5 mg/L) [42], a strain in Argentina in 2014 (MIC = 0.5 mg/L) [43], and FC428 (Japan 2015, MIC = 0.5 mg/L) [44]. While the first superbugs and subsequently identified ceftriaxone-resistant *N. gonorrhoeae* strains did not appear to result in any sustained transmission, further spread of FC428 or genetically closely related strains has been reported in Canada [45], Denmark [46] and Australia [47] in 2017, and France in 2018 [48], predominately associated with travel to Asia. This indicates an adequate fitness of FC428 and related ceftriaxone-resistant *N. gonorrhoeae* strains, despite the remodeled mosaic PBP2 resulting in the ceftriaxone resistance. The FC428 and related ceftriaxone-resistant *N. gonorrhoeae* strains have similar genetic ESC resistance determinants as the first Japanese ceftriaxone-resistant superbug H041 [37]. Accordingly, the FC428 mosaic PBP2 sequence contains the A311V and T483S key ceftriaxone resistance amino acid substitutions, but not the T316P substitution, found in H041 [37, 44]. FC428 and related strains have a lower ceftriaxone MIC (0.5 mg/L) due to the lack of the PBP2 T316P mutation. However, it cannot be excluded that the lack of this mutation restores the fitness of these ceftriaxone resistant strains. It is important to stress that also nonmosaic PBP2s can cause ceftriaxone resistance in *N. gonorrhoeae*, which has been shown particularly in Asia (e.g., in China, Korea, and Vietnam) but also in Argentina [5, 43]. Accordingly, *N. gonorrhoeae* can develop ceftriaxone resistance using different molecular pathways and only one or a few amino acid substitutions in PBP2 are required for development of ceftriaxone resistance in a large proportion of strains spreading globally. Nevertheless, all these mentioned ceftriaxone-resistant *N. gonorrhoeae* strains were either susceptible, intermediate or had a very low-level resistance to azithromycin.

The first global failure of treating pharyngeal gonorrhea with recommended dual antimicrobial therapy (ceftriaxone 500 mg plus azithromycin 1 g) was reported in the UK in 2016, caused by an XDR *N. gonorrhoeae* strain with resistance to both ceftriaxone and azithromycin [49]. Finally, it is a grave concern that the first reported gonorrhea case globally with combined ceftriaxone resistance (MIC = 0.5 mg/L) and high-level azithromycin resistance

(MIC > 256 mg/L) was reported in early 2018 in UK [50]. This case was a 50-year-old male diagnosed with gonorrhea in February 2018 following sexual intercourse with a female in Thailand. The urethral gonorrhea was cured but pharyngeal gonorrhea failed treatment with ceftriaxone 1 g plus doxycycline 100 mg twice daily for 7 days, as well as subsequent treatment with spectinomycin 2 g. Ertapenem 1 g intravenously daily for 3 days finally eradicated the pharyngeal infection. A comparison with all available *N. gonorrhoeae* WGS sequences elucidated that this strain belonged to a clade of high-level azithromycin resistant strains (containing the A2059G mutation in all four alleles of the 23S rRNA gene) but it had additionally through a transformation and recombination event acquired the identical mosaic PBP2 as in FC428 [44]. The genetically most closely related genome sequenced *N. gonorrhoeae* isolates were from China, Japan, and the UK [50]. The genomic backbone of this isolate [50] was also relatively closely related to the eighth most prevalent *N. gonorrhoeae* multiantigen sequence typing (NG-MAST) genogroup G4995 in Europe in 2013 [19]. Only some month(s) later, two similar cases with ceftriaxone resistance and high-level azithromycin resistance, due to identical AMR determinants, were identified in Australia [51]. The first of these isolates was cultured in Western Australia from a man with urethral discharge who was returning from Southeast Asia, where he had sexual contact with a local woman. The second isolate was from a woman in Queensland, who had no record of recent overseas travel [51]. The combination of ceftriaxone resistance and high-level azithromycin resistance in *N. gonorrhoeae* poses a major public health threat globally.

3 Detection of Antimicrobial Resistance or Susceptibility in *Neisseria gonorrhoeae*

To perform a complete AMR testing, culture of *N. gonorrhoeae* is essential. Standardized, quality-assured, and quantitative MIC-based methods, that is, agar dilution method or MIC gradient strip test (e.g., Etest), are preferred. The performance characteristics and quality of different MIC gradient strip tests can significantly differ [52]. Several qualitative disc diffusion tests are also used, especially in less-resourced settings. However, because of the suboptimal correlation with MIC-based methods, these disc diffusion methods are only recommended for use when agar dilution or MIC gradient strip tests are not available and any new or rare AMR should then always be confirmed by MIC-based testing [2, 53].

Culture of *N. gonorrhoeae* is being replaced by nucleic acid amplification tests (NAATs) for diagnosis of gonorrhea in many more-resourced settings. In these settings, it is essential to have adequate GASPs. However, it would additionally be valuable to use

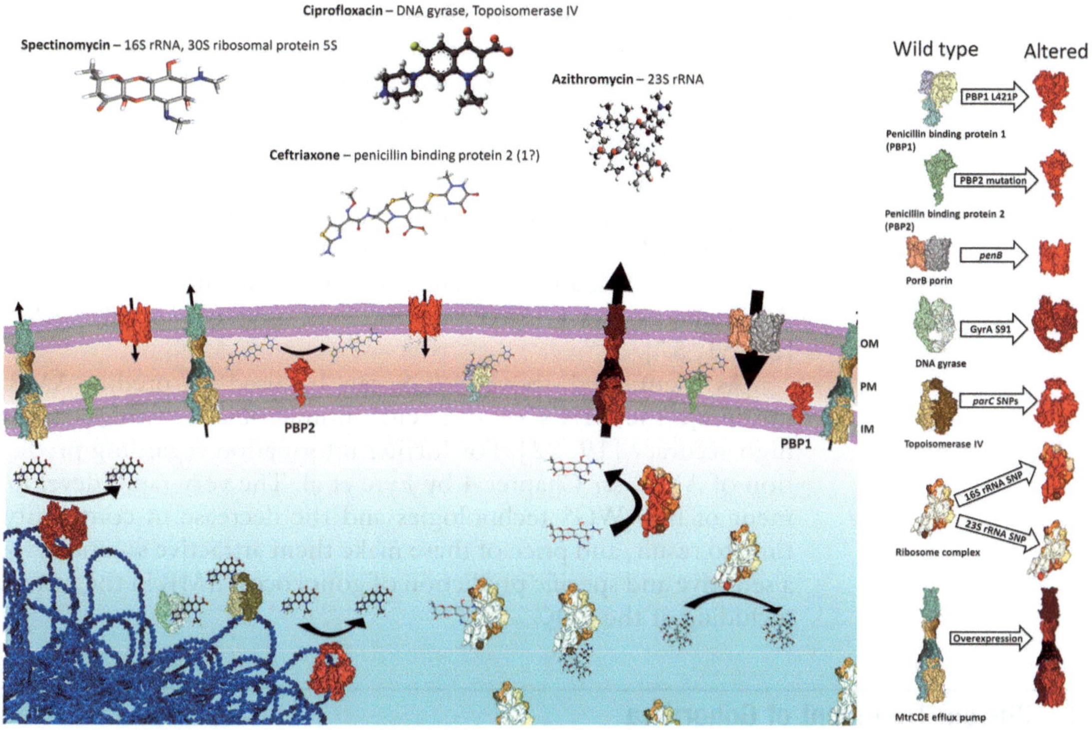

Fig. 1 Main antimicrobial targets and mechanisms/determinants resulting in resistance to extended-spectrum cephalosporins (ceftriaxone, cefixime), azithromycin, spectinomycin, and ciprofloxacin. Larger mutations or single nucleotide polymorphisms (SNPs): in *ponA* coding for PBP1 (PBP1 L421P), *penA* coding for PBP2 (PBP2 mosaic, or PBP2 A501 and/or G542/P551 alterations), *porB1b* coding for PorB1b ("*penB* alteration"; G120 $\pm$ A121 alterations of PorB1b), *gyrA* coding for GyrA (alterations in S91[$\pm$D95]), *parC* coding for ParC (alterations in D86, S87, S88, and/or E91), 16S rRNA gene (C1192T SNP), and 23S rRNA gene (C2611T or A2059G SNP). *IM* inner membrane, *PS* periplasmic space, *OM* outer membrane

molecular AMR surveillance. The number of molecular tests and use of these mainly real-time PCRs for detection of AMR determinants to predict AMR in *N. gonorrhoeae* is increasing [2, 15, 29, 54–56]. Most of these assays target the main AMR determinant for a single antimicrobial ([15]; Fig. 1). Several of these assays are promising (particularly for prediction of ciprofloxacin susceptibility/resistance), but there are important shortcomings for direct testing of clinical, especially extragenital specimens, and few assays have been adequately validated and quality assured [15]. Unfortunately, for most antimicrobials the sensitivity and/or specificity of these molecular AMR assays in their prediction of AMR and particularly the MICs of given antimicrobials are suboptimal [2, 15, 29, 54–56]. Currently, no commercial gonococcal NAAT detects any AMR determinants, but assays are under development. In the future, sensitive and specific point-of-care (POC) tests with simultaneous detection of *N. gonorrhoeae* and molecular AMR determinants for multiple antimicrobials will hopefully be available. These

POC tests could be used for immediate diagnosis, AMR surveillance as well as to guide individualized treatments. Nevertheless, molecular AMR prediction will never totally replace phenotypic AMR testing, which detects also AMR due to unknown AMR determinants. The *N. gonorrhoeae* Sequence Typing for Antimicrobial Resistance (NG-STAR; https://ngstar.canada.ca) molecular typing scheme that uses the DNA sequences of seven AMR determinants (*penA*, *mtrR*, *porB*, *ponA*, *gyrA*, *parC*, and 23S rRNA) has been shown valuable to standardize the nomenclature of AMR determinants, track AMR strains, and indirectly predict AMR [50, 51, 57].

As mentioned above, WGS can be used to predict AMR, including the MICs of different antimicrobials, with relatively high accuracy [19, 22]. For further information regarding prediction of AMR, *see* Chapter 4 by Eyre et al. The very rapid development of new WGS technologies and the decrease in complexity, time to result, and price of these make them attractive solutions for a sensitive and specific prediction of gonococcal AMR in the future, including at the POC.

4 Current Treatment of Gonorrhea

Empirical therapy at the first health care visit using evidence-based treatment guidelines is usually applied. In many countries, dual antimicrobial therapy (mainly ceftriaxone 250–500 mg × 1 intramuscularly plus azithromycin 1–2 g × 1 single oral dose) is currently the only option for empirical first-line therapy [7–12]. Dual antimicrobial therapies were introduced as a response to the emerging ESC resistance and also have activity against concurrent *Chlamydia trachomatis* and many *Mycoplasma genitalium* infections. However, in some countries (e.g., Japan, China, the Netherlands, Azerbaijan, Belarus, and Ukraine), ceftriaxone monotherapy (500–1 g × 1 IM/IV) has continued to be used. In countries using monotherapy, it is important with regular, comprehensive, local, quality-assured AMR surveillance, TOC for all patients, and a high proportion of patients returning for TOC. The therapies currently recommended for uncomplicated gonorrhea in adults by the WHO (global recommendations), and in Europe, Germany, the UK, Australia, the USA, and Canada are recapitulated in Table 2 [7–13].

All these guidelines mainly recommend ceftriaxone plus azithromycin as first-line therapy. Nevertheless, the ceftriaxone doses vary from 250 mg × 1 (WHO, USA and Canada) to 1 g × 1 (Germany). The azithromycin doses vary from 1 g × 1 (WHO, USA, Canada, UK and Australia), to 1.5 g × 1 (Germany), and finally to 2 g × 1 (Europe) (Table 2) [7–13]. Notably, the evidence-based rationale for the dual therapies and the different doses of ceftriaxone and azithromycin for currently spreading

Table 2

Antimicrobial treatment recommended for uncomplicated gonorrhoea in adults by the WHO (global recommendations), in Europe, Germany, the UK, Australia, the USA, and Canada

	WHO [11]	Europe [7]	Germany [13]	UK [12]	Australia [10]	USA [8]	Canada [9]
Recommended (first-line) regimens for anogenital infections[a]	Ceftriaxone 250 mg × 1 IM PLUS azithromycin 1 g × 1 p.o. OR Cefixime 400 mg × 1 p.o. PLUS Azithromycin 1 g × 1 p.o. *Single therapy* (one of the following, based on recent local resistance data confirming susceptibility) • Ceftriaxone 250 mg × 1 IM • Cefixime 400 mg × 1 p.o. • Spectinomycin 2 g × 1 IM	Ceftriaxone 500 mg × 1 IM PLUS Azithromycin 2 g × 1 p.o.[b]	Ceftriaxone 1 g × 1 IM/IV PLUS Azithromycin 1.5 g × 1 p.o.	Ceftriaxone 1 g × 1 IM PLUS Azithromycin 1 g × 1 p.o.	Ceftriaxone 500 mg × 1 IM PLUS Azithromycin 1 g × 1 p.o.	Ceftriaxone 250 mg × 1 IM PLUS Azithromycin 1 g × 1 p.o.	Ceftriaxone 250 mg × 1 IM PLUS Azithromycin 1 g × 1 p.o. OR Cefixime 800 mg × 1 p.o. PLUS Azithromycin 1 g × 1 p.o.
Recommended treatment for pharyngeal infections	• Ceftriaxone 250 mg × 1 IM PLUS azithromycin 1 g × 1 p.o. • Cefixime 400 mg × 1 p.o. PLUS azithromycin 1 g × 1 p.o. *Single therapy* (based on recent local resistance data confirming susceptibility) • Ceftriaxone 250 mg × 1 IM The WHO STI guideline suggests dual therapy over single therapy	As for anogenital infections	As for anogenital infections OR if *N. gonorrhoeae* known to be susceptible: – Ciprofloxacin 500 mg × 1 p.o. or Ofloxacin 400 mg × 1 p.o. – Azithromycin 1.5 g × 1 p.o.	As for anogenital infections OR if *N. gonorrhoeae* known to be fluoroquinolone susceptible – Ciprofloxacin 500 mg × 1 p.o. or Ofloxacin 400 mg × 1 p.o.	As for anogenital infections	As for anogenital infections	Ceftriaxone 250 mg × 1 IM PLUS Azithromycin 1 g × 1 p.o. ***Alternatives***: Cefixime 800 mg × 1 p.o. PLUS Azithromycin 1 g × 1 p.o. OR Azithromycin 2 g × 1 p.o.

(continued)

Table 2
(continued)

	WHO [11]	Europe [7]	Germany [13]	UK [12]	Australia [10]	USA [8]	Canada [9]
Recommended regimen when extended-spectrum cephalosporin resistance identified or failure with recommended dual regimen	• If reinfection is suspected, re-treat with a WHO-recommended regimen, reinforce sexual abstinence or condom use, and provide partner treatment • If treatment failure occurred after treatment with a regimen not recommended by WHO, re-treat with a WHO-recommended regimen • If treatment failure occurred and resistance data are available, re-treat according to susceptibility • If treatment failure occurred after treatment with a WHO-recommended single therapy, re-treat with WHO-recommended dual therapy • If treatment failure occurred after a WHO-recommended dual therapy, re-treat with one of the following dual therapies: – Ceftriaxone 500 mg × 1 IM PLUS azithromycin 2 g × 1 p.o. – Cefixime 800 mg × 1 p.o. PLUS azithromycin	– Ceftriaxone 1 g × 1 IM PLUS Azithromycin 2 g × 1 orally – Gentamicin 240 mg × 1 IM PLUS Azithromycin 2 g × 1 orally[b]	No recommendation	No recommendation	No recommendation	– Re-treatment with recommended dual regimen – Gemifloxacin 320 mg × 1 orally PLUS Azithromycin 2 g × 1 OR Gentamicin 240 mg × 1 IM PLUS Azithromycin 2 g × 1 can be considered	It is strongly recommended that treatment be guided by antimicrobial susceptibility test results to determine the appropriate antimicrobial agent in consultation with an expert in infectious diseases and local public health authorities

(continued)

Table 2
(continued)

	WHO [11]	Europe [7]	Germany [13]	UK [12]	Australia [10]	USA [8]	Canada [9]
	2 g × 1 p.o. – Gentamicin 240 mg × 1 IM PLUS azithromycin 2 g × 1 p.o. – Spectinomycin 2 g × 1 IM (if not an oropharyngeal infection) PLUS azithromycin 2 g × 1 p.o.						

IM intramuscularly, *p.o.* per os

[a]Uncomplicated gonococcal infections of the cervix, urethra, and rectum

[b]Azithromycin tablets may be taken with or without food but gastrointestinal side effects can be less if taken after food

N. gonorrhoeae strains are not well established, with randomized clinical controlled trials (RCTs) absent. Instead, these dual antimicrobial therapies, including selection of antimicrobials and doses, were instead initially introduced based on old clinical trials, AMR surveillance data, predicted AMR trends, case reports of treatment failures using ESCs [5, 16], pharmacokinetic/pharmacodynamic simulations [58], and expert opinions. Ceftriaxone and azithromycin, included in the dual antimicrobial therapy, might not protect each other from initial resistance induction/selection [59]. However, these dual therapies are currently eradicating mainly all gonorrhea cases; concomitant resistance to ceftriaxone and azithromycin remains sporadic; and the transmission of any emerged ceftriaxone resistance is consequently prevented. Finally, in settings where up-to-date, local, comprehensive, quality-assured AMR surveillance exist WHO recommends also monotherapy with ceftriaxone 250 mg × 1, cefixime 400 mg × 1, or spectinomycin 2 g × 1 [11]. Notably, due to the very low cure rates, spectinomycin monotherapy should only be used if pharyngeal gonorrhea has been excluded and otherwise, azithromycin should also be given [2, 7].

In general, the dual antimicrobial therapies mentioned above are currently effective and they have likely been involved in decreasing the level of ESC resistance internationally [19]. Nevertheless, it is a main concern that the susceptibility of gonococcal isolates to ceftriaxone is still decreasing in many settings, the resistance to azithromycin has increased in many countries, and concomitant resistance to ceftriaxone and azithromycin has been found in several WHO regions [2, 5, 29, 49–51]. As mentioned above, the first global treatment failure with recommended dual therapy was reported in 2016 from the UK [49] and the first global case with combined ceftriaxone resistance and high-level azithromycin resistance was reported in 2018 in the UK [50]. The latter strain had emerged from *N. gonorrhoeae* strains with high-level azithromycin that had acquired a mosaic *penA* allele [50]. *N. gonorrhoeae* strains with high-level resistance to azithromycin (MIC $\geq$ 256 mg/L) have been verified in many countries, and a sustained national transmission of such strains is ongoing in the UK [2, 28]. Thus, the currently available dual antimicrobial treatments may not be long-term solutions and they are too expensive for wide-scale use in many less-resourced settings, which limit the prevention of emergence and spread of gonococcal AMR worldwide. Accordingly, new cost-effective and accessible antimicrobials are essential.

5 Future Treatment of Gonorrhea

5.1 Repurposing or Use of "Old" Antimicrobials

The susceptibility to spectinomycin is high globally [2, 5, 7, 11] and it would be valuable to have spectinomycin widely available worldwide again. However, the cure rate of pharyngeal gonorrhea is low [2–5]. Therefore, spectinomycin should ideally be used in dual therapy, for example, with azithromycin, which might also mitigate the emergence or spread of AMR. Repurposing of antimicrobials, such as gentamicin, rifampicin, ertapenem, and fosfomycin, have been suggested for future therapy of gonorrhea, but all of these have shortcomings. Briefly, several have not been formally assessed in a clinical trial for treatment of urogenital and extragenital gonorrhea (rifampicin, ertapenem, and fosfomycin); rifampicin is reserved for other infections (e.g., tuberculosis); AMR determinants already exist in the gonococcal population internationally (ertapenem and rifampicin) or resistance might emerge rapidly (fosfomycin); <95% clinical cure rate (gentamicin and fosfomycin); and for all of them a lack of appropriate pharmacokinetic/pharmacodynamic parameters for gonorrhea and known correlates between MIC and treatment outcome [2, 5, 16, 60–66]. These shortcomings most likely prohibit a widespread use of these antimicrobials in empirical monotherapy. However, they might be valuable in cases of ceftriaxone resistance, ESC allergy, and particularly in novel dual antimicrobial regimens.

5.2 New Antimicrobials or Other Therapeutics Compounds with Mainly In Vitro and/or Nonhuman Data Available

In recent decades, several new antimicrobials (derivatives of earlier developed antimicrobials or new antimicrobial classes) have proven relatively potent in vitro against *N. gonorrhoeae* strains, but clinical data regarding treatment of gonorrhea are mainly lacking. These antimicrobials include the fluoroquinolones avarofloxacin (JNJ-Q2), delafloxacin (RX-3341), sitafloxacin (DU-6859), and WQ-3810; bicyclic macrolides (bicyclolides) modithromycin (EDP-420/EP-013420/S-013420) and EDP-322; tetracycline derivatives eravacycline (TP-434) and tigecycline (fluorocycline and glycylcycline, respectively); 2-acyl carbapenems SM-295291 and SM-369926; aminomethyl spectinomycin; lipoglycopeptide dalbavancin; pleuromutilin lefamulin (BC-3781); boron-containing inhibitor AN3365; LpxC inhibitors; FabI inhibitor (e.g., MUT056399); tricyclic topoisomerase inhibitor REDX05931; topoisomerase II inhibitors closthioamide (CTA) and VXc-486 (VT12-008911); and SMT-571 (DDS-01), which was identified through high density transposon mutant profiling combined with WGS and machine learning tools. Notably, several other nonantimicrobial therapeutic compounds, using new targets and/or mechanisms of action, have been developed and suggested for future prevention or treatment of gonorrhea. These include, for example, noncytotoxic nanomaterials; inhibitors of efflux pumps, particularly coadministered with appropriate antimicrobials, that increase the susceptibility to certain antimicrobials, the innate host defense, and toxic metabolites; molecules mimicking host defensins; and host defense peptides, immune modulators or therapeutic vaccine (LL-37 (multifunctional cathelicidin peptide), IDR1, immunobodies [Fc], factor H-Fc immunotherapeutic molecule). However, many of these are in very early development for gonorrhea treatment. These antimicrobials and therapeutic compounds have been recently described elsewhere [2, 3, 5, 16, 60, 67–70].

5.3 Novel Antimicrobials in Clinical Trial Evaluation

Solithromycin (CEM-101), zoliflodacin (AZD0914/ETX0914), and gepotidacin (GSK2140944) are novel orally administered antimicrobials in clinical evaluation for uncomplicated gonorrhea treatment [3].

5.3.1 Solithromycin

The fluoroketolide solithromycin, as macrolides and ketolides, binds to the 50S ribosomal subunit and inhibits protein synthesis [71]. Solithromycin has shown a high in vitro activity against geographically, temporally, and genetically diverse wild-type, MDR, and XDR *N. gonorrhoeae* reference strains and clinical AMR isolates, that is, with MIC_{50} and MIC_{90} of 0.064–0.125 mg/L and 0.125–0.25 mg/L, respectively [72]. Nevertheless, strains with high-level resistance to azithromycin (MIC $\geq$ 256 mg/L; due to the A2059G SNP in 3–4 of the four 23S rRNA gene alleles) were shown early to be resistant to solithromycin (MICs = 4–32 mg/L) [72].

The phase II RCT (two-center, open-label, noncomparative) evaluating the efficacy of solithromycin 1 g $\times$ 1 or 1.2 g $\times$ 1 orally in the treatment of adults ($\geq$19 years) with uncomplicated urogenital gonorrhea included 46 subjects who could be evaluated for microbiological cure (1 g $\times$ 1 ($n = 22$) and 1.2 g $\times$ 1 ($n = 24$)). All (100%) subjects were culture negative at TOC. Side effects were dose-dependent and with the 1 g $\times$ 1 dose the most common were mild diarrhea (42%), nausea (26%), and fatigue/asthenia (10%) [73].

The phase III clinical trial evaluating solithromycin (1 g $\times$ 1 orally) in a multicenter, open-label noninferiority RCT (SOLITAIRE-U; NCT02210325) for treatment of uncomplicated urogenital gonorrhea in men and women has ended recently. The final results have not been presented. However, it is a major concern that analysis of the data from the initial patient cohort of 262 patients (study initially aimed to enroll 300 subjects) showed that solithromycin had only a 91.3% success in the population that could be evaluated microbiologically (despite 100% success for women). Consequently, solithromycin showed inferiority to the standard-of-care treatment (ceftriaxone 500 mg $\times$ 1 plus azithromycin 1 g × 1). The *N. gonorrhoeae* isolates from the treatment failures did not show solithromycin resistance pretreatment or at test-of-cure. Consequently, treatment failures were likely associated with a suboptimal duration of solithromycin exposure at the infectious site and optimizations of the dosing regimen (or possibly formulation, or both) appear essential [3, 74]. This is a major concern because the dose-dependent side effects observed in the phase II study were concerning and a higher solithromycin dose might not be tolerated.

5.3.2 Gepotidacin

Gepotidacin is the first-in-class triazaacenaphthylene antibacterial (bacterial topoisomerase II inhibitor). It inhibits DNA replication in bacteria by interactions with DNA gyrase (GyrA subunit) and topoisomerase IV (ParC subunit) [75–77]. However, gepotidacin has slightly different binding sites and mechanism of action compared to fluoroquinolones [77]. Gepotidacin has shown a high in vitro activity against geographically, temporally and genetically diverse resistant, including fluoroquinolone-resistant, MDR and XDR, gonococcal isolates [77]. The gepotidacin MIC_{50}, MIC_{90}, and MIC range was 0.5 mg/L, 1 mg/L, and 0.032–4 mg/L, respectively. No significant consistent cross-resistance between gepotidacin and any other antimicrobials, including the fluoroquinolone ciprofloxacin, was found. Nevertheless, the ParC D86N amino acid substitution, associated with ciprofloxacin resistance, was associated also with enhanced MICs of gepotidacin [77]. Notably, 5% of genome sequenced European gonococcal isolates from 2013 ($n = 1054$) contained the ParC D86N amino acid substitution [19, 77].

A multicenter, open-label phase II RCT recently evaluated gepotidacin (1.5 g and 3 g single oral dose, respectively) for treatment of uncomplicated gonorrhea [78]. Microbiological cure was achieved by 97% (29/30) and 95% (37/39) of subjects, respectively, in the treatment of urogenital gonorrhea. The most common side effects were gastrointestinal and the majority of these were mild or moderate [78]. It is a major concern that all isolates from these treatment failures ($n = 3$) were resistant to ciprofloxacin with the preexisting D86N amino acid substitution in ParC [79]. Test-of-cure isolates from the two treatment failures with the gepotidacin 3 g dose also demonstrated resistance induction/selection to gepotidacin (MIC increased ≥32-fold to ≥32 mg/L) and these isolates had induced/selected an additional resistance mutation in the second target for gepotidacin, that is, an A92T mutation in GyrA [79]. Accordingly, due to the dual targeting mechanism-of-action of gepotidacin, mutations in the quinolone resistance-determining region (QRDR) of both ParC and GyrA are required for high-level resistance. The frequency of single-step resistance mutations, when *N. gonorrhoeae* (without the ParC D86N amino acid substitution) is exposed to 4× and 8× MIC of gepotidacin, has been shown to be low [80]. However, resistance mutations are induced at a higher frequency if mutations are present in one of the targets (e.g., the ParC D86N mutation) [77]. It might be possible to optimize the formulation, dosing regimen and/or the PK/PD parameters of gepotidacin. However, the in vitro findings of the preexisting prevalent ParC D86N resistance mutation, the induction/selection of the GyrA A92T resistance mutation during treatment, and the rates of treatment failures in the phase II RCT are worrying.

5.3.3 Zoliflodacin

The first-in-class spiropyrimidinetrione zoliflodacin targets particularly the GyrB subcomponent of DNA gyrase. Zoliflodacin is a topoisomerase II inhibitor; however, it has a novel target as well as mechanisms of action compared to fluoroquinolones and other previously developed antimicrobials [81–84]. Thus, zoliflodacin inhibits DNA biosynthesis and causes an accumulation of double-strand cleavages [83]. Zoliflodacin initially showed a high in vitro activity against geographically, temporally and genetically diverse wild type, MDR and XDR *N. gonorrhoeae* reference strains and clinical isolates [85]. Subsequent examinations of contemporary, consecutive and selected clinical isolates in Europe (873 isolates from 21 European countries), the USA (100 isolates), and China (187 isolates) have further confirmed the high in vitro potency of, and lack of resistance to, zoliflodacin with MIC_{50}, MIC_{90}, and MIC range of 0.06–0.125 mg/L, 0.125–0.25 mg/L, and <0.002–0.25 mg/L, respectively [85–88]. The frequency of induced/selected zoliflodacin resistance mutations in *gyrB* has also been proven to be very low [82, 84].

A multicenter, open-label phase II RCT evaluating the efficacy, tolerability and safety of zoliflodacin 2 g × 1 or 3 g × 1 orally, compared to ceftriaxone 500 mg × 1 intramuscularly, for treatment of uncomplicated urogenital gonorrhea in males and females (18–55 years) has ended recently [89]. The RCT included 117 subjects who could be evaluated microbiologically, treated with zoliflodacin 2 g, zoliflodacin 3 g, or ceftriaxone 500 mg. The cure rates were 98% (48/49), 100% (47/47), and 100% (21/21), respectively. In general, zoliflodacin was well-tolerated and only 12% (21/179) of subjects reported any side effects, that is, 20 mild and one moderate (mostly gastrointestinal side effects) [89]. Accordingly, zoliflodacin remains promising for future treatment of gonorrhea, but a larger number of patients and particularly additional pharyngeal and rectal infections need to be examined. A phase III RCT is now under final planning, in collaboration between the manufacturer Entasis, the Global Antibiotic Research and Development Partnerships (GARDP) and WHO.

6 Conclusions

Gonorrhea with its severe complications and sequelae and the development of AMR in *N. gonorrhoeae*, are major public health concerns. Dual antimicrobial therapy (ceftriaxone 250–500 mg × 1 plus azithromycin 1–2 mg × 1 [7–13]) is currently implemented for treatment in many countries. Dual therapy should be used particularly in settings where up-to-date, local, and high-quality AMR data do not support other therapy and TOC is not mandatory. These recommended dual therapies currently have high cure rates; they have likely been involved in decreasing the level of ESC resistance internationally [19], and inhibit spread of AMR gonococcal strains. These dual therapies also eradicate concomitant *C. trachomatis* and many *M. genitalium* infections. The susceptibility of gonococcal isolates to ceftriaxone is still decreasing in many settings, resistance to azithromycin has increased in many countries, and concomitant resistance to ceftriaxone and azithromycin has been found in several settings [5, 29, 49–51]. The first global treatment failure with recommended dual therapy has been verified [49], and the first cases with combined ceftriaxone resistance and high-level azithromycin resistance were reported in 2018 in the UK and Australia [50, 51]. Consequently, the currently available dual therapies with ceftriaxone plus azithromycin are no long-term solutions and new antimicrobials for future treatment of gonorrhea are essential. Many new antimicrobials have proven relatively potent in vitro activity against gonococcal strains, but clinical data regarding gonorrhea treatment are mostly completely lacking [2, 5, 16, 60–66]. Solithromycin, gepotidacin, and zoliflodacin are novel antimicrobials in clinical evaluation for treatment of gonorrhea [71–89], and particularly zoliflodacin appears very promising and requires more attention.

Until we have new effective treatments for gonorrhea (urogenital and extragenital), as stated in the "WHO Global Action Plan to Control the Spread and Impact of Antimicrobial Resistance in *Neisseria gonorrhoeae*" [90], holistic actions are essential. These include an enhanced advocacy and awareness; prevention, early diagnosis, contact tracing, treatment (patient and sexual contacts), TOC, and additional measures for effective management of anogenital and pharyngeal gonorrhea; antimicrobial stewardship resulting in more rational drug use (e.g., use of evidence-based updated treatment guidelines and monitor antimicrobial use (i.e., including misuse and overuse)); surveillance of infection, AMR and treatment failures; and intensified research (e.g., regarding novel AMR determinants, molecular methods for AMR prediction, new antimicrobials and a gonococcal vaccine) [90].

In the near future, new molecular technologies, combined with phenotypic methods, will revolutionize the detection of

N. gonorrhoeae; knowledge of the induction/selection, evolution, fitness, and epidemiology of *N. gonorrhoeae* and its AMR (including in noncultured samples) [18–27, 40, 91]; rapid POC detection of *N. gonorrhoeae* and AMR, and personalized therapy at the first health care visit [54–56]; antimicrobial discovery and design; our understanding of pharmacokinetics/pharmacodynamics of antimicrobials in the gonorrhea treatment (urogenital and extragenital anatomical sites); and the *N. gonorrhoeae* vaccine field [92–95]. Ultimately, a gonococcal vaccine is the only sustainable solution for management and control of gonorrhea.

References

1. Newman L, Rowley J, Vander Hoorn S et al (2015) Global estimates of the prevalence and incidence of four curable sexually transmitted infections in 2012 based on systematic review and global reporting. PLoS One 10:e0143304
2. Unemo M, Shafer WM (2014) Antimicrobial resistance in *Neisseria gonorrhoeae* in the 21st century: past, evolution, and future. Clin Microbiol Rev 27:587–613
3. Unemo M, Bradshaw CS, Hocking JS et al (2017) Sexually transmitted infections: challenges ahead. Lancet Infect Dis 17:e235–e279
4. Unemo M, Jensen JS (2017) Antimicrobial-resistant sexually transmitted infections: gonorrhoea and *Mycoplasma genitalium*. Nat Rev Urol 14:139–152
5. Unemo M (2015) Current and future antimicrobial treatment of gonorrhoea – the rapidly evolving *Neisseria gonorrhoeae* continues to challenge. BMC Infect Dis 15:364
6. Lewis DA (2014) Global resistance of *Neisseria gonorrhoeae*: when theory becomes reality. Curr Opin Infect Dis 27:62–67
7. Bignell C, Unemo M (2013) 2012 European guideline on the diagnosis and treatment of gonorrhoea in adults. Int J STD AIDS 24:85–92
8. Workowski KA, Bolan GA (2015) Sexually transmitted diseases treatment guidelines, 2015. MMWR Recomm Rep 64 (RR-03):1–137
9. Public Health Agency of Canada (2013) Canadian guidelines on sexually transmitted infections. Gonococcal infections. Public Health Agency of Canada, Ottawa, ON. www.phac-aspc.gc.ca/std-mts/sti-its/cgsti-ldcits/assets/pdf/section-5-6-eng.pdf. Accessed 9 Aug 2018
10. Australasian Sexual Health Alliance (ASHA) Australian STI management guidelines for use in primary care. www.sti.guidelines.org.au/sexually-transmissible-infections/gonorrhoea#management. Accessed 9 Aug 2018
11. World Health Organization (WHO), Department of Reproductive Health and Research (2016) WHO guidelines for the treatment of *Neisseria gonorrhoeae*. WHO, Geneva, pp 1–64
12. Bignell C, Fitzgerald M (2011) UK national guideline for the management of gonorrhoea in adults, 2011. Int J STD AIDS 22:541–547
13. AWMF-Register. Nr. 059/004 – S2k-Leitlinie: Gonorrhoe bei Erwachsenen und Adoleszenten aktueller Stand: 08/2013. 1–31 [In German]
14. Lewis DA (2010) The Gonococcus fights back: is this time a knock out? Sex Transm Infect 86:415–421
15. Donà V, Low N, Golparian D et al (2017) Recent advances in the development and use of molecular tests to predict antimicrobial resistance in *Neisseria gonorrhoeae*. Expert Rev Mol Diagn 17:845–859
16. Unemo M, Nicholas RA (2012) Emergence of multidrug-resistant, extensively drug-resistant and untreatable gonorrhea. Future Microbiol 7:1401–1422
17. Tapsall JW, Ndowa F, Lewis DA et al (2009) Meeting the public health challenge of multidrug- and extensively drug-resistant *Neisseria gonorrhoeae*. Expert Rev Anti-Infect Ther 7:821–834
18. De Silva D, Peters J, Cole K et al (2016) Whole-genome sequencing to determine transmission of *Neisseria gonorrhoeae*: an observational study. Lancet Infect Dis 16:1295–1303
19. Harris SR, Cole MJ, Spiteri G et al (2018) Public health surveillance of multidrug-resistant clones of *Neisseria gonorrhoeae* in Europe: a genomic survey. Lancet Infect Dis 18:758–768

20. Jacobsson S, Golparian D, Cole M et al (2016) WGS analysis and molecular resistance mechanisms of azithromycin-resistant (MIC > 2 mg/L) *Neisseria gonorrhoeae* isolates in Europe from 2009 to 2014. J Antimicrob Chemother 71:3109–3116
21. Peters J, Cresswell F, Amor L et al (2018) Whole genome sequencing of *Neisseria gonorrhoeae* reveals transmission clusters involving patients of mixed HIV serostatus. Sex Transm Infect 94:138–143
22. Eyre DW, De Silva D, Cole K et al (2017) WGS to predict antibiotic MICs for *Neisseria gonorrhoeae*. J Antimicrob Chemother 72:1937–1947
23. Demczuk W, Martin I, Peterson S et al (2016) Genomic epidemiology and molecular resistance mechanisms of azithromycin-resistant *Neisseria gonorrhoeae* in Canada from 1997 to 2014. J Clin Microbiol 54:1304–1313
24. Demczuk W, Lynch T, Martin I et al (2015) Whole-genome phylogenomic heterogeneity of *Neisseria gonorrhoeae* isolates with decreased cephalosporin susceptibility collected in Canada between 1989 and 2013. J Clin Microbiol 53:191–200
25. Grad YH, Kirkcaldy RD, Trees D et al (2014) Genomic epidemiology of *Neisseria gonorrhoeae* with reduced susceptibility to cefixime in the USA: a retrospective observational study. Lancet Infect Dis 14:220–226
26. Ryan L, Golparian D, Fennelly N et al (2018) Antimicrobial resistance and molecular epidemiology using whole-genome sequencing of *Neisseria gonorrhoeae* in Ireland, 2014–2016: focus on extended-spectrum cephalosporins and azithromycin. Eur J Clin Microbiol Infect Dis 37:1661-1672
27. Grad YH, Harris SR, Kirkcaldy RD et al (2016) Genomic epidemiology of gonococcal resistance to extended-spectrum cephalosporins, macrolides, and fluoroquinolones in the United States, 2000–2013. J Infect Dis 214:1579–1587
28. Fifer H, Cole M, Hughes G et al (2018) Sustained transmission of high-level azithromycin-resistant *Neisseria gonorrhoeae* in England: an observational study. Lancet Infect Dis 18:573–581
29. Wi T, Lahra MM, Ndowa F et al (2017) Antimicrobial resistance in *Neisseria gonorrhoeae*: global surveillance and a call for international collaborative action. PLoS Med 14:e1002344
30. Spiteri G, Cole M, Unemo M et al (2013) The European Gonococcal Antimicrobial Surveillance Programme (Euro-GASP) – a sentinel approach in the European Union (EU)/European Economic Area (EEA). Sex Transm Infect 89(Suppl 4):iv16–iv18
31. Cole MJ, Spiteri G, Jacobsson S et al (2015) Is the tide turning again for cephalosporin resistance in *Neisseria gonorrhoeae* in Europe? Results from the 2013 European surveillance. BMC Infect Dis 15:321
32. Kirkcaldy RD, Kidd S, Weinstock HS et al (2013) Trends in antimicrobial resistance in *Neisseria gonorrhoeae* in the USA: The Gonococcal Isolate Surveillance Project (GISP), January 2006 – June 2012. Sex Transm Infect 89(Suppl 4):iv5–i10
33. Kirkcaldy RD, Harvey A, Papp JR et al (2016) *Neisseria gonorrhoeae* antimicrobial susceptibility surveillance – The Gonococcal Isolate Surveillance Project, 27 sites, United States, 2014. MMWR Surveill Summ 65:1–19
34. Martin I, Sawatzky P, Liu G et al (2016) Decline in decreased cephalosporin susceptibility and increase in azithromycin resistance in *Neisseria gonorrhoeae*, Canada. Emerg Infect Dis 22:65–67
35. Lahra MM (2015) Australian Gonococcal Surveillance Programme. Australian Gonococcal Surveillance Programme annual report, 2014. Commun Dis Intell Q Rep 39:E347–E354
36. Town K, Obi C, Quaye N et al (2017) Drifting towards ceftriaxone treatment failure in gonorrhoea: risk factor analysis of data from the Gonococcal Resistance to Antimicrobials Surveillance Programme in England and Wales. Sex Transm Infect 93:39–45
37. Ohnishi M, Golparian D, Shimuta K et al (2011) Is *Neisseria gonorrhoeae* initiating a future era of untreatable gonorrhea? Detailed characterization of the first strain with high-level resistance to ceftriaxone. Antimicrob Agents Chemother 55:3538–3545
38. Unemo M, Golparian D, Nicholas R et al (2012) High-level cefixime- and ceftriaxone-resistant *N. gonorrhoeae* in France: novel *penA* mosaic allele in a successful international clone causes treatment failure. Antimicrob Agents Chemother 56:1273–1280
39. Cámara J, Serra J, Ayats J et al (2012) Molecular characterization of two high-level ceftriaxone-resistant *Neisseria gonorrhoeae* isolates detected in Catalonia, Spain. J Antimicrob Chemother 67:1858–1860
40. Vincent LR, Kerr SR, Tan Y et al (2018) *In vivo*-selected compensatory mutations restore the fitness cost of mosaic *penA* alleles that confer ceftriaxone resistance in *Neisseria gonorrhoeae*. MBio 9(2). https://doi.org/10.1128/mBio.01905-17

41. Lahra MM, Ryder N, While DM (2014) A new multidrug-resistant strain of *Neisseria gonorrhoeae* in Australia. N Engl J Med 371:1850–1851
42. Deguchi T, Yasuda M, Hatazaki K et al (2016) New clinical strain of *Neisseria gonorrhoeae* with decreased susceptibility to ceftriaxone in Japan. Emerg Infect Dis 22:142–144
43. Gianecini R, Oviedo C, Stafforini G et al (2016) *Neisseria gonorrhoeae* resistant to ceftriaxone and cefixime, Argentina. Emerg Infect Dis 22:1139–1141
44. Nakayama S, Shimuta K, Furubayashi K et al (2016) New ceftriaxone- and multidrug-resistant *Neisseria gonorrhoeae* strain with a novel mosaic *penA* gene isolated in Japan. Antimicrob Agents Chemother 60:4339–4341
45. Lefebvre B, Martin I, Demczuk W et al (2018) Ceftriaxone-resistant *Neisseria gonorrhoeae*, Canada, 2017. Emerg Infect Dis 24:381–383
46. Terkelsen D, Tolstrup J, Johnsen CH et al (2017) Multidrug-resistant *Neisseria gonorrhoeae* infection with ceftriaxone resistance and intermediate resistance to azithromycin, Denmark, 2017. Euro Surveill 22:1273
47. Lahra MM, Martin I, Demczuk W et al (2018) Cooperative recognition of internationally disseminated ceftriaxone-resistant *Neisseria gonorrhoeae* strain. Emerg Infect Dis 24:735–740
48. Poncin T, Fouere S, Braille A et al (2018) Multidrug-resistant *Neisseria gonorrhoeae* failing treatment with ceftriaxone and doxycycline in France, November 2017. Euro Surveill 23 (21). https://doi.org/10.2807/1560-7917.ES.2018.23.21.1800264
49. Fifer H, Natarajan U, Jones L et al (2016) Failure of dual antimicrobial therapy in treatment of gonorrhoea. New Engl J Med 374:2504–2506
50. Eyre DW, Sanderson ND, Lord E et al (2018) Gonorrhoea treatment failure caused by a *Neisseria gonorrhoeae* strain with combined ceftriaxone and high-level azithromycin resistance, England, February 2018. Euro Surveill 23 (27). https://doi.org/10.2807/1560-7917.ES.2018.23.27.1800323
51. Whiley DM, Jennison A, Pearson J et al (2018) Genetic characterization of *Neisseria gonorrhoeae* resistant to both ceftriaxone and azithromycin. Lancet Infect Dis 18:717–718
52. Jönsson A, Jacobsson S, Foerster S et al (2018) Performance characteristics of newer MIC gradient strip tests compared with the Etest for antimicrobial susceptibility testing of *Neisseria gonorrhoeae*. APMIS 126:822–827
53. Unemo M, Ison C (2013) Gonorrhoea. In: Ballard R, Ison C, Lewis D, Ndowa F, Peeling R (eds) Laboratory diagnosis of sexually transmitted infections, including human immunodeficiency virus. World Health Organization (WHO), Geneva
54. Low N, Unemo M (2016) Molecular tests for the detection of antimicrobial resistant *Neisseria gonorrhoeae*: when, where, and how to use? Curr Opin Infect Dis 29:45–51
55. Goire N, Lahra MM, Chen M et al (2014) Molecular approaches to enhance surveillance of gonococcal antimicrobial resistance. Nat Rev Microbiol 12:223–229
56. Sadiq ST, Mazzaferri F, Unemo M (2017) Rapid accurate point-of-care tests combining diagnostics and antimicrobial resistance prediction for *Neisseria gonorrhoeae* and *Mycoplasma genitalium*. Sex Transm Infect 93:S65–S68
57. Demczuk W, Sidhu S, Unemo M et al (2017) *Neisseria gonorrhoeae* sequence typing for antimicrobial resistance, a novel antimicrobial resistance multilocus typing scheme for tracking global dissemination of *N. gonorrhoeae* strains. J Clin Microbiol 55:1454–1468
58. Chisholm SA, Mouton JW, Lewis DA et al (2010) Cephalosporin MIC creep among gonococci: time for a pharmacodynamic rethink? J Antimicrob Chemother 65:2141–2148
59. Rice LB (2015) Will use of combination cephalosporin/azithromycin therapy forestall resistance to cephalosporins in *Neisseria gonorrhoeae*? Sex Transm Infect 91:238–240
60. Ison CA, Deal C, Unemo M (2013) Current and future treatment options for gonorrhoea. Sex Transm Infect 89(Suppl 4):iv52–iv56
61. Hauser C, Hirzberger L, Unemo M et al (2015) In vitro activity of fosfomycin alone and in combination with ceftriaxone or azithromycin against clinical *Neisseria gonorrhoeae* isolates. Antimicrob Agents Chemother 59:1605–1611
62. Dowell D, Kirkcaldy RD (2013) Effectiveness of gentamicin for gonorrhoea treatment: systematic review and meta-analysis. Sex Transm Infect 89:142–147
63. Hathorn E, Dhasmana D, Duley L et al (2014) The effectiveness of gentamicin in the treatment of *Neisseria gonorrhoeae*: a systematic review. Syst Rev 3:104
64. Olsen B, Månsson F, Camara C et al (2012) Phenotypic and genetic characterisation of bacterial sexually transmitted infections in Bissau, Guinea-Bissau, West Africa: a prospective cohort study. BMJ Open 2:e000636

65. Ross JD, Harding J, Duley L et al (2017) LB1.5 the efficacy and safety of gentamicin for the treatment of genital, pharyngeal and rectal gonorrhoea: a randomised controlled trial. Sex Transm Infect 93(Suppl 2):A42–A43
66. Rodríguez A, Gallego A, Olay T et al (1977) Bacteriological evaluation of fosfomycin in clinical studies. Chemotherapy 23(Suppl 1):247–258
67. Bruhn DF, Waidyarachchi SL, Madhura DB et al (2015) Aminomethyl spectinomycins as therapeutics for drug-resistant respiratory tract and sexually transmitted bacterial infections. Sci Transl Med 7:288ra75
68. Savage VJ, Charrier C, Salisbury AM et al (2016) Biological profiling of novel tricyclic inhibitors of bacterial DNA gyrase and topoisomerase IV. J Antimicrob Chemother 71:1905–1913
69. Jönsson A, Foerster S, Golparian D et al (2018) In vitro activity and time-kill curve analysis of sitafloxacin against a global panel of antimicrobial-resistant and multidrug-resistant *Neisseria gonorrhoeae* isolates. APMIS 126:29–37
70. Jacobsson S, Mason C, Khan N, Meo P, Unemo M (2019) In vitro activity of the novel oral antimicrobial SMT-571, with a new mechanism of action, against MDR and XDR Neisseria gonorrhoeae: future treatment option for gonorrhoea? J Antimicrob Chemother pii: dkz060. https://doi.org/10.1093/jac/dkz060. [Epub ahead of print]
71. Llano-Sotelo B, Dunkle J, Klepacki D et al (2012) Binding and action of CEM-101, a new fluoroketolide antibiotic that inhibits protein synthesis. Antimicrob Agents Chemother 54:4961–4970
72. Golparian D, Fernandes P, Ohnishi M et al (2012) In vitro activity of the new fluoroketolide solithromycin (CEM-101) against a large collection of clinical *Neisseria gonorrhoeae* isolates and international reference strains including those with various high-level antimicrobial resistance-potential treatment option for gonorrhea? Antimicrob Agents Chemother 56:2739–2742
73. Hook EW III, Golden M, Jamieson BD et al (2015) A phase 2 trial of oral solithromycin 1200 mg or 1000 mg as single-dose oral therapy for uncomplicated gonorrhea. Clin Infect Dis 61:1043–1048
74. Cempra (2017) Cempra provides corporate update and reports fourth quarter and full year 2016 financial results. http://investor.cempra.com/releasedetail.cfm?ReleaseID=1014807. Accessed 2 Mar 2017
75. Biedenbach DJ, Bouchillon SK, Hackel M et al (2016) In vitro activity of gepotidacin, a novel triazaacenaphthylene bacterial topoisomerase inhibitor, against a broad spectrum of bacterial pathogens. Antimicrob Agents Chemother 60:1918–1923
76. Scangarella-Oman NE, Dixon P, Koeth LM et al (2016) Analysis of agar dilution MIC testing methods and variables and *in vitro* activity of gepotidacin (GSK2140944) against *Neisseria gonorrhoeae.* ASM Microbe. Boston, MA, Poster 462
77. Jacobsson S, Golparian D, Scangarella-Oman N et al (2018) In vitro activity of the novel triazaacenaphthylene gepotidacin (GSK2140944) against MDR *Neisseria gonorrhoeae.* J Antimicrob Chemother 73:2072–2077
78. Taylor SN, Morris DH, Avery AK et al (2018) Gepotidacin for the treatment of uncomplicated urogenital gonorrhea: A phase 2, randomized, dose-ranging, single-oral dose evaluation. Clin Infect Dis 67:504–512
79. Scangarella-Oman NE, Hossain M, Dixon PB et al (2018) Microbiological analysis from a phase 2 randomized study in adults evaluating single oral doses of gepotidacin in the treatment of uncomplicated urogenital gonorrhea caused by Neisseria gonorrhoeae. Antimicrob Agents Chemother 62(12)
80. Farrell DJ, Sader HS, Rhomberg PR et al (2017) *In vitro* activity of gepotidacin (GSK2140944) against *Neisseria gonorrhoeae.* Antimicrob Agents Chemother 61:pii:e02047-16
81. Huband MD, Bradford PA, Otterson LG et al (2015) In vitro antibacterial activity of AZD0914: a new spiropyrimidinetrione DNA Gyrase/Topoisomerase inhibitor with potent activity against Gram-positive, fastidious Gram-negative, and atypical bacteria. Antimicrob Agents Chemother 59:467–474
82. Foerster S, Golparian D, Jacobsson S et al (2015) Genetic resistance determinants, in vitro time-kill curve analysis and pharmacodynamic functions for the novel topoisomerase II inhibitor ETX0914 (AZD0914) in *Neisseria gonorrhoeae.* Front Microbiol 6:1377
83. Basarab GS, Kern GH, McNulty J et al (2015) Responding to the challenge of untreatable gonorrhea: ETX0914, a first-in-class agent with a distinct mechanism-of-action against bacterial type II topoisomerases. Sci Rep 5:11827
84. Alm RA, Lahiri SD, Kutschke A et al (2015) Characterization of the novel DNA gyrase inhibitor AZD0914: Low resistance potential

and lack of cross-resistance in *Neisseria gonorrhoeae*. Antimicrob Agents Chemother 59:1478–1486

85. Jacobsson S, Golparian D, Alm RA et al (2014) High in vitro activity of the novel spiropyrimidinetrione AZD0914, a DNA gyrase inhibitor, against multidrug resistant *Neisseria gonorrhoeae* isolates suggests a new effective option for oral treatment of gonorrhea. Antimicrob Agents Chemother 58:5585–5588
86. Unemo M, Ringlander J, Wiggins C et al (2015) High in vitro susceptibility to the novel spiropyrimidinetrione ETX0914 (also known as AZD0914) among 873 contemporary clinical *Neisseria gonorrhoeae* isolates in 21 European countries during 2012–2014. Antimicrob Agents Chemother 59:5220–5225
87. Su XH, Wang BX, Le WJ et al (2015) Multidrug-resistant *Neisseria gonorrhoeae* isolates from Nanjing, China, are sensitive to killing by a novel DNA Gyrase inhibitor, ETX0914 (AZD0914). Antimicrob Agents Chemother 60:621–623
88. Papp JR, Lawrence K, Sharpe S et al (2016) In vitro growth of multidrug-resistant *Neisseria gonorrhoeae* isolates is inhibited by ETX0914, a novel spiropyrimidinetrione. Int J Antimicrob Agents 48:328–330
89. Taylor SN, Marrazzo J, Batteiger BE et al (2018) Single-dose zoliflodacin (ETX0914) for treatment of urogenital gonorrhea. N Engl J Med 379:1835–1845
90. World Health Organization (WHO) (2012) Global action plan to control the spread and impact of antimicrobial resistance in *Neisseria gonorrhoeae*. WHO, Geneva. http://www.who.int/reproductivehealth/publications/rtis/9789241503501/en/
91. Graham RM, Doyle CJ, Jennison AV (2017) Epidemiological typing of *Neisseria gonorrhoeae* and detection of markers associated with antimicrobial resistance directly from urine samples using next generation sequencing. Sex Transm Infect 93:65–67
92. Gottlieb SL, Deal CD, Giersing B et al (2016) The global roadmap for advancing development of vaccines against sexually transmitted infections: update and next steps. Vaccine 34:2939–2947
93. Jerse AE, Deal CD (2013) Vaccine research for gonococcal infections: where are we? Sex Transm Infect 89(Suppl 4):iv63–iv68
94. Zielke RA, Wierzbicki IH, Baarda BI et al (2016) Proteomics-driven antigen discovery for development of vaccines against gonorrhea. Mol Cell Proteomics 15:2338–2355
95. Rice PA, Shafer WM, Ram S et al (2017) *Neisseria gonorrhoeae*: drug resistance, mouse models, and vaccine development. Annu Rev Microbiol 71:665–686

Chapter 4

Prediction of Minimum Inhibitory Concentrations of Antimicrobials for *Neisseria gonorrhoeae* Using Whole-Genome Sequencing

David W. Eyre, Daniel Golparian, and Magnus Unemo

Abstract

Bacterial whole-genome sequencing is now increasingly available to researchers, reference laboratories and individual healthcare institutions. It can be possible to predict antimicrobial minimum inhibitory concentrations (MICs) for *Neisseria gonorrhoeae* using sequencing data, for many antimicrobials within one or two MIC doubling dilutions of the phenotypic value. With emerging rapid sequencing technologies, it may be possible in future to predict antimicrobial resistance faster than existing culture-based methods. Sequencing also provides insights into the genetic mechanisms underlying antimicrobial resistance, their spread in time and space, as well as the molecular epidemiology of the gonococcal strains.

Key words Antibiotic, Antimicrobial, Minimum inhibitory concentrations, Resistance, Whole-genome sequencing

1 Introduction

Antimicrobial resistance in *Neisseria gonorrhoeae* is a major public health concern. Treatment with dual antimicrobial therapy (ceftriaxone plus azithromycin) is widely recommended and the last remaining option for empirical first-line treatment of gonorrhoea internationally, but emerging resistance to these antimicrobials threatens to leave few treatment options available [1, 2]. The genetic mechanisms underlying resistance to antimicrobials in *N. gonorrhoeae* are generally well characterized [2]. It is therefore possible to use whole-genome sequencing (WGS) data to predict antimicrobial resistance (or susceptibility), including the antimicrobial minimum inhibitory concentrations (MICs) [3]. Prediction at the level of MIC allows the development and spread of resistance to be monitored with more precision than a categorical "susceptible" or "resistant" classification. Additionally, it can provide more

Myron Christodoulides (ed.), *Neisseria gonorrhoeae: Methods and Protocols*, Methods in Molecular Biology, vol. 1997,
https://doi.org/10.1007/978-1-4939-9496-0_4,

robust predictions where strains have antibiotic MICs near susceptibility or resistance breakpoints.

Potential applications include characterization of the genetic basis of resistance or predisposition to develop full resistance, both in individual outbreaks and as part of national and international surveillance programs, in combination with the WGS-based epidemiology of the gonococcal strains. WGS-based predictions also allow strains without detailed phenotypic data to be included in surveillance. In future, there is the prospect of rapid and portable sequencing technologies being applied direct to clinical samples, such that predictions may also be used in real-time to inform individual patient management.

A range of mechanisms are responsible for antimicrobial resistance including point mutations and small insertions and deletions in the bacterial chromosome in key target genes, their regulators, efflux pump genes, or porin genes. Additionally, acquisition of novel genetic material can confer resistance, for example via recombination or via mobile genetic elements such as plasmids. Different bioinformatic approaches are required to optimally detect and characterize each class of genetic variant, and these are described in detail below.

Applying WGS, it is possible to predict *N. gonorrhoeae* antimicrobial MICs with relatively high precision. Using a multivariable logistic regression-based approach, in a dataset of 681 samples from national antimicrobial resistance surveillance programs in the UK, the USA, and Canada, MICs for five antimicrobials were predicted to the nearest MIC doubling dilution 53% of the time, and 93% were within ±1 and 98% within ±2 MIC doubling dilutions [3]. Some antimicrobial resistance phenotypes are conferred by a single genetic mechanism (e.g., high-level resistance to azithromycin), whereas MIC prediction for others requires the effect of multiple mechanisms involving a multitude of genes to be considered together.

Combinations of genetic resistance determinants may also help to identify key lineages for surveillance. This is the focus of the *N. gonorrhoeae* Sequence Typing for Antimicrobial Resistance (NG-STAR) typing Scheme [4]. This scheme is based on seven well-characterized resistance genes and creates a nomenclature for tracking the global dissemination of *N. gonorrhoeae* strains, together with a web interface (https://ngstar.canada.ca/) for researchers, clinicians and public health practitioners to identify NG-STAR types.

2 Materials

A complete protocol for WGS-based MIC prediction is presented, starting from a cultured isolate. The workflow is based around the most widely available sequencing platform for bacterial WGS, the Illumina MiSeq. The steps before and after sequencing are also applicable to all other Illumina platforms. Detailed library preparation and MiSeq sequencing steps are omitted as these follow the manufacturer's instructions. More extensive modifications are required for long-read sequencing technologies, such as Oxford Nanopore Technologies MinION, but as these are rapidly evolving they are not considered further here.

The bioinformatic methods assume familiarity with running software from a linux-based command line and a basic knowledge of Python. The software tools described can be installed on most linux distributions and macOS. A variety of installation methods can be used, but the use of a package manager with bioinformatic tool support such as brew or conda is recommended. Additionally, some understanding of the National Center for Biotechnology Information (NCBI) Basic Local Alignment Search Tool (BLAST) for searching for sequences of interest is also assumed but could be gained from a number of available online introductions to BLAST.

2.1 Isolate Preparation for DNA Extraction

1. *Neisseria gonorrhoeae* isolate (e.g., from stock previously inoculated in commercially available 20% (v/v) glycerol solution, frozen at −80 °C).
2. VCAT (vancomycin, colistin sulfate, amphotericin B, trimethoprim) selective-agar, or equivalent commercially available selective gonococcal agar.
3. 10 μL inoculating loops.
4. Humidified incubator set to 37 °C with 5% (v/v) CO_2.

2.2 DNA Extraction

2.2.1 Mechanical Disruption

1. MP Biomedicals FastPrep24 homogenizer or equivalent.
2. 2 mL Tubes of MP Biomedicals Lysing Matrix B.
3. 0.9% sterile saline: commercially available or add 0.9 g of NaCl to 100 mL of distilled H_2O and sterilize by autoclaving at 121 °C and 2.68 kg/cm^2 for 15 min.
4. QuickGene DNA tissue kit S (DT-S), including EDT-01 Proteinase K, LDT-01 Lysis Buffer, WDT-01 Wash Buffer, CDT-01 Elution Buffer, Cartridges, Wash Tubes, Collection Tubes.
5. Vortex mixer.
6. Fridge.
7. Microtube centrifuge.
8. Heat block.

2.2.2 DNA Preparation and Extraction

1. 2 mL Tubes of MP Biomedicals Lysing Matrix B.
2. QuickGene Mini80 nucleic acid isolation device.

3. QuickGene DNA tissue kit S (DT-S), including EDT-01 Proteinase K, LDT-01 Lysis Buffer, WDT-01 Wash Buffer, CDT-01 Elution Buffer, Cartridges, Wash Tubes, Collection Tubes.
4. 100% Ethanol.
5. Microtube centrifuge.
6. 1.5 mL microcentrifuge tubes.
7. Qubit Fluorometer.

2.3 Library Preparation

1. Nextera XT DNA Library Preparation Kit (Illumina).
2. Nextera Index Kit (96 indexes, 384 samples) (Illumina).
3. 10 mM Tris–HCl buffer, pH 7.8, with no detergent present.
4. Microtube centrifuge.
5. Thermocycler.
6. Qubit Fluorometer.
7. Agilent TapeStation (e.g., 2200 TapeStation).

2.4 Illumina MiSeq Sequencing

1. Illumina MiSeq.
2. MiSeq Reagent Kit (e.g. v2 300-cycle).
3. Illumina Sequence Analysis Viewer quality control software.
4. FastQC post-sequencing quality control software (http://www.bioinformatics.babraham.ac.uk/projects/fastqc/).

2.5 Sequence Data Assembly

1. Input sequence data: two FASTQ files containing forward and reverse reads (e.g., the output from a MiSeq run for each sample: samplename_S1_L001_R1_001.fastq.gz and samplename_S1_L001_R2_001.fastq.gz).
2. De novo assembly software (e.g., Velvet (https://www.ebi.ac.uk/~zerbino/velvet/) with VelvetOptimiser (https://github.com/tseemann/VelvetOptimiser)).
3. A locally installed copy of BLAST from NCBI (e.g., from ftp://ftp.ncbi.nlm.nih.gov/blast/executables/blast+/LATEST/).
4. Mapping software, BWA mem (https://github.com/lh3/bwa).
5. A FASTA file containing the sequence of the 23S RNA gene, reference_23s.fa (available at https://github.com/davideyre/gc_mic_prediction_chapter).
6. Samtools (https://github.com/samtools/samtools).

2.6 Identification of Resistance Determinant Presence

2.6.1 Determining 23S rRNA Gene Mutation Copies

1. A bam file containing sequence reads mapped to a 23S rRNA gene.
2. Python 2.7 and the pysam python library and its dependencies (http://pysam.readthedocs.io/en/latest/installation.html).

2.6.2 BLAST Searches to Identify Resistance Determinants

1. A list of genes or promoters, including their specific resistance mutations, to be searched for; these can be obtained from available reference genomes based on available literature. An example list sufficient to predict MICs for cefixime, azithromycin, ciprofloxaxin, benzylpenicillin, and tetracycline, gene_targets.fa, is provided at https://github.com/davideyre/gc_mic_prediction_chapter and contains *penA*, *penB*, *mtrR*, *mtrR* promoter, *gyrA*, *parC*, *ponA*, *tetM*, *blaTEM*, *ermB*, *ermC*, *rpsJ*.
2. A list of variants in these genes to be searched for (Table 1).
3. A FASTA file containing assembled contigs for the sample to be tested.
4. A locally installed copy of BLAST from NCBI (e.g., from ftp://ftp.ncbi.nlm.nih.gov/blast/executables/blast+/LATEST/).
5. Optionally, the Biopython.Bio.BLAST library from the Biopython package (http://biopython.org/DIST/docs/api/Bio.Blast-module.html).

2.6.3 Identification of penA Allele

1. A locally installed copy of BLAST from NCBI (e.g., from ftp://ftp.ncbi.nlm.nih.gov/blast/executables/blast+/LATEST/).
2. A list of *penA* alleles for which MIC prediction coefficients have been previously calculated; penA_alleles.fa, available at https://github.com/davideyre/gc_mic_prediction_chapter.
3. Python 2.7 and the pysam python library and its dependencies (http://pysam.readthedocs.io/en/latest/installation.html).
4. Sequence alignment viewer (e.g., SeaView (http://doua.prabi.fr/software/seaview)).

2.7 MIC Prediction

1. A list of resistance determinants present.
2. A list of regression coefficients for each resistance determinant for each antimicrobial of interest: an example is provided in Table 2.

3 Methods

3.1 Isolate Preparation for DNA Extraction

1. Scrape the frozen gonococcal isolate stock solution using a 10 μL loop and inoculate the VCAT agar plate(s) by spreading the contents of the loop over the agar surface.

Table 1
Susceptibility-modifying genetic elements required for MIC prediction for five antimicrobials

Gene/ element	Characteristic	Summary	AZM	CFX	CIP	PEN	TET
penA	Allele	Reduced β-lactam acylation of penicillin binding protein (PBP) 2		✓		✓	
mtrR promoter	Deletion of A in repeat (−35A) A → C in repeat (−38) 2 bp insertion *mtr*120 (at position 79)	Over-expression of MtrCDE efflux pump	✓ ✓	✓ ✓		✓ ✓	✓ ✓
mtrR	A39T G45D Truncation	Over-expression of MtrCDE efflux pump (A39T is included in the scheme, but experimental data for this action are lacking)	✓	✓		✓	✓
penB (*porB1b*)	G120K A121D/N	Reduced influx		✓		✓	✓
ponA (*ponA*1 allele)	L421P	Reduced β-lactam acylation of PBP1				✓	
bla_{TEM}	Presence/ absence	Penicillinase				✓	
23S rRNA	C2611T A2059G	Four copies of these genes present, increasing resistance with increased number of copies with SNPs via decreased binding to 50S ribosome	✓				
erm(B), *erm (C)*	Presence/ absence	Methylate 23S rRNA to block binding	✓				
gyrA	S91F D95N/G	Reduce quinolone binding to DNA gyrase (GyrA)			✓		
parC	D86N S87R/I/W S88P E91K	Reduce quinolone binding to topoisomerase IV (ParC)			✓		
rpsJ	V57M	Reduces affinity of 30S ribosome for tetracycline					✓
tet(M) plasmid	Presence/ absence	TetM resembles elongation factor G, binds 30S ribotype and prevents tetracycline binding					✓

The *mtrR* promoter variants can be determined by inspecting the *mtrR* promoter sequence (provided at https://github.com/davideyre/gc_mic_prediction_chapter in gene_targets.fa) between positions 145 and 156. An A at position 79 of the *mtrR* promoter sequence provided is a marker for the *mtr*120 variant. Truncation of *mtrR* can be determined by translating the *mtrR* gene sequence in Seaview and looking for the presence of a premature stop codon. Based on a table in ref. 3, please see this publication for details of the original description of each resistance mechanism

AZM azithromycin, *CFX* cefixime, *CIP* ciprofloxacin, *PEN* penicillin, *TET* tetracycline

Table 2
MIC prediction coefficients

Cefixime		Benzylpenicillin		Azithromycin		Ciprofloxacin		Tetracycline	
Predictor	Multivariate coefficient	Predictor	Multivariate coefficient	Predictor	Multivariate coefficient	Predictor	Multivariate coefficient	Predictor	Multivariate coefficient
Country: England	0	Country: England	0	Country: England	0	Country: England	0	Country: England	0
Country: Canada	0.6	Country: Canada	1.33	Country: Canada	2.28	Country: Canada	−0.14	Country: Canada	−0.77
Country: USA	1.36	Country: USA	1.57	Country: USA	1.74	Country: USA	−0.44	Country: USA	−1.42
mtrR 045	0.01	bla_{TEM}	4.55	*ermB* or *ermC*	4.48	*gyrA* 91	2.89	*mtrR* 039	
*mtr*120	1.28	*mtrR* 039	0.11	*mtrR* 039	0.37	*gyrA* 95	4.07	*mtrR* 045	−0.07
penA: M32091	0	*mtrR* 045	0.7	*mtrR* promoter: A-38C	1.97	*parC* 86	2.31	*mtrR* promoter: A-38C	0.66
penA: II	0.65	mtr120	0.98	*mtrR* promoter: deletion	1.31	*parC* 87	2.9	*mtrR* promoter: deletion	0.17
penA: KP677512	1.92	*mtrR* promoter: A-38C	0.83	*mtrR* truncation	−0.97	*parC* 91	2.3	*mtrR* truncation	−0.39
penA: KP721218	2.01	*mtrR* promoter: deletion	0.31	rRNA A2059G: 4 copies	8.55	Interaction: *parC* 87 + *parC* 91	−2.96	*penB* 120	0.72
penA: V	1.35	*mtrR* truncation	−0.36	rRNA C2611T: per copy [range 0–4]	1.52	Multivariate constant	−5.8	*rpsJ* 57	2.66
penA: X07470	1.52	*penA*: M32091	0	Interaction: *ermB* or *ermC* + *mtrR* promoter: deletion	−3.6			*tetM*	6.98

(continued)

Table 2
(continued)

Cefixime		Benzylpenicillin		Azithromycin		Ciprofloxacin		Tetracycline	
Predictor	Multivariate coefficient	Predictor	Multivariate coefficient	Predictor	Multivariate coefficient	Predictor	Multivariate coefficient	Predictor	Multivariate coefficient
penA: XII	1.79	*penA*: II	−0.68	Interaction: *mtrR* promoter: deletion + rRNA C2611T per copy	−0.25			Interaction: *mtrR* 045 + *tetM*	−0.9
penA: XIX	0.96	*penA*: KP677512	−0.55	Interaction: *mtrR* promoter: deletion + Country: USA	−0.77			Interaction: *mtrR* promoter: deletion + *penB* 120	0.55
penA: XV	−0.44	*penA*: KP721218	0.36	Interaction: rRNA C2611T per copy + Country: Canada	−0.6			Interaction: *mtrR* promoter: deletion + *tetM*	−1.02
penA: XXII	0.57	*penA*: V	0.59	Interaction: rRNA C2611T per copy + Country: USA	−0.45			Interaction *rpsJ* 57 + *tetM*	−2.69
penA: XXXIV	3.08	*penA*: X07470	−0.61	Multivariate constant	−3.13			Multivariate constant	−1.34
penA: XXXV	0.89	*penA*: XII	0.45						
penA: A501P	4.74	*penA*: XIX	−0.27						
penA: A501T	0.97	*penA*: XV	−0.85						
penA: A501V	0.96	*penA*: XXII	−0.29						
penA: N512Y	1.33	*penA*: XXXIV	0.4						
penB 120	0.07	*penA*: XXXV	−0.34						

penB 121	−1.33	*penA*: F504 L	1.18
Interaction: *mtrR* 045 + *penA*: II	0.67	*penB* 120	1
Interaction: *mtrR* 045 + Country: Canada	0.72	*ponA* 421	0.74
Multivariate constant	−7.83	Interaction: bla_{TEM} + *mtrR* 39	1.5
		Interaction: bla_{TEM} + *mtrR* promoter: deletion	2.42
		Interaction: bla_{TEM} + *penA*: XXXIV	−4.72
		Interaction: bla_{TEM} + *ponA* 421	−1.67
		Interaction: bla_{TEM} + Country: USA	−1.26
		Interaction: *penA*: X07470 + Country Canada	1.06
		Multivariate constant	−3.87

Unless otherwise specified, the presence of a non-wild-type residue at a given site should be taken to be associated with the coefficient (e.g., *gyrA* 95) could represent either D95N or D95G. Based on a table in ref. 3

2. Place the agar plate in a humid incubator at 37 °C with 5% (v/v) CO_2 atmosphere for 24 h.
3. Transfer a single colony from the growth after 24 h to a second VCAT agar plate, using a 10 μL loop. Aim to achieve profuse growth on most of the plate, but leave an area of the plate with sparse streaks to allow the purity of growth to be assessed (*see* **Notes 1** and **2**).
4. Repeat **step 2**.

3.2 DNA Extraction

This protocol uses proprietary reagents and equipment for a two-step process involving mechanical disruption of the bacteria followed by DNA preparation and extraction (*see* **Note 3**).

3.2.1 Mechanical Disruption

1. Inspect the gonococcal growth on the VCAT agar plates for purity before proceeding (*see* **Note 4**).
2. Add 200 μL of sterile saline and 480 μL of LDT-01 and 20 μL of EDT-01 solutions provided by the QuickGene DNA tissue kit (DT-S) (*see* **Note 5**) to a Lysing Matrix B tube.
3. Add a 10 μL loop-full of pure bacterial growth to the Lysing Matrix B tube.
4. Replace lid and vortex for 5 s.
5. Homogenize the bacterial cells for 2 cycles in the FastPrep24 homogenizer using settings of speed 6 m/s and 40 s per cycle. Rest the samples between cycles in the fridge at 4 °C for 5 min.
6. Centrifuge the Lysing Matrix B tubes at ~15,000 × *g* for 10 min.
7. Heat the Lysing Matrix B tubes at 70 °C for 10 min in the heat block.

3.2.2 DNA Preparation and Extraction

1. Prepare 2.25 mL of fresh Wash Solution by adding an equal volume of 100% ethanol to WDT-01 solution provided in the QuickGene DNA tissue kit (DT-S).
2. Add one cartridge per sample to the QuickGene Mini80 Cartridge Holder.
3. Prepare the QuickGene Mini80 Tube Holder with sufficient Waste Tubes and Collection Tubes.
4. Add 450 μL of 100% ethanol to a 1.5 mL microcentrifuge tube.
5. Transfer the supernatant from the Lysing Matrix B tube described in **step** 7 of Subheading 3.2.1 to the microcentrifuge tube containing the 100% ethanol.
6. Pipette up and down once to mix.
7. Transfer the sample to a QuickGene Filter Cartridge held in the QuickGene Mini80 Cartridge Holder.

8. Place the Cartridge Holder rack in the QuickGene Mini80 in the wash position and pressurize until there is no lysate left in the Filter Cartridge.
9. Add 750 μL of Wash Solution into the Filter Cartridge to start the Wash step. Pressurize until there is no Wash Solution left in the Filter Cartridge.
10. Repeat **step 9** twice more.
11. Transfer the Cartridge Holder rack to the elution position.
12. Add 400 μL of CDT-01 Elution Buffer into the Filter Cartridge (*see* **Note 6**).
13. Incubate at room temperature for 2 min.
14. Pressurize until there is no CDT-01 Elution Buffer left in the Filter Cartridge.
15. Store the Collection tube at −20 °C until required.
16. Measure the extracted DNA by Qubit Fluorometer, according to the manufacturer's instructions (*see* **Note 7**).

3.3 Library Preparation

1. Use the Nextera XT DNA Library Preparation Kit (Illumina) according to the manufacturer's instructions (*see* **Notes 8–10**).
2. The specified starting DNA requirement is 1 ng (e.g., 5 μL of 0.2 ng/μL DNA preparation).
3. An example sample dilution spreadsheet is provided at https://github.com/davideyre/gc_mic_prediction_chapter, normalization_spreadsheet.xlsx. Dilute DNA samples in 10 mM Tris–HCl buffer, pH 7.8, using a two-step procedure (e.g., first dilute input DNA samples to 3 ng/μL, and then dilute to 0.2 ng/μL). A suggested post-PCR cleanup normalization is also included as an alternative to the standard magnetic bead normalization. Finally, calculations for sample dilutions and sample handling for loading onto the MiSeq flowcell are also provided.
5. Run 2 μL of the library on an Agilent 2200 TapeStation (or e-quivalent) to check the size distribution for all of libraries. Examine for the presence of a single symmetrical peak around ~500 bp, demonstrating an appropriate size distribution for the DNA fragments, with smaller fragments removed (*see* **Note 10**).
6. Run multiple samples simultaneously in one run using the Nextera Index Kit.
7. Determine the total sequencing output according to the chosen MiSeq Reagent Kit (e.g., the MiSeq Reagent Kit v2 (300-cycle) generates a maximum of 4.5 Gb of output). Each genome is around 2.1 Mb in length, and it is suggested that ~100× coverage is targeted, such that approximately 4500/(2.1 × 100) samples can be sequenced simultaneously, that is, ~20 samples.

3.4 Illumina MiSeq Sequencing

1. Set up and run the MiSeq according to the manufacturer's instructions, using the relevant settings for the sequencing kit used (*see* **Notes 11** and **12**).
2. Select "Use Adaptor Trimming" and set the "Application" to "FASTQ Only," when using Illumina Experiment Manager to create a sample sheet. This will ensure that the output data produced is demultiplexed into two individual FASTQ files per sample with adapter sequences removed.
3. Assess the sequencing quality for the sequencing run with Sequence Analysis Viewer. Ensure that the sequencing yield meets the expected range for the sequencing kit chosen.
4. Use FastQC to make a "per sample" and "per run" quality control assessment. FastQC contains a traffic light system to make interpretation accessible: green (acceptable), orange (warning), and red (unacceptable) (*see* **Note 13**).

3.5 Sequence Data Assembly

1. Install Velvet (`git clone` http://github.com/dzerbino/velvet.github) (*see* **Note 14**). Edit the Makefile to set `MAXKMERLENGTH` to an odd number around the maximum read length generated (e.g., `MAXKMERLENGTH =301`). Type `make` and the resulting executable binaries velvetg and velveth will be made.
2. Download and install VelvetOptimiser, BLAST, BWA mem and samtools using the authors' instructions.
3. Run VelvetOptimiser on the input files. For example:

```
VelvetOptimiser.pl -f '-shortPaired -fastq.gz -separate samplename_S1_L001_R1_001.fastq.gz samplename_S1_L001_R2_001.fastq.gz' -s 35 -e 127 -x 4 -t 8 -p samplename_output (see Note 15).
```

   VelvetOptimiser determines the optimal settings to use for the final assembly, including optimizing the length of the original reads retained to build the assembly, and the k-mer size, which is typically around two thirds of the original read length. In the example, the input files are samplename_S1_L001_R1_001.fastq.gz and samplename_S1_L001_R2_001.fastq.gz. We search for the optimal k-mer size to use between 35 and 127 in increments of 4, and use eight threads and an output prefix of samplename_output (*see* **Note 16**).

4. Within the output directories, the contigs.fa file contains the assembled contigs and the stats.txt, a summary of the quality of the assembly. Assess the quality of the assembly by examining the following parameters:
   (a) The "Total bases in contigs > 1k" should be approximately the size of the genome of 2.1 Mb. Much longer is suggestive of misassembly (e.g., due to contamination).

   (b) Ideally, the "Number of contigs > 1k" should be a few hundred rather than thousands, which suggests a fragmented assembly, (e.g., due to sequencing errors).
   (c) Larger values of "n50" reflect better quality assemblies with fewer (and larger) contigs needed to span the genome; at least half of the nucleotides in the assembly belong to contigs of length n50 or longer.

5. Create a BLAST database for each set of contigs, here called contigs.fa, so that they can be searched for resistance determinants:

```
makeblastdb -dbtype nucl -in contigs.fa.
```

6. Assess for mutations in the four copies of the 23S rRNA genes by mapping the reads against a 23S gene reference, as follows:
   (a) Save a local copy of reference_23s.fa and create a bwa index for it:

```
bwa index reference_23s.fa.
```

   (b) Map the sequence reads against the 23S reference:

```
bwa mem reference_23s.fa samplename_S1_L001_R1_001.
fastq.gz samplename_S1_L001_R2_001.fastq.gz > sample-
name.sam.
```

   (c) Store the mapped reads in a sorted bam file:

```
samtools view -uS -F 4 samplename.sam | samtools sort - -o
samplename.bam.
```

   (d) Index the sorted bam file:

```
samtools index samplename.bam.
```

3.6 Identification of Resistance Determinant Presence

3.6.1 Determining 23S rRNA Gene Mutation Copies

Determine the number of the four 23S rRNA alleles that contain the A2059G and C2611T mutations for A2059G (*E. coli* numbering) by calculating the ratio of A and G base calls at site A2045 (using the reference_23s.fa file suggested) and for C2611T, the ratio of C to T base calls at site C2597.

1. Use the mapped bam file created in **step 6** of Subheading 3.5 as input, and use pysam to determine the ratio of base counts at the two sites. An example python function that contains the necessary commands for mapping the data and running pysam, get23S.py, is provided at https://github.com/davideyre/gc_mic_prediction_chapter.
2. The proportion of the total bases at the two sites that are G (for 2045) or T (for 2597) should be approximately 0, 0.25, 0.5, 0.75, or 1, corresponding to 0, 1, 2, 3, or 4 copies of the A2059G or C2611T mutation.

3.6.2 BLAST Searches to Identify Resistance Determinants

Run BLAST to search for each of the genes of interest in the contigs.fa file and the BLAST database created in **steps 3** and **5** of Subheading 3.5 (*see* **Notes 17–19**).

1. Obtain the top match for each gene present in the core genome (*penA*, *penB*, *mtrR*, *mtrR* promoter, *gyrA*, *parC*, *ponA*, *rpsJ*) (*see* **Note 20**).
2. Determine if any match is present (e.g., requiring 90% identity as a threshold for identifying presence) for each gene that is variably present (*tetM*, *blaTEM*, *ermB*, *ermC*).
3. Various strategies are available for obtaining these BLAST results, including running BLAST directly from the command line. For example:

```
blastn -db contigs.fa -query gyrA.fa -outfmt 5.
```

   This command runs a search within contigs.fa for the *gyrA* gene (a copy of which is contained in gyrA.fa), and the results of the search are provided in XML format (omit the -outfmt 5 for a standard BLAST output). Separate FASTA files can be created for each of the genes of interest from gene_targets.fa, or the gene_targets.fa file used to search for all of these simultaneously.

4. Manually search the outputs from the BLAST for known variants at a particular position. For example, in the case of *gyrA* S91F, using the XML formatted BLAST output, take the text from the <Hsp_hseq> section (this is the match to *gyrA* found in contigs.fa) and create a new contigs_gyrA.fa file with this sequence. It is suggested also that the <Hsp_qseq> (the reference *gyrA* sequence) is included in the contigs_gyrA.fa. The final file should look like the example provided here, where Hsp_hseq has been renamed test and Hsp_qseq renamed reference, https://github.com/davideyre/gc_mic_prediction_chapter.
5. Open contigs_gyrA.fa in SeaView and from the Props menu select "view as proteins," then browse to position 91. Where insertions or deletions have occurred, relative to the reference, this needs to be accounted for when identifying the site of interest.
6. Automation of the BLAST searches can be achieved by running BLAST using the Biopython Bio.Blast library, which allows BLAST to be called from Python and for XML output to be parsed. Refer to the Biopython documentation for further details.

3.6.3 Identification of penA *Allele*

For the MIC prediction method presented, the closest *penA* allele with a described MIC coefficient must be identified.

1. Obtain the *penA* sequence from the BLAST search described in Subheading 3.6.2, and save it, removing any dashes (e.g., penA_example_gene.fa).
2. Use SeaView (or another tool, *see* **Note 21**) to translate the sequence to a protein sequence, and save the file (e.g., penA_example_protein.fa).
3. Run a protein-based BLAST search to identify the closest match in the penA_alleles.fa file. First create BLAST indexes

```
makeblastdb -dbtype prot -in penA_alleles.fa.
```

4. Run a protein BLAST search using the list of known alleles to identify the match with the highest identity

```
blastp -db penA_alleles.fa -query penA_example_protein.fa
(see Note 22).
```

5. The closest matching *penA* alleles are displayed as a list.

3.7 MIC Prediction (See Note 23)

1. Determine which of the relevant resistance determinants are present for each antimicrobial from the output of the final step in Subheadings 3.6.1–3.6.3.
2. Look up the associated regression coefficient for each determinant l in Table 2.
3. Calculate the MIC as follows: first of all, obtain the sum of the relevant coefficients for resistance mechanisms that are present and a constant representing the wild-type MIC. Secondly, obtain the inverse $\log_2$ transformation of the sum, that is,

$$\log_2(\text{MIC}) = \text{coeff\_A} + \text{coeff\_B} + \text{const}.$$
$$\text{MIC} = 2^{(\text{coeff\_A}+\text{coeff\_B}+\text{const})},$$

where coeff_A and coeff_B are two examples of coefficients associated with two resistance mechanisms from Table 2 and const is the multivariate constant provided in Table 2.
4. The country coefficient allows for variation in the MIC as measured in different countries' surveillance programmes. Use the coefficients for the USA or Canada if you wish to predict the MIC based on phenotypic methods in these countries, or omit the coefficient if you wish to predict the MIC using UK phenotypic methods.

4 Notes

1. If the frozen stock was made from a single colony or sub-culture of a single colony, the subculture step may be omitted, but it is still recommended if possible to ensure pure growth from a single colony for the extraction.

2. Assess the purity of growth by checking that all colonies present share the same morphology.
3. Alternative DNA extraction procedures for bacteria can be used, provided that sufficient high quality DNA is generated (e.g., kits such as the Wizard Genomic DNA Purification Kit Protocol or an automated DNA extraction on for example, the QIAsymphony instrument).
4. If the culture is not pure, then subculture a single colony on to a fresh VCAT agar plate as described in **step 3** of Subheading 3.1.
5. The QuickGene DNA Tissue Kit S (DT-S) is specifically designed for the QuickGene-Mini480 and the QuickGene-Mini80 devices (Kurabo Industries Ltd.). All consumables required for a DNA isolation are contained within the kit.
6. For mucoid organisms, warm the CDT-01 Elution Buffer in a water bath to 50 °C before use.
7. Quantification of DNA concentration using spectrophotometric methods (e.g., NanoDrop) is not sufficiently precise, but can provide additional information on DNA purity. DNA concentration can also be measured by PicoGreen using a plate-reader and standard curve for high throughput applications.
8. Refer to manufacturer's website (https://emea.illumina.com/products/by-type/sequencing-kits/library-prep-kits/nextera-xt-dna.html).
9. Any available Illumina library kit can be used.
10. An AMPure (Beckman) bead-based clean-up of the DNA is recommended prior to library preparation (https://www.beckmancoulter.com/wsrportal/techdocs?docname=B37419).
11. The recommended sequencing kit assumes a relatively low throughput sequencing scenario, where the desired batch size is small.
12. Alternative sequencing kits can be used, including MiSeq Reagent Kit v3 (600-cycles) and MiSeq Reagent Kit v2 (500-cycles) which have greater sequencing outputs.
13. When using FastQC with data generated from a Nextera library preparation, it is expected that the per base sequence content will be biased for the beginning of reads (https://sequencing.qcfail.com/articles/positional-sequence-bias-in-random-primed-libraries/). This is not a reason to trim these reads or reject the sequence.
14. Instructions are provided for Velvet, but Spades can be used alternatively and effectively. Spades software includes the iteration for optimal kmer size, etc. (http://cab.spbu.ru/software/spades/).

15. samplename_S1_L001_R1_001.fastq.gz and samplename_S1_L001_R2_001.fastq.gz are the default names generated from a MiSeq run: samplename is the sample name provided before starting sequencing; S1 indicates the row of the sample sheet, here row 1; L001 the sequencing lane – always 1 on an MiSeq run; R1 the forward read and R2 the reverse read; and 001 a fixed last segment.
16. When running VelvetOptimiser, the optimal k-mer length is frequently around two thirds of the total read length, so the *–s* and *–e* values can be set closer to this to reduce run time if needed. Conversely, using the default step size for the k-mer of 2 (rather than × 4) is more resource intensive but may produce marginally more optimal assemblies.
17. The protocol provided can be modified to determine NG-STAR resistance profiles. A list of alleles as DNA sequences can be downloaded from the NG-STAR website, https://ngstar.canada.ca/welcome/home (e.g., for *gyrA*, gyrA.fasta.txt.). These can then be compared directly to a FASTA file of DNA sequence contigs (in the example provided, this is called `contigs.fa`) to identify the closest matches. For example:

```
blastn -db contigs.fa -query gyrA.fasta.txt.
```

18. Several introductions and tutorials to using BLAST are available online (https://www.ncbi.nlm.nih.gov/home/tutorials/).
19. The BLAST approach presented is basic, deliberately so, to allow the reader to understand from first principles how BLAST outputs are generated and can be analyzed. One approach to automation using Biopython is suggested, but alternative solutions also exist. For example, BIGSdb is freely available software that automates BLAST searches and identification of specific alleles for each gene (http://bigsdb.readthedocs.io/en/latest/). Pathogen.watch also allows resistance determinants and other genetic markers to be identified from assemblies, and has specific analytical tools for *N. gonorrhoeae*. Other commercially available software is also able to automate BLAST searches (e.g., BioNumerics, CLC Genomics Workbench, or SeqSphere).
20. If no BLAST match is found for more divergent genes (e.g., *penA*), and the assembly is of adequate quality, consider using one or more alternative alleles as the BLAST query, instead of the example provided.
21. DNA sequences can be translated using SeaView. However, if adopting an approach using Biopython, this can also be used for sequence translation and comparison.
22. In more fragmented assemblies it is possible that a particular gene is partially present on two different contigs. In the case

where the top hit contains only part of the gene of interest, review further hits to see if the remainder of the gene is present on another contig, or obtain an assembly using a different tool (e.g., Spades).

23. There are limitations to the current MIC prediction scheme, notably that it does not include prediction of ceftriaxone susceptibility. Ceftriaxone MICs of ≥0.5 mg/L have been associated frequently with two substitutions in *penA*, A311V and T483S, and the presence of an additional substitution, T316P, with higher MICs, ~2 mg/L. These mutations could be searched for within the *penA* protein sequence determined above, but do not form part of the previously published scheme. Furthermore, unknown resistance determinants affecting the MICs of antimicrobials are missed in any scheme that genetically predicts resistance.

    More generally, the scheme itself could be and likely will be, updated. This would require parallel phenotyping and WGS sequencing of a collection of isolates. An updated regression model could be fitted to the dataset, in a similar approach to that previously described, using multivariable linear regression and $\log_2$ transformed MIC values and accounting for interactions.

References

1. Unemo M (2015) Current and future antimicrobial treatment of gonorrhoea – the rapidly evolving *Neisseria gonorrhoeae* continues to challenge. BMC Infect Dis 15:364. https://doi.org/10.1186/s12879-015-1029-2
2. Unemo M, Shafer WM (2014) Antimicrobial resistance in *Neisseria gonorrhoeae* in the 21st century: past, evolution, and future. Clin Microbiol Rev 27(3):587–613. https://doi.org/10.1128/CMR.00010-14
3. Eyre DW, De Silva D, Cole K et al (2017) WGS to predict antibiotic MICs for *Neisseria gonorrhoeae*. J Antimicrob Chemother 72 (7):1937–1947. https://doi.org/10.1093/jac/dkx067
4. Demczuk W, Sidhu S, Unemo M et al (2017) *Neisseria gonorrhoeae* sequence typing for antimicrobial resistance, a novel antimicrobial resistance multilocus typing scheme for tracking global dissemination of *N. gonorrhoeae* strains. J Clin Microbiol 55(5):1454–1468. https://doi.org/10.1128/JCM.00100-17

Chapter 5

Gonococcal Colony Typing

John E. Heckels

Abstract

Gonococcal colony typing is part a science and part an art that has been central to studies which have identified crucial virulence antigens and also demonstrated the ability of the bacteria to undergo rapid phase and antigenic variation. Without this fundamental work, modern molecular biological studies of gonococcal pathogenesis would not have been possible. Indeed colony typing is still essential when performing biological experiments with clinical and laboratory isolates and for monitoring their outcome. In this chapter, the methods for performing colony typing and techniques to optimize the process are described.

Key words Colony typing, *Neisseria gonorrhoeae*, Pili, Opa protein

1 Introduction

In early studies of gonococcal virulence Kellogg and colleagues in 1963 [1] described that fresh gonococcal isolates exhibited four distinct colony types (T1–4) when cultured on solid laboratory media. T1 and T2 colony types were small and domed, T3 and T4 were larger and flat. Additionally, types T1 and T2 were dark in color, T3 was lighter and T4 was colorless. Cultures of fresh primary isolates from acute gonorrhea of males produced mainly the T1 colony type, but nonselective laboratory subculture resulted in an increased proportion of types 3 and 4. The T1 phenotype was rapidly lost after 3–4 rounds of subculture with T4 becoming the majority colony type. As a result, most laboratory-maintained strains produced only T4 colonies. However, each of the types could be maintained from fresh isolates by selective subculture of individual colonies.

The importance of these observations was immediately apparent as selective subculture maintaining the T1 colony type retained virulence in urethral challenge of human volunteers. In contrast, the T4 colony type lost virulence and did not cause infection [1]. The basis for this difference in morphology and virulence became apparent through independent studies by Jephcott et al.,

Myron Christodoulides (ed.), *Neisseria gonorrhoeae: Methods and Protocols*, Methods in Molecular Biology, vol. 1997,
https://doi.org/10.1007/978-1-4939-9496-0_5,

who reported that T1 colonies, unlike T4, produced "filamentous appendages" [2] and by Swanson et al. who identified these as pili [3]. These were among the first studies to identify bacterial pili as important virulence factors and soon the role of pili in cell adherence was demonstrated [4].

The early studies of Kellogg et al. [1] noted that colony differentiation required observation by transmitted rather than reflected light, but it proved difficult to reliably reproduce the conditions in other laboratories. Subsequently, Walstad et al. [5] reported that "dark colonies" (T3 and dark variants of T1 and T2) had markedly increased amounts of a single protein of molecular mass (*Mr*) 28 kDa, located in the cell envelope, the apparent *Mr* of which on SDS-PAGE varied depending on the method of solubilization, and which was unrelated to piliation. They suggested its surface location "could be important to pathogenicity or antigenicity" [5]. At the same time, Swanson embarked on a series of detailed studies of colony morphology and opacity using defined media and lighting conditions [6] and demonstrated that opaque colony types from different strains expressed a protein of *Mr* 24–30 kDa, which he designated "colony opacity-associated protein" [7]. This work and the methods involved, which required clear solid growth media and the use of stereomicroscopy with colonies illuminated by a substage reflecting mirror, were reported at the First International Pathogenic Neisseria Meeting (San Francisco, California, USA, 1978) and greatly facilitated the selection of gonococcal variants for studies in other laboratories.

Previously, Heckels had reported that gonococcal outer membranes (OM) contained two major proteins, protein I (PI) and a protein II (PII) of apparent *Mr* 29.5 kDa, which could be characterized by a mobility shift in apparent *Mr* on SDS-PAGE when membranes were derivatized at 37 or 100 °C [8]. While expression of PI (now PorB) was stable, variants lacking PII arose spontaneously. It soon became apparent that PII corresponded to the opacity proteins described by Walstad et al. and Swanson. Using the method described by Swanson, Lambden, and Heckels were able to select a series of opacity variants from a single strain that produced a series of different opacity proteins of apparent *Mr* of 27–29 kDa, either singly or in combination [9] and also with differences in the *Mr* of the pilus subunit protein (pilin) [10]. Thus, by careful selection of colony types (Fig. 1), it was possible to select matched pairs of colonies that expressed (a) the same PII opacity protein (or none) but differed in their piliation state, (b) lacked pili but differed in the opacity protein expressed and (c) lacked opacity protein but differed in pilin expression. These variants demonstrated that expression of opacity proteins was associated with increased adhesion of gonococci to epithelial cells [11].

Following this early work, the importance of the PII/opacity related proteins in gonococcal/cell interactions was soon

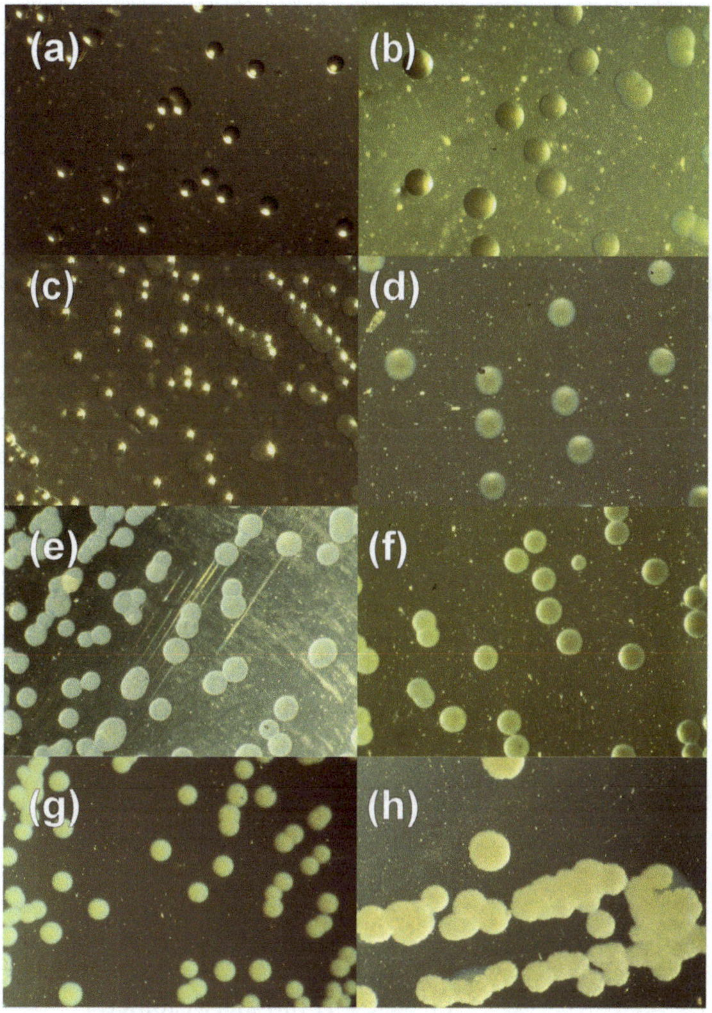

Fig. 1 Colony variants selected from a single strain of gonococcus (P9): (**a**) Pil^{+}, transparent (Tr), (**b**) Pil^{-}, transparent, (**c**) Pil^{+}, intermediate opacity, (**d**) Pil^{-}, intermediate opacity, (**e–h**) Pil^{-}, opaque differing in opacity [9]

established in several laboratories and has become a major field in the study of gonococcal and meningococcal pathogenesis. At the Sixth International Pathogenic Neisseria Meeting (Pine Mountain, Georgia, USA, 1988), a unified nomenclature was agreed that these proteins responsible for colony opacity be designated Opa proteins in both gonococci and meningococci and the encoding genes *opa* [12]. Countless subsequent studies have studied the role of Opa proteins in cellular interactions, virulence and other key biological properties. These all stem from the original studies of gonococcal colony opacity.

In this chapter, we describe methods for preparing clear gonococcal typing media, colony typing using a stereomicroscope, and selecting, growing, and storing new colony variants.

2 Materials

2.1 Clear Gonococcal Typing Media (Adapted from Swanson [6])

1. Proteose Peptone Agar (per liter): 10 g of Bacto-agar No. 1 (Oxoid, *see* **Note 1**), 10 g of Proteose Peptone (Difco), 1 g of KH_2PO_4, 5.24 g of $K_2HPO_4 \cdot 3H_2O$, 5 g of sodium chloride, 1 g of soluble starch.
2. Supplement A (per 800 mL): 100 mg of cocarboxylase, 10 mg of cyanocobalamin, 20 mg of ferric nitrate, 100 g of glucose, 10 g of L-glutamine, 13 mg of p-amino benzoic acid, 3 mg of thiamine hydrochloride, 250 mg of β-nicotinamide adenine dinucleotide.
3. Supplement B (per 200 mL): 26 g of L-cysteine hydrochloride, 1 g of adenine, 30 mg of guanidine hydrochloride, 320 mg of hypoxanthine, 800 mg of uracil, 100 mL of 0.1 M hydrochloric acid.
4. Deionized distilled H_2O.
5. Glass bottle (e.g., Schott Duran).
6. Autoclave.
7. 0.45 μm filter and syringe.
8. Hot plate.
9. Water bath.
10. 90 mm × 16 mm sterile petri plates (hold 20 mL of agar).
11. Class II biosafety cabinet.

2.2 Growing Neisseria gonorrhoeae

1. Clear gonococcal typing agar plates.
2. *Neisseria gonorrhoeae* strains/isolates stored at −80 °C or in liquid nitrogen vapor phase.
3. Sterile gouger.
4. 10 μL sterile inoculating loops.
5. Humidified incubator set at 37 °C with 5% (v/v) CO_2.

2.3 Colony Typing

1. *Neisseria gonorrhoeae* agar plate cultures.
2. Binocular dissecting microscope equipped with a flat substage reflecting mirror and overhead illumination. The stage should have a circular aperture of approximately 2 cm (Fig. 2).
3. 1 μL sterile inoculating loop.

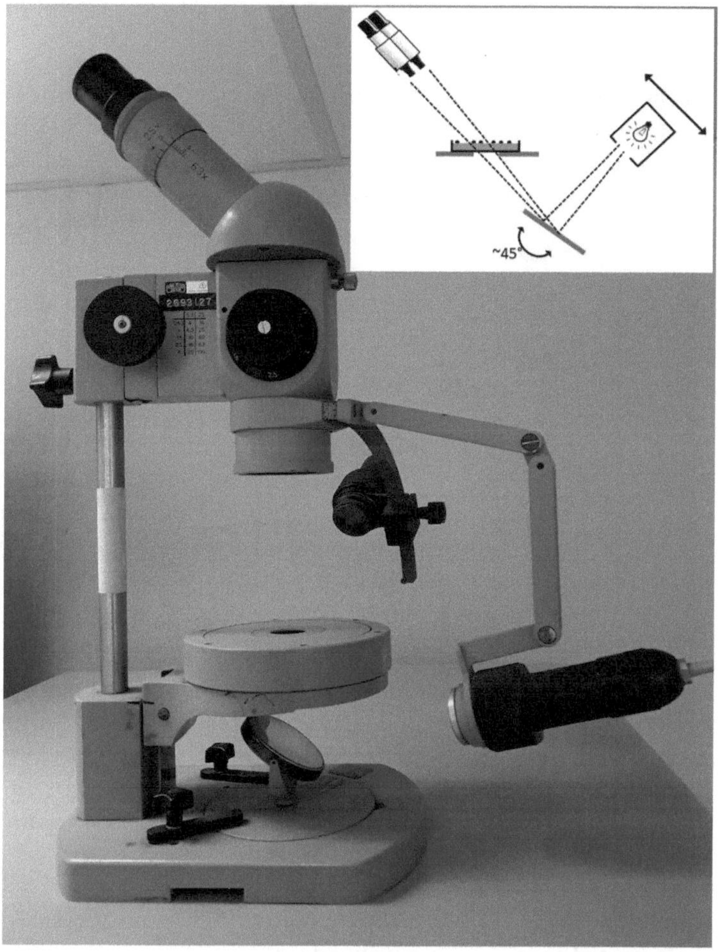

Fig. 2 Binocular dissecting microscope equipped with a flat substage reflecting mirror and overhead illumination suitable for colony typing. Inset showing adjustment of illumination and mirror to obtain optimal colony differentiation

2.4 Growth and Storage of New Colony Variants

1. Storage medium: 1% (w/v) proteose peptone, 10% (v/v) glycerol. Dissolve the ingredients in deionized H_2O, dispense into small aliquots (~4 mL) in glass bottles and sterilize at 2.68 kg/ cm^2 at 121 °C for 15 min. Store the bottles at room temperature for several weeks.
2. Cryovials.
3. Liquid nitrogen.
4. Clear gonococcal typing agar plates.
5. 1 and 10 μL inoculating loops.
6. Binocular dissecting microscope.

3 Methods

3.1 Clear Gonococcal Typing Media (Adapted from Swanson [6])

1. Add the components of the proteose peptone agar to deionized H_2O to a total volume of 1 L in a strong glass bottle and autoclave at 2.68 kg/cm^2, 121 °C for 15 min. Allow to cool to ~45–50 °C in a water bath.
2. Dissolve the components of Supplement A in 800 mL of deionized H_2O. Sterilize the solution by syringe passage through a 0.45 μm filter and dispense into aliquots of 8 mL, and keep at −20 °C.
3. Heat the 0.1 M HCl solution to boiling point with a hot plate and with the exception of the cysteine, add the components of Supplement B to dissolve. Cool the solution to room temperature and then add the cysteine. Adjust the final volume to 200 mL with deionized H_2O and sterilize the solution by syringe passage through a 0.45 μm filter. Store in aliquots of 2 mL, at −20 °C.
4. When the proteose peptone agar has cooled to 45–50 °C, add 8 mL of Supplement A and 2 mL of Supplement B per liter, mix gently and pour the molten agar into sterile petri dishes. Allow the agar to set in a Class II biosafety cabinet with the lids removed to dry. Use the plates immediately or store them in sealed plastic bags at 4 °C for up to 1 month. Stored plates must be allowed to dry again if any condensation has accumulated.

3.2 Growing *Neisseria gonorrhoeae*

1. Prepare plates of clear gonococcal typing media as described in Subheading 2.1.
2. Use a sterile gouger to transfer gonococci from frozen vial onto the typing media and streak with an inoculating loop to obtain growth on isolated colonies.
3. Incubate the plates for 16–20 h in an incubator at 37 °C with a humidified atmosphere containing 5% (v/v) CO_2.

3.3 Colony Typing

Colony typing is carried out using a binocular dissecting microscope equipped with a flat substage reflecting mirror and overhead illumination, with the stage having a circular aperture of ~2 cm (Fig. 2).

1. Place the plates with gonococcal colonies on the stage of the binocular microscope (with the lid removed if possible).
2. Adjust the angle of the reflecting mirror to approximately 45°.
3. Adjust the position of the illumination so that the angle of incident light on the mirror is approximately 90°.

4. Observe the colonies through the binocular eyepiece and adjust the angle of the mirror (and if necessary the illumination) so that any highlights on the colonies are off-center. This enables optimal differentiation of colony opacity (Fig. 2).
5. Colonies of Pil^+ gonococci have a morphology which appears small and domed with a bright highlight (Fig. 1a, c) while colonies of Pil^- gonococci have a large flat appearance without highlight (Fig. 1b, d–h. In addition colonies may appear with a range of opacities ranging from transparent (Fig. 1a, b) through intermediate opacity (Fig. 1c, d) to highly opaque with a granular appearance (Fig. 1h).
6. Using a 1 μL inoculation loop to pick off single colonies differing in morphology for subculture or storage.

3.4 Growth and Storage of New Colony Variants

1. When performing primary isolation of new colony variants, pick a single colony for immediate subculture onto a fresh agar plate. Incubate for 16–20 h in an incubator at 37 °C with a humidified atmosphere containing 5% (v/v) CO_2. On the following day, suspend the resulting growth in 1–2 mL of storage medium in a cryo-vial, and snap-freeze in liquid nitrogen to produce a Master Stock.
2. To check that the Master Stock bacteria have the expected colony morphology, scrape off material from the top of the Master Stock using a sterile gouger and streak across a fresh agar plate to produce a First Subculture. Check the subsequent growth as described in Subheading 3.3.
3. Produce several vials of the First Subculture material to act as Submaster Stocks, which can be used to produce inocula for further experiments. The original Master Stock should not be used except in emergency. One Submaster Stock can be retained to produce further stocks when required.
4. When performing routine experiments, use a stock culture to inoculate one plate. Examine the resulting growth for colony morphology and then pick ~10 identical colonies with 1 μL inoculating loops to inoculate a single plate to provide the experimental material. This ensures that faster growing colony types do not outgrow the desired colony type (*see* **Note 2**).
5. Where large-scale cultures are being prepared for experiments, agar plates should be inoculated to produce confluent growth over approximately two-thirds of the plate and the remaining third streaked to produce isolated single colonies so that the colony type of the experimental material can be confirmed (*see* **Notes 3** and **4**).
6. While variants differing in Opa proteins can be readily isolated on the basis of colony opacity, variants differing in the pilin expressed cannot be selected by differences in morphology.

One successful strategy is to culture a Pil^+ variant, select a single colony of a Pil^- variant for subculture and then select one of the minority Pil^+ revertants for further study. Such revertants will frequently express pilin differing in apparent *Mr* (and hence antigenic and biological properties) from the original variant. This procedure can be repeated to obtain a panel of gonococci differing in pilin expression.

7. Isolation of similar colonial variants is very much more difficult with the related *N. meningitidis*. The expression of the capsular polysaccharide makes differentiation of colony morphology impossible. Noncapsulated strains show much less clear differences in morphology, although nonpiliated variants can be selected by repeated subculture of larger colonies and differences on Opa expression detected through small differences in colony opacity.

4 Notes

1. The source and batch of the agar used to produce the typing media may be crucial. Batches of agar from a single manufacturer of apparently the same product may vary wildly in the degree of opacity observed. Occasional batches fail to produce any discernible difference in colony opacity between known transparent, intermediate and opaque variants. In our laboratory, Oxoid Bacto Agar No.1 has proved successful [9]; equally successful results were obtained by Swanson using Difco Noble Agar [6].

   Once distinct variants have been isolated they can be used to check new batches of agar, particularly the differentiation of a transparent and the least "opaque" intermediate variant. It is useful to retain samples of "good" agar to be used for primary isolation of opacity variants from fresh clinical isolates.
2. Growth of cultures from frozen stocks often contains a higher proportion of undesired colony types. It is therefore important to select colonies of the desired morphology for a second subculture to produce experimental inocula. Given the relatively rapid rate of colony variation, rather than select a single colony, it is wise to pick multiple colonies of the desired morphology to minimize the risk of inadvertent selection of an unwanted variant.
3. If liquid cultures are used in experiments, the resulting growth should be plated out on colony typing agar to ensure that the bulk population is of the desired colony type.
4. As any experimental procedure may select a minority colony variant, it is important at the end of the experiment to check the predominant colony type in the surviving population.

Acknowledgments

I would like to dedicate this chapter to the memory of John Swanson (1936–2013), whose meticulous observational studies formed the basis of all subsequent gonococcal colony typing and who was the foundation of future studies of the critical role of pili and Opa proteins in gonococcal (and subsequently meningococcal) virulence.

References

1. Kellogg DS Jr, Peacock WL Jr, Deacon WE et al (1963) *Neisseria gonorrhoeae*. I. Virulence genetically linked to clonal variation. J Bacteriol 85:1274–1279
2. Jephcott AE, Reyn A, Birch-Anderson A (1971) Brief report: *Neisseria gonorrhoeae*. iii. Demonstration of presumed appendages to cells from different colony types. Acta Pathol Microbiol Scand B Microbiol Immunol B79:437–439
3. Swanson J, Kraus SJ, Gotschlich EC (1971) Studies on gonococcus infection. I. Pili and zones of adhesion: their relation to gonococcal growth patterns. J Exp Med 134:886–906
4. Swanson J (1973) Studies on gonococcal infection. IV. Pili: their role in attachment of gonococci to tissue culture cells. J Exp Med 137:571–589
5. Walstad DL, Guymon LF, Sparling PF (1977) Altered outer membrane protein in different colonial types of *Neisseria gonorrhoeae*. J Bacteriol 129(3):1623–1627
6. Swanson J (1978a) Studies on gonococcus infection. XII. Colony color and opacity variants of gonococci. Infect Immun 19:320–321
7. Swanson J (1978b) Studies on gonococcus infection. XIV. Cell wall protein differences among color/opacity colony variants of *Neisseria gonorrhoeae*. Infect Immun 21:292–302
8. Heckels JE (1977) The surface properties of *Neisseria gonorrhoeae*: isolation of the major components of the outer membrane. J Gen Microbiol 99:333–341
9. Lambden PR, Heckels JE (1979) Outer membrane protein composition and colonial morphology of *Neisseria gonorrhoeae* strain P9. FEMS Microbiol Lett 5:263–265
10. Lambden PR, Robertson JN, Watt PJ (1981) The preparation and properties of alpha-pili and beta-pili from variants of *Neisseria gonorrhoeae* P9. J Gen Microbiol 124:109–117
11. Lambden PR, Heckels JE, James LT et al (1979) Variations in surface protein composition associated with virulence properties in opacity types of *Neisseria gonorrhoeae*. J Gen Microbiol 114:305–312
12. Hitchcock PJ (1989) Unified nomenclature for pathogenic *Neisseria* species. Clin Microbiol Rev 2:S64–S65

Chapter 6

Preparation of Lipooligosaccharide (LOS) from *Neisseria gonorrhoeae*

Myron Christodoulides

Abstract

Neisseria gonorrhoeae is a gram-negative obligate human pathogen that contains lipooligosaccharide (LOS) as a major constituent within the outer membrane. LOS plays a major role in pathogenesis by inducing host inflammatory responses and also enabling evasion of host innate immunity through sialylation. Epitopes within LOS are also potential vaccine candidates. In this chapter, we describe a general method based on the Westphal hot phenol extraction process to purify whole LOS from *N. gonorrhoeae* for structural analyses and for use in in vivo and in vitro biological assays.

Key words Lipooligosaccharide, LOS, Phenol extraction, Silver staining, Electrodialysis

1 Introduction

Neisseria gonorrhoeae lipooligosaccharide (LOS) is located in the outer membrane of the pathogen and is composed of a lipid A, an inner and outer core oligosaccharide and is structurally distinct from lipopolysaccharide (LPS) of gram-negative enteric bacilli due to the lack of a repeating polysaccharide O-side chain [1]. *N. gonorrhoeae*, like its sister *N. meningitidis*, harbors the *lgt* genes encoding LOS with high genetic diversity [2, 3] and exhibits antigenic variation, which is mostly due to phase variation of the related gene expression. Sialylation of LOS enables gonococci to mimic host cell surfaces that also express sialic acid [4]; consequently, gonococci become more resistant to antibody and complement-mediated killing and phagocytosis and sialylated LOS plays an important role in establishing infection in humans and experimental mouse models [5]. In common with LPS, LOS is recognized by TLR4, which leads to activation of the innate immune system [6, 7] and release of inflammatory cytokines [8, 9].

Purification of LOS from gonococci has allowed several aspects of its structure and biology to be examined. The structure of gonococcal LOS has been characterized using a variety of chemical,

Myron Christodoulides (ed.), *Neisseria gonorrhoeae: Methods and Protocols*, Methods in Molecular Biology, vol. 1997,
https://doi.org/10.1007/978-1-4939-9496-0_6,

enzymatic, two-dimensional NMR, HPLC, and mass spectrometry methods [10–16]. In addition, purified LOS is useful for studying the mechanisms of innate immune system activation and identifying other putative non-TLR4 molecules that can act as LOS receptors. Understanding the structure and biology of LOS is also important in the development of gonorrhea vaccines [17].

In this chapter, we describe the method that is routinely followed for gonococcal whole LOS extraction, which is essentially the Westphal process using hot phenol extraction [18], with the modifications introduced by Johnson and Perry [19], Stead et al. [20], and Lambden and Heckels [21].

2 Materials

2.1 Growth of N. gonorrhoeae

1. Supplemented GC agar for growth of *N. gonorrhoeae*: dissolve the following chemicals in 500 mL of distilled H_2O—5 g of Bacto-agar (no. 1, Oxoid), 0.5 g of KH_2PO_4, 2.62 g of $K_2HPO_4 \cdot 3H_2O$, 5 g of proteose peptone, 2.5 g of NaCl, and 0.5 g of starch.
2. Supplement A: dissolve the following in 800 mL of distilled H_2O—100 mg of cocarboxylase, 10 mg of cyanocobalamin, 20 mg of ferric nitrate, 100 mg of glucose, 10 g of L-glutamine, 13 mg of para-amino-benzoic acid, 3 mg of thiamine hydrochloride, and 250 mg of β-nicotinamide adenine dinucleotide. Sterilize the solution through a 0.2 μM filter and store in aliquots at −20 °C (*see* **Note 1**).
3. Supplement B: dissolve the following in 100 mL of boiling 0.1 M hydrochloric acid—1 g of adenine, 30 mg of guanidine hydrochloride, 320 mg of hypoxanthine, and 800 mg of uracil. Allow the solution to cool to room temperature and then add 26 g of cysteine and add distilled H_2O to make a final volume of 200 mL. Sterilize the solution through a 0.2 μM filter and store in aliquots at −20 °C (*see* **Note 1**).
4. GC agar plates: sterilize the GC medium agar by autoclaving for 15 min at 120 °C with 2.68 kg/cm^2 pressure. Allow sterile medium to cool to ~50 °C, and under aseptic conditions, add 4 mL of Supplement A and 1 mL of Supplement B per 500 mL and pour into sterile petri dishes and allow setting at room temperature.
5. *N. gonorrhoeae* strains stored at −80 °C or under liquid nitrogen in sterile proteose peptone broth (10% w/v) containing 10% (v/v) glycerol.
6. Sterile plastic Petri dishes (20 mL volume, 90 mm dish; 100 mL, 145 mm dish).
7. Humidified incubator set to 37 °C with 5% (v/v) CO_2.

2.2 DNase, RNase, and Proteinase K Treatment

1. Solubilizing buffer: 0.1 M Tris–HCl, pH 6.8, buffer containing 2% (w/v) sodium dodecyl sulfate (SDS), 2% (v/v) β-mercaptoethanol, and 10% (v/v) glycerol.
2. 10 mg/mL DNase I, 10 mg/mL RNase, and 10 mg/mL Proteinase K solutions in sterile distilled H_2O.

2.3 Phenol Treatment

1. 90% (w/v) phenol.
2. Vortex homogenizer.
3. Benchtop centrifuge, with polypropylene pots.

2.4 Dialysis and LOS Precipitation

1. Dialysis tubing: 20 mm flat width diameter.
2. 5 mM (final concentration) Sodium ethylenediaminetetraacetic acid (Na-EDTA).
3. 1 mM $MgCl_2$ solution and $MgCl_2$ crystals.
4. 60% (v/v) acetone solution and 100% acetone.
5. 0.025 M Tris–acetate buffer, pH 7.2.
6. Equipment: glass pipettes, vacuum desiccator, magnetic stirrer, benchtop and high-speed centrifuges, water bath.

2.5 Quality Control of LOS

2.5.1 Construction of Linear Gels

1. Acrylamide monomer solution: 48% (w/v) acrylamide, 1.5% (w/v) *N,N′*-methylene bisacrylamide, stored at 4 °C.
2. 20% (w/v) SDS in distilled H_2O, stored at room temperature.
3. Tetramethylethylenediamine (TEMED).
4. 10% (w/v) ammonium persulfate (APS) in distilled H_2O, prepared fresh for each use.
5. Gel buffer: 3 M Tris–HCl, pH 8.5 buffer containing 0.3% (w/v) SDS, stored at 4°C.
6. Anode buffer: 0.2 M Tris–HCl buffer, pH 8.9, prepared fresh for each use.
7. Cathode buffer: 0.1 M Tris–HCl buffer, pH 8.2, containing 0.1 M Tricine and 0.1% (w/v) SDS, prepared fresh for each use.
8. Sample buffer: 0. 05 M Tris–HCl, pH 6.8, containing 4% (w/v) SDS, 12% (v/v) glycerol, 2% (v/v) 2-mercaptoethanol, and 0.02% (w/v) Brilliant Blue G.
9. 100% methanol.
10. H_2O-saturated 100% butanol.
11. Vacuum pump.

2.5.2 Electrophoresis

1. Life Technologies glass gel casting plates and V15.17 gel tanks or equivalent.
2. Electrical power packs.

2.5.3 Silver Staining of LOS

1. Fixing solution: 25% (v/v) isopropanol and 7% (v/v) glacial acetic acid in distilled H_2O.
2. Oxidizing solution: 2.67% (v/v) fixing solution with 0.7% (w/v) periodic acid in distilled H_2O.
3. Silver staining solution: prepare fresh for each use. Add 0.56 mL of a 35% (w/v) ammonia solution slowly to 18.8 mL of 0.1 M sodium hydroxide solution with constant stirring. Use a fume hood and a white background. A 20% (w/v) silver nitrate solution is then added dropwise until the saturation point is reached visibly, and the solution diluted with distilled H_2O up to a final volume of 100 mL (*see* **Note 2**).
4. Developing solution: 0.05% (v/v) formalin solution containing 0.05 mg/mL of citric acid, prepared fresh for each use.
5. Stop solution: 0.1% (v/v) glacial acetic acid.
6. Equipment: rotary shaker, warm room, clean glass dishes.

2.5.4 Staining for Protein

1. PAGE Blue 83 staining solution: 10% (v/v) glacial acetic acid, 20% (v/v) isopropanol, and 0.5 mg/mL of PAGE Blue 83 in distilled H_2O.
2. PAGE Blue 83 destaining solution: 10% (v/v) glacial acetic acid and 10% (v/v) isopropanol in distilled H_2O.
3. Coomassie Blue staining solution: 50% (v/v) methanol, 10% (v/v) glacial acetic acid and 0.25% (w/v) Coomassie Blue in distilled H_2O.
4. Coomassie Blue destaining solution: 5% (v/v) methanol and 7.5% (v/v) glacial acetic in distilled H_2O.

3 Methods

3.1 Growth of N. gonorrhoeae

1. Take a stab of *N. gonorrhoeae* from a frozen stock culture kept either at −80 °C or under liquid nitrogen and swab onto a GC agar plate (9 cm) and incubate overnight at 37 °C with 5% (v/v) CO_2.
2. Prepare lawn growth of bacteria by culturing from this overnight growth onto six fresh GC agar plates (9 cm) and incubate overnight at 37 °C with 5% (v/v) CO_2.
3. In order to produce sufficient wet weight of bacteria for LOS extraction, remove the total bacterial growth from a single agar plate and suspend in ~1 mL of sterile proteose peptone broth (10% w/v, no glycerol) and spread ~100 μL per large GC agar plate (ten plates in total). Repeat for the other overnight plates to prepare ~50–60 large GC agar plates in total. Incubate the large plates overnight at 37 °C with 5% (v/v) CO_2.

3.2 DNase, RNase, and Proteinase K Treatment

The method of Davis and Goldberg [22] is used, with minor modification to volumes.

1. Scrape the growth from the 50–60 plates into a steel beaker containing a small volume of ~6 mL solubilizing buffer. Suspend the pellet completely without vortex.
2. Boil the suspended bacteria in a water bath for 15 min and allow the solution to cool at room temperature (15 min.).
3. Add 5 μL of both DNase I and RNase solutions and incubate the sample at 37 °C for 30 min (*see* **Note 3**).
4. Add 10 μL of Proteinase K solution and incubate in a water bath at 60 °C for 3 h.

3.3 Phenol Treatment

1. Prewarm 90% (w/v) phenol solution (*see* **Note 4**) to 70 °C in a water bath and add it at a ratio of 1:1 to the nuclease and protease-treated sample prepared in **step 4** of Subheading 3.2. Attach the beaker to a blender, immerse the beaker into a water bath at 70 °C and shear the bacteria by vortex mixing for 15 min.
2. Remove the blended suspension into a polypropylene pot (*see* **Note 5**) and with a bench-top machine, centrifuge at 10,000 × *g* for 30 min to separate the upper aqueous and lower phenol phases.
3. Remove the upper aqueous phase with a glass pipette and retain.
4. Extract the remaining phenol layer with water as described in **step 1** in Subheading 3.3 and repeat **step 2** in Subheading 3.3.
5. Combine the two aqueous phase volumes and discard the phenol as hazardous waste (*see* **Note 6**).

3.4 Dialysis and LOS Precipitation

1. Prepare a length of dialysis tubing and if necessary, bring to the boil in distilled H_2O for 10 min to aid removal of contaminating glycerol and sulfur compounds. Wash tubing with distilled H_2O, tie one end and add aqueous phase volume. Use 50% more tubing than the volume of sample. Tie the other end and dialyze against running tap water for 48 h.
2. Open the dialysis sac and add Na-EDTA to a final concentration of 5 mM, close the tubing and dialyze against a minimum of 2 L of deionised distilled H_2O overnight. On the following day, replace the water several times during the day.
3. Remove the liquid to a glass bijou bottle and add a few crystals of $MgCL_2$, followed by 2 volumes of cold acetone (100%). Leave the sample overnight at 4 °C.
4. Centrifuge the resulting precipitate in a glass universal in a bench-top centrifuge at 10,000 × *g* for 10 min. Remove the

supernatant fluid and add a volume of 60% (v/v) acetone to wash the precipitate. Centrifuge again, remove the supernatant fluid and place the universal in a vacuum desiccator for 30 min to remove the residual acetone.

5. Suspend the precipitate in a volume of 0.025 M Tris–acetate buffer, pH 7.2 and homogenize until completely dissolved. This may take some time.
6. Add a solution of $MgCl_2$ to give a final concentration of 1 mM and transfer the solution to a fresh dialysis bag and dialyze overnight at 37 °C against 0.025 M Tris–acetate buffer, pH 7.2 containing 1 mM $MgCl_2$ (2–4 L).
7. On the following day, centrifuge the solution at 1000 × *g* for 15 min and keep the supernatant fluid, discarding the pellet.
8. Centrifuge the supernatant at 100,000 × *g* (high-speed centrifuge) for 2 h and then discard the final supernatant fluid. Wash the pellet with distilled H_2O and repeat the centrifugation step.
9. Weigh an Eppendorf tube and dry the sample under vacuum to estimate dry weight of extracted LOS. At this stage, purified LOS can be examined using low molecular weight gel electrophoresis with silver-staining and potential protein contamination with Coomassie blue staining (*see* **Note** 7). If a homogeneous preparation is required, the LOS can be subject to electrodialysis with appropriate base neutralization (*see* **Note 8**).

3.5 Quality Control of LOS

3.5.1 Construction of Linear Gels

The gels are constructed according to the method of Schagger and von Jagow [23] with a discontinuous buffer system and with tricine as the trailing ion, using the composition volumes shown in Table 1.

1. Clean glass gel casting plates with methanol (100%), separate with rubber spacers (1 mm) and hold together with clips.
2. Prepare the separating gel as shown in Table 1. Add APS and TEMED last and deaerate the gel solution by connecting to a vacuum before pouring into the casting mold, to avoid inhibition of polymerization by oxygen. Pour to a height of 10 cm in large gel plate system and allow polymerization for 1–2 h at room temperature. Next, add the spacer gel solution to a height of 3 cm and overlay with H_2O-saturated butanol and leave to polymerize for 1 h at room temperature. Remove the H_2O-saturated butanol with a Pasteur pipette and overlay with the stacking gel solution with the appropriate comb inserted. Leave to polymerize for 30 min at room temperature.
3. Prepare the samples to a final volume of 20 μL with sample buffer, heat to 100 °C for 10 min in a heater bath and load into the comb wells.

Table 1
Composition of low molecular weight gels

	Volume (mL)		
Solution	**Separating gel**	**Spacer gel**	**Stacking gel**
Acrylamide	6.7	2	1
Gel buffer	6.7	3.33	3.1
Glycerol	2.1	–	–
Distilled H_2O	4.5	4.63	8.4
APS	0.067	0.033	0.1
TEMED	0.0067	0.0033	0.01

3.5.2 Electrophoresis

1. Load the cast gels into the Life technologies V15.17 gel tanks or equivalent and fill the reservoirs with the anode and cathode buffers. Run the electrophoresis for 1 h at room temperature at 30 V to concentrate the samples and then increase voltage to 90 V for a further 16 h.

3.5.3 Silver Staining of LOS

The staining method is from Hitchcock and Brown [24]. All reactions are done at 37 °C on a rotary shaker, using solutions prewarmed to 37 °C.

1. Following electrophoresis, gently remove the gel, place into a glass dish and immerse in fixing solution for 30 min.
2. Remove the fixing solution and add oxidizing solution for 5 min.
3. Wash eight times with distilled H_2O (4 min each wash) and then incubate the gel with silver staining solution for 10 min.
4. Remove excess stain and wash four times with distilled H_2O (10 min each wash) and add the developing solution until the LOS bands become visible and reach the desired intensity. Any protein contamination will also appear with silver staining.
5. Stop the reaction by incubating the stained gel in stop solution for 1 h followed by washing with distilled H_2O.
6. Scan the gel to record a permanent image and store in the dark in distilled H_2O.

3.5.4 Staining for Protein

Staining can be done with either PAGE Blue 83 or Coomassie Blue stains (*see* **Note 9**).

1. Place the gel in a glass tray with PAGE Blue 83 staining solution and shake gently for 1 h or for 2–4 h with Coomassie Blue stain.

2. Remove the stain(s) and add either PAGE Blue 83 or Coomassie Blue destaining solutions and shake gently for up to 4 h, with several changes of solutions, until the background is clear.
3. Store the destained gels in distilled H_2O with 7% (v/v) acetic acid.

4 Notes

1. Supplement A and B can be substituted with commercial enrichment reagents (e.g., Vitox or Isovitalix) and used at concentrations recommended by the manufacturers.
2. It is critical that fresh ammonia is used for silver staining of low molecular weight gels; ammonia solutions stored at 4 °C for long periods are labile.
3. Treatment with DNase and RNase is optional; Davis and Goldberg [22] report that there is minimal difference in the quality of LPS between nuclease-treated and nontreated samples. The inclusion of Proteinase K is recommended.
4. Care must be taken when using phenol as the chemical is highly corrosive and acts as an anesthetic, such that burns are initially painless. Personal Protective Equipment must be worn at all times: use chemical splash goggles and/or a face shield, wear impervious clothing, including close-toed shoes, lab coat and butyl rubber or neoprene gloves. Avoid heat, flames and ignition sources. The use of phenol requires a full risk assessment and a decontamination kit must be available for potential spillages.
5. Centrifugation of phenol-containing solutions must be done only in polypropylene pots. Phenol will destroy centrifuge heads, so ensure that the outside of the containers are clean and place inside plastic bagging (if space permits) before inserting into rotor.
6. Phenol must be disposed of as hazardous waste.
7. High pressure liquid chromatography (HPLC) can be used to separate and purify LOS further from any contaminating polypeptides.
8. Regardless of the extraction method use, LOS will invariably contain inorganic cations (e.g., Na^+, K^+, Mg^{2+}, Ca^{2+}) and low molecular weight basic amines like putrescine, spermine, spermidine, and ethanolamine. Electrodialysis can be used to separate a large proportion of the cations and basic amines from LOS molecules, but this produces acidic LOS with reduced solubility, thereby requiring neutralization with an appropriate base to produce defined salt forms. Electrodialysis can be done

as described by Galanos and Luderitz [25], by dissolving dry LOS (*see* **step 9** in Subheading 3.4) in distilled H_2O and subjecting the LOS solution to electrodialysis with a commercially available electrodialyzer. The insoluble free acid obtained can then be neutralized with 1% (v/v) triethylamine solution to give a clear, nonviscous solution that contains no detectable nucleic acid or protein. Losses throughout the extraction and electrodialysis procedures can be monitored using the Limulus Amoebocyte Lysate (LAL) assay, widely available from several manufacturers.

9. Gels can be stained with either PAGE Blue 83 or Coomassie Blue, depending on availability and laboratory practice.

References

1. Kahler CM, Stephens DS (1998) Genetic basis for biosynthesis, structure, and function of meningococcal lipooligosaccharide (Endotoxin). Crit Rev Microbiol 24(4):281–334
2. Arking D, Tong Y, Stein DC (2001) Analysis of lipooligosaccharide biosynthesis in the *Neisseriaceae*. J Bacteriol 183(3):934–941. https://doi.org/10.1128/JB.183.3.934-941.2001
3. Zhu P, Klutch MJ, Bash MC et al (2002) Genetic diversity of three lgt loci for biosynthesis of lipooligosaccharide (LOS) in *Neisseria* species. Microbiology 148:1833–1844. https://doi.org/10.1099/00221287-148-6-1833
4. Mandrell RE, Griffiss JM, Macher BA (1988) Lipooligosaccharides (LOS) of *Neisseria gonorrhoeae* and *Neisseria meningitidis* have components that are immunochemically similar to precursors of human blood group antigens. Carbohydrate sequence specificity of the mouse monoclonal antibodies that recognize crossreacting antigens on LOS and human erythrocytes. J Exp Med 168(1):107–126
5. Ram S, Shaughnessy J, de Oliveira RB et al (2017) Gonococcal lipooligosaccharide sialylation: virulence factor and target for novel immunotherapeutics. Pathog Dis 75(4). https://doi.org/10.1093/femspd/ftx049
6. Zhou X, Gao X, Broglie PM et al (2014) Hexa-acylated lipid A is required for host inflammatory response to *Neisseria gonorrhoeae* in experimental gonorrhea. Infect Immun 82(1):184–192. https://doi.org/10.1128/iai.00890-13
7. Liu M, John CM, Jarvis GA (2010) Phosphoryl moieties of lipid A from *Neisseria meningitidis* and *N. gonorrhoeae* lipooligosaccharides play an important role in activation of both MyD88- and TRIF-dependent TLR4-MD-2 signaling pathways. J Immunol 185(11):6974–6984. https://doi.org/10.4049/jimmunol.1000953
8. John CM, Liu M, Jarvis GA (2009) Profiles of structural heterogeneity in native lipooligosaccharides of *Neisseria* and cytokine induction. J Lipid Res 50(3):424–438. https://doi.org/10.1194/jlr.M800184-JLR200
9. Patrone JB, Stein DC (2007) Effect of gonococcal lipooligosaccharide variation on human monocytic cytokine profile. BMC Microbiol 7:7. https://doi.org/10.1186/1471-2180-7-7
10. Takayama K, Qureshi N, Hyver K et al (1998) Characterisation of a structural series of lipid a obtained from the LPS of *Neisseria gonorrhoeae*. J Biol Chem 261:10624–10631
11. Takayama K, Qureshi N, Hyver K et al (1986) Characterization of a structural series of lipid A obtained from the lipopolysaccharides of *Neisseria gonorrhoeae*. Combined laser desorption and fast atom bombardment mass spectral analysis of high performance liquid chromatography-purified dimethyl derivatives. J Biol Chem 261(23):10624–10631
12. Leavell MD, Leary JA, Yamasaki R (2002) Mass spectrometric strategy for the characterization of lipooligosaccharides from *Neisseria gonorrhoeae* 302 using FTICR. J Am Soc Mass Spectrom 13(5):571–576. https://doi.org/10.1016/S1044-0305(02)00360-4
13. Kerwood DE, Schneider H, Yamasaki R (1992) Structural analysis of lipooligosaccharide produced by *Neisseria gonorrhoeae*, strain ms11mk (variant-a) - a precursor for a gonococcal lipooligosaccharide associated with virulence. Biochemist 31:12760–12768
14. Yamasaki R, Bacon BE, Nasholds W et al (1991) Structural determination of

oligosaccharides derived from lipooligosaccharide of *Neisseria gonorrhoeae* f62 by chemical, enzymatic, and 2-dimensional NMRmethods. Biochemist 30:10566–10575

15. Swanson KV, Griffiss JM (2006) Separation and identification of *Neisserial* lipooligosaccharide oligosaccharides using high-performance anion-exchange chromatography with pulsed amperometric detection. Carbohydr Res 341(3):388–396. https://doi.org/10.1016/j.carres.2005.11.017
16. Phillips NJ, John CM, Reinders LG et al (1990) Structural models for the cell-surface lipooligosaccharides of *Neisseria gonorrhoeae* and *Haemophilus influenzae*. Biomed Environ Mass Spec 19:731–745
17. Rice PA, Shafer WM, Ram S et al (2017) *Neisseria gonorrhoeae*: Drug Resistance, Mouse Models, and Vaccine Development. Annu Rev Microbiol 71:665–686. https://doi.org/10.1146/annurev-micro-090816-093530
18. Westphal O, Westphal O, Jann K (1965) Bacterial lipopolysaccharides: extraction with phenol-water and further applications of the procedure. In: Whistler RW, Wolfrom ML (eds) Methods in carbohydrate chemistry. Academic, New York, NY, pp 83–91
19. Johnson KG, Perry MB (1976) Improved techniques for the preparation of bacterial lipopolysaccharides. Can J Microbiol 22(1):29–34
20. Stead A, Main JS, Ward ME et al (1975) Studies on the lipopolysaccharides isolated from strains of *Neisseria gonorrhoeae*. J Gen Microbiol 88:123–131
21. Lambden PR, Heckels JE (1982) Synthesis of immunogenic oligosaccharide-protein conjugates from the lipopolysaccharide of *Neisseria gonorrhoeae* P9. J Immunol Meth 48:233–240
22. Davis MR Jr, Goldberg JB (2012) Purification and visualization of lipopolysaccharide from Gram-negative bacteria by hot aqueous-phenol extraction. J Vis Exp 63. https://doi.org/10.3791/3916
23. Schagger H, von Jagow G (1987) Tricine-sodium dodecyl sulphate-polyacrylamide gel electrophoresis for the separation of proteins in the range from 1 to 100 kDa. Anal Biochem 166:368–379
24. Hitchcock PJ, Brown TM (1983) Morphological heterogeneity among *Salmonella* lipopolysaccharide chemotypes in silver-stained polyacrylamide gels. J Bacteriol 154:269–277
25. Galanos C, Luderitz O (1975) Electrodialysis of lipopolysaccharides and their conversion to uniform salt forms. Eur J Biochem 54(2):603–610

Chapter 7

Purification of Type IV Pili and Pilin Subunits

Lisa Craig and Tuba Altindal

Abstract

The Type IV pili are displayed peritrichously on the surfaces of *Neisseria gonorrhoeae* cells. Here we present protocols for isolating and purifying Type IV pili and dissociating them into PilE pilin subunits. Pilus filaments are isolated from the bacterial cell surface by mechanical shearing and purified by differential precipitation and centrifugation. PilE subunits are extracted by treating the purified pili with detergent to disrupt the hydrophobic interactions holding them together in the filaments. Purified pili and pilin subunits can be used for structural, biophysical, or biochemical characterization and as antigens for antibody production.

Key words *Neisseria gonorrhoeae*, Type IV pili, Pilin, PilE

1 Introduction

Type IV pili are long thin polymers displayed on the surfaces of *Neisseria gonorrhoeae* and many other bacterial pathogens. They mediate a range of bacterial behaviors, including twitching motility, natural transformation, adhesion, and microcolony formation [1, 2]. Type IV pili are comprised of thousands of copies of the major pilin, PilE, the building block of the pilus, plus a few minor pilin subunits that are likely located at the tip of the pilus [3–6]. Hyper-variability in the amino acid sequences of the major pilins allows them to change the antigenicity of the pilus from one cell generation to the next to evade the immune response [7]. Pilus filaments are several microns in length but only ~6 nm in diameter. They are remarkably robust, withstanding pulling forces of >100 pN [8], and are resistant to denaturation by chemicals and heat [9]. These properties serve *N. gonorrhoeae* well in colonizing the genitourinary tract and other tissues.

The major pilin, PilE, is a small ladle-shaped protein with a hydrophobic N-terminal α-helix and a globular C-terminal domain [3, 4]. In the intact pilus filaments, the N-terminal α-helices wind around each other, forming a hydrophobic core that anchors the

Myron Christodoulides (ed.), *Neisseria gonorrhoeae: Methods and Protocols*, Methods in Molecular Biology, vol. 1997, https://doi.org/10.1007/978-1-4939-9496-0_7,

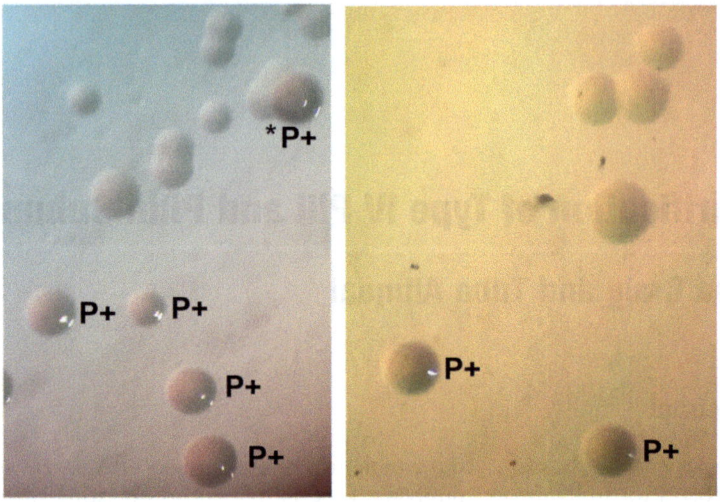

Fig. 1 Morphology of pilus-expressing (P+) and nonexpressing (P−) *N. gonorrhoeae* bacteria. Pilus-positive (P+) colonies are indicated. *P+ is a P + colony from which 2 P− colonies have grown

C-terminal domains on the surface of the filament [5]. Since subunits are held together primarily by hydrophobic interactions they can be disrupted using detergent to produce soluble pilin subunits.

Pili are purified from *N. gonorrhoeae* bacteria grown in 5% (v/v) CO_2 on agar plates with complex medium ("GC" plates, for *go*nococcal). *N. gonorrhoeae* use phase variation to turn pilus expression on or off [10, 11], so pilus-positive (P+) colonies must be selected at each passage step when growing the bacteria [12, 13]. P+ colonies are distinguished from nonpiliated (P−) colonies by their glossy raised spherical morphology (Fig. 1). This is presumably because of the close contacts made by pilus–pilus interactions and pilus retraction, which pull the cells into tight aggregates. In contrast, P− colonies tend to be flatter and duller in appearance (Fig. 1). Here, we describe a stepwise protocol for purifying pili from *N. gonorrhoeae* by mechanical shearing and differential precipitation. This method has been used to obtain *N. gonorrhoeae* pili for structure determination by cryoelectron microscopy [3, 5]. Purified pili can be dissociated into pilin subunits by detergent solubilization for high resolution structure determination [3, 4].

2 Materials

The number of plates of *N. gonorrhoeae* used for the pilus preparation will depend on the application. The protocol described below uses 100 plates, which should yield >5 mg of purified pili. The protocol may be scaled as needed.

2.1 Pilus Purification

2.1.1 Large Batch Culture of Gonococci

1. *N. gonorrhoeae* glycerol stock stored at −80 °C (e.g., strain MS11 or C30, *see* **Note 1**).
2. GC medium base (Difco).
3. Supplement I stock, 500 mL: 40% (w/v) glucose, 68 mM L-glutamine, 75 μM thiamine. Add 200 g of glucose (dextrose) to ~400 mL of distilled H_2O and heat until dissolved. Do not boil. Cool the solution to room temperature and add 5 g of L-glutamine and 10 mg of thiamine. Mix until dry ingredients are dissolved. Bring the solution to 500 mL with distilled H_2O. Dispense 15 mL aliquots into 15 mL Falcon tubes. Store stock solutions at −20 °C, keeping two tubes aside at 4 °C.
4. Supplement II stock, 50 mL: 21 mM $Fe(NO_3)_3{\cdot}9H_2O$. Add 0.42 g of $Fe(NO_3)_3{\cdot}9H_2O$ to 50 mL of distilled H_2O. Stir until dissolved. Dispense 1.5 mL aliquots into 2 mL cryotubes and store at −20 °C, keeping two tubes aside at 4 °C.
5. Noble agar.
6. Sterile 50 mL Falcon tubes (conical bottom, polypropylene).
7. 2 mL cryotubes.
8. 3 L Erlenmeyer flasks.
9. 0.2 μm Acrodisc syringe filters (25 mm diameter).
10. 12 mL syringes.
11. Sterile pipette tips.
12. Sterile 10 μL loops.
13. Plastic sterile petri plates with covers.
14. Humidified CO_2 incubator set at 37 °C, 5% (v/v) CO_2. Pre-equilibrate to ensure that the CO_2 level is stable at 5%.
15. Fyrite CO_2 analyzer.
16. Metal pan.
17. Large stir bar.
18. Aluminum foil.
19. Stir plate.
20. Bunsen burner.

2.1.2 Pilus Production

1. GC broth, 100 mL: 1.5% (w/v) proteose peptone #3, 23 mM K_2HPO_4, 7.4 mM KH_2PO_4, 86 mM NaCl. Add 1.5 g of proteose peptone #3, 0.4 g of K_2HPO_4, 0.1 g of KH_2PO_4, 0.5 g of NaCl to 100 mL of distilled H_2O. Autoclave in 50 mL volumes for 30 min at 121 °C and 2.7 kg/cm^2 (liquid cycle). This solution is stable for several months at 4 °C.
2. Sterile 50 mL Falcon tubes (conical bottom, polypropylene).
3. Sterile cotton swabs.

4. GC agar plates.
5. Repeat pipettor (100 μL).
6. Glass spreader.
7. Small hand-operated inoculating turntable.
8. 70% (v/v) ethanol.
9. 4 L beakers.
10. Bunsen burner.

2.1.3 Harvest Pili

1. Dithiothreitol (DTT), 1 M, 10 mL. Add 1.54 g of DTT to 10 mL of distilled H_2O. Stir to dissolve. Dispense into 100 μL aliquots and freeze at −20 °C.
2. Pilus Buffer, 50 mL: 50 mM CHES (2-[cyclohexylamino] ethanesulfonic acid), pH 9.5. Store at 4 °C.
3. 20% (w/v) sodium azide (NaN_3) stock solution, 10 mL. Add 2 g of NaN_3 to 10 mL of distilled H_2O (*see* **Note 2**).
4. Chilled Dialysis Buffer (Tris-buffered saline (TBS)-NaN_3), 8 L: 50 mM Tris–HCl, pH 7.4 containing 150 mM NaCl and 0.02% (w/v) NaN_3. Prepare two 4 L solutions of Dialysis Buffer: add 24.2 g of Tris–HCl and 35 g of NaCl to each of two 4 L beakers. Bring to 4 L with distilled H_2O and adjust pH to 7.4. Add 0.8 g of NaN_3 (*see* **Note 2**). Cover beakers and store at 4 °C.
5. 6000–8000 Da molecular weight cutoff (MWCO) dialysis tubing (Spectra/Por 1, 40 mm flat width).
6. 50 mL Falcon tubes (conical bottom, polypropylene).
7. Rubber spatula.
8. Metal spatula.
9. Small hand-operated inoculating turntable.
10. 70% (v/v) ethanol.
11. Dialysis tubing clamps.
12. Long-necked funnel.
13. Benchtop vortexer.
14. Refrigerated centrifuge with 50 mL fixed angle rotor; capable of 12,000 × *g*.
15. Stir bar.
17. Plastic wrap.

2.1.4 Purify Pili

1. 4 L of chilled Dialysis Buffer.
2. 50 mL centrifuge tubes (round bottom, polypropylene).
3. Chilled Pilus Buffer/1 mM DTT.
4. Pasteur pipette.

5. 15 mL conical centrifuge tube.
6. 50 mM CHES buffer (2-[cyclohexylamino]ethanesulfonic acid), pH 9.5.
7. 15% (w/v) acrylamide–sodium dodecyl sulfate (SDS) gels.
8. Polyacrylamide gel electrophoresis (PAGE) apparatus.

2.2 Pilus Dissociation into PilE Subunits

1. Amicon filtration membranes (Biomax 50,000 Molecular Weight Cut Off, 25 mm, Ultracel PL-10, 44.5 mm).
2. Pilin Buffer, 100 mL: 100 mM Tris–HCl, pH 8, 20 mM NaCl, 0.02% (w/v) NaN_3, 1.5% (w/v) *n*-octyl-β-D-glucopyranoside (βOG). Add 1.21 g of Tris–HCl and 117 mg of NaCl to ~98 mL of distilled H_2O. Stir to dissolve dry ingredients and adjust pH to 8.0. Add 100 μL of 20% (w/v) sodium azide (NaN_3) and 1.5 g of βOG. Bring volume to 100 mL with distilled H_2O. Store at 4 °C.
3. 1 M DTT, frozen in 100 μL aliquots.
4. 10 mL Amicon Stirred Cell.
5. 50 mL Amicon Stirred Cell.
6. Compressed nitrogen cylinder and regulator (90–99.9% purity).
7. 15 mL Falcon tubes (conical bottom, polypropylene).
8. 100 mL beaker.
9. 15% (w/v) acrylamide–SDS gels.
10. SDS-PAGE apparatus.
11. 1.5 mL microcentrifuge tubes.
12. Liquid nitrogen.

3 Methods

Colonies are passed several times to obtain dense lawns of P+ *N. gonorrhoeae* that are harvested and suspended in a high pH buffer. The pili are released from the bacteria by mechanical shearing using a vortexer, and the bacteria are removed from the pili in solution by medium-speed centrifugation. The high pH buffer prevents the pili from aggregating. The pili are purified and concentrated by one or two rounds of precipitation using a neutral pH buffer, which causes the pili to aggregate, allowing them to be pelleted by medium-speed centrifugation. Purified, concentrated pili are suspended in a high pH buffer to disaggregate them. A flow chart is shown in Fig. 2. Sterile technique should be used for all steps for Days 1–5 and biosafety precautions should be taken at all times when handling *N. gonorrhoeae* (*see* **Notes 3** and **4**).

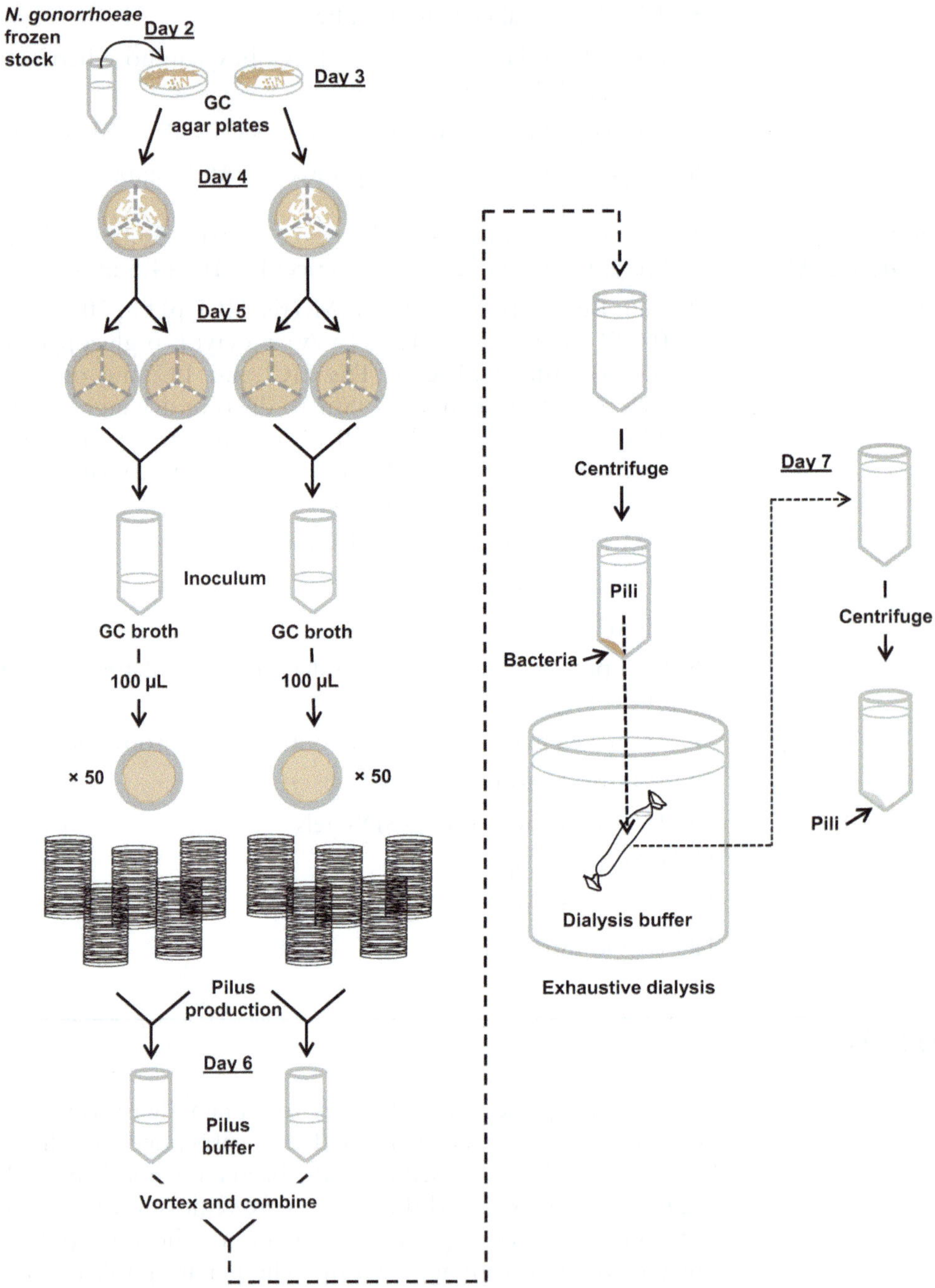

Fig. 2 Flowchart for pilus preparation

3.1 Pilus Purification

3.1.1 Large Batch Culture of Gonococci

1. On Day 1, prepare two 1.5 L solutions of GC agar, enough for approximately 100–110 plates. One 1.5 L solution of GC agar yields 50–60 plates. Weigh out 54.4 g of GC medium base and 1.9 g of Noble agar and add these to two 3-L Erlenmeyer flasks. Place a large stir bar into each flask and add 1.5 L of distilled

H_2O. Cover the flasks with aluminum foil and stir the solutions until the dry ingredients are completely dissolved. Place the flasks in a large metal pan with several cm of distilled H_2O at the bottom. Autoclave the flasks for 45 min on liquid cycle, slow exhaust.

2. Gently stir the GC agar on a stir plate, allowing it to cool until you can hold your hand on the side of the flask (~65 °C). This takes less than an hour if done in a cold room.
3. Turn on a Bunsen burner. Filter-sterilize one tube each of Supplement I (15 mL) and Supplement II (1.5 mL) directly into each flask of cooled GC agar using a 12 mL syringe and 0.2 μm Acrodisc syringe filter attachment using sterile technique (pull the foil lid back, flame the rim of the flask before and after adding supplements, replace lid). Swirl to mix but do not over-mix to cause bubbles.
4. Arrange 100 Petri plates upright in stacks of 10. Remove the foil lid from one flask and flame the rim. Begin pouring plates from the bottom of the stack, lifting the lid on the bottom plate along with all plates above it. Pour enough GC agar to generously cover the bottom of the plate. Replace the lid, move up to the next lid and repeat. Maintain the flask at approximately the same angle to avoid washing solution from the rim back into the flask. Pour gently to avoid generating bubbles. When a stack is complete, gently move it away and pull the next empty stack close to the burner, keeping the flask at approximately the same level at all times. Flame the rim of the flask and pour the next set of plates. Repeat until both flasks have been emptied (*see* **Note 5**). Leave the plates on the counter undisturbed overnight to allow the agar to become solid (*see* **Note 6**).
5. Set the incubator for 5% (v/v) CO_2. After several hours use a Fyrite CO_2 analyzer to confirm that the CO_2 level is at 5% (v/v) (*see* **Note 7**). Place a pan of distilled H_2O in the incubator to maintain humidity.
6. On Day 2, scrape a piece of the *N. gonorrhoeae* frozen stock using a sterile plastic pipette tip, and transfer it to a GC agar plate using sterile technique (*see* **Note 8**). Streak the solution using a sterile loop on the plate to obtain single colonies (*see* Fig. 3a for desired growth pattern). Repeat for a second plate. These bacteria will be sub-cultured twice, with the goal of obtaining multiple colonies of P+ *N. gonorrhoeae* bacteria to be used to prepare an inoculum for large scale growth to harvest pili.
7. Place the agar plates upside-down in a humidified 37 °C incubator with 5% (v/v) CO_2 and grow for exactly 24 h.

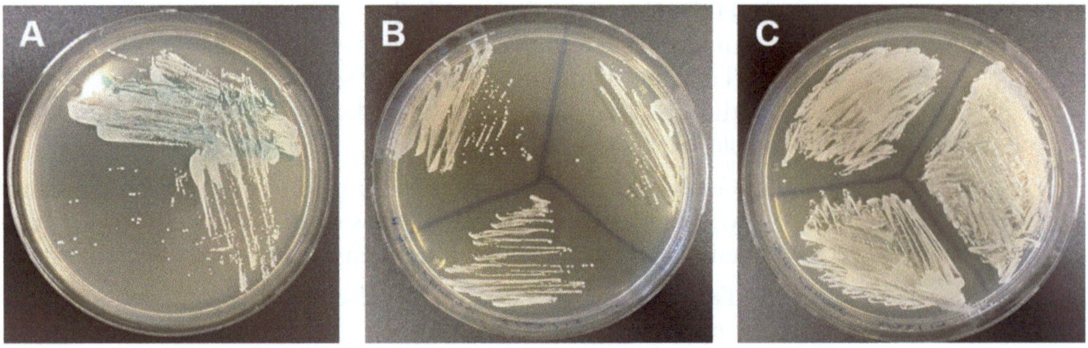

Fig. 3 *N. gonorrhoeae* plates as they should appear on (**a**) Day 3, (**b**) Day 4, and (**c**) Day 5. Note that the colonies may appear darker in color than those shown here

8. On Day 3, examine colonies under a light microscope at 4× or 10× magnification. *N. gonorrhoeae* bacteria form tan- or amber-colored circular colonies. Pilus-positive (P+) colonies reflect light and have a raised, almost spherical morphology, whereas P- colonies are dull and flat (Fig. 1, *see* **Note 9**). Lightly touch an isolated P+ colony with a sterile loop and streak it onto one third of a GC agar plate to obtain single colonies (*see* Fig. 3b for desired growth pattern). Repeat this step with five additional P+ colonies, streaking a total of two GC plates.
9. Repeat **step** 7.
10. On Day 4, examine colonies under a light microscope at 4× or 10× magnification to identify P+ colonies. Using a sterile loop pick an entire P+ colony and streak it onto one-third of a GC plate. The goal is to obtain a dense lawn of *N. gonorrhoeae* bacteria (*see* Fig. 3c for desired growth pattern) that will be used as the inoculum to grow the bacteria from which the pili will be harvested. Prepare four GC agar plates from 12 P+ colonies.
11. Repeat **step** 7.

3.1.2 Pilus Production

1. On Day 5, transfer 8 mL of GC broth into each of two 50 mL Falcon tubes.
2. Remove two plates from incubator and examine them to confirm that the colonies are P+ and to identify any contamination (*see* **Note 10**). If many of the colonies are P− or plates are contaminated, go back to the plates from **step 10** in Subheading 3.1.1, and prepare a new inoculum. Scrape up as many bacteria as possible from both plates with a sterile cotton swab and swirl them off into the GC broth in one of the tubes. Gently vortex the solution to suspend the bacteria.

Work quickly to minimize the time the bacteria spend in the GC broth.

3. Place 50 GC agar plates right-side up in stacks of 10 near a Bunsen burner flame. Place a glass spreader in 70% (v/v) ethanol. Dispense 100 μL aliquots of the bacterial suspension with a repeat pipettor onto each plate within a stack. Spread each inoculum evenly around the plate using a flame-sterilized glass spreader and a turntable.
4. Place the plates upside down in the humidified incubator, with no more than four plates per stack. Ensure that the temperature is 37 °C and the CO_2 is at 5% (v/v).
5. Take out the second pair of inoculum plates and prepare the next 50 plates of culture, as described in **step 3**. Grow bacteria for 23–24 h.

3.1.3 Harvest Pili

Sterile technique is not needed from Day 6 onward (*see* **Note 11**), since the bacteria are not grown further, but all solutions and suspensions should be kept on ice. It is a good idea to save a sample at each step for SDS-PAGE analysis.

1. On Day 5, prepare two 4 L solutions of Dialysis Buffer. Store these at 4 °C overnight.
2. On Day 6, prepare a 50 mL solution of Pilus Buffer/DTT. Thaw a 1 mL tube of 1 M DTT and add 50 μL to the Pilus Buffer. Transfer 20 mL of Pilus Buffer/DTT to each of two 50 mL Falcon tubes. Place on ice.
3. Add 0.62 g of DTT to one of the 4 L beakers containing Dialysis Buffer. Keep cold.
4. Soak a ~16 cm length of 40 mm dialysis tubing in distilled H_2O for 30 min.
5. Examine colonies under a light microscope at 4× or 10× magnification of the single *N. gonorrhoeae* colonies to confirm that they are P+. Examine the plates by eye for contamination (*see* **Note 10**).
6. Scrape up the bacteria from each plate with a rubber spatula and an inoculating turntable, and then use a metal spatula to scrape them into one of the Falcon tubes containing Pilus Buffer/DTT, avoiding any areas of contamination on the plates. Bacteria from ~50 plates can be added to each tube.
7. Once all of the bacteria have been harvested, shear the pili from the bacteria by vigorously mixing the suspension with a benchtop vortexer at high speed. Mix each tube for three 1 min intervals, alternating the tubes, which are stored on ice between each interval.

8. Combine the bacterial suspensions into one of the tubes. Prepare a balance tube and centrifuge at 12,000 × *g* for 20 min at 4 °C to pellet the *N. gonorrhoeae* bacteria, leaving the sheared pili suspended in the supernatant. Decant the supernatant into a single clean beaker.
9. Seal one end of the soaked dialysis tubing with a clamp and transfer the supernatant into the tubing using a funnel. Remove air and seal the other end, leaving a couple of centimeters for expansion. Place the dialysis bag into the chilled Dialysis Buffer/DTT solution. Place a stir bar in the beaker and cover it with plastic wrap. Dialyze for 4 h or overnight at 4 °C, stirring gently (*see* **Note 12**).

3.1.4 Purify Pili

1. On Day 7, add 0.62 g of DTT to the second 4 L beaker that contains Dialysis Buffer. Transfer the dialysis bag to this solution and dialyze for another 4 h at 4 °C. The buffer exchange from the basic Pilus Buffer (pH 9.5) to the neutral Dialysis Buffer (pH 7.5) causes the pili to aggregate, allowing them to be pelleted by centrifugation. The pilus solution in the dialysis bag (dialysate) will become slightly opaque.
2. Transfer the dialysate to a 50 mL Falcon tube, prepare a balance tube and centrifuge at 12,000 × *g* for 20 min at 4° C. The pili should appear as a small white pellet at the bottom of the tube. Decant the supernatant from the tube into a single bottle and store at 4 °C. Leave the tube upside down over a paper towel for a few minutes to drain residual supernatant.
3. Add 3–5 mL of Pilus Buffer/DTT to the tube and very gently loosen the pellet from the tube wall by pipetting with a Pasteur pipette or by agitation. Avoid foam/froth. Store the tube at 4 °C for 4–24 h, agitating occasionally, to solubilize the pili.
4. Once the pellet is fully dissolved, transfer the solution to a 15 mL conical centrifuge tube and centrifuge at 10,000 × *g* for 20 min to remove insoluble contaminants and residual bacteria. Transfer the supernatant containing the pili into a clean 15 mL tube. Store the pellet at 4 °C (*see* **Note 13**).
5. Repeat the dialysis (**step 9** of Subheading 3.1.3 and **steps 1–4** of Subheading 3.1.4) to further purify the pili (*see* **Note 14**), then suspend the final pellet in ~1–2 mL of Pilus Buffer/DTT.
6. Determine the protein concentration by UV absorbance. Blank the UV spectrophotometer with Pilus Buffer/DTT. Prepare a 1/50 dilution of the pilus solution in Pilus Buffer/DTT and measure the UV absorbance at $\lambda_{280\ \mathrm{nm}}$ (A_{280}) and $\lambda_{260\ \mathrm{nm}}$ (A_{260}). Calculate the amount of protein using the following equation:

$$[(A_{280})(1.5) - (A_{260})(0.76)] \times (\text{dilution factor [e.g.50]}) = \text{protein concentration (mg/mL)}.$$

If the readings are low, use a less-diluted sample. Pilus preparations from *N. gonorrhoeae* C30 bacteria have yielded 250 μg of pili per plate, so it is reasonable to expect 25 mg of protein from 100 plates.

7. Analyze the pili by SDS-PAGE using a 15% (w/v/) acrylamide gel: load 10 μL of a 1/100 dilution, a 1/10 dilution and undiluted to show the protein band (~17 kDa) and estimate the purity of the preparation (*see* **Note 15**).
8. Always store intact pili at 4 °C (*see* **Note 16**).

3.2 Pilus Dissociation into PilE Subunits

If desired, the purified pili can be dissociated with the detergent *n*-octyl-β-D-glucopyranoside (βOG), which disrupts the hydrophobic interactions between the PilE subunits, yielding solubilized monomers. The pili are washed exhaustively with βOG solution in a stirred cell concentrator with a Biomax 50,000 MWCO membrane that retains the intact pilus filaments but allows the dissociated pilin subunits to pass through into the filtrate. This process can take days to weeks and is somewhat unpredictable. Once all of the pili are dissociated, the pilin subunits in the filtrate are concentrated on an Ultracel PL-10 membrane.

1. Soak a Biomax 50,000 MWCO membrane (25 mm) in distilled H_2O for 30 min (*see* **Note 17**).
2. Prepare 100 mL of Pilin Buffer solution and add 100 μL of 1 M DTT (*see* **Note 18**).
3. Assemble a 10 mL Amicon Stirred Cell concentrator with the Biomax 50,000 MWCO membrane and place on a stir plate at 4 °C. Connect the concentrator to a compressed nitrogen cylinder. Add solid βOG to the purified pilus solution to bring it to ~1.5% (w/v). Add this solution to the Stirred Cell. Bring the volume up to 10 mL with Pilin Buffer/DTT.
4. Concentrate the pilus solution to ~0.5 mL with pressure from nitrogen gas (5.1 kg/cm^2) while stirring and collect the filtrate (flow-through) in a clean 100 mL beaker. Do not allow the membrane to dry.
5. Bring the volume back to 10 mL with Pilin Buffer/DTT and concentrate to 1 mL, collecting the filtrate in the same beaker used in **step 4**. Repeat 5–10 times. Determine the concentration of the retentate using $A_{280/260}$ measurements as described in **step 6** of Subheading 3.1.4, to see if the pili are dissociating (*see* **Note 19**). Prepare fresh Pilin Buffer/DTT and continue until the protein concentration in the retentate approaches zero.
6. Combine all pilin-containing filtrates and concentrate to 0.5–1 mL on the 50 mL Amicon Stirred Cell concentrator using an Ultracel PL-10 membrane (*see* **Note 20**).

7. Carefully remove the concentrated protein solution from the Stirred Cell using a micropipettor without touching the filter and transfer it to a microcentrifuge tube. Wash the filter/cell with 50–100 μL of Pilin Buffer/DTT and add this solution to the microcentrifuge tube.
8. Vortex lightly to mix. Determine the protein concentration using $A_{280/260}$ measurements as described in **step 6** of Subheading 3.1.4. Run an aliquot (~5–10 μg) on a 15% (w/v) acrylamide-SDS gel to confirm the presence of a 17 kDa band.
9. Dispense the purified pilin (PilE) into microcentrifuge tubes and flash-freeze in liquid nitrogen. Store at −80 °C.

4 Notes

1. Frozen glycerol stocks of *N. gonorrhoeae* bacteria can be prepared as follows: Scrape several P+ colonies at once from an overnight growth plate using a sterile loop and streak these onto a fresh GC agar plate. Grow the bacteria for 24 h at 37 °C with 5% (v/v) CO_2. Prepare a 10 mL solution containing 0.5 g of trypticase soy broth (TSB) and 2 mL of 20% (v/v) glycerol in distilled H_2O. The viscous glycerol can be dispensed with a 5 mL pipettor using a tip that has 2–3 mm of its end cut off. Mix the TSB/glycerol solution to dissolve the TSB then autoclave on liquid cycle for 15 min at 121 °C and 2.78 kg/cm^2 and cool to room temperature. Examine the plate of *N. gonorrhoeae* bacteria to confirm that the colonies are mostly P+. Scrape the bacteria from the plate using a sterile cotton swab and suspend the cells in the TSB/glycerol solution. Dispense 1 mL aliquots of the cell suspension into labeled sterile 2 mL cryotubes, which are immediately capped. Insert the tubes with long metal tongs upright into liquid nitrogen to flash-freeze. Store the tubes at −80 °C.
2. Use a mask when weighing NaN_3 to avoid inhaling the powder.
3. Sterile containers and techniques should be used from Days 1–5 when growing *N. gonorrhoeae*. At Day 6 the pili are sheared from the bacteria and sterility is not necessary from this point forward.
4. *N. gonorrhoeae* bacteria are very sensitive, so instructions should be followed carefully. *N. gonorrhoeae* are Biosafety Level 2 pathogens. Work should be performed in a Biosafety Cabinet wherever possible and Personal Protective Equipment should be worn at all times. All materials that come into contact with the bacteria must be decontaminated or treated as biohazardous waste.
5. Work rapidly to avoid the agar polymerizing while in the flask.

6. If plates are not to be used immediately they can be placed in a plastic bag and stored at 4 °C for several weeks.
7. If the CO_2 level is not at or close to 5% (v/v), adjust the setting appropriately and wait for an hour or more before checking level again. Repeat as necessary until the level reaches 5% (v/v). The CO_2 level should be checked each day to ensure that it is stable at 5% (v/v).
8. The glycerol stocks are rather dilute, so a substantial amount of the frozen suspension should be transferred to the GC plate to ensure that viable *N. gonorrhoeae* bacteria are present.
9. *N. gonorrhoeae* bacteria from glycerol stocks grow slowly and colonies are quite small after 24 h growth, which can make it difficult to distinguish P+ and P− colonies. It is advisable to practice identifying P+ colonies using bacteria that have grown for 2 days. View the colonies under a light microscope, 4× to 10× magnification, preferably with illumination on the top and bottom. On the first day the colonies will likely be mostly P+, but on the second day P− colonies will be evident and the colonies will be larger so they can easily be distinguished. The P + colonies appear raised and almost spherical, like a bead of water on a hydrophobic surface. If the colonies are illuminated from the top as well as from the bottom they will reflect light and appear much shinier than the P− colonies, which are duller and flatter. Occasionally you will see P− colonies growing out of P+ colonies, as in Fig. 3. Once you can distinguish P+ and P− colonies you can begin your experiment and confidently select P+ colonies even though they are small.
10. Contaminants will differ from *N. gonorrhoeae* in colony morphology and color. Typically, yeast will contaminate the plates and appear as large flat, very opaque white or yellow colonies.
11. It is not necessary to use any sterile containers from this point onward.
12. Make sure that the stir bar does not touch the tubing. The tubing clamps can be tied/taped to the sides of the beaker to keep the bag near the top of the solution.
13. If pilus yield is low, suspend this pellet in ~2 mL of Pilus Buffer/DTT to try and recover more pili.
14. This second dialysis step is optional.
15. The % purity can be estimated for overloaded lanes where there is a strong pilin band: estimate the combined densities of the nonpilin bands, that is, those migrating above or below the 17 kDa pilin band, relative to the density of the pilin band. Thus, if the pilin band represented ~95% of the total density in a single lane, the protein is ~95% pure.

16. Do not freeze the pili. We have found that freezing can distort the pilus structure and introduce heterogeneity into the population.
17. The membrane filter is very fragile. Handle with care.
18. It will be necessary to prepare this solution several times to fully dissociate the pili into pilin subunits.
19. Since the filtrate will become more dilute with each wash, it is more accurate to follow dissociation progress by quantifying the reduction in protein concentration in the retentate, rather than the increase in concentration of the filtrate. Eventually, as the concentration of pilin in the retentate decreases, you should see a corresponding increase in protein concentration in the filtrate.
20. A final concentration of 10–15 mg/mL is desirable for crystallization and X-ray structure determination.

Acknowledgments

This work was supported by the Canadian Institutes of Health Research (MOP125959). We thank Katrina Forest, University of Wisconsin-Madison, Wisconsin, the USA, for helpful discussions.

References

1. Merz AJ, So M (2000) Interactions of pathogenic *Neisseriae* with epithelial cell membranes. Annu Rev Cell Dev Biol 16:423–457
2. Naumann M, Rudel T, Meyer TF (1999) Host cell interactions and signalling with *Neisseria gonorrhoeae*. Curr Opin Microbiol 2:62–70
3. Craig L, Volkmann N, Arvai AS et al (2006) Type IV pilus structure by cryo-electron microscopy and crystallography: implications for pilus assembly and functions. Mol Cell 23:651–662
4. Parge HE, Forest KT, Hickey MJ et al (1995) Structure of the fibre-forming protein pilin at 2.6 Å resolution. Nature 378:32–38
5. Wang F, Coureuil M, Osinski T et al (2017) Cryoelectron microscopy reconstructions of the *Pseudomonas aeruginosa* and *Neisseria gonorrhoeae* type IV pili at sub-nanometer resolution. Structure 25:1423–1435 e1424
6. Winther-Larsen HC, Wolfgang M, Dunham S et al (2005) A conserved set of pilin-like molecules controls type IV pilus dynamics and organelle-associated functions in *Neisseria gonorrhoeae*. Mol Microbiol 56:903–917
7. Seifert H, So M (1984) Genetic mechanisms of bacterial antigenic variation. Microbiol Rev 52:327–336
8. Maier B, Potter L, So M et al (2002) Single pilus motor forces exceed 100 pN. Proc Natl Acad Sci U S A 99:16012–16017
9. Li J, Egelman EH, Craig L (2012) Structure of the *Vibrio cholerae* type IVb pilus and stability comparison with the *Neisseria gonorrhoeae* type IVa pilus. J Mol Biol 418:47–64
10. Hagblom P, Segal E, Billyard E et al (1985) Intragenic recombination leads to pilus antigenic variation in *Neisseria gonorrhoeae*. Nature 315:156–158
11. Segal E, Billyard E, So M et al (1985) Role of chromosomal rearrangement in *N. gonorrhoeae* pilus phase variation. Cell 40:293–300
12. Swanson J (1978) Studies on gonococcus infection. XII. Colony color and opacity variants of gonococci. Infect Immun 19:320–331
13. Vink C, Rudenko G, Seifert HS (2012) Microbial antigenic variation mediated by homologous DNA recombination. FEMS Microbiol Rev 36:917–948

Chapter 8

Peptidoglycan Composition in *Neisseria*

Ryan E. Schaub and Joseph P. Dillard

Abstract

The composition of *Neisseria* peptidoglycan has been of scientific interest for over four decades. Initial investigations focused on discovering the mechanisms causing rising rates of antibiotic resistance in *N. gonorrhoeae* by determining differences in peptidoglycan composition in penicillin susceptible and resistant strains. The discovery that cytotoxic peptidoglycan fragments are also released by *Neisseria* furthered the interest in peptidoglycan composition. This method describes the purification, enzymatic degradation, and separation of peptidoglycan fragments by high-performance liquid chromatography (HPLC). It also describes the preparation of samples so that they can be positively identified by mass spectrometry.

Key words Peptidoglycan, HPLC, Cell wall, *O*-Acetylation

1 Introduction

The murein sacculus is composed of chains of repeating disaccharide units with the chains linked together by peptide stems. This structural component of the cell is nearly ubiquitous in bacteria. The general architecture of peptidoglycan and the addition of disaccharide pentapeptide units are similar in nearly all bacteria. Following the insertion of glycan strands, the peptidoglycan is continuously modified to allow for cell growth, division, and separation. Peptidoglycan needs to be degraded to allow for the insertion of new peptidoglycan strands without disrupting the structural integrity of the cell. Despite iterative addition of disaccharide pentapeptide units, when broken down, the macromolecular structure is composed of dozens of unique constituents.

Gram-negative peptide stems contain a diaminopimelic acid (DAP) at the third position. DAP is used to link to other DAP residues or, more commonly, to the fourth position D-alanines (Ala) on other peptide stems. DAP-containing peptidoglycan has long been known to be an adjuvant [1, 2]. More recently, it has become clear that human nucleotide-binding oligomerization

Myron Christodoulides (ed.), *Neisseria gonorrhoeae: Methods and Protocols*, Methods in Molecular Biology, vol. 1997,
https://doi.org/10.1007/978-1-4939-9496-0_8,

domain-containing protein 1 (NOD1) is capable of detecting peptidoglycan-derived muramyl tripeptides [3]. Some *Neisseria* species are unusual in that they release a significant proportion of their peptidoglycan that would have been otherwise recycled. Host cells are then able to sense these released fragments generated from the peptidoglycan sacculus. Understanding the structure of the sacculus can allow for a better understanding of how immunogenic peptidoglycan fragments are generated.

The previous method for identifying muropeptides relied on paper chromatography and large quantities of purified peptidoglycan that could be enzymatically digested and visualized using ninhydrin [4]. This method was capable of separating eleven types of peptidoglycan fragments. Just prior to the development of this method, studying the nature of peptidoglycan required radiolabeled murein sugars and/or peptide stems to detect peptidoglycan fragments that could be separated by paper or size-exclusion chromatography [5, 6], but they could only identify a small subset of the total complexity. The method for separating peptidoglycan fragments by High Performance Liquid Chromatography (HPLC) was developed by Glauner and Schwarz and first published in 1983 [7]. The unexpected result from these experiments was just how complex macromolecular peptidoglycan structures are. Glauner and Schwarz could identify 80 different peptidoglycan-derived compounds in *Escherichia coli* which included amino acid substitutions and DAP—lipoprotein linkages [7].

Soon after this method was pioneered, the peptidoglycan composition of *Neisseria gonorrhoeae* was determined [8]. *Neisseria* peptidoglycan was, and continues to be, of special interest because of its association to β-lactam resistance. Mutations in *penA*, encoding the biosynthetic transpeptidase PBP2, can cause resistance to penicillin and other β-lactam antibiotics. One of the initial findings that were made using this method was that *penA* mutants have different peptidoglycan composition in the presence and absence of antibiotics in *N. gonorrhoeae* [9]. Similar results were also reported for *N. meningitidis* [10]. The separation technique was later used in combination with mass spectrometry that has now mostly replaced the chemical identification of sugars and amino acids. The new mass spectrometry techniques, including MS/MS, confirmed the sequence of amino acids and the location of *O*-acetyl groups [11].

In order to determine the peptidoglycan composition, peptidoglycan first needs to be isolated from other cellular components by boiling cells in buffered sodium dodecyl sulfate (SDS). The insoluble peptidoglycan is then washed and treated with α-amylase to remove any bacterial glycogen present. A protease is then added to remove any peptidoglycan-associated proteins. Once the peptidoglycan is isolated, mutanolysin is used to cut the β-1,4 bond between *N*-acetylmuramic acid and *N*-acetylglucosamine.

The peptidoglycan fragments are isolated from any insoluble peptidoglycan and treated with sodium borohydride to prevent anomerization of the reducing sugar. The peptidoglycan fragments are separated by reversed-phase HPLC using a phosphate buffered pH gradient and methanol. Separated fragments can be visualized as peaks on a chromatogram. The identity of these peaks can be positively identified by collecting corresponding fractions and determining the mass by mass spectrometry.

2 Materials

2.1 Preparation of Peptidoglycan

1. *Neisseria gonorrhoeae* strains or isolates stored in GCBL with 15% (v/v) glycerol in cryovials at −80 °C or under liquid nitrogen.
2. *Micrococcus lysodeikticus* ATCC No. 4698.
3. Sterile polyester-tipped applicator swabs.
4. GCB agar plates: add 15 g/L of proteose peptone no. 3, 1 g/L of corn starch, 4 g/L of dipotassium phosphate (K_2HPO_4), 1 g/L of monopotassium phosphate (KH_2PO_4), 5 g/L of sodium chloride (NaCl), and 10 g/L of agar to 1 L of distilled H_2O. Stir to mix and autoclave at 121 °C and 2.68 kg/cm^2 for 30 min. Allow medium to cool to 55 °C and then add 10 mL of Kellogg's Supplements.
5. 100× Kellogg's Supplements: add 40 g of glucose, 1 g of glutamine, 2 mg of cocarboxylase, and 5 mg of iron(III) nitrate nonahydrate ($Fe(NO_2)_3 \cdot 9H_2O$) to 100 mL of distilled H_2O. Sterilize by filtration with through a 0.22 μm filter with a syringe.
6. GCB Liquid medium (GCBL): add 15 g/L of proteose peptone no. 3, 1 g/L of corn starch, 4 g/L of dipotassium phosphate (K_2HPO_4), 1 g/L of monopotassium phosphate (KH_2PO_4), and 5 g/L of sodium chloride (NaCl) to 1 L of distilled H_2O. Stir to mix and autoclave at 121 °C and 2.68 kg/cm^2 for 30 min. Allow medium to cool to 55 °C and then add 10 mL of Kellogg's Supplements.
7. Fernbach baffled culture flasks.
8. UV-VIS Spectrophotometer.
9. Sorvall RC 6+ refrigerated superspeed centrifuge, or equivalent.
10. 500 mL Nalgene PPCO centrifuge bottles, or equivalent.
11. Sorvall SLA-3000 fixed angle rotor, or equivalent.
12. 50 mL PPCO Oak Ridge centrifuge tubes, or equivalent.
13. Sorvall SS-34 fixed angle rotor, or equivalent.

14. Phosphate buffered saline (PBS): add 8 g/L of NaCl, 0.2 g/L of potassium chloride (KCl), 1.44 g/L of dibasic sodium phosphate (Na_2HPO_4), 0.24 g/L of monobasic potassium phosphate (KH_2PO_4) and adjust pH to 7.4 with HCl.
15. 25 mM sodium phosphate buffer, pH 6: add 3 mL of 1.0 M dibasic sodium phosphate (Na_2HPO_4) and 22 mL of monobasic sodium phosphate (NaH_2PHO_4) and bring the volume up to 1 L and autoclave at 121 °C and 2.68 kg/cm^2 for 30 min.
16. 25 mM sodium phosphate buffer, pH 6.0 containing 4% (w/v) or 8% (w/v) sodium dodecyl sulfate (SDS).
17. Boiling water bath.
18. Pipettes and pipette tips.
19. α-Amylase buffer: 10 mM Tris–HCl buffer, pH 8.0 containing 30 mM NaCl, 1 mM $CaCl_2$, 320 mM imidazole.
20. α-Amylase. Prepare a fresh 2 mg/mL (~20 U/mL) solution in 25 mM Tris–HCl buffer, pH 7.5 before adding to samples.
21. Pronase. Stored in aliquots at −20 °C at a concentration of 2 mg/mL.
22. Quartz cuvette.
23. Humidified incubator set at 37 °C with 5% (v/v) CO_2.
24. Shaking platform inside an incubator set at 37 °C.
25. Tube roller drum inside the 37 °C incubator.
26. pH meter.
27. Sodium azide (NaN_3) solution in distilled H_2O. Dilute to final concentration of 0.02% (w/v) from stock solution for use.

2.2 Peptidoglycan Digestion and Sodium Borohydride Treatment

1. Microcentrifuge tubes.
2. Muramidase (EC 3.2.1.17) mutanolysin.
3. 10 kDa MWCO spin column.
4. 0.5 M sodium borate, pH 8.0: add 7.729 g of boric acid to 150 mL water and adjust pH to 8 with 10 N of sodium hydroxide (NaOH). Bring the volume up to 250 mL and filter-sterilize through a 0.22 μm filter.
5. Sodium borohydride ($NaBH_4$) solution: add $NaBH_4$ to a final concentration of 2 mg/mL to 0.5 M borate buffer, pH 8.0, immediately before adding solution to samples.
6. 85% (w/w) phosphoric acid (H_3PO_4).
7. pH paper.
8. Vortex mixer.
9. Tube roller drum inside a 37 °C incubator.

2.3 HPLC Separation of Peptidoglycan Fragments

1. High-performance liquid chromatography (HPLC) system with UV detection ability, a 500 μL loop, a column oven and a fraction collector.
2. 4.6 mm × 250 mm, 5 μm particle, ODS-Hypersil chromatography column.
3. 50 mM sodium phosphate Buffer A, pH 4.33 (*see* **Note 1**).
4. 50 mM sodium phosphate Buffer B, pH 5.10 with 15% (v/v) methanol.

2.4 Mass Spectrometry

1. Microcentrifuge tubes.
2. SpeedVac.
3. HPLC system with UV detection ability, a 500 μL loop, a column oven and a fraction collector.
4. 4.6 mm × 250 mm, 5 μm particle, ODS-Hypersil chromatography column.
5. 0.05% (v/v) trifluoroacetic acid in distilled H_2O: Desalting Buffer A.
6. 50% (v/v) acetonitrile with 0.05% (v/v) trifluoroacetic acid in distilled H_2O: Desalting Buffer B.
7. Access to a positive ion electrospray ionization time-of-flight mass spectrometer.

3 Methods

3.1 Preparation of Peptidoglycan

Macromolecular peptidoglycan is separated from the majority of cellular components by boiling in buffered SDS. Covalently attached proteins are then digested with a proteinase, and the peptidoglycan is purified again using boiling SDS.

1. Scrape some *Neisseria gonorrhoeae* from a cryovial of bacteria stored at −80 °C or under liquid nitrogen onto eight GCB agar plates using sterile polyester-tipped applicator swabs. Incubate overnight at 37 °C in a humid atmosphere with 5% (v/v) CO_2.
2. Use a sterile swab to remove the growth from 8 GCB agar plates and suspend into 25 mL of GCBL and measure the optical density at $\lambda_{540\ nm}$ (OD_{540}) using a UV-VIS spectrophotometer. Add the appropriate amount of suspended cells to 1 L of GCBL medium at OD_{540} of 0.25 in a Fernbach baffled culture flask. Incubate with shaking at 125 rpm at 37 °C with aeration until the bacteria grow to an OD_{540} of 1.0 (*see* **Note 2**).
3. Harvest the bacteria by centrifugation at 7000 × *g* for 10 min at 4 °C in 500 mL centrifuge bottles.
4. Suspend the bacteria in 25 mL of cold PBS.

5. Pellet the bacteria by centrifugation at 7000 × *g* for 10 min at 4 °C. At this point, the bacteria can be stored at −80 °C for later use.
6. Suspend the pelleted bacteria in 10 mL of 25 mM sodium phosphate buffer, pH 6.0.
7. Place 10 mL of 25 mM sodium phosphate buffer, pH 6.0 containing 8% (w/v) SDS in an Oak Ridge tube and place the tube upright in a boiling water bath.
8. Add the suspended bacteria drop-wise using a pipette into the Oak Ridge tube with boiling SDS and boil for 30 min.
9. Remove the tubes from the boiling water bath and let them cool to room temperature.
10. Centrifuge the samples at 45,000 × *g* for 30 min at 15 °C and discard the supernatant into a biohazardous liquid waste container.
11. Wash the pellet by suspending it in 10 mL of 25 mM sodium phosphate buffer, pH 6.0, followed by centrifugation at 45,000 × *g* for 30 min at 15 °C and discard the supernatant. Repeat this step until all the SDS has been removed (*see* **Note 3**).
12. After the last wash, discard the supernatant and suspend the pellet in 1 mL of α-amylase buffer.
13. Add 200 μg (or approximately 5 U) of α-amylase suspended in 25 mM Tris-HCl pH 7.5 and incubate on a roller drum at 37 °C for 1 h.
14. Add 200 μg (~200 U) of pronase and incubate on a roller drum at 37 °C for 2 h or overnight.
15. Place 10 mL of 25 mM sodium phosphate buffer, pH 6.0 with 4% (w/v) SDS in an Oak Ridge tube in a boiling water bath.
16. Repeat **steps 7–11**.
17. Suspend the insoluble peptidoglycan in 1 mL of 25 mM sodium phosphate buffer, pH 6.0. Peptidoglycan can be stored at −20 °C or at 4 °C with 0.02% (w/v) sodium azide solution.
18. Quantify the sample concentration by measuring the absorbance in a UV spectrophotometer at $\lambda_{206\ nm}$ in a quartz cuvette by creating a standard curve with either lyophilized *Micrococcus lysodeikticus* or lyophilized peptidoglycan to determine the sample concentration.

3.2 Peptidoglycan Digestion and Sodium Borohydride Treatment

Purified macromolecular peptidoglycan is broken down using the muramidase mutanolysin that has both lysozyme-like 1,4-*N*-acetylmuramidase activity and *N,O*-diacetylmuramidase activity that are necessary to cleave *O*-acetylated peptidoglycan. The muramic acid of the resulting disaccharides is reduced to muramitol with $NaBH_4$, in order to eliminate the confusion of having two anomers with different dissociation constants on an ODS column.

1. Add 200 μg of mutanolysin to 500 μL of peptidoglycan suspended in 25 mM sodium phosphate pH 6. Incubate with rotation at 37 °C for 2 h or overnight.
2. Separate the soluble peptidoglycan fragments by filtering them through a 10 kDa MWCO spin column and collecting the flow-through.
3. Add 500 μL of freshly prepared $NaBH_4$ solution (0.5 M sodium borate pH 8 with 2 mg/mL $NaBH_4$) to soluble peptidoglycan fragments and leave for 20 min at room temperature in uncapped tubes. Briefly cap and vortex the samples after 10 min (*see* **Note 4**).
4. Quench the reaction by adding approximately 20 μL of 85% (w/w) phosphoric acid.
5. Check the pH and add additional 85% (w/w) phosphoric acid to bring the samples to between pH 3 and 4, if necessary.
6. Store the samples at 4 °C.

3.3 HPLC Separation of Peptidoglycan Fragments

1. Set the column oven to 30 °C and allow the column and detector lamp to warm up.
2. Create or select the following HPLC method: Detector at $\lambda_{206\ nm}$, 0.5 mL/min flow rate, 10 min of Buffer A, a linear gradient of 0–100% of Buffer B over 120 min, 100% of Buffer B for 20 min.
3. Equilibrate the column to Buffer A and then inject approximately 200 μL of sample and begin collecting fractions.
4. After the run, use the HPLC software to compare chromatograms. The area under the curve can be calculated to compare peaks corresponding to different peptidoglycan species.

3.4 Mass Spectrometry

Mass spectrometry is recommended for identification of HPLC peaks (Fig. 1, *see* **Note 5**). Analyze the chromatographs and collect the fractions corresponding to peaks of interest and pool fractions if necessary (*see* **Note 6**).

1. Use a SpeedVac concentrator to remove liquid from the samples.
2. Suspend the samples in 200 μL of H_2O.
3. Create or select an HPLC method: Detector at $\lambda_{206\ nm}$, 1 mL/min flow rate, 10 min of Desalting Buffer A, a linear gradient of 0–100% of Desalting Buffer B for 10 min, 10 min of Desalting Buffer B.
4. Inject the concentrated sample.

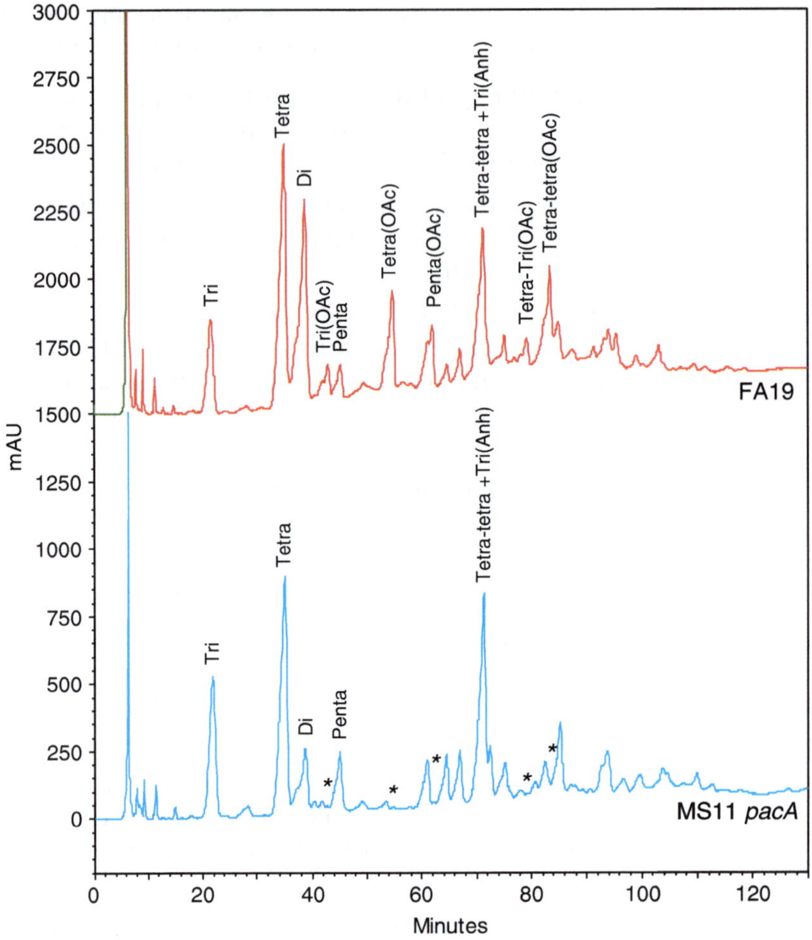

Fig. 1 Chromatograms showing the peptidoglycan composition of two strains of *N. gonorrhoeae*. A comparison of strain FA19 and an MS11 *pacA* mutant defective in *O*-acetylation. An asterisk shows the locations where peaks containing acetylated PG fragments would appear in a wild-type strain. Some peaks contain two different peptidoglycan fragments, such as the one containing Penta(OAc) and the one containing cross-linked Tetra-tetra monomers and Tri(Anh). FA19 is known to contain more dipeptide than other strains, but it is otherwise similar to other wild-type strains [12]. Labels indicate disaccharides with two (di), three (tri), four (tetra), or five (penta) amino acids. *OAc, O*-acetylation; *Anh,* 1,6-anhydromuramyl moiety

5. Collect the peak(s) of interest by comparing the chromatogram to a blank run and/or a high salt run.
6. Concentrate the samples into a single microcentrifuge tube by SpeedVac.
7. Run the sample on a mass spectrometer.
8. Identify peaks by comparing the acquired mass to that of the anticipated peptidoglycan fragment or to other published values.

4 Notes

1. Due to the strong pH dependency of this technique and small variations that can occur making these buffers, make sure there is enough buffer for all needed runs before starting. Before running samples, prepare Buffers A and Buffer B. These buffers are made by first adding sodium hydroxide (NaOH) to create a 50 mM solution, then adding phosphoric acid to obtain the desired pH. To prevent growth, 1 μg/mL sodium azide can be added to Buffer A. Filter and degas all the buffers.
2. To accurately compare peptidoglycan profiles it is important to grow cultures to the same optical density. During different growth phases the proportion of peptide stems changes [13]. An alternative method is to take cells directly from 40 GCB plates, but the profiles from plates will be different than those from liquid.
3. A simple and effective, although somewhat crude, way to determine if there is SDS in the 20 mM sodium phosphate buffer, pH 6.0 is to take advantage of the detergent property of SDS and look for bubbles after vortexing samples. Although it is not necessary, a more exact concentration of SDS can be determined using colorimetric assays with methylene blue [14] or the carbocyanine dye "stains-all" [15].
4. This reaction violently releases hydrogen gas; do not cap the solution.
5. The identity of some peaks should be apparent from comparing chromatographs. In some cases, mutants can be used to identify some peaks, like those of acetylated fragments in an *O*-acetylation mutant (Fig. 1). The certainty of other peaks may be in question due to small variations in HPLC setup or by overlapping peaks. Other factors such as buffer composition and column temperature can cause variations. It is best practice to always confirm the identity of any peaks of interest by mass spectrometry. Samples will likely need to be desalted before mass spectrometry because phosphate buffer is incompatible with many systems.
6. It is beneficial to take peaks from multiple HPLC runs, especially if minor peaks are being collected. Sufficient material is required to obtain an accurate mass.

Acknowledgments

This work was supported by NIH grant R01AI097157.

References

1. Adam A, Ciorbaru R, Ellouz F et al (1974) Adjuvant activity of monomeric bacterial cell wall peptidoglycans. Biochem Biophys Res Commun 56(3):561–567
2. Nauciel C, Fleck J, Martin JP et al (1973) Adjuvant activity of Gram-negative bacterial peptidoglycans in delayed hypersensitivity. C R Acad Sci Hebd Seances Acad Sci D 276 (26):3499–3500
3. Girardin SE, Boneca IG, Carneiro LA et al (2003) Nod1 detects a unique muropeptide from gram-negative bacterial peptidoglycan. Science 300(5625):1584–1587. https://doi.org/10.1126/science.1084677
4. Primosigh J, Pelzer H, Maass D et al (1961) Chemical characterization of mucopeptides released from the *E. coli* B cell wall by enzymic action. Biochim Biophys Acta 46:68–80
5. Rosenthal RS, Wright RM, Sinha RK (1980) Extent of peptide cross-linking in the peptidoglycan of *Neisseria gonorrhoeae*. Infect Immun 28(3):867–875
6. de Pedro MA, Schwarz U (1981) Heterogeneity of newly inserted and preexisting murein in the sacculus of *Escherichia coli*. Proc Natl Acad Sci U S A 78(9):5856–5860
7. Glauner B, Schwarz U (1983) The analysis of murein composition with high pressure liquid chromatography. The targets of penicillin. Walter de Gruyter, Inc, Berlin
8. Dougherty TJ (1985) Analysis of *Neisseria gonorrhoeae* peptidoglycan by reverse-phase, high-pressure liquid chromatography. J Bacteriol 163(1):69–74
9. Dougherty TJ (1985) Involvement of a change in penicillin target and peptidoglycan structure in low-level resistance to beta-lactam antibiotics in *Neisseria gonorrhoeae*. Antimicrob Agents Chemother 28(1):90–95
10. Antignac A, Boneca IG, Rousselle JC et al (2003) Correlation between alterations of the penicillin-binding protein 2 and modifications of the peptidoglycan structure in *Neisseria meningitidis* with reduced susceptibility to penicillin G. J Biol Chem 278 (34):31529–31535. https://doi.org/10.1074/jbc.M304607200
11. Martin SA, Rosenthal RS, Biemann K (1987) Fast atom bombardment mass spectrometry and tandem mass spectrometry of biologically active peptidoglycan monomers from *Neisseria gonorrhoeae*. J Biol Chem 262(16):7514–7522
12. Schaub RE, Perez-Medina KM, Hackett KT et al (2019) Neisseria gonorrhoeae PBP3 and PBP4 facilitate NOD1 agonist peptidoglycan fragment release and survival in stationary phase. Infect Immun 87:e00833-18. https://doi.org/10.1128/IAI.00833-18
13. Pisabarro AG, de Pedro MA, Vazquez D (1985) Structural modifications in the peptidoglycan of *Escherichia coli* associated with changes in the state of growth of the culture. J Bacteriol 161(1):238–242
14. Hayashi K (1975) A rapid determination of sodium dodecyl sulfate with methylene blue. Anal Biochem 67(2):503–506
15. Rusconi F, Valton E, Nguyen R et al (2001) Quantification of sodium dodecyl sulfate in microliter-volume biochemical samples by visible light spectroscopy. Anal Biochem 295 (1):31–37. https://doi.org/10.1006/abio.2001.5164

Chapter 9

Isolation of Naturally Released Gonococcal Outer Membrane Vesicles as Vaccine Antigens

Ian P. Francis, Xiuping Lui, and Lee M. Wetzler

Abstract

The emergence and spread of fully antimicrobial resistant *Neisseria gonorrhoeae* (GC) highlights a clear need for next-generation antigonococcal therapeutics. A broadly reactive anti-GC vaccine would best address this global public health threat. Polyantigenic outer membrane vesicles (OMVs) derived from GC can overcome the challenges posed by GC's high rate of phase and antigen variation. In fact, GC OMVs have already shown promise as a vaccine antigen; however, all previous studies have utilized vesicles contaminated by RMP, a bacterioprotective antigen known to entirely abrogate vaccine-induced bactericidal activity in vivo. Additionally, these studies primarily utilized vesicles isolated through techniques like membrane disruption with detergents, which are known to increase contamination of cytoplasmic components as compared to naturally released OMVs (nOMVs). This chapter describes the isolation and characterization of naturally released nOMVs through sequential size and weight restrictive filtration. nOMVs are characterized by morphology, proteomics, and bioactivity via various methods. Herein we also describe methods for further evaluation of the innate and induced immunogenicity of *rmp*-deficient GC nOMVs by cell stimulation and murine vaccination. Per these methods, nOMVs are found to be largely homogenous spherical structures approximately 70 nm in diameter containing a consistent subset of GC outer membrane proteins. The *rmp*-deficient vesicles demonstrate a morphology and, with the exception of RMP, antigenic profile consistent with that of nOMVs derived from wild time *N. gonorrhoeae*. Additionally, vesicles lacking RMP are able to engage and strongly activate a diverse array of pattern recognition receptors in vitro. These methods lay the groundwork for future experiments examining the in vivo protective efficacy of the anti-GC response induced by these nOMVs as well as studies examining the mechanism of vaccine induced female genital tract immunity.

Key words *Neisseria gonorrhoeae*, Outer membrane, Vesicles, Vaccine, Antigens

1 Introduction

As a gram-negative bacterium, *Neisseria gonorrhoeae*'s (GC) envelope is composed of three distinct layers: an inner lipid bilayer, a "periplasmic" space populated principally by a thin peptidoglycan wall, and an outer lipid bilayer [1]. It is at this outer lipid bilayer that outer membrane vesicles (OMVs) are produced [2]. OMVs were first described in a 1967 study of *Vibrio cholera*

Myron Christodoulides (ed.), *Neisseria gonorrhoeae: Methods and Protocols*, Methods in Molecular Biology, vol. 1997, https://doi.org/10.1007/978-1-4939-9496-0_9, © Springer Science+Business Media, LLC, part of Springer Nature 2019

growth and were thought to be an artifact of laboratory culture [3]. Since then, structures similar to these have been consistently observed from a variety of gram-negative bacteria [4]. While the exact mechanisms remain unclear, the reproducibility of these observations suggests the production of OMVs is an active and conserved process in gram-negative bacteria [4].

The body of OMV literature remains thin, particularly for any one pathogen, however a general understanding of the structure of vesicles can be appreciated by examining all findings together. OMVs are single lipid bilayer spheres that carry an array of outer membrane derived lipids, proteins, and lipopolysaccharide [5, 6], periplasmic peptidoglycan and proteins [7], and finally cytoplasmic proteins and nucleic acids [8, 9] (Fig. 1). Several studies have examined the proteome of vesicles and compared it to both the bacterial proteome and outer-membrane proteome at the time of blebbing [10–12]. In each of these studies, including Zielke and colleagues' examination of GC OMVs, it was shown that OMVs contain a predictable subset of proteins in different relative concentrations from the source outer membrane and bacterium. This suggests active sorting of proteins to curate the OMV proteome, likely to facilitate their function.

One of the challenges that has slowed the investigation of OMVs, is the lack of definitive identifying characteristics [13] resulting in OMVs being defined by the rough criteria of size and morphology [4]. The study of OMVs is only further complicated by the extreme difficulty associated with the pure isolation of the particles, especially in quantities large enough for experimentation [13].

The most commonly used method of OMV isolation is the induction of vesiculation by chemically disrupting the lipid bilayer using the bile acid, deoxycholic acid resulting in detergent derived vesicles (dOMVs) [14]. Several others methods of OMV isolation have also been described including mechanical shearing of OMVs from cells by passing them through a narrow-gauge needle (mOMVs) [15], or using chelators to disrupt the association of the outer bilayer with the underlying cell wall thereby promoting vesicle (cOMVs) release [16]. In addition to these methods, some researchers continue to isolate and characterize vesicles naturally released by bacteria (nOMVs) [17].

Detailed molecular characterization and comparison of cOMVs, dOMVs, mOMVs, and nOMVs shows that these vesicles are very different from one another in composition [17–19]. It was shown that dOMVs contained much greater quantities of cytoplasmic compounds than the other vesicle forms [18]. In contrast, cOMVs and nOMVs are both enriched with outer membrane lipoproteins but contain different cohorts in ratios distinct from each other [18]. It has also been shown, though not as completely, that there is significant batch variability with all OMVs, especially when isolated through secondary mechanisms (cOMVs, mOMVs, dOMVs) [20, 21].

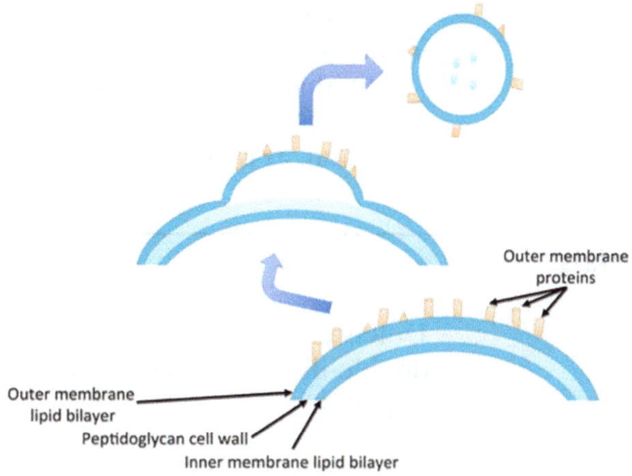

Fig. 1 A graphical representation of Gram-negative bacterial envelope structure and outer membrane vesicle genesis. Gram-negative associated outer membrane vesicles (OMVs) are released naturally in response to an unidentified signal. As represented here, the vesicles carry a subset of outer membrane associated proteins as well as some periplasmic debris within their single-lipid-bilayer-bound lumen

Like with any pharmaceutical, consistency in composition and effect are characteristics required of a vaccine antigen. Variations in formulation can result in batch dependent differences in protection efficacy, or even unpredictable adverse events. The body of OMV literature suggests that nOMVs demonstrate the most consistency in composition, provided they are prepared under similar and reproducible conditions [4, 17, 20].

While there is variability in described OMV formation, composition, and function, their potential as complex vaccine antigens is clear. The target of a neutralizing vaccine, like one that would prevent gonorrhea by blocking initial adhesion to the mucosal surface, is most frequently an externally exposed membrane structure. In OMVs there are is broad array of outer membrane structures that are presented in their native form since they are derived from the outer membrane of the actual pathogen. This should ensure that the faces of antigens that the immune system initially comes into contact with should be externally exposed. This will hopefully ensure, following booster exposure, that the trained immunological memory is focused on the externally exposed epitopes of membrane-bound structures. This will select for the type of antigen that typically mediates binding. As demonstrated by previous studies, the induction of a unidimensional immune memory does not result in protection against GC. OMVs, in addition to likely inducing some degree of neutralizing immunity, will induce a remarkably broad immune response due to their molecular diversity. It is possible that this broad immunological pressure could overwhelm the gonococcus and successfully induce protective immunity [21].

In this chapter we present a method for isolating naturally released OMVs from gonococci and the protocols for their characterization, as prerequisites for eventual use as potential vaccines.

2 Materials

Unless otherwise stated, all buffers and reagents are prepared with double-distilled water (ddH_2O). For cell culture, sterile ddH_2O is required.

2.1 Growth of Neisseria gonorrhoeae

1. *Neisseria gonorrhoeae* strain MS11 with a genetic deletion of the Reduction Modifiable Protein (RMP) gene (MS11Δ*rmp*).
2. Chocolate agar plates (commercially sourced).
3. GC liquid broth: dissolve 1.5% (w/v) protease peptone 3, 22.96 mM K_2HPO_4, 7.35 mM KH_2PO_4, 85.62 mM NaCl, in ddH_2O. Autoclave at 121 °C with 2.68 kg/cm^2, cool to 56 °C and then add 1% (v/v) Isovitalex (Becton Dickinson).
4. Sterile synthetic swabs.
5. 50 mL vented conical tubes (Corning mini bioreactor or equivalent).
6. 125 mL DeLong flasks (Bellco or equivalent).
7. 2 L vented top Fernbach flasks.
8. Class II biosafety cabinet.
9. Shaking incubator.
10. Humidified incubator set at 37 °C and 5% (v/v) CO_2.
11. Spectrophotometer (Bio-Rad SmartSpec 3000 or equivalent).

2.2 nOMV Isolation

1. 5× nOMV buffer: add 15% (w/v) sucrose and 1 M glycine to ddH_2O, and adjust pH to 8.0 with NaOH.
2. 500 mL sealable centrifuge bottles (Beckman Coulter or equivalent).
3. 0.45 μm polyethersulfone (PES) filters.
4. Chocolate agar plates.
5. Ultrafiltration cell (Millipore Amicon cell or similar) and appropriate 100,000 NMWL membranes for the cell.
6. Medical grade N_2 gas source.
7. Reinforced high-speed ultracentrifuge tubes (Beckman Coulter or equivalent).
8. 0.05% (w/v) sodium azide, sterilized by filtration (0.22 μm filter).

9. Beckman Coulter Avanti J-E floor centrifuge, (or equivalent) and rotor (Beckman Coulter or equivalent).
10. Beckman Coulter Optima XPN 100 k floor ultracentrifuge (or equivalent) and rotor (Beckman Coulter SW-28 or equivalent).
11. Pasteur pipettes.
12. Externally vented chemical hood.
13. Class II safety cabinet.
14. Humidified incubator set at 37 °C and 5% (v/v) CO_2.
15. Sterile screw-top cryogenic vials.

2.3 Antigenic Characterization of nOMVs

2.3.1 Protein Quantification

1. nOMV.
2. Cold (−20 °C) acetone.
3. Lowry protein assay (Pierce Modified kit or equivalent).
4. Sterile phosphate-buffered saline (PBS), pH 7.4.
5. Microcentrifuge tubes.
6. Benchtop centrifuge.
7. Vortex mixer.

2.3.2 Polyacrylamide Gel Electrophoresis Separation of nOMV Associated Structures

1. Cold acetone-prepared nOMV.
2. 15% (w/v) polyacrylamide–sodium dodecyl sulfate (SDS) gel.
3. Loading buffer: 100 mM Tris–HCl pH 6.8 containing 5% (v/v) β-mercaptoethanol, 4% (w/v) SDS, 0.2% (w/v) bromophenol blue.
4. 10× SDS-PAGE running buffer: add 144 g of glycine, 30 g of Tris, and 10 g of SDS in 1 L ddH_2O.
5. 1× dilution of 4× SDS-Sample Buffer (Boston Bioproducts or equivalent).
6. Precision Plus Pre-Stained Protein Dual Colored Standard (Bio-Rad or equivalent).
7. Gel electrophoresis system (e.g., Bio-Rad Mini PROTEAN 3 Cell system with Amersham Pharmacia Biotech electrophoresis power supply (EPS 601) used in the optimization of this protocol).
8. Coomassie Brilliant Blue.
9. Coomassie destain solution: 10% (v/v) methanol and 10% (v/v) isopropanol in ddH_2O.
10. Incubation chamber.
11. Plate rocker.
12. Digital imaging system (e.g., Bio-Rad Gel Doc XR or equivalent).

2.4 Structural Characterization of nOMVs

2.4.1 Evaluation of Isolated nOMV Effective Diameter by Dynamic Light Scattering (DLS)

1. nOMV.
2. 1.5mL polystyrene cuvette.
3. Nanoparticle Analyzer (Brookhaven Instruments 90Plus or equivalent).
4. Brookhaven Instrument Corporation's Particle Sizing Software.
5. Personal computer.

2.4.2 nOMV Visualization by Negative Staining Transmission Electron Microscopy (TEM)

1. nOMV.
2. 300 mesh copper grids.
3. Negative action "Self-holding" Forceps (Dumont or similar).
4. #1 Whatman filter paper cut into eight even slices.
5. 0.2 μm filtered ultrapure deionized H_2O.
6. Pasteur pipettes and bulb.
7. 2% (w/v) uranyl acetate stain solution prepared from radioactively depleted powder (Ted Pella or equivalent) in H_2O.
8. Vacuum Evaporator (e.g., Denton Vacuum Evaporator or equivalent).
9. Glow Discharge Apparatus (e.g., Balzers/Baltech Glow Discharge Apparatus or equivalent).
10. Phillips CM12 Transmission Electron Microscope configured with a lanthanum hexaboride (LaB_6) crystal electron source (or equivalent TEM).
11. Image capture CCD cameras, or built-in plate camera and EM film.

2.4.3 Morphological Evaluation of nOMVs

1. Digital micrographs of nOMV samples.
2. ImageJ software (Version 2.0.0-rc-65/1.51u, Build 961c5f1b7f).

2.5 In Vitro Characterization of nOMV Interactions with Innate Immune Receptors

1. Human Embryonic Kidney (HEK) cell lines overexpressing human Toll-like receptors (TLRs) 2 and TLR4 (InvivoGen).
2. Negative control HEK line (InvivoGen).
3. RPMI medium with 10% (v/v) fetal bovine serum (FBS) for cell culture.
4. 24-well sterile cell culture plates.
5. nOMV.
6. Cell culture grade Lipopolysaccharide (LPS).
7. Purified gonococcal P1B (when unavailable, the TLR2 agonist Pam_3CSK_4 can be used).
8. Cell culture grade purified TNF-α.

9. Human IL-8 ELISA (e.g., Becton Dickinson OptEI or equivalent).
10. Humidified incubator set to 37 °C and 5% (v/v) CO_2.

2.6 Immunogenicity of nOMVs

2.6.1 Mice

1. Standard, Association for Assessment and Accreditation of Laboratory Animal Care-accredited animal facility (or equivalent federal government body).
2. Female, 6- to 8-week-old C57/BL6 mice.
3. Mouse cages.

2.6.2 Preimmunization Bleed

1. C57/BL6 mice, housed in cages.
2. Tailveiner restrainer.
3. Anesthesia liquid vaporizer and anesthesia induction chamber.
4. Sterile razor blade.
5. 1000 U/mL heparin sulfate.
6. Sterile gauze.
7. Microcentrifuge tubes.
8. Benchtop centrifuge.
9. Pipettor and sterile pipette tips.

2.6.3 Vaginal Secretion Washings

1. C57/BL6 mice, housed in cages.
2. Nose cone.
3. Insulin syringes.
4. Cage CO_2 diffuser.
5. 2% (v/v) and 5% (v/v) Isoflurane mixtures.
6. Heating pad (chemical or electric).
7. Sterile phosphate buffered saline (PBS), pH 7.4.
8. cOmplete EDTA-free protease inhibitor cocktail (Roche or equivalent).
9. Pipettor and sterile pipette tips.
10. Microcentrifuge tubes.
11. Benchtop centrifuge.
12. CO_2 source.

2.6.4 Immunization and Sampling

1. Concentrated pure sterile isolated RMP-deficient gonococcal nOMVs.
2. Pure LPS.
3. C57/BL6 mice, housed in cages.
4. Nose cone.
5. Insulin syringes.

6. Cage CO_2 diffuser.
7. 2% (v/v) and 5% (v/v) isoflurane mixtures.
8. Tailveiner restrainer.
9. Heating pad (chemical or electric).
10. Sterile phosphate-buffered saline (PBS), pH 7.4.
11. cOmplete EDTA-free protease inhibitor cocktail (Roche or equivalent).
12. Pipettor and sterile pipette tips.
13. CO_2 source.
14. Microcentrifuge tubes.
15. Benchtop centrifuge.

3 Methods

3.1 Growth of *Neisseria gonorrhoeae*

1. Streak Δ*rmp* MS11 strain of *N. gonorrhoeae* on chocolate agar plates and grow overnight in a humidified incubator at 37 °C and 5% (v/v) CO_2 (*see* **Note 1**).
2. Approximately 12–15 h after streaking (*see* **Note 2**) collect bacterial colonies using a sterile synthetic swab and transfer to approximately 15 mL of prewarmed liquid GC broth in a sterile conical tube. Once this initial bacterial suspension reaches an optical density (OD) at $\lambda_{600\ \mathrm{nm}}$ of approximately 0.6 as measured by spectrophotometry, prepare four 20 mL starter cultures in four sterile, vented 50 mL bioreactors by adding approximately 3.2 mL of the bacterial suspension to 16.8 mL of prewarmed liquid GC broth. Place these 20 mL starter cultures in a shaking incubator set for 37 °C and 180 rmp.
3. After cultures reach an OD_{600} of approximately 0.6 (around 4 h of growth) use each 20 mL culture to seed two 60 mL cultures to an OD_{600} of approximately 0.1 by adding 10 mL of starter culture to 50 mL of prewarmed complete GC media in sterile 125 mL vented DeLong flasks. Grow these eight 60 mL cultures in a shaking incubator at 37 °C and 180 rpm until they reach an OD_{600} of approximately 0.6 (~4 h growth).
4. Once the 60 mL cultures reach an OD_{600} of approximately 0.6, pool the cultures and then evenly split them to seed 1500 mL cultures in two separate sterile 2 L vented top Fernbach flasks, that is, 240 mL of bacterial suspension plus 1260 mL of complete GC media. Grow this final culture overnight in a shaking incubator set to 37 °C and 160 rpm.

3.2 nOMV Isolation (See Note 3)

1. Collect the overnight (8–12 h) growth of 3-L of bacterial culture and split it evenly among six 500 mL sealable centrifuge

tubes. Remove the bacteria and large debris from the solution through centrifugation at 4000 × *g* for 30 min at 4 °C.

2. Following centrifugation, remove carefully the low speed supernatant with a Pasteur pipette and filter with a 0.45 μL PES filter to sterilize and further remove large debris.
3. Next, add ~600 mL of sterile 5× nOMV buffer to the low speed supernatant to provide osmotic support for the vesicles.
4. Confirm the sterility of the low speed supernatant by plating 200 μL on a chocolate agar plate and overnight culture in a humidified incubator at 37 °C and 5% (v/v) CO_2. Keep the low speed supernatant in a sealed sterile bottle(s) at 4 °C while sterility is confirmed.
5. Once sterility of the low speed supernatant is confirmed, concentrate the high molecular weight structures by stirred ultrafiltration.
6. Assemble a 400 mL Amicon ultrafiltration cell with a 100,0000 NMWL membrane following the manufacturer's instructions.
7. In a cell culture hood, fill the ultrafiltration cell with the low speed supernatant and seal. Move the cell to an externally vented chemical hood and attached to a source of medical grade gaseous nitrogen. Pressurize the cell pressurized and maintain at 1.78 kg/cm^2 and under continuous stirring, concentrate the low speed supernatant until only ~100 mL of the original 3000 mL remains (*see* **Note 4**).
8. Once the low speed supernatant has been concentrated into ~100 mL of ultrafiltration retentate, stop the ultrafiltration process and depressurize the cell.
9. Divide the ultrafiltration retentate evenly between reinforced high-speed ultracentrifuge tubes, which are weighed to ensure balance. Finally, remove the high molecular weight structures smaller than 45 μm, presumably the isolated vesicles, from the retentate by ultracentrifugation at 100,000 × *g* for 2 h at 4 °C.
10. Following centrifugation, remove gently the supernatant from the presumed vesicle pellet with a pipette and discard. Place a small volume of 1× nOMV buffer on top of the nOMV pellet and allow the vesicles to enter solution while stored overnight at 4 °C (*see* **Note 5**).
11. Consolidate the nOMV colloid into one sterile screw-top cryogenic vial and <0.001% (w/v) sodium azide to prevent fungal growth.
12. Store the nOMV preparation at 4 °C and maintained sterile during all subsequent uses. In these conditions, vesicle integrity is maintained at least a year after isolation (Fig. 2).

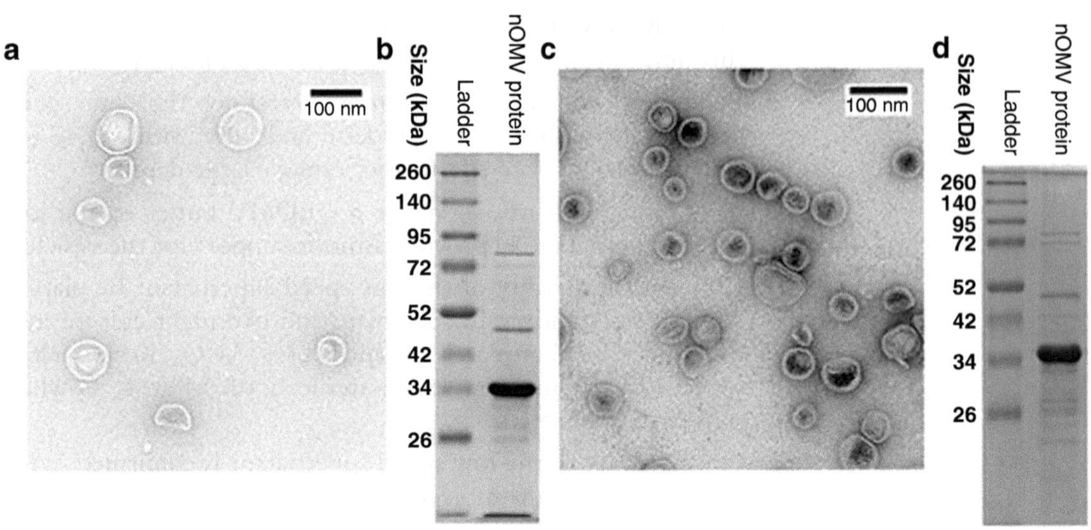

Fig. 2 nOMVs lacking RMP, display stability for up to 1 year in standard 4 °C storage conditions. Vesicles were isolated from a Δ*rmp* strain of GC and evaluated at 2 weeks (**a** and **b**) and 12 months (**c** and **d**) after isolation for morphology by UA negative stain TEM (**a** and **c**) and protein composition by PAGE and Coomassie blue staining (**b** and **d**). PAGE was run using 5 μg of nOMV associated protein (in indicted lanes). Images were processed and scale bars generated in ImageJ. Contrast differences are the result of different image capture methods (CCD camera versus film-based plate camera)

3.3 Antigenic Characterization of nOMVs

Protein concentration of nOMV associated protein is determined by Pierce Modified Lowry protein assay of the suspended nOMV protein following cold acetone extraction/purification.

3.3.1 Protein Quantification

1. Place a small volume (~5 μL) of nOMV suspension in a microcentrifuge tube.
2. Add approximately four-times the volume of clean cold (−20 °C) acetone to the microcentrifuge tube, close the tube and briefly vortex to mix.
3. Place the acetone-nOMV solution in a −20 °C freezer for at least 2 h.
4. Remove sample from the freezer and centrifuge for 10 min at 4 °C and 10,000 × *g* to pellet the protein.
5. Decant the supernatant and allow the protein pellet to air dry.
6. Once dry, suspend the protein pellet in a known volume of PBS and quantify the protein concentration by Pierce Modified Lowry assay, following the manufacturer's instructions.

3.3.2 Polyacrylamide Gel Electrophoresis Separation of nOMV Associated Structures

1. Suspend 10 μg of pelleted cold acetone-prepared nOMV associated protein in 10 μL of loading buffer.
2. Boil the samples for 5 min immediately before loading into wells of a 15% (v/v) polyacrylamide, 10% w/v) SDS gel secured in a properly assembled gel electrophoresis box containing the

manufacturer's recommended volume of electrophoresis buffer (prepared from a 1/10 dilution of the 10× stock).

3. Run the electrophoresis at 60 V for 20 min to concentrate sample in the stacking gel, and then for 90 min at 120 V to separate components in the resolving gel.
4. Disassemble the gel electrophoresis box, remove the gel from its casting-mold and place it in a clean plastic incubation container.
5. Add enough preprepared Coomassie blue stain solution (follow manufacturer's directions for the preparation of the stain) to the incubation chamber to completely cover the gel. Place the chamber on a rocker and allow the gel to stain for 10–20 min (*see* **Note 6**).
6. After staining, drain the Coomassie stain from the gel (*see* **Note 7**) and wash the gel gently four times with ddH_2O.
7. Cover the gel with destaining solution and incubate on a rocker until background staining is removed and the protein bands are clear (*see* **Note 8**).
8. Image gels using digital caption techniques available to you. Example images shown in Figs. 2 and 3, which were captured with a Bio-Rad Gel Doc XR imaging system (*see* **Note 9**).

3.4 Structural Characterization of nOMVs

Directions are for a Brookhaven 90Plus Nanoparticle Analyzer. Refer to operating instructions for the specific equipment available to you.

3.4.1 Evaluation of Isolated nOMV Effective Diameter by Dynamic Light Scattering (DLS)

1. Per manufacturer's directions, turn the analyzer on and allow the laser to warm up for ~15 min. While waiting, start the Brookhaven Instrument Corporation's Particle Sizing Software to collect data.
2. Transfer ~1 mL of nOMV suspension to a cleaned, 1.5 mL polystyrene cuvette.
3. Use the following run parameters: Temperature, 25 °C; Suspension, Water; Viscosity, 0.89 cp; Ref. Index Fluid, 1.330; Angle, 90.00; Wavelength, 657 nm; Dust cutoff, 90; Run duration, 30 s (*see* **Note 10**).
4. After the runs, export a table of observed effective diameters as well as a histogram showing diameter distribution and frequency from the software to a personal computer, as shown in Fig. 4.
5. Collect the sample for future use and shut down the machine.

3.4.2 nOMV Visualization by Negative Staining Transmission Electron Microscopy (TEM)

Methods below are for the preparation of TEM grids for subsequent analysis by electron microscopy (*see* **Note 11**).

1. Produce the grids "in-house" by adhering Formvar support film to 300-mesh copper grids and then coating with atomized

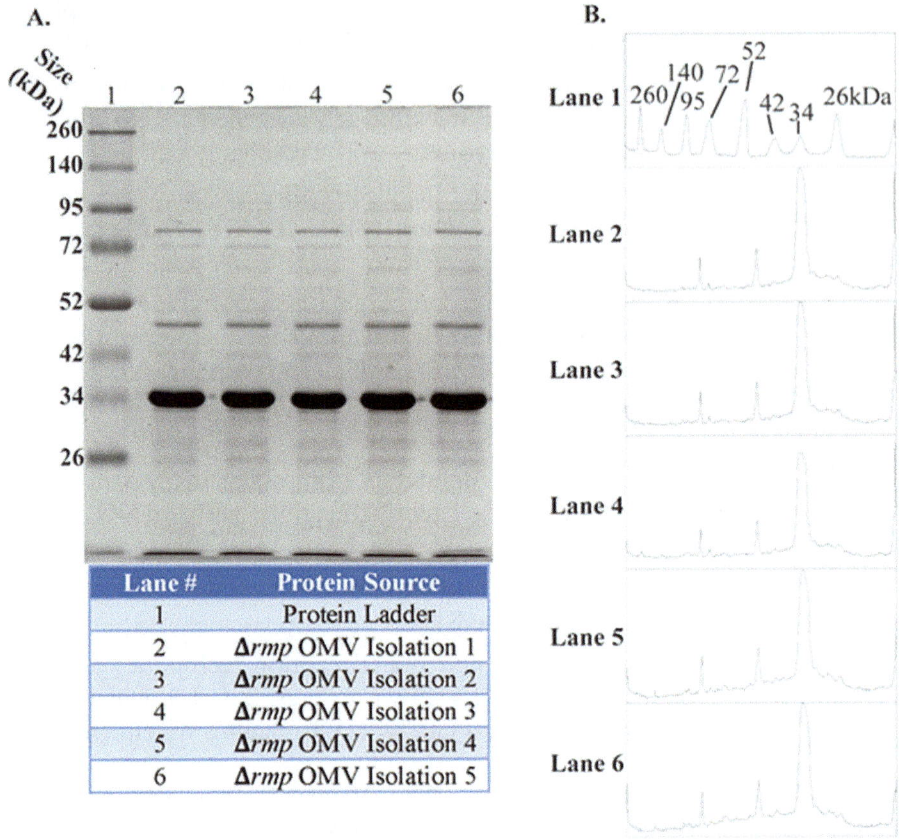

Fig. 3 Protein composition of *Δrmp* nOMVs appears to be consistent between isolations. Proteins (5 μg as measured by modified Lowry protein assay) from Δrmp nOMVs were separated by denaturing polyacrylamide gel-electrophoresis (PAGE). Lane 1 contains the SpectraTM broad range protein ladder. Lanes 2–6 each contain protein from different nOMV isolations. (**a**) Coomassie blue stained polyacrylamide gel. (**b**) The protein banding pattern of each lane from the stained gel was evaluated by ImageJ and visualized by a histogram of pixel density. Moving left to right along the histogram corresponds with high to low molecular weight protein bands

carbon from carbon rods using a vacuum evaporator and store until needed (*see* **Note 12**).

2. Immediately prior to grid use, render the grids hydrophilic by placing them carbon side up in a glow discharge apparatus set at 180 V and expose them to two 30 s plasma discharges under a near vacuum (~0.08 mBar). Use the grids within 1 h (*see* **Note 13**).
3. To adhere and stain the nOMV sample to the hydrophilic carbon surface, first dilute a small volume of the sample of interest (OMVs from a single isolation), if necessary, to obtain a protein concentration of no more than 2 μg protein per 1 μL of suspension.

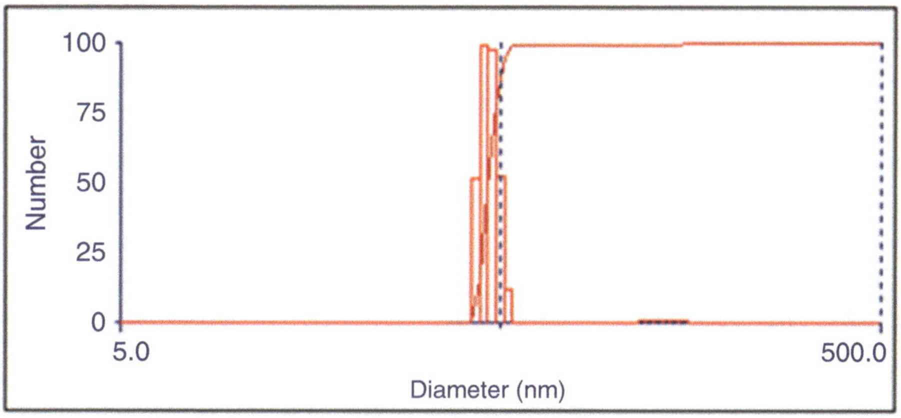

d(nm)	G(d)	C(d)	d(nm)	G(d)	C(d)	d(nm)	G(d)	C(d)
36.8	0	0	63.7	0	99	110.2	0	99
38.7	0	0	66.9	0	99	115.8	0	99
40.6	0	0	70.3	0	99	121.7	1	100
42.7	51	16	73.9	0	99	128.0	1	100
44.9	100	48	77.7	0	99	134.5	1	100
47.2	96	79	81.7	0	99	141.4	0	100
49.6	52	96	85.9	0	99	148.6	0	100
52.2	12	99	90.3	0	99	158.2	0	100
54.8	0	99	94.9	0	99	164.2	0	100
57.6	0	99	99.7	0	99	172.6	0	100
60.6	0	99	104.8	0	99	181.4	0	100

Fig. 4 nOMVs isolated from Δ*rmp Neisseria gonorrhoeae* appear to be a single population with a narrow range of diameters by dynamic light scattering. Approximately 1 mL of resulting material from a single isolation protocol, suspended in a sterile 3% (w/v) sucrose, 0.2 M glycine solution, was analyzed by dynamic light scattering (DLS). Based on light scattering, DLS predicted a single population of particles with a narrow normal distribution of diameters centered around ~45 nm. *d*(nm) is diameters in nanometers, *G*(d) is relative percentage contribution of the size range, *C*(d) is cumulated percentage contribution

4. With the grid held carbon side up using self-holding forceps, gently place 4 μL of the sample on the face of the grid and incubate for 1 min at room temperature.
5. Remove the sample from the grid by gently placing the edge of a piece of clean filter paper perpendicular to the grid's face against the grid edge for approximately 5 s or until most liquid is removed from the grid.
6. Immediately wash the grid 12 times by gently placing a series of drops of 0.2 μm filtered ultrapure deionized H_2O on the grid face and then immediately removing them via gentle suction from a Pasteur pipette (*see* **Note 14**).
7. Following the final wash, remove any remaining water from the grid, using the same perpendicular blotting technique used above, with a clean piece of filter paper.
8. Immediately stain the grid with one drop of a ~2% (w/v) uranyl acetate stain for 1 min.

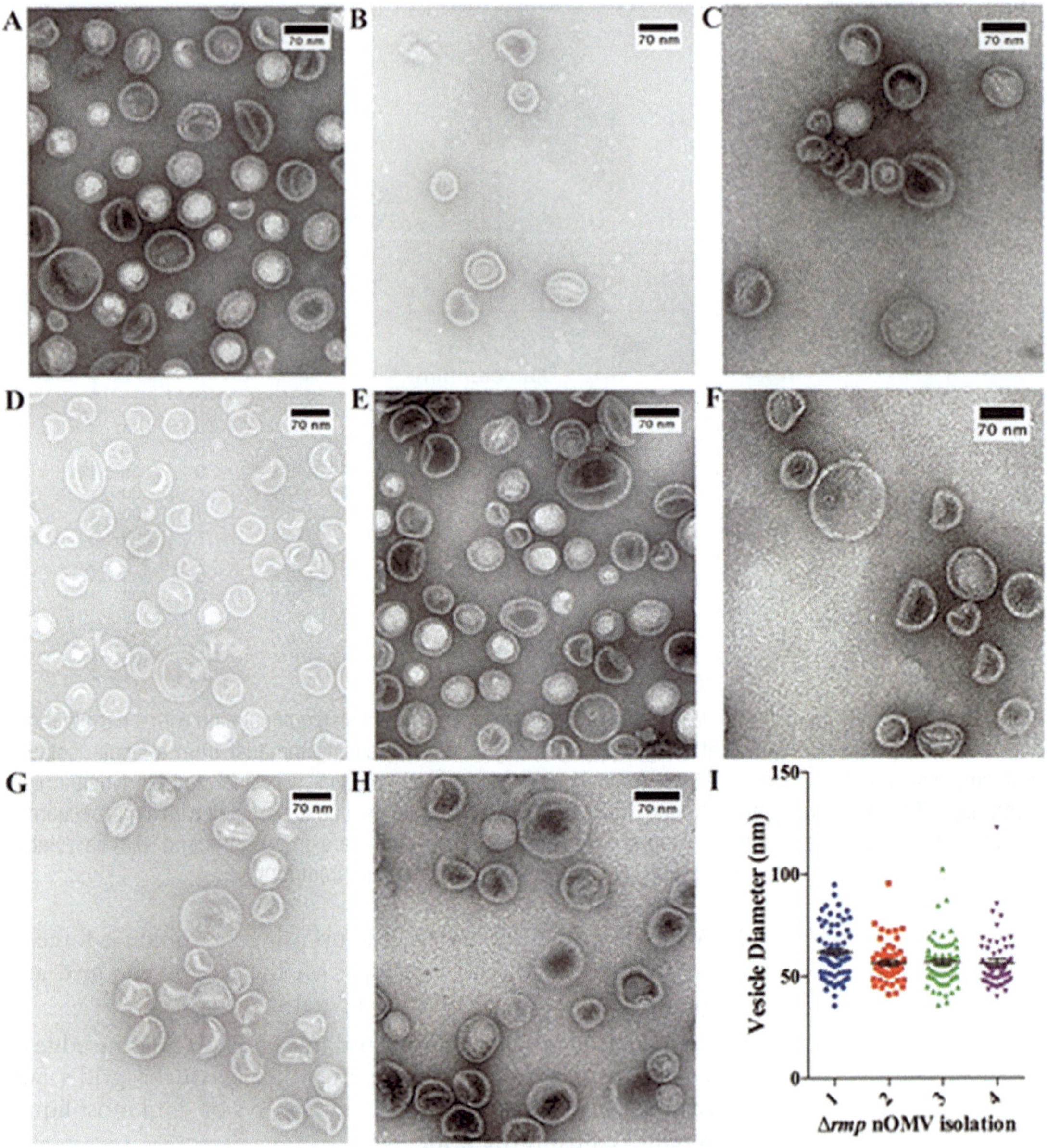

Fig. 5 nOMVs isolated from Δ*rmp* GC have a consistent structure, morphology and size between separate isolations. Vesicles were isolated from large volume growths of Δ*rmp Neisseria gonorrhoeae* in late log growth phase. Electron micrographs (**a**–**h**) are representative images of vesicles each from a different isolation and growth. All samples were negative stained with ~2% UA and 43,750× EM images were captured by CCD camera (**a**, **c**, **e**, **f**, **h**) or built in plate camera (**b**, **d**, **g**). Images were processed and scale bars generated in ImageJ. (**i**) the diameters of 53–67 representative vesicles from four separate isolations were measured by ImageJ; data were analyzed and graphed in Prism. Plot whiskers indicate mean and standard error

9. Remove the stain solution from the grid with a clean piece of filter paper via the same perpendicular blotting technique described in **step 5** above, this time blotting for 10 s.
10. Allow the grid to fully air dry for approximately 3 min before placing in a protective grid case for subsequent examination and imaging by TEM (*see* **Note 15**).
11. Capture the TEM images using proprietary CCD cameras, or by built-in plate camera on EM film (*see* **Note 16**). Typical results demonstrating consistency of isolations of nOMV are shown in Fig. 5.

3.4.3 Morphological Evaluation of nOMVs

1. Identify several digital micrographs from a single nOMV sample taken from different grid squares of the same EM grid (*see* **Note 17**).
2. Using a systematic grid pattern, measure every vesicle with clear, in-focus outer membrane frontiers and do not demonstrate alterations in shape due to external structures (i.e., other vesicles or grid bars) in a given visual field.
3. For each vesicle, take two perpendicular measurements from the outer membrane limit to outer membrane limit (forming an "X" across the vesicle). Average the two diameter measurements to accurately estimate individual vesicle diameter.

3.5 In Vitro Characterization of nOMV Interactions with Innate Immune Receptors

The ability of nOMVs to interact and stimulate cells through the Toll-Like-Receptor (TLR) homodimer TLR4/TLR4 and heterodimer TLR1/TLR2 was evaluated using HEK cells designed to overexpress those receptors in isolation and a control cell line that was transfected with a noncoding vector (pcDNA) [22].

1. For each genotype of HEK cells, seed 24-well plates with TLR 2/TLR1 heterodimer overexpressing, TLR4 homodimer overexpressing and pcDNA transformed control cells, each at a density of 0.5×10^5 cells per well.
2. Incubate the cells overnight at 37 °C with 5% (v/v) CO_2 overnight until confluency is reached.
3. Once confluent, treat each monolayer in the wells for 12 h with a single compound at a given concentration (e.g., 10 μg/mL of nOMV, 1 μg/mL of nOMV, 0.1 μg/mL of nOMV, 100 ng/mL of LPS, 10 μg/mL of P1B, 20 ng/mL of TNF-α, and an untreated control).
4. Following the 12 h stimulation, collect the supernatants and evaluate IL-8 production by ELISA, per manufacturer's protocol. Results demonstrating the innate immune stimulating ability of nOMVs are shown in Fig. 6.

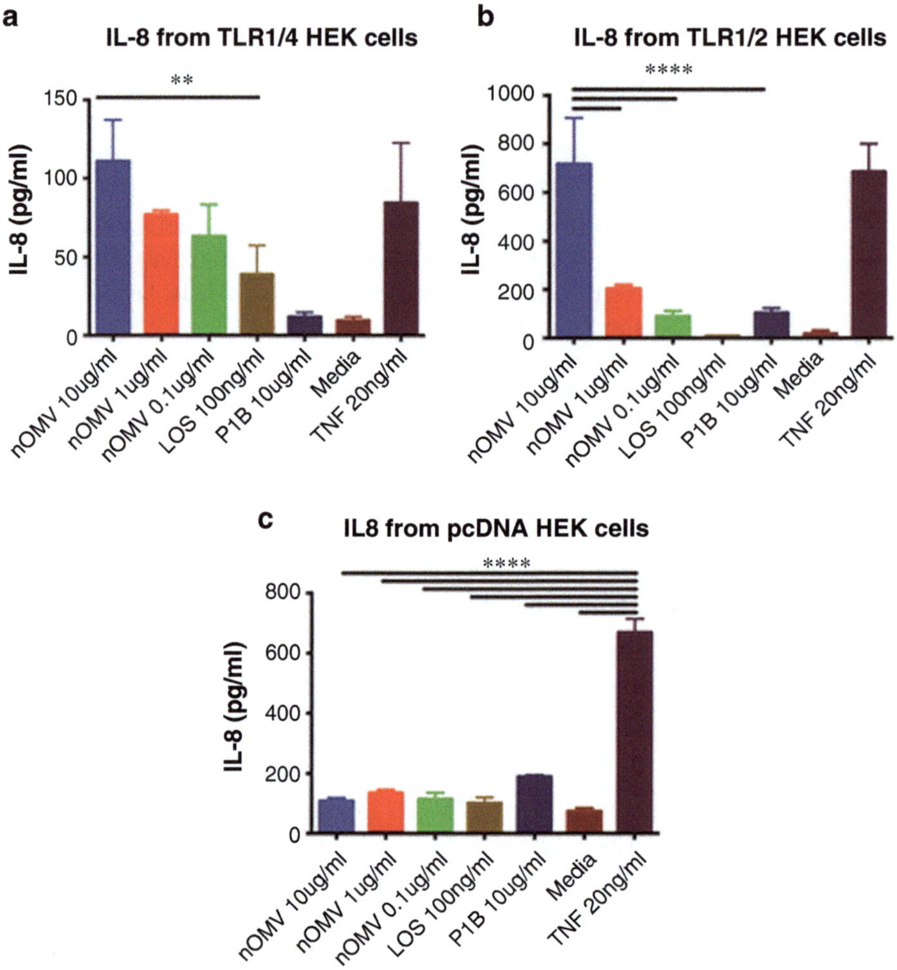

Fig. 6 Δ*rmp* nOMVs bind to and strongly activate TLR1/2 and TLR4 signaling pathways. PRR activation was indirectly measured by secretion of IL-8 by HEK cells overexpressing TLR4 (**a**), TLR1/TLR2 (**b**) or control cells after a 12 h stimulation with 10, 1 or 0.1 μg nOMVs, 100 ng LOS, 10 μg P1B, or 20 ng TNF-α. Histogram shows mean with standard deviation, significance was calculated by one way ANOVA between each condition. *$p < 0.05$, **$p < 0.01$, ***$p < 0.001$, ****$p < 0.0001$ by ANOVA. Data were analyzed and graphs were generated in Prism. Supernatant IL-8 levels were measured by ELISA

3.6 Immunogenicity of nOMVs

3.6.1 Mice

1. Acquire 6- to 8-week-old female wild-type C57/BL6 mice (*see* **Note 18**).
2. Allow mice to acclimate to housing in standard animal facility for 2 weeks after arrival before being used in experiments.
3. Separate the 8- to 10-week-old female C57/BL6 mice into cages based on treatment group (*see* **Note 19**).

3.6.2 Preimmunization Bleed

1. On the day prior to the first immunization, collect sera from all mice from tail veins.

2. Place mouse in a tailveiner restrainer and make a small cut with a sterile blade on the dorsal side of the tail as distal from the base as possible. Collect ~200 μL of blood in a microcentrifuge tube containing 10 μL of 1000 U/mL heparin sulfate.
3. Apply pressure on the incision site with sterile gauze for ~15 s to encourage clotting and then return mouse to their cage.
4. Process whole blood for sera by subsequent centrifugation for 10 min at 10,000 × *g* at 4 °C. Pipette sera from the pellet, dispense into aliquots and freeze at −20 °C for future use.

3.6.3 Vaginal Secretion Washings

1. Anesthetize mouse with isoflurane. Induction should be done in an anesthesia induction box with a 5% (v/v) isoflurane O_2 mixture.
2. Maintain anesthesia with a 2% (v/v) isoflurane mixture delivered through a nosecone at ~0.8 L/min. Make sure to place the anesthetized mouse on a covered heating pad to maintain body temperature throughout the procedure.
3. To collect vaginal secretions, place the anesthetized mouse in a supine position and immobilize with gentle downward pressure on the tail.
4. Gently place the tip of a sterile P20 pipette tip at the dorsal aspect of the external vaginal os and wash the vagina by pipetting up and down three times with 30 μL of sterile PBS containing 1× protease inhibitors.
5. Collect the initial 30 μL of wash into a sterile microcentrifuge tube and use a new clean tip to wash the vagina in the same manner with another 30 μL of sterile PBS containing 1× protease inhibitors.
6. Repeat the washing procedure for a total of five washes resulting in ~150 μL of wash fluid in the sterile microcentrifuge tube. Store on ice until frozen for future analysis.
7. Allow the mice to recover from anesthesia alone in a clean cage before being returned to their home cage.

3.6.4 Immunization and Sampling

On the next day, immunize mice according to their group (*see* **Note 20**).

1. For nOMV vaccines, dilute isolated sterile Δ*rmp* nOMVs in sterile PBS to the desired concentrations. Immunize control mice with normal sterile PBS.
2. For subcutaneous (SC) immunizations, immobilize the mouse using the tailveiner and inject the desired dose of nOMVs in 100 μL just below the skin along the back above the tail. Return the mouse to its home cage.

3. For intranasal (IN) immunizations, anesthetize the mouse with isoflurane exactly as described in **step 1** of Subheading 3.6.3. Immobilize the anesthetized mouse gently in the hand by grabbing the skin at the neck and the lower body. As the mouse is held at a ~30° angle (head up), use a P20 to deliver 10 μL of nOMV vaccine into each nostril (20 μL total per dose). Hold the mouse at this angle until the solution fully enters the nasal passages. Place the mouse in a clean cage by itself to recover and then moved to its home cage.
4. Collect serum and vaginal secretions from all mice, 2 weeks following the final immunization, as described in Subheadings 3.6.2 and 3.6.3 respectively (*see* **Note 21**).
5. Following sample collection, sacrifice mice in the manner consistent with the standards set by the laboratory animal use guidance body of your institute (*see* **Note 22**).

4 Notes

1. Culture of *Neisseria gonorrhoeae* must take place in biosafety level 2+ biosafety cabinets. Lab personnel must also wear appropriate personal protective equipment (double-layered gloves, goggles, full coverage laboratory coat, and mask) at all times.
2. Do not grow bacteria for longer than 18 h, as *N. gonorrhoeae* will start to lyse due to overgrowth.
3. The process of nOMV isolation should be completed in a biosafety level 2+ biosafety cabinet for steps prior to confirmation of sterility and in a tissue culture hood for steps following, in order to maintain a clean vesicle preparation.
4. This will require periodic pauses in the process to depressurize the cell, transfer to a clean culture hood, refill with low speed supernatant and finally repressurize and restart the filtration process.
5. It is not unusual for gentle pipetting to be necessary to fully suspend the vesicle pellet after overnight incubation.
6. To facilitate dissolving the Coomassie stain particles, the solution can be sonicated and then remaining particles removed by filtration with Whatman 3MM filter paper.
7. Dispose of waste in accordance to your institutions regulations.
8. This may require periodic replacement of old destaining solution with fresh solution.
9. Alternatively, gels can be dried on a frame (Owl Separation System, NH) between two sheets of drying paper (Promega, WI, USA), per manufacturer's protocol.

10. At least five runs are recommended with a final average documented.
11. We encourage researchers to work with institutional cores or individuals with experience for final visualization of structures by electron microscopy.
12. Several commercial products with formvar supported carbon layers are available; discuss with your EM core facility to determine the best option for you.
13. Grids need to be used within 1 h because the hydrophilic ability of the grid will fade over time after treatment.
14. Be sure not to let the grid completely dry between drops.
15. We achieved best results with a Phillips CM12 Transmission Electron Microscope configured with a lanthanum hexaboride (LaB6) crystal electron source set to an accelerating voltage of 120 kV and liquid Nitrogen cold finger to visualize the uranyl acetate stained samples.
16. Images can be captured using either TVIPS TEITZ (Gauting, Germany) 1Kx1K CCD camera using EMmenu software (Version 4.0.9.83), TVIPS TEITZ TemCam F216 2Kx2K CCD camera using the same software, or by the built-in plate-camera (Phillips CM12) on electron image film (Kodak, SO-163 Film) developed according to manufacturer protocol and solutions.
17. Exported images from the EM software should have accurate image scale embedded. Despite this we recommend double checking that the scale is accurate and ImageJ's measure function is working properly by drawing a line along the length of the image scale bar and directing ImageJ to "measure" (function is found under the "analyze" menu tab).
18. Mice used in our studies were purchased from Jackson Laboratories; mice can also be bred "in-house" or purchased from sources appropriate to country.
19. No more than five mice per cage should be allowed, and multiple cages per treatment group should be used if necessary.
20. Based on our previous studies of P1B immunization, we utilize a 3-immunization schedule during which immunizations are separated by 2 weeks and sera is collected the day prior to all immunizations.
21. Frozen sera and vaginal secretions can be utilized in a variety of assays including ELISA, western blotting, and multiplex assay, among others.
22. In our studies, mice are sacrificed by CO_2 asphyxiation followed by cervical dislocation.

Acknowledgments

The authors want to thank the lab of Dr. Esther Bullitt and Donald Gantz for their help with electron microscopy. Thank you to Neelou Etesami for her help with editing. Thank you to the NIH/-NIAID for the funding to support this work (2R01AI103400-05 awarded to L.M.W.).

References

1. Bos MP, Robert V, Tommassen J (2007) Biogenesis of the gram-negative bacterial outer membrane. Annu Rev Microbiol 61:191–214. https://doi.org/10.1146/annurev.micro.61.080706.093245
2. Dorward DJ, Judd RC (1988) The isolation and partial characterization of naturally-evolved outer membrane blebs of *Neisseria gonorrhoeae*. In: Poolman JZ, Meyer TF, Heckels JE, PRH M, Smith H, Beuvery EC (eds) Gonococci and meningococci. Kluwer Academic Publishers, Dordrecht, pp 349–356
3. Chatterjee SN, Das J (1967) Electron microscopic observations on the excretion of cell-wall material by *Vibrio cholerae*. J Gen Microbiol 49(1):1–11. https://doi.org/10.1099/00221287-49-1-1
4. Kulp A, Kuehn MJ (2010) Biological functions and biogenesis of secreted bacterial outer membrane vesicles. Annu Rev Microbiol 64:163–184. https://doi.org/10.1146/annurev.micro.091208.073413
5. Vanaja SK, Russo AJ, Behl B et al (2016) Bacterial outer membrane vesicles mediate cytosolic localization of LPS and caspase-11 activation. Cell 165(5):1106–1119. https://doi.org/10.1016/j.cell.2016.04.015
6. Pettit RK, Judd RC (1992) The interaction of naturally elaborated blebs from serum-susceptible and serum-resistant strains of *Neisseria gonorrhoeae* with normal human serum. Mol Microbiol 6(6):729–734
7. Kaparakis M, Turnbull L, Carneiro L et al (2010) Bacterial membrane vesicles deliver peptidoglycan to NOD1 in epithelial cells. Cell Microbiol 12(3):372–385. https://doi.org/10.1111/j.1462-5822.2009.01404.x
8. Lindmark B, Rompikuntal PK, Vaitkevicius K et al (2009) Outer membrane vesicle-mediated release of cytolethal distending toxin (CDT) from *Campylobacter jejuni*. BMC Microbiol 9:220. https://doi.org/10.1186/1471-2180-9-220
9. Koeppen K, Hampton TH, Jarek M et al (2016) A novel mechanism of host-pathogen interaction through sRNA in bacterial outer membrane vesicles. PLoS Pathog 12(6):e1005672. https://doi.org/10.1371/journal.ppat.1005672
10. Zielke RA, Wierzbicki IH, Weber JV et al (2014) Quantitative proteomics of the *Neisseria gonorrhoeae* cell envelope and membrane vesicles for the discovery of potential therapeutic targets. Mol Cell Proteomics 13(5):1299–1317. https://doi.org/10.1074/mcp.M113.029538
11. Olofsson A, Vallstrom A, Petzold K et al (2010) Biochemical and functional characterization of *Helicobacter pylori* vesicles. Mol Microbiol 77(6):1539–1555. https://doi.org/10.1111/j.1365-2958.2010.07307.x
12. McMahon KJ, Castelli ME, Garcia Vescovi E et al (2012) Biogenesis of outer membrane vesicles in *Serratia marcescens* is thermoregulated and can be induced by activation of the Rcs phosphorelay system. J Bacteriol 194(12):3241–3249. https://doi.org/10.1128/JB.00016-12
13. Yanez-Mo M, Siljander PR, Andreu Z et al (2015) Biological properties of extracellular vesicles and their physiological functions. J Extracell Vesicles 4:27066. https://doi.org/10.3402/jev.v4.27066
14. Claassen I, Meylis J, van der Ley P et al (1996) Production, characterization and control of a *Neisseria meningitidis* hexavalent class 1 outer membrane protein containing vesicle vaccine. Vaccine 14:1001–1008
15. Zhu W, Thomas CE, Chen CJ et al (2005) Comparison of immune responses to gonococcal PorB delivered as outer membrane vesicles, recombinant protein, or Venezuelan equine encephalitis virus replicon particles. Infect Immun 73(11):7558–7568. https://doi.org/10.1128/iai.73.11.7558-7568.2005
16. Keiser PB, Biggs-Cicatelli S, Moran EE et al (2011) A phase 1 study of a meningococcal native outer membrane vesicle vaccine made from a group B strain with deleted lpxL1 and synX, over-expressed factor H binding protein,

two PorAs and stabilized OpcA expression. Vaccine 29(7):1413–1420. https://doi.org/10.1016/j.vaccine.2010.12.039

17. Post DM, Zhang D, Eastvold JS et al (2005) Biochemical and functional characterization of membrane blebs purified from *Neisseria meningitidis* serogroup B. J Biol Chem 280 (46):38383–38394. https://doi.org/10.1074/jbc.M508063200. M508063200 [pii]
18. van de Waterbeemd B, Mommen GP, Pennings JL et al (2013) Quantitative proteomics reveals distinct differences in the protein content of outer membrane vesicle vaccines. J Prot Res 12(4):1898–1908
19. Lappann M, Otto A, Becher D et al (2013) Comparative proteome analysis of spontaneous outer membrane vesicles and purified outer membranes of *Neisseria meningitidis*. J Bact 195(19):4425–4435. https://doi.org/10.1128/jb.00625-13
20. van de Waterbeemd B, Streefland M, van der Ley P et al (2010) Improved OMV vaccine against *Neisseria meningitidis* using genetically engineered strains and a detergent-free purification process. Vaccine 28(30):4810–4816. https://doi.org/10.1016/j.vaccine.2010.04.082
21. van der Pol L, Stork M, van der Ley P (2015) Outer membrane vesicles as platform vaccine technology. Biotechnol J 10(11):1689–1706. https://doi.org/10.1002/biot.201400395
22. Chow JC, Young DW, Golenbock DT et al (1999) Toll-like receptor-4 mediates lipopolysaccharide-induced signal transduction. J Biol Chem 274(16):10689–10692

Chapter 10

Transformation in *Neisseria gonorrhoeae*

Melanie M. Callaghan and Joseph P. Dillard

Abstract

Laboratory techniques for transformation of the pathogenic *Neisseria* are well developed, and take advantage of the natural transformability of these species. More recently, these techniques have been successfully applied to some nonpathogenic species of *Neisseria* as well. This chapter provides foundational information on the mechanism of *Neisseria* transformation, considerations for DNA transformation substrate design, two methods for transforming *Neisseria* in the laboratory, and guidelines for identifying successful transformants.

Key words Natural transformation, Type IV pili, Screening, Competence

1 Introduction

The natural competence of the *Neisseria* spp. facilitates their extensive horizontal gene transfer, especially intragenus genetic exchange [1]. In *Neisseria gonorrhoeae*, natural transformation contributes to their panmictic population structure and the spreading of antibiotic resistance observed in the clinic. Gonococci are capable of transformation during all growth phases [2], which makes them easy to genetically manipulate in the laboratory. The techniques used to transform gonococci have also been successfully applied to *Neisseria meningitidis* and some commensal *Neisseria*, allowing for more extensive research into these species as well [3–6]. The cellular transformation process can be conceptualized into three steps: binding, uptake, and recombination [7].

Like many other competent bacteria, *N. gonorrhoeae* rely on type IV pili to bind single- and double-stranded exogenous DNA (*see* **Note 1**) [8]. These filamentous surface structures are primarily polymers of the major pilin subunit PilE, with a variety of minor pilin subunits [7, 9, 10]. One of these minor pilins, ComP, is the cognate receptor for the DNA Uptake Sequence (DUS), a 10–12 bp sequence that ComP binds specifically. This interaction creates a degree of species specificity in neisserial DNA uptake, as

Myron Christodoulides (ed.), *Neisseria gonorrhoeae: Methods and Protocols*, Methods in Molecular Biology, vol. 1997, https://doi.org/10.1007/978-1-4939-9496-0_10,

some different species within the *Neisseria* have and recognize variations of the uptake sequence [11–13] (*see* **Note 2**).

After binding DNA, the pilus motor PilT is required to forcefully pull DNA into the proximity of the cell, occurring at random locations around the cell envelope [2, 14–16]. Nonspecific linearization of plasmid DNA by putative extracellular endonuclease activity has been observed at this stage [17]. PilT retraction makes the cavity of the outer membrane pore, PilQ, available to bind a small DNA loop, which enters the periplasm [18]. Many copies of the soluble periplasmic protein ComE bind the incoming DNA. These proteins compact DNA, pulling it in through PilQ and continually binding and compacting the DNA molecule as it enters the cell. The amount of DNA that can be held in the periplasm at any given time is dependent and proportional to the amount of ComE protein present, and the maximal carrying capacity of the gonococcal periplasm has been estimated to be about 40 kb [16]. From the periplasm, a single strand of DNA is driven through the inner membrane pore ComA to enter the cytosol [18].

After the DNA has been shuttled through both membranes, the newly acquired genetic information is available for chromosomal integration by homologous recombination. *N. gonorrhoeae* has multiple recombination systems; while *recA* is the only identified essential gene for homologous recombination, *recBCD* mutations also decrease the frequency of homologous recombination events [9, 19].

Transformation can be accomplished in the laboratory on solid (Method 1) and in liquid (Method 2) medium. Plate transformations, also called spot transformations, are done by spreading bacteria over a dried spot of DNA on an agar plate [20]. This method is very simple and the most efficient [21]. Liquid transformations are done by exposing concentrated bacteria to the DNA of interest in magnesium-supplemented media, then diluting and allowing for recovery in tissue culture flasks for several hours before plating. This method is slightly more involved and less efficient than spot transformation [21], but takes fewer days and allows for quantitative comparisons of transformation efficiency. Both of these techniques use the same preparations and screening techniques.

2 Materials

2.1 Spot Transformation

1. DNA Substrate: Purified chromosomal DNA, circular or linear plasmid DNA, Polymerase Chain Reaction (PCR) products, or synthesized DNA fragment(s).
2. GCB agar plates: Add 36.25 g of GC Medium Base and 1.25 g of agar to 1 L of ddH_2O, add a stir bar to dissolve the solids on a magnetic stir plate, and autoclave for 40 min at 121 °C and

Table 1
Recommended antibiotic concentrations given in μg/mL for selection in several species of *Neisseria*

Antibiotic	*N. gonorrhoeae* MS11	*N. gonorrhoeae* FA1090	*N. meningitidis* ATCC 13102	*N. musculi* AP2031	*N. sicca* ATCC 29256	*N. mucosa* ATCC 25996
Erythromycin	10	2	10 [3]			
Chloramphenicol	10	2	10 [3]	2.5 [6]		
Spectinomycin	75	75				
Streptomycin	100	100	100 [3]	100 [4]		
Kanamycin	80	80	80	50 [4]	40 [5]	40 [5]
Tetracycline	2	0.2				
Nalidixic acid	3	1				
Rifampicin				50 [6]		

$2.68\ kg/cm^2$. The medium may appear flaky. After cooling to ~50–55 °C, add 10 mL of Kellogg's Supplements I and II, combined. Add any desired antibiotic (Table 1). Stir until mixed completely and pour the plates immediately. Allow the agar to set at room temperature and store at 4 °C.

3. Kellogg's Supplements I and II, combined: Add 40 g of glucose (dextrose), 1 g of glutamine, 2 mg of cocarboxylase (thiamine pyrophosphate), and 5 mg of $Fe(NO_3)_3{\cdot}9H_2O$ to 100 mL of ddH_2O. Filter sterilize through a 0.22 μm filter with a syringe. Store at −20 °C until needed. Although Kellogg's supplements were originally designed as two separate solutions, both supplements can be made and stored as a single solution (*see* **Note 3**).
4. Dacron swabs.
5. Humidified incubator set to 37 °C with 5% (v/v) CO_2.
6. 10 μL sterile inoculating loops.
7. Glycerol (to make stocks of successful transformants).
8. SpeedVac (optional).
9. Class II Biosafety Cabinet.
10. Cryovials.

2.2 Liquid Medium Transformation

1. DNA Substrate.
2. GCBL medium: add 15 g of proteose peptone #3, 4 g of K_2HPO_4, 1 g of KH_2PO_4, and 1 g of NaCl to 1 L of double-distilled (dd)H_2O. Check that the pH is 7.2 and autoclave the medium for 30–35 min at 121 °C and $2.68\ kg/cm^2$. For every 100 mL of medium, add 1 mL of 0.5 M sodium bicarbonate solution (to give a final concentration of sodium bicarbonate of 0.005 M, 0.042% w/v) and 1 mL of Kellogg's Supplements I and II, combined, before use.
3. GCB agar plates (*see* Subheading 2.1.2).
4. Kellogg's Supplements I and II, combined (*see* Subheading 2.1.3).
5. Dacron swabs.
6. Humidified incubator set to 37 °C with 5% (v/v) CO_2.
7. 37 °C water bath or heat block (optional).
8. 1.5 mL microcentrifuge tubes or 15 mL screw-cap tubes.
9. T25 cm^2 cell culture flasks.
10. 100 mM magnesium sulfate ($MgSO_4$) stock solution in distilled H_2O.
11. 5 mM $MgSO_4$ in GCBL medium.
12. Glycerol (to make stocks of successful transformants).

2.3 Colony PCR Screening and Selection

1. Colony lysis solution: 20 mM Tris–HCl buffer, pH 8.5, containing 1% (v/v) Triton X-100 and 2 mM of ethylenediaminetetraacetic acid (EDTA).
2. 10 μL sterile inoculating loops.
3. Sterile toothpicks.
4. PCR materials: sterile ddH_2O, any polymerase and corresponding buffer, 20 mM dNTP stock, template DNA from colony lysis, forward and reverse primers, thermocycler, PCR tubes (*see* **Note 4**).
5. Agarose gel electrophoresis materials: 0.8% (w/v) agarose in TBE buffer (adjust percentage accordingly if PCR fragments of interest are very small or very large), 1× TBE buffer, ethidium bromide for visualization, linear DNA standard ladder (*see* **Note 5**).

3 Methods

There are two common approaches for transforming *Neisseria*: spot transformation on agar plates and liquid transformation. *See* **Note 6** for more information about growing gonococci in the laboratory.

3.1 Spot Transformation

The primary transformation method used in the laboratory is spot transformation, wherein bacteria are grown on an agar plate in the presence of concentrated DNA. This technique is done most efficiently using 100–1000 ng of total DNA in a ~1 cm diameter spot. The entire transformation process, from the first revival of bacteria to the final stock of the new strain takes 6 days (Fig. 1a).

1. On day 1, prepare the DNA substrate. Purified chromosomal DNA, PCR products, and synthesized DNA fragments can all be used without modification (*see* **Note 7**). Plasmids under 8 kb in size should be linearized before transformation to achieve genetic modification by a double-crossover linear recombination event (*see* **Note 8**). Aim to maximize the final concentration of prepared DNA. *See* **Note 9** for information regarding transformation substrate design.
2. Revive bacteria from −80 °C stocks by scraping a small amount of a frozen stock of piliated *N. gonorrhoeae* with a pipet tip and dab onto a prewarmed, nonselective GCB agar plate (*see* **Notes 10** and **11**). Use a small amount of frozen stock, as single colonies are desired (*see* **Note 12**). Use a Dacron swab or 10 μL inoculating loop to spread the suspension across the plate surface. Incubate for approximately 20 h in a humidified atmosphere at 37 °C with 5% (v/v) CO_2.
3. On day 2, spot the substrate by drawing a small circle on the bottom of a GCB plate and pipetting 10–20 μL of DNA

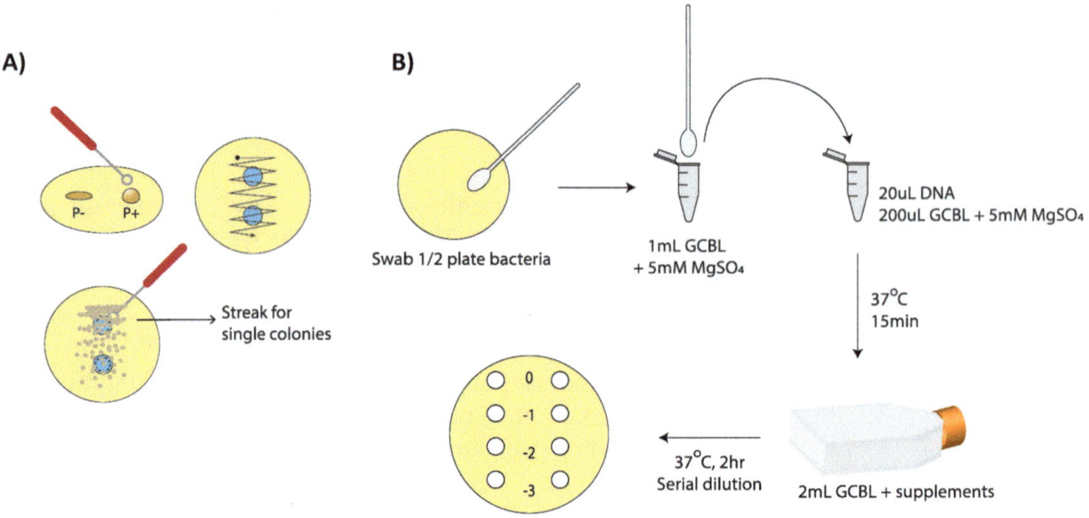

Fig. 1 Schematics of (**a**) spot and (**b**) liquid transformation protocols

substrate solution into the circle. Place the plate(s) in a biosafety cabinet to dry with the lids removed. If you are doing multiple transformations, they can be done on a single sectored plate. Maximal DNA concentration and purity are preferred, with a final quantity of DNA in the 100–1000 ng range (*see* **Note 13**). If the prepared substrate concentration is too low, concentrate it using a SpeedVac concentrator, or pipette aliquots of 20 μL multiple times, allowing each spot to dry before applying a fresh aliquot on top of it.

4. Streak a single, piliated colony of *N. gonorrhoeae* using an inoculating loop onto the new plate directly over the circle of DNA. Avoid nonpiliated colonies, as the pilus is required for transformation (*see* **Note 11**). Incubate for approximately 20 h in a humidified atmosphere at 37 °C with 5% (v/v) CO_2.
5. On day 3, streak the bacteria growing within the circles of DNA onto a new GCB plate, using an inoculating loop to streak for single colonies. If the transformants are selectable, streak onto a GCB plate containing the antibiotic or other selection agent. If transformants are not selectable, streak onto a nonselective GCB plate, taking extra care to avoid taking any bacteria from an area of the overnight plate that did not contain DNA. Incubate for approximately 20 h in a humidified atmosphere at 37 °C with 5% (v/v) CO_2.
6. On day 4, patch single transformant colonies onto a fresh nonselective GCB or GCB + selection plate using sterile toothpicks. Patching is necessary, as a single *N. gonorrhoeae* colony does not contain enough bacteria to screen the transformants reliably. Grow the patches overnight in a humidified atmosphere at 37 °C with 5% (v/v) CO_2.

7. On day 5, screen the patches as described in Subheading 3.3 below (*see* **Note 14**). Bacterial patches provide sufficient DNA for PCR-based screening and sequencing. Streak the successful, confirmed transformants with an inoculating loop onto at least one half of a new GCB plate.
8. On day 6, transfer the overnight plate growth of the new transformed *N. gonorrhoeae* strain into 2 mL of prewarmed GCBL medium using a Dacron swab. Add glycerol to the bacterial stock to a final concentration of 15% (v/v) in GCBL medium, place into labeled cryovials, and freeze immediately. Store transformed bacteria at −80 °C or under liquid nitrogen.

3.2 Liquid Medium Transformation

In some cases it may be more practical to transform *Neisseria* in liquid culture instead of on plates, for example when aiming to quantify transformation frequency. For liquid transformation, supplemental magnesium is advantageous and a selectable transformation marker is highly recommended. The entire transformation process, from the first revival of bacteria to the final stock of the new strain takes 5 days (Fig. 1b).

1. On day 1, prepare the DNA substrate and revive the gonococci as described in **steps 1** and **2** of Subheading 3.1.
2. On day 2, combine prepared DNA substrate with 200 μL of GCBL medium containing 5 mM $MgSO_4$. This solution should contain between 200 and 1000 ng of DNA, and no more than 20 μL of DNA solution should be added in total. Warm this solution to 37 °C in an incubator or water bath.
3. Add 1 mL of prewarmed GCBL medium containing 5 mM $MgSO_4$ into a 1.5 mL microcentrifuge tube or a 15 mL screw-cap tube. Ensure that it remains at approximately 37 °C by either pre-warming the tube or incubating the aliquot again until warm. Transfer about one-half of the growth of piliated *Neisseria* from the GCB agar plate using a Dacron swab (*see* **Note 11**). Less bacterial growth (down to one-quarter of the growth from a GCB agar plate) can be used, if overnight growth is very dense. Add 20 μL of this cell suspension to the DNA-medium preparation.
4. Incubate the DNA–bacteria mixture at 37 °C in a 1.5 mL microcentrifuge tube for 15 min without aeration. After this initial incubation, add the transformation mixture to a pre-warmed T25 cm^2 cell culture flask containing 2 mL of GCBL medium with Supplements I and II ($MgSO_4$ supplementation is optional) and incubate at 37 °C in an atmosphere of 5% (v/v) CO_2 for 2 h.
5. To isolate single colonies of transformants, serially dilute the transformation mixture ten-fold in GCBL medium (in microcentrifuge tubes containing 900 μL of GCBL

medium, add 100 μL of transformation mixture) and plate 100 μL of the appropriate dilutions onto GCB agar plates. For antibiotic selection, ten-fold serial dilution between 10^0 and 10^{-4} is generally sufficient, whereas for plating on GCB agar without a selectable marker, ten-fold serial dilution between 10^{-3} and 10^{-6} is often necessary to isolate single colonies.

6. Follow **steps 6–8** from Subheading 3.1.

3.3 Colony PCR Screening and Selection

After patching single colonies of *Neisseria* transformants, colony PCR (and restriction digestion, if necessary) can be used to determine which transformants have the desired mutation (Fig. 2).

1. Swipe about half of a patch of overnight bacterial growth using an inoculating loop or sterile toothpick into 50–100 μL of Colony Lysis Solution [22] to lyse the bacteria. Incubate the cell suspension for 15 min at 94 °C in a heat block, thermocycler, or water bath.
2. Set up a PCR reaction with the desired polymerase and specific primers for the target gene, using the lysed bacterial suspension as the DNA template for amplification. The lysed bacterial suspension should constitute no more than 1/10 of the total reaction volume (*see* **Note 4**).
3. For detecting insertions or deletions, analysis of PCR products by size is usually sufficient (*see* **Note 5**). For insertions large enough to be detected by agarose gel electrophoresis, the use of primers flanking the insertion site is appropriate. To detect smaller insertions, combinations of one flanking and one internal primer give clearer results. For detecting point mutations, design the mutation to either add or remove a restriction site. If the point mutation introduces an antibiotic resistance, transformants can be selected for as described in Subheading 3.1 and 3.2.
4. Sequence to confirm the expected mutation. If the point mutation does not confer a resistance or other apparent phenotype, do a PCR with primers that flank the mutation and surrounding region. Digest the PCR products with the enzyme corresponding to the altered restriction site, and screen for cutting (or lack thereof) accordingly. Include controls for PCR and digest using the parent strain for comparison.

4 Notes

1. The importance of the DNA substrate being single-stranded (SS) or double-stranded (DS) is an unresolved question in *Neisseria* transformation. Studies have had variable results

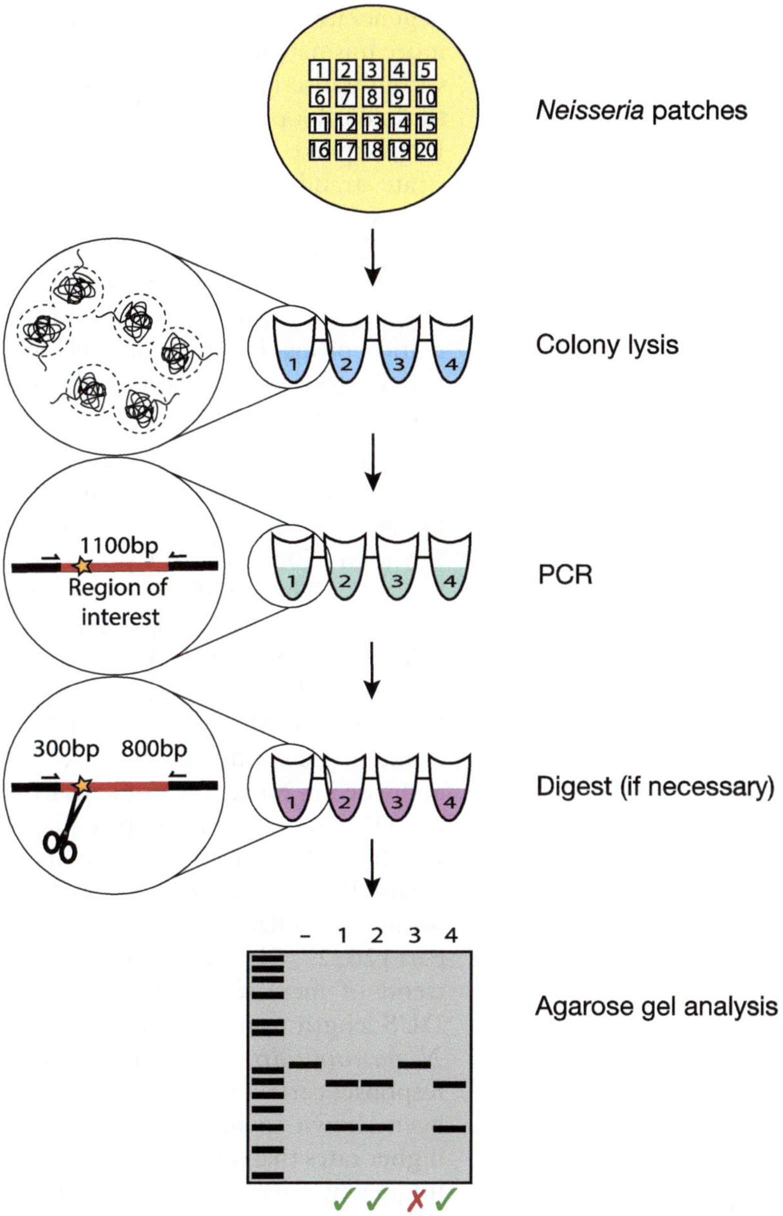

Fig. 2 Schematic of the screening method

when comparing transformation efficiency by SS- and DS-DNA; while in some cases they transformed with comparable efficiency, others observed vastly more efficient transformation by DS-DNA [23–25]. These studies were all done with DNA of different lengths derived from variable sources, which could explain the different results. Additionally, approximately half of DUSs are found in inverted repeats (IR), wherein two proximal DUSs can pair, or a single DUS is found near a degenerate counterpart that still allows pairing with one or more mismatches. A vast majority of these IR-DUS are located within transcription terminators [26, 27]. The potential role of this stem loop formation within transcription terminators for DNA uptake is speculative and reports regarding DNA substrate stranded-ness have variable findings. Although the presence of DUSs existing as IRs or direct repeats on double-stranded plasmid DNA did not alter transformation over a single DUS in *N. meningitidis* [27], the addition of a DS-DUS to a 300 bp SS-oligonucleotide did increase transformation probability in *N. gonorrhoeae* [25], indicating a potential for pairing of repeats only when the DNA substrate is single-stranded.

2. The nature of the DNA substrates that gonococci take up has been a topic of study for many years. *Neisseria* preferentially uptake DNA from the same or similar species, as facilitated by the pilin ComP-DNA Uptake Sequence (DUS) interaction [8, 12]. The DUS is found frequently throughout the genome; in *N. gonorrhoeae* and *N. meningitidis* this sequence is 5′-ATGCCGTCTGAA-3′ [26, 27] and it occurs on average once per 1.1 kb, comprising approximately 1% of the genome [13, 28–30]. Multiple "dialects" of this sequence have been found in preferential use by different *Neisseria* species, which is addressed in **Note 9** (Commensals) [13]. The first discovery of the DUS reported a 10 bp sequence, lacking the leading AT, which is sufficient for DNA uptake [11]. Around 76% of *Neisseria* DUS contain the additional 2 bp at the beginning, which somewhat enhances uptake frequency over its 10 bp counterpart [26, 27, 31]. Testing by Ambur et al. [27] found a general trend of increased transformation frequency with increased DUS length from 8 bp to 12 bp for both *N. gonorrhoeae* and *N. meningitidis*. Species and strain differences may alter DUS response; certain gonococcal strains (FA19 and FA1090) have been shown to uptake DNA lacking a DUS at significantly higher rates than others (MS11 and 1291), and transformation frequency enhancement with a 12 bp DUS over a 10 bp DUS varies from 0-fold to >30-fold in studies using different *Neisseria* strains [27, 31]. For maximal transformation success in the laboratory setting, it is recommended to use transforming DNA containing the species-appropriate 12 bp DUS. For *N. gonorrhoeae*, plasmids such as pUP1 and pIDN1 containing the 12 bp DUS have already been made [32, 33].

3. The large quantity of glucose is slow to dissolve; leave ample time for stirring. This solution should be stored at −20 °C; the iron and glucose will react over time unless frozen. Add 1 mL of the Supplements I and II solution for every 100 mL of GCBL medium.

4. A sample colony PCR reaction (total volume 25 μL) contains 16 μL of ddH_2O, 2.5 μL of DNA from colony lysis, 2.5 μL of Taq polymerase buffer, 2.5 μL of 20 mM dNTPs, 0.5 μL of forward primer, 0.5 μL of reverse primer, and 0.5 μL of Taq polymerase.
5. Use agarose gel electrophoresis to determine the sizes of PCR products obtained from each potential transformants. 0.8% (w/v) agarose gels are recommended for screening, although in rare cases of extremely small or large PCR products this concentration should be adjusted. Run a linear DNA standard ladder, PCR reaction using the same primers and the parent strain or construct as a template, and all colony PCR reactions from the putative transformants to allow for easier size determination and comparison. Any standard DNA visualization method is appropriate.
6. *Neisseria* are fastidious bacteria that are sensitive to changes in temperature and nutrient availability. For maximal growth in a laboratory setting, prewarm all media, plates, and tubes to 37 °C before inoculation. Minimize the amount of time that plates and cultures are at room temperature and return them to 37 °C as soon as possible. In liquid culture, *Neisseria* do not survive in stationary phase for an extended period of time. For overnight growth, dab a small amount of the glycerol stock onto a GCB agar plate and spread it uniformly across the surface. Revival from frozen stocks is best done without antibiotics. The ideal time period for growth on GCB agar plates for *N. gonorrhoeae* is between 18–20 h; after 24 h viability begins to decrease. These growth patterns may be shifted later in slower growing species, such as *N. musculi* AP2031, or when grown on certain antibiotics, such as chloramphenicol.
7. The most common substrates used to create laboratory constructs of *Neisseria* are linearized plasmids and chromosomal DNA. However, PCR-amplified DNA fragments and synthesized gene blocks are also effective.
   (a) Linear plasmids: Mutations and constructs are often built on plasmids, and these can be used immediately for transformation. The steric limitations of circular plasmids make single-crossover events between the plasmid and the chromosome possible, whereas linear DNA facilitates double-crossover events, which are most often the desired outcome. Restriction digest is the most common method for linearizing plasmids. Proximity of the cut site to the mutation of interest is an important consideration, because sufficient flanking DNA is required for a recombination event.

(b) Chromosomal DNA: this may be preferred over plasmids because of the extensive homology regions available for crossover, and additionally chromosomal DNA preparations can be used for transformation immediately without additional processing. However, this approach may require additional screening steps to ensure that mutations other than the mutation of interest that are present on the transforming DNA are not incorporated by the recipient strain, and that mutations already made to the recipient strain are not corrected by recombination with the transforming DNA.

As an example to compare directly the transformation efficiency of different substrates, we transformed wild type *N. gonorrhoeae* strain MS11 in liquid culture with a variety of pAH15 substrate derivatives. The plasmid pAH15 contains *rpsE*, which encodes the 30S ribosomal protein S5, containing an A to C point mutation of base pair 70. This mutation confers resistance to the antibiotic spectinomycin. We used 250 ng of circular or linearized pAH15 plasmid, or chromosomal DNA from strain MS11 containing the pAH15 point mutation in *rpsE* as transformation substrates (Fig. 3a). All of these substrates transformed MS11 with similar efficiency (One-way ANOVA, $p = 0.072$). It is impressive that 250 ng of chromosomal DNA transformed with comparable efficiency to 250 ng of the plasmid DNA, as there is a much higher molarity of the DNA region of interest in the

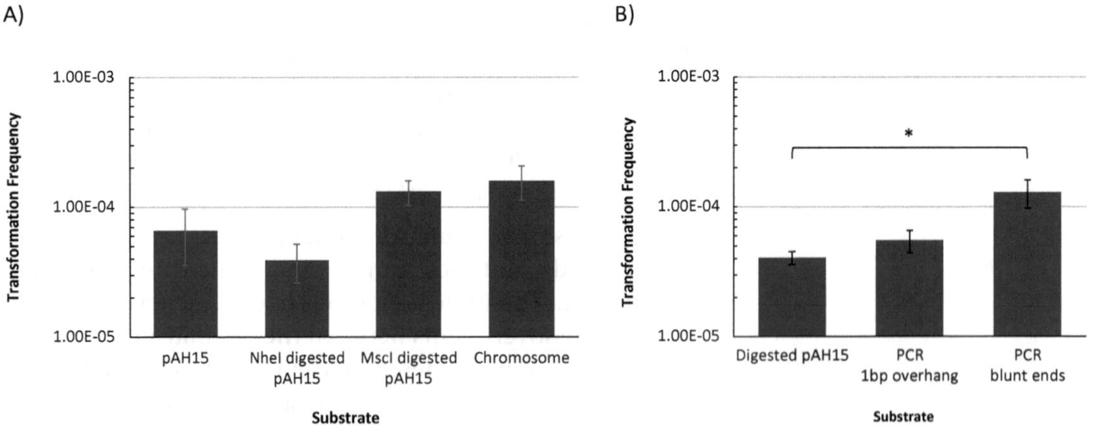

Fig. 3 Transformation efficiency of different DNA substrates. (**a**) Circular and linearized plasmids, as well as chromosomal DNA, all transform *N. gonorrhoeae*. Substrates are not differently efficient (one-way ANOVA, $p = 0.072$). (**b**) The same DNA sequence derived from either plasmid digestion or PCR transform at different frequencies. Substrates are differently efficient (one-way ANOVA, $p = 0.022$). Asterisks indicate Student's *t*-test $p < 0.05$. Error bars are the standard errors of the mean. Frequencies reported as resistant colony forming units (CFU)/total CFU

plasmid samples. This may be a testament to the value of large homology regions found on the chromosome.

(c) PCR products: These potential transformation substrates contain no off-target mutations and can easily be purified at high concentrations. As an example to compare the transformation efficiency of a PCR product and a digested plasmid we digested pAH15 with NheI and MscI enzymes, purified and blunted the plasmid to leave a 1249 bp product that contained both the mutant *rpsE* and a 12 bp-DUS. We compared this digestion product to a PCR product of identical sequence generated with either Taq or Phusion polymerase enzyme (Fig. 3b). All of these substrates transformed MS11 and are efficient for laboratory use. The Phusion polymerase PCR product was significantly more efficient for transformation than the pAH15 digest (Student's *t*-test, $p = 0.032$). These two products are identical except for methylation of the plasmid DNA by *E. coli*. Therefore, it is possible that this difference is a result of a restriction modification system (RMS), many of which have been identified in pathogenic *Neisseria* and shown to affect transformation [20, 34, 35].

(d) Synthesized DNA: Synthetic fragments of DNA can be used directly as a substrate for transformation. To test the efficiency of this approach, we used a synthesized a 500 bp DNA fragment that contained a portion of MS11 *rpsE* and the upstream region, with the A to C mutation from pAH15 at the center of the homologous sequence. At the 3′ end, a single Crick DUS was added. 150 ng of this fragment transformed MS11 with a frequency of 2.68×10^{-5} ($\pm 1.06 \times 10^{-5}$), comparable to the other substrates tested. This DNA fragment could also be used as a PCR template, and the resultant PCR product used for transformation.

8. Plasmids can be linearized by restriction enzyme digestion. Be sure to choose a restriction enzyme that does not cut within the region of interest, regions of homology, or the DUS, as this will prevent successful uptake and homologous recombination.

9. Understanding the biology of the *Neisseria* requires the ability to genetically manipulate members of the genus. Different mutations require specific considerations for design, which are summarized in Table 2.

(a) Point mutation: A double-crossover homologous recombination event is required to replace the chromosomal sequence of *Neisseria* without adding material. For point mutations, only a very small region of the chromosome needs to be recombined to achieve mutation. Plasmids for

Table 2
Types of mutations that can be made using transformation of *Neisseria* and the constructs required to make them

Mutation type	Description	Transformation substrate	Diagram
Point mutation	Alter a single base pair	Linear DNA containing mutation and surrounding region (≥250 bp on each side)	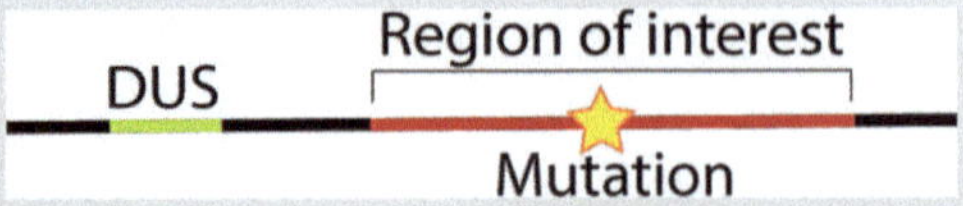
Insertion/ allelic replacement	Add a sequence into a target region	Linear DNA containing inserted sequence flanked by ≥500 bp of target sequence on each side	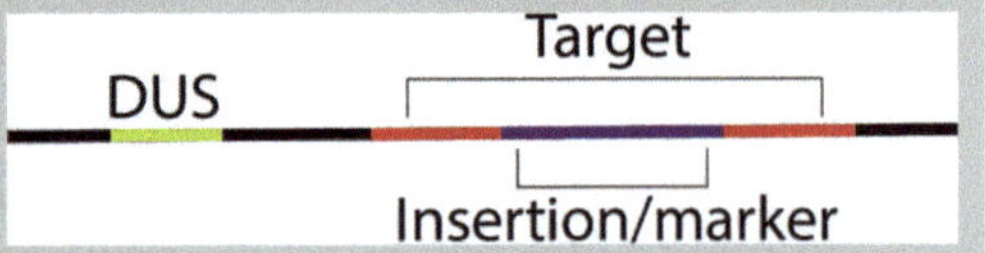
Insertion duplication mutagenesis	Interrupt a gene with a selectable marker	Circular DNA containing resistance marker and target region	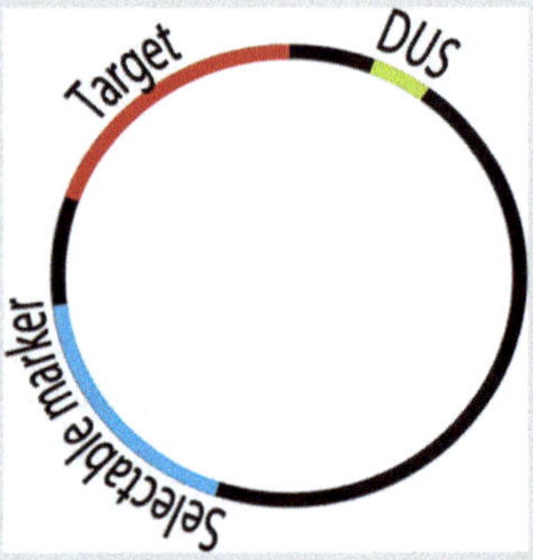
Unmarked deletion	Delete a gene	Linear DNA containing ≥500 bp flanking DNA of target gene	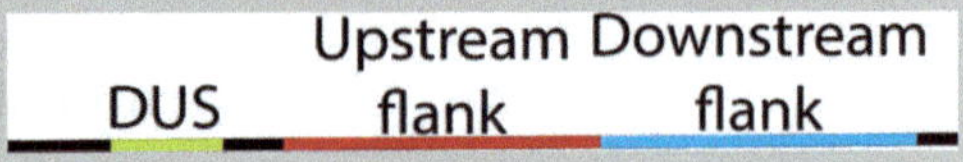

delivering a point mutation should contain the DUS and at least 500 bp of chromosomal DNA sequence containing the mutation, although as little as 73 bp have been shown to transform *N. meningitidis* [21, 36, 37]. The plasmid should be linearized to favor double crossover unless it is quite large (over 8 kb). If this point mutation does not cause a selectable phenotype, design the mutation to insert or remove a restriction enzyme site to allow for screening of potential transformants. These mutations can also be achieved by using a 500 bp (or larger) synthesized DNA fragment.

(b) Allelic replacements: Replacing an existing *Neisseria* gene with an antibiotic resistance cassette creates selectable deletion mutants. This common process can be done using a plasmid containing the resistance cassette flanked by the regions on either side of the target gene, in addition to the DUS. Assembly cloning all three inserts into the vector will expedite substrate creation. After linearization and transformation with the plasmid, double crossover recombination on either side on the resistance cassette/ target gene will swap these genes to yield selectable transformants.

(c) Targeted insertions: Inserting a selectable marker into a locus to create an interruption can be done exactly as allelic replacement, wherein the flanking regions of the marker are designed to match the loci on either side of the intended interruption site, instead of the regions flanking a region to be replaced.

(d) Insertion-duplication mutagenesis: This is an efficient approach for gene interruption that has been used in both *N. gonorrhoeae* and *N. meningitidis* [33, 38]. This technique utilizes an undigested plasmid substrate containing an antibiotic resistance cassette, the target of interest, and a DUS [33]. Transformation with the uncut plasmid facilitates a single crossover event with the chromosome, inserting the entire plasmid contents into the targeted site of recombination and creating selectable mutants. Propagation of the plasmid in a $RecA^{+}$ *E. coli* strain prior to transformation of *Neisseria* will generate dimer plasmids that are more efficient at generating insertion-duplication mutants [39]. Target fragments as small as 290 bp of homologous DNA have been successfully reported [33].

(e) Unmarked deletions: Deleting a gene without otherwise altering the genome can be achieved using the same approach as gene replacement, by using a transformation substrate lacking the antibiotic resistance cassette and containing only the flanking regions adjacent to one another. Use PCR to screen transformants containing a clean deletion. A restriction site can be added between the two flanking regions to allow for addition restriction digest screening, or to make future introduction of an antibiotic resistance cassette into the plasmid construct quick and easy.

(f) In vitro transposon mutagenesis: *Himar1 mariner* transposition is a method for both targeted and random mutagenesis of *Neisseria*. In this approach, the DUS, a purified hyperactive transposase, and a mini-transposon with a kanamycin resistance marker flanked by the *Himar1* inverted repeat target are combined with the DNA of interest. The mutagenized DNA is then used to transform the parent *Neisseria* strain, which uptakes and incorporates the mutations by homologous recombination onto the chromosome [40].

(g) Commensals: The genus *Neisseria* includes many nonpathogenic *Neisseria*, which are often found as nasopharyngeal colonizers [41]. All investigated *Neisseria* have a 12 bp DUS, although several variants have been identified. The variations in DUS have been termed "dialects," reflecting the idea that each species (or group of species) use their own version of the genetic "word" for DNA uptake to select for genetic material from their own species and exclude the DNA of others [13]. The DUS with sequence 5′-ATGCCGTCTGAA-3′ has been found in pathogens *N. gonorrhoeae* and *N. meningitidis*, as well as the human commensals *N. lactamica*, *N. cinerea*, and *N. polysaccharea*. The DUS variant identified in the human commensals *N. subflava*, *N. flavescens*, and others differs by a single base pair, with sequence 5′-A**G**GCCGTCTGAA-3′. Another variant, 5′-A**G**G**T**CGTCTGAA-3′, is found in the human commensal *N. sicca* and the macaque colonizer *N. macacae*. In some cases, the DUS variation most prevalent in the genome is strain dependent; *N. mucosa* C102 uses the single base pair variant, whereas *N. mucosa* ATCC25996 uses the two base pair variant [13]. As in the pathogens, interaction between the DUS and its cognate receptor ComP in the commensals appears to be specific, as demonstrated by replacing meningococcal ComP with *N. subflava* ComP to change the DUS preference [42]. The extent of transformation exclusivity that this interaction provides may be species or strain dependent, but preference for transformation of DNA with a species-specific DUS has been demonstrated [13, 42].

Laboratory transformation of many commensal species, including *N. musculi*, *N. sicca*, *N. mucosa*, *N. elongata*, and *N. macacae*, has been done using the same general protocols described in this chapter [4, 5, 43, 44]. Allelic replacement is greatly favored over clean deletion of genes for these species, although unmarked deletions are theoretically possible. Recall that different *Neisseria* species recognize variants of the DUS [13]. Additionally, some commensal species may require longer growth periods or notably higher quantities of DNA (~20 μg) for optimal transformation.

10. To make piliated bacterial stocks, streak a piliated colony onto a GCB agar plate and prepare the stock after 18–20 h of growth. If you are using stocks with a mix of piliated and nonpiliated bacteria, an extra day is required for the transformation protocol. Revive your bacteria from a frozen stock onto a GCB agar plate on the day before the transformation and grow overnight. Then, on day 1, streak a single piliated colony and grow the plate overnight and continue from day 2.
11. Type IV pili are essential for transformation in the pathogenic *Neisseria*, and have been identified in most nonpathogenic species as well [8, 45, 46]. Uniquely, the pathogenic *Neisseria* undergo antigenic variation of the major pilin gene, expressing variants of the gene by recombining one or more alleles from different loci in the genome [1, 47]. Antigenic variation of the pilus in *N. gonorrhoeae* alters colony morphology visibly, and this change can be observed using a stereo microscope. Piliated colonies appear raised and smaller in diameter, as they are able to build the colony upward due to pilus interactions. Nonpiliated colonies appear flatter with larger diameters, as they grow outward instead of upward. Spontaneous loss of pilus expression is common, and these events are visible as blebs of nonpiliated, flat growth protruding from the piliated colony base [48]. *N. meningitidis* also undergoes pilin antigenic variation, although this is not very visible on plates. Likewise, the pathogenic *Neisseria* antigenically vary surface opacity proteins. This change is also visible using a stereo microscope with a mirror; colonies expressing opacity proteins appear more granulated and coruscant than those that are not. Opacity proteins appear to be lacking from most nonpathogenic *Neisseria*, and have only been identified in *N. polysaccharea*, *N. flavescens*, and *N. lactamica* [1].
12. If growth is too thick and a single colony cannot be isolated for the transformation, streak again from this plate onto a fresh GCB plate for single colonies and grow overnight as before.
13. It is recommended that the amount of DNA solution spotted be varied (e.g., spot 2 and 10 μL of DNA in the two circles, sized appropriately) and that the bacteria be distributed such that different circles represent different bacteria–DNA ratios. This also dilutes out residual components of the DNA preparation that may inhibit gonococcal growth to allow bacteria to survive in contact with the transforming DNA.
14. If the screening process takes more than one day, patches will need to be streaked again approximately every 24 h, or *N. gonorrhoeae* bacteria begin to die. When the bacteria are not in use, keep the agar plates in a humidified atmosphere at 37 °C with 5% (v/v) CO_2. Gonococci do not survive well at room temperature for an extended period of time.

References

1. Marri PR, Paniscus M, Weyand NJ et al (2010) Genome sequencing reveals widespread virulence gene exchange among human *Neisseria* species. PLoS One 5(7):e11835. https://doi.org/10.1371/journal.pone.0011835
2. Biswas GD, Sox T, Blackman E et al (1977) Factors affecting genetic transformation of *Neisseria gonorrhoeae*. J Bacteriol 129 (2):983–992
3. Woodhams KL, Benet ZL, Blonsky SE et al (2012) Prevalence and detailed mapping of the gonococcal genetic island in *Neisseria meningitidis*. J Bacteriol 194(9):2275–2285. https://doi.org/10.1128/JB.00094-12
4. Weyand NJ, Ma M, Phifer-Rixey M et al (2016) Isolation and characterization of *Neisseria musculi* sp. nov., from the wild house mouse. Int J Syst Evol Microbiol 66:3585–3593. https://doi.org/10.1099/ijsem.0.001237
5. Chan JM, Dillard JP (2016) *Neisseria gonorrhoeae* crippled its peptidoglycan fragment permease to facilitate toxic peptidoglycan monomer release. J Bacteriol 198 (21):3029–3040. https://doi.org/10.1128/JB.00437-16
6. Ma M, Powell DA, Weyand NJ et al (2018) A natural mouse model for *Neisseria* colonization. Infect Immun 86(5):e00839–e00817. https://doi.org/10.1128/IAI.00839-17
7. Hamilton HL, Dillard JP (2006) Natural transformation of *Neisseria gonorrhoeae*: from DNA donation to homologous recombination. Mol Microbiol 59(2):376–385. https://doi.org/10.1111/j.1365-2958.2005.04964.x
8. Sparling PF (1966) Genetic transformation of *Neisseria gonorrhoeae* to streptomycin resistance. J Bacteriol 92(5):1364–1371
9. Koomey JM, Falkow S (1987) Cloning of the *recA* gene of *Neisseria gonorrhoeae* and construction of gonococcal *recA* mutants. J Bacteriol 169(2):790–795. https://doi.org/10.1128/jb.169.2.790-795.1987
10. Giltner CL, Habash M, Burrows LL (2010) *Pseudomonas aeruginosa* minor pilins are incorporated into type IV pili. J Mol Biol 398 (3):444–461. https://doi.org/10.1016/J.JMB.2010.03.028
11. Goodman SD, Scocca JJ (1988) Identification and arrangement of the DNA sequence recognized in specific transformation of *Neisseria gonorrhoeae*. Proc Natl Acad Sci U S A 85 (18):6982–6986. https://doi.org/10.1073/pnas.85.18.6982
12. Cehovin A, Simpson PJ, McDowell MA et al (2013) Specific DNA recognition mediated by a type IV pilin. Proc Natl Acad Sci 110 (8):3065–3070. https://doi.org/10.1073/pnas.1218832110
13. Frye SA, Nilsen M, Tønjum T et al (2013) Dialects of the DNA uptake sequence in *Neisseriaceae*. PLoS Genet 9(4):e1003458. https://doi.org/10.1371/journal.pgen.1003458
14. Wolfgang M, Lauer P, Park HS et al (1998) PilT mutations lead to simultaneous defects in competence for natural transformation and twitching motility in piliated *Neisseria gonorrhoeae*. Mol Microbiol 29(1):321–330. https://doi.org/10.1046/j.1365-2958.1998.00935.x
15. Maier B, Potter L, So M et al (2002) Single pilus motor forces exceed 100 pN. Proc Natl Acad Sci U S A 99(25):16012–16017. https://doi.org/10.1073/pnas.242523299
16. Gangel H, Hepp C, Müller S et al (2014) Concerted spatio-temporal dynamics of imported DNA and ComE DNA uptake protein during gonococcal transformation. PLoS Pathog 10(4):e1004043. https://doi.org/10.1371/journal.ppat.1004043
17. Biswas GD, Burnstein KL, Sparling PF (1986) Linearization of donor DNA during plasmid transformation in *Neisseria gonorrhoeae*. J Bacteriol 168(2):756–761. https://doi.org/10.1128/jb.168.2.756-761.1986
18. Krüger NJ, Stingl K (2011) Two steps away from novelty – principles of bacterial DNA uptake. Mol Microbiol 80(4):860–867. https://doi.org/10.1111/j.1365-2958.2011.07647.x
19. Mehr IJ, Seifert HS (1998) Differential roles of homologous recombination pathways in *Neisseria gonorrhoeae* pilin antigenic variation, DNA transformation and DNA repair. Mol Microbiol 30(4):697–710. https://doi.org/10.1046/j.1365-2958.1998.01089.x
20. Gunn JS, Stein DC (1996) Use of a non-selective transformation technique to construct a multiply restriction/modification-deficient mutant of *Neisseria gonorrhoeae*. Mol Gen Genet 251:509–517. https://doi.org/10.1007/BF02173639
21. Dillard JP (2011) Genetic manipulation of *Neisseria gonorrhoeae*. Curr Protoc Microbiol 4A(2):1–4A.2.23
22. Wright CJ, Jerse AE, Cohen MS et al (1994) Nonrepresentative PCR amplification of variable gene sequences in clinical specimens

containing dilute, complex mixtures of microorganisms. J Clin Microbiol 32(2):464–468

23. Stein DC, Danaher RJ, Cook TM (1991) Characterization of a *gyrB* mutation responsible for low-level nalidixic acid resistance in *Neisseria gonorrhoeae*. Antimicrob Agents Chemother 35(4):622–626. https://doi.org/10.1128/AAC.35.4.622
24. Duffin PM, Seifert HS (2012) Genetic transformation of *Neisseria gonorrhoeae* shows a strand preference. FEMS Microbiol Lett 334(1):44–48. https://doi.org/10.1111/j.1574-6968.2012.02612.x
25. Hepp C, Gangel H, Henseler K et al (2016) Single-stranded DNA uptake during gonococcal transformation. J Bacteriol 198(18):2515–2523. https://doi.org/10.1128/JB.00464-16
26. Smith HO, Gwinn ML, Salzberg SL (1999) DNA uptake signal sequences in naturally transformable bacteria. Res Microbiol 150:603–616. https://doi.org/10.1016/S0923-2508(99)00130-8
27. Ambur OH, Frye SA, Tønjum T (2007) New functional identity for the DNA uptake sequence in transformation and its presence in transcriptional terminators. J Bacteriol 189(5):2077–2085. https://doi.org/10.1128/JB.01408-06
28. Davidsen T, Rødland EA, Lagesen K et al (2004) Biased distribution of DNA uptake sequences towards genome maintenance genes. Nucleic Acids Res 32(3):1050–1058. https://doi.org/10.1093/nar/gkh255
29. Hamilton HL, Domínguez NM, Schwartz KJ et al (2005) *Neisseria gonorrhoeae* secretes chromosomal DNA via a novel type IV secretion system. Mol Microbiol 55(6):1704–1721. https://doi.org/10.1111/j.1365-2958.2005.04521.x
30. Spencer-Smith R, Roberts S, Gurung N et al (2016) DNA uptake sequences in *Neisseria gonorrhoeae* as intrinsic transcriptional terminators and markers of horizontal gene transfer. Microb Genomics 2(8):1–11. https://doi.org/10.1099/mgen.0.000069
31. Duffin PM, Seifert HS (2010) DNA uptake sequence-mediated enhancement of transformation in *Neisseria gonorrhoeae* is strain dependent. J Bacteriol 192(17):4436–4444. https://doi.org/10.1128/JB.00442-10
32. Elkins C, Thomas CE, Seifert HS et al (1991) Species-specific uptake of DNA by gonococci is mediated by a 10-base-pair sequence. J Bacteriol 173(12):3911–3913
33. Hamilton HL, Schwartz KJ, Dillard JP (2001) Insertion-duplication mutagenesis of *Neisseria*: Use in characterization of DNA transfer genes in the gonococcal genetic island. J Bacteriol 183(16):4718–4726. https://doi.org/10.1128/JB.183.16.4718-4726.2001
34. Stein DC, Gunn JS, Radlinska M et al (1995) Restriction and modification systems of *Neisseria gonorrhoeae*. Gene 157:19–22. https://doi.org/10.1016/0378-1119(94)00649-D
35. Budroni S, Siena E, Hotopp JCD et al (2011) *Neisseria meningitidis* is structured in clades associated with restriction modification systems that modulate homologous recombination. Proc Natl Acad Sci U S A 108(11):4494–4499. https://doi.org/10.1073/pnas.1019751108
36. Ambur OH, Frye SA, Nilsen M et al (2012) Restriction and sequence alterations affect DNA uptake sequence-dependent transformation in *Neisseria meningitidis*. PLoS One 7(7):e39742. https://doi.org/10.1371/journal.pone.0039742
37. Rotman E, Seifert HS (2014) The genetics of *Neisseria* species. Annu Rev Genet 48(1):405–431. https://doi.org/10.1146/annurev-genet-120213-092007
38. van der Ley P, Kramer M, Steeghs L et al (1996) Identification of a locus involved in meningococcal lipopolysaccharide biosynthesis by deletion mutagenesis. Mol Microbiol 19(5):1117–1125. https://doi.org/10.1046/j.1365-2958.1996.464992.x
39. Michel B, Niaudet B, Ehrlich SD (1983) Intermolecular recombination during transformation of *Bacillus subtilis* competent cells by monomeric and dimeric plasmids. Plasmid 10:1–10. https://doi.org/10.1016/0147-619X(83)90052-5
40. Pelicic V, Morelle S, Lampe D et al (2000) Mutagenesis of *Neisseria meningitidis* by *in vitro* transposition of *Himar1 mariner*. J Bacteriol 182(19):5391–5398. https://doi.org/10.1128/JB.182.19.5391-5398.2000
41. Liu G, Tang CM, Exley RM (2015) Non-pathogenic *Neisseria*: members of an abundant, multi-habitat, diverse genus. Microbiology 161:1297–1312. https://doi.org/10.1099/mic.0.000086
42. Berry JL, Cehovin A, McDowell MA et al (2013) Functional analysis of the interdependence between DNA uptake sequence and its cognate ComP receptor during natural transformation in *Neisseria* species. PLoS Genet 9(12):1–10. https://doi.org/10.1371/journal.pgen.1004014
43. Weyand NJ, Wertheimer AM, Hobbs TR et al (2013) *Neisseria* infection of rhesus macaques as a model to study colonization, transmission,

persistence, and horizontal gene transfer. Proc Natl Acad Sci 110(8):3059–3064. https://doi.org/10.1073/pnas.1217420110

44. Rendón MA, Hockenberry AM, McManus SA et al (2013) Sigma factor RpoN (σ54) regulates *pilE* transcription in commensal *Neisseria elongata*. Mol Microbiol 90(1):103–113. https://doi.org/10.1111/mmi.12350
45. Aho EL, Keating AM, McGillivray SM (2000) A comparative analysis of pilin genes from pathogenic and nonpathogenic *Neisseria* species. Microb Pathog 28(2):81–88. https://doi.org/10.1006/mpat.1999.0325
46. Cahoon LA, Seifert HS (2011) Focusing homologous recombination: pilin antigenic variation in the pathogenic *Neisseria*. Mol Biol Evol 81(5):1136–1143. https://doi.org/10.3174/ajnr.A1256.Functional
47. Criss AK, Kline KA, Seifert HS (2005) The frequency and rate of pilin antigenic variation in *Neisseria gonorrhoeae*. Mol Microbiol 58(2):510–519. https://doi.org/10.1111/j.1365-2958.2005.04838.x
48. Kellogg DS, Peacock WL, Deacon WE et al (1963) *Neisseria gonorrhoeae*. I. Virulence genetically linked to clonal variation. J Bacteriol 85:1274–1279

Chapter 11

Strategies for Global RNA Sequencing of the Human Pathogen *Neisseria gonorrhoeae*

Ryan McClure and Caroline A. Genco

Abstract

Over the last several years transcriptomic analysis of bacterial pathogens has become easier and less expensive. This technique is used to determine expression levels for all genes of a particular species or collection of species under a given condition, including genes that are not yet known to exist. While transcriptomics can be a powerful tool to better understand bacterial regulatory responses to specific host environments, the experimental approach and data analysis must be performed correctly to ensure robust, accurate, and translational results. Here, we describe experimental protocols related to transcriptomic analysis of the sexually transmitted disease pathogen *Neisseria gonorrhoeae*. Methods are described for the extraction of high-quality RNA, examination of RNA to ensure quality, the generation of cDNA libraries for sequencing and the downstream analysis of raw sequencing data to determine gene expression levels. Much of this work can be carried out with equipment and reagents that are readily available, and the methods can be performed by a majority of laboratory groups. RNA-seq and transcriptomic analyses are set to become even more common in the coming years. The protocols described here will provide a standardized set of methods for applying this powerful technique to the study of *N. gonorrhoeae* under a variety of conditions.

Key words *Neisseria gonorrhoeae*, RNA-seq, Transcriptomic, RNA, cDNA

1 Introduction

Over the last 10 years the emergence of global sequencing analysis has revolutionized our ability to view the complete transcriptomic profile of a number of human pathogens. This methodology (also termed RNA-seq) involves the collection of RNA from a biological system (cells cultured planktonically, whole tissues, live animals, etc.) followed by RNA sequencing to view the expression level of thousands of genes simultaneously. The most common use of RNA-seq is to identify genes that are differentially expressed due to some external stimuli or genetic alteration. The ability to apply this analysis to the complete genome allows for the identification of entire pathways that may respond to stimuli, as well as RNA transcripts that are not yet known to exist, such as bacterial sRNAs.

Myron Christodoulides (ed.), *Neisseria gonorrhoeae: Methods and Protocols*, Methods in Molecular Biology, vol. 1997,
https://doi.org/10.1007/978-1-4939-9496-0_11,

The ability of RNA-seq to collect large amounts of information on a biological system has led to its application to a number of pathogenic bacterial species, including *Neisseria gonorrhoeae*. *N. gonorrhoeae* is the causative agent of gonorrhea, a sexually transmitted infection that affects approximately 500,000 people each year in the USA and over 100 million worldwide. The incidence of gonorrhea has been on the rise since 2010 and, significantly, *N. gonorrhoeae* strains that are resistant to a number of antibiotics, including current antibiotic treatments recommended by the Centers for Disease Control, have emerged. It is well established that during infection of the human genital tract this pathogen must contend with a complex and dynamic environment including changing iron and oxygen levels as well as the presence of resident microflora and cells of the innate immune system, and, in some cases, antibiotic treatment. This environment requires a corresponding response from *N. gonorrhoeae* that includes the proper timing of gene expression patterns, mediated by a number of DNA binding transcription factors as well as regulatory RNAs. Over the last several years, several groups have used global RNA-seq to query the gene expression profiles of *N. gonorrhoeae* during growth under a variety of in vitro conditions and our group has expanded this analysis to include the response of the gonococcus to natural mucosal infection in men and women [1–10]. The global aspects of RNA-seq make it ideally suited for examining human pathogens as the response of multiple organisms (host and pathogen cells) can be queried at once. In addition, transcripts that have not yet been annotated, such as sRNAs, can be identified and studied with this method, an advantage that is especially important in *N. gonorrhoeae* as it contains relatively few DNA binding proteins and may rely heavily on regulatory sRNAs. Finally, RNA-seq can be used to identify highly expressed outer membrane proteins that may represent target vaccine candidates or targets for new drug development for this pathogen.

While RNA-seq is a powerful tool to understand the global bacterial response to stimuli, it is important that methods used to carry out sequencing, as well as the analysis of sequencing results, is performed properly to ensure robust scientific interpretation. In this chapter we describe a number of strategies related to RNA-seq analysis of *N. gonorrhoeae*. Methods for the isolation of high quality RNA are discussed, as well as the preparation of RNA for sequencing, the construction of cDNA libraries, and the subsequent computational analysis and interpretation of raw RNA-seq data. As RNA-seq becomes faster, more accurate, and less expensive, its use as a tool to examine a number of bacterial species, including *N. gonorrhoeae*, will only increase. This demand will benefit from the application of standardized methods and protocols to interrogate gene regulatory patterns in pathogenic bacteria.

2 Materials

2.1 Collection of RNA from Neisseria gonorrhoeae

1. *N. gonorrhoeae* cells/samples (*see* **Note 1**).
2. Microcentrifuge capable of RCF of 10,000 × *g*.
3. 1.5 mL Eppendorf tubes.
4. TRIzol (*see* **Note 2**).
5. 100% chloroform.
6. 100% isopropanol.
7. 75% (v/v) ethanol.
8. RNase-free H_2O (*see* **Note 3**).

2.2 Cleaning RNA

1. 3 M sodium acetate.
2. 5 μg/mL glycogen.
3. 75% (v/v) and 100% ethanol.
4. Microcentrifuge capable of RCF of 10,000 × *g*.

2.3 RNA Quality and mRNA Enrichment

2.3.1 Measuring RNA Concentration

1. Purified RNA.
2. 15 mL conical tubes.
3. Qubit™ 4 Fluorometer (*see* **Note 4**).
4. Qubit™ RNA HS Assay Kit.
5. Qubit™ Assay Tubes.
6. Vortex mixer.

2.3.2 Removing Contaminating DNA

1. Purified RNA.
2. 1.5 mL Eppendorf tubes.
3. DNA-free DNA removal kit, containing 10× Sample Buffer, DNase I, and DNase Inactivation Reagent.
4. Heat block.
5. Microcentrifuge capable of RCF of 10,000 × *g*.

2.3.3 Determining RNA Quality

1. Purified RNA.
2. RNase-free H_2O.
3. RNaseZap.
4. Agilent 2100 Bioanalyzer, including Agilent Chip Priming Station, Agilent Syringe Kit and IKA vortex mixer.
5. Agilent RNA 6000 Pico Kit.
6. Heat block.
7. MicrobeEXPRESS Kit.
8. RiboMinus™ Eukaryote Kit (*see* **Note 5**).
9. Magnetic stand (e.g., Ambion P/NAM10026, or equivalent).

2.4 Preparation of cDNA Libraries from RNA

1. Purified mRNA.
2. 37 and 70 °C heat block or water bath incubators.
3. Microcentrifuge capable of RCF of 10,000 × *g*.
4. Vortex mixer.
5. 200 μL polymerase chain reaction (PCR) tubes.
6. Magnetic stand (e.g., Ambion P/NAM10026, or equivalent).
7. RapidSeq High Yield Directional mRNA Sample Prep Kit (*see* **Note 6**).
8. Thermal cycler.
9. Applied Biosystems Real-Time PCR Machine (*see* **Note** 7).
10. 96-well plates for Applied Biosystems Real-Time PCR (*see* **Note** 7).
11. 20× EVAGREEN Dye in H_2O (Biotium) (*see* **Note** 7).
12. 15 mL conical tubes.
13. Qubit™ 4 Fluorometer.
14. Qubit™ dsDNA HS Assay Kit.
15. Qubit™ Assay Tubes.
16. Agilent 2100 Bioanalyzer.
17. Agilent High Sensitivity DNA Kit.

2.5 Bioinformatics

2.5.1 Sequencing of cDNA Libraries and Quality Check

1. Sample .fastq sequencing files.
2. Mac or PC.
3. FastQC program.

2.5.2 Alignment of Reads to N. gonorrhoeae Genome

1. Sample .fastq sequencing files.
2. Mac or PC.
3. Rockhopper program.

3 Methods

3.1 Collection of RNA from Neisseria gonorrhoeae

The input RNA for RNA-seq will be isolated from *N. gonorrhoeae* after growth of the organism in isolation in vitro (e.g., querying response to iron or oxygen levels) or as part of a multispecies experiment (e.g., examining response of *N. gonorrhoeae* to invasion of epithelial cells or from infected patients). The isolation of RNA from *N. gonorrhoeae* is generally the same regardless of the specifics of a particular experiment (*see* **Note 8** for general approaches when working with RNA). There are a number of different methods for collecting RNA; some employ kits with silica membranes, while others use a form of phenol–chloroform extraction. Extraction with phenol–chloroform is generally slower and includes the use of hazardous chemicals, but does collect all RNA in the sample, including sRNAs of *N. gonorrhoeae*. Many of the silica membrane kits are specific for RNA species larger than 200 nucleotides, thus missing regulatory RNAs in *N. gonorrhoeae* (*see* **Note 2** for a discussion of these methods). The protocol below uses TRIzol, a proprietary approach based on phenol–chloroform extraction.

1. Chill 2 mL of 75% (v/v) ethanol per sample in a −20 °C freezer.
2. Collect samples of *N. gonorrhoeae* for RNA extraction by centrifugation for 5 min at 6000 × *g*. Generally, TRIzol will efficiently extract RNA from ~10^6 cells. For *N. gonorrhoeae* this corresponds to approximately 2 mL at an optical density (OD) at $\lambda_{600\ \mathrm{nm}}$ of 0.5 (*see* **Note 9**).
3. Discard the supernatant and proceed to RNA isolation or store cell pellets at −80 °C (*see* **Note 10**).
4. Add 1 mL of TRIzol to each cell pellet sample and pipet gently to mix. If samples are removed from −80 °C for RNA removal, then add TRIzol immediately upon removal, do not wait for the pellet to thaw (*see* **Note 11**).
5. Incubate samples at room temperature (RT) for 5 min.
6. Add 0.2 mL of chloroform to each sample and invert several times to mix.
7. Incubate samples at RT for 3 min, whereupon the sample should partially separate into an aqueous layer (top clear layer) and a phenol–chloroform layer (bottom pink layer).
8. Centrifuge at 12,000 × *g* for 15 min at 4 °C to complete this separation.
9. The layers should be clearly separated now. The top aqueous layer contains RNA, the bottom pink layer contains proteins. Precipitated DNA is also visible in the sample as a white layer between the aqueous and phenol layers (*see* **Note 12**).
10. Remove the upper aqueous layer containing RNA with a pipette and move to a new 1.5 mL tube. Approximately 400 μL of aqueous layer can be collected easily using two rounds of pipetting of 200 μL each. Do not attempt to gather all 400 μL at once. Once the 400 μL is collected, an additional 100–200 μL may be collected as well, but this must be done gently (*see* **Note 13**).
11. Add 0.5 mL of 100% isopropanol to the collected aqueous layer containing RNA (*see* **Note 14**).
12. Incubate at RT for 10 min.
13. Centrifuge at 12,000 × *g* for 10 min at 4 °C (*see* **Note 15**).

14. Carefully discard the supernatant, making every attempt to avoid touching the bottom or sides of the tube with the pipette tip. This could draw up RNA, leading to loss of sample. Leave a few microliters of supernatant in the bottom of the tube rather than risk losing RNA as the sample will be washed vigorously in the next steps.
15. Add 1 mL of chilled 75% (v/v) ethanol to the sample and mix by vortexing.
16. Centrifuge at 7500 × *g* for 5 min at 4 °C. At this point, RNA from *N. gonorrhoeae* is often visible as a white pellet at the bottom of the tube. However, this is not always the case (*see* **Note 16**).
17. Even if a pellet is not visible, remove the supernatant and repeat **steps 15** and **16** for a second wash (*see* **Note 17**).
18. Carefully, remove the supernatant, as much as possible, and invert the tube to allow the RNA to dry. Remove as much ethanol as possible with a 200 μL followed by a 20 μL pipet tip (20 μL tips are better at gathering very small amounts of ethanol at the bottom of the tube) and then invert the tube and allow it to dry to for 5 min. Remove ethanol on the sides of the tube and well above the pellet carefully with a Kimwipes (*see* **Note 18**).
19. Dissolve the RNA in 20 μL of RNase-free H_2O (*see* **Note 19**).
20. Estimate the concentration of the RNA with a NanoDrop or other spectrophotometric method. With 10^6 bacteria, the total amount of RNA collected for *N. gonorrhoeae* should be at least 500 ng. Take note of the ratio of absorbance at $\lambda_{260\ \mathrm{nm}}/\lambda_{280\ \mathrm{nm}}$. This number is reported if a NanoDrop is used and should be at least 2.0 if the RNA is pure. If the number is below 1.8 then the RNA is impure and should be cleaned (*see* Subheading 3.2).
21. Store the RNA (*see* **Note 20**).

3.2 Cleaning RNA

1. Add 0.4 μL of 5 μg/mL glycogen, 2 μL of 3 M sodium acetate and 50 μL of chilled 100% ethanol to the RNA preparation from **step 19** in Subheading 3.1.
2. Store at −80 °C for 2 h or at −20 °C overnight.
3. Centrifuge at 10,000 × *g* for 25 min at 4 °C.
4. Remove the supernatant and wash with 1 mL of chilled 75% (v/v) ethanol.
5. Centrifuge at 10,000 × *g* for 10 min at 4 °C.
6. Remove the supernatant, dry the RNA, dissolve it in RNase-free H_2O and estimate concentration again as described in **step 20** of Subheading 3.1.

3.3 RNA Quality and mRNA Enrichment

Before RNA can be used for cDNA libraries any contaminating DNA must be removed via enzymatic degradation. Next, the quality of the RNA should be checked using an Agilent 2100 Bioanalyzer. Finally, it is often advisable to enrich the sample for mRNAs and deplete rRNAs and tRNAs. The latter noncoding RNAs can account for 90–95% of the RNA in the cell, making interrogation of the more interesting mRNAs more difficult. While depletion of these noncoding RNAs is not absolutely required, it can lead to far more data on gene expression profiles in bacteria, including *N. gonorrhoeae*, and is almost always the suggested course. The methods used for depleting RNA from *N. gonorrhoeae* cultured by itself will be different from the methods used if *N. gonorrhoeae* is cocultured, for example, with host cells. If *N. gonorrhoeae* is cultured with host cells then both eukaryotic (host cell) and prokaryotic (gonococcal cell) rRNA/tRNA should be depleted. Two separate approaches must be used for this dual depletion and the protocol below describes both. It is critical to treat samples from control conditions identically to those from experimental conditions regarding rRNA/tRNA depletion. RNA-seq data from samples that were depleted of rRNA using different approaches (only depleted prokaryotic rRNA vs. depleted of both prokaryotic and eukaryotic RNA) should never be compared. All RNA-seq data sets that will be compared to each other should have rRNA depleted in the same manner. This may mean depleting both eukaryotic and prokaryotic rRNA from samples of *N. gonorrhoeae* cultured alone, if, at some point, they will be compared to *N. gonorrhoeae* samples cultured with host cells that were depleted of both prokaryotic and eukaryotic rRNA.

3.3.1 Measuring RNA Concentration

1. We generally use a Qubit™ 4 Fluorometer and the RNA HS kit to measure the concentration of RNA in samples.
2. For each sample, add 200 μL of RNA buffer from the RNA HS kit to a 15 mL conical tube, include enough for two additional samples representing the RNA standards. Multiple 200 μL aliquots can be combined into the same 15 mL conical tube.
3. For each sample, add 1 μL of Qubit RNA Reagent to the RNA buffer in the 15 mL conical tube. Add an additional 2 μL for the RNA standards. Mix by inverting.
4. Add 199 μL of this mixture to a Qubit Assay tube, one tube for each sample. Add 190 μL of this mixture to two additional tubes for the RNA standards.
5. Add 1 μL of the RNA sample to be measured to each assay tube with a buffer/reagent mixture. Add 10 μL of the standards to the standard tubes. For the samples, other amounts of RNA can be used (up to 10 μL), but the total volume in each tube should always be 200 μL.

6. Vortex each tube for 5 s.
7. Measure each standard first, followed by each RNA sample. Calculate the stock concentration based on the dilutions. The Qubit also has an option to calculate this.

3.3.2 Removing Contaminating DNA

1. Use a commercial DNA-free Kit to remove contaminating DNA (*see* **Note 21**).
2. If necessary, dilute samples to less than 200 μg/mL of RNA (*see* **Note 22**).
3. Add 0.1 volumes of 10× Sample Buffer and 1 μL of DNaseI from the DNA-free removal kit.

   The total volume of the reaction should not exceed 50 μL. Split the sample if necessary.
4. Incubate at 37 °C for 30 min on a heat block.
5. Add 0.1 volumes of the DNase Inactivation Reagent component from the DNA-free removal kit. If 0.1 volumes would be less than 2 μL, then add 2 μL of the Inactivation Reagent.
6. Mix by inverting.
7. Incubate at RT for 3 min, mixing by inverting once every min.
8. Centrifuge at 10,000 × *g* for 2 min at 4 °C.
9. Remove the supernatant containing the RNA to a clean 1.5 mL Eppendorf tube, taking care not to collect any of the Inactivation Reagent.

3.3.3 Determining RNA Quality

The best way to determine RNA quality is by using an Agilent 2100 Bioanalyzer or similar machine to visualize the 16s/23s rRNA bands. These can serve as a proxy for the total RNA in the sample. If these RNA species are intact, the sample is likely of high quality, if not, then the sample is at least partially degraded. *See* **Note 23** for using degraded RNA for RNA-seq. Because 16s/23s rRNA species are used to determine quality, this step must be done before depletion of these rRNAs and mRNA enrichment. We have had the most success with the Agilent RNA 6000 Pico Kit. This kit is designed for a range of RNA concentrations that are commonly isolated with the experiments described here. It comes with a RNA ladder as well as 25 RNA chips to be used with the Bioanalyzer.

1. Remove the Agilent RNA 6000 Pico Kit from 4 °C and allow it to equilibrate to RT for 30 min.
2. Prepare the RNA ladder as described in the kit instructions. The ladder can be stored in 2 μL aliquots at −80 °C and is sufficient for more than 25 chips.
3. Prepare the gel–dye mix as described in the kit instructions.
4. Prepare aliquots of each RNA sample to be tested: dilute to 1 ng/μL in RNase-free H_2O with a total volume of 3–5 μL.

5. Wash the machine by using a cleaning chip with 300 μL of RNaseZap. Place this chip in the machine and close the lid for 30 s.
6. Rinse the machine by using a second cleaning chip with 300 μL of RNase-free H_2O. Place this chip in the machine and close the lid for 30 s.
7. Repeat **step 6** once.
8. Allow the machine to dry by opening the lid for 5 min.
9. Open a new RNA chip and place in the chip priming station.
10. Add 9 μL of the gel–dye ladder to the well indicated in the instructions.
11. Close the chip priming station and press down the syringe until it can be locked on the lowest setting. The syringe should give some resistance but if it is difficult to depress then this may indicate old gel–dye mixture in the chip priming station. *See* **Note 24** on cleaning the priming station.
12. Wait 30 s.
13. Release the syringe and verify that is moves back to the 0.3 mL mark within 3 s (*see* **Note 25**).
14. Add gel–dye mixture and pico conditioning solution into the wells indicated in the instructions.
15. Add 5 μL of marker to each well that will contain a sample, including the ladder well. One chip is sufficient for eleven samples. However, it is good practice to run a control of high quality RNA (this can be isolated or purchased) in the 11th well leaving room for ten samples. Fill any well that will not contain a sample with 6 μL of marker. No well should ever be empty.
16. Denature the diluted RNA samples to be tested by heating them at 70 °C for 2 min. on a heat block. Once denatured, place the tubes on ice.
17. Add 1 μL of each RNA sample to each well.
18. Add 1 μL of diluted ladder to the ladder well.
19. Vortex the chip using the Agilent IKA vortex mixer for 1 min. Hold the chip gently in place to ensure it is not shaken from the vortex mixer.
20. Within 5 min of vortex mixing, place the chip in the Bioanalyzer, close the lid, and select the appropriate run ("Total RNA Pico"). Click Start.
21. The run takes approximately 45 min to complete. At the end of the run immediately remove the chip and discard it into a biological waste bin.

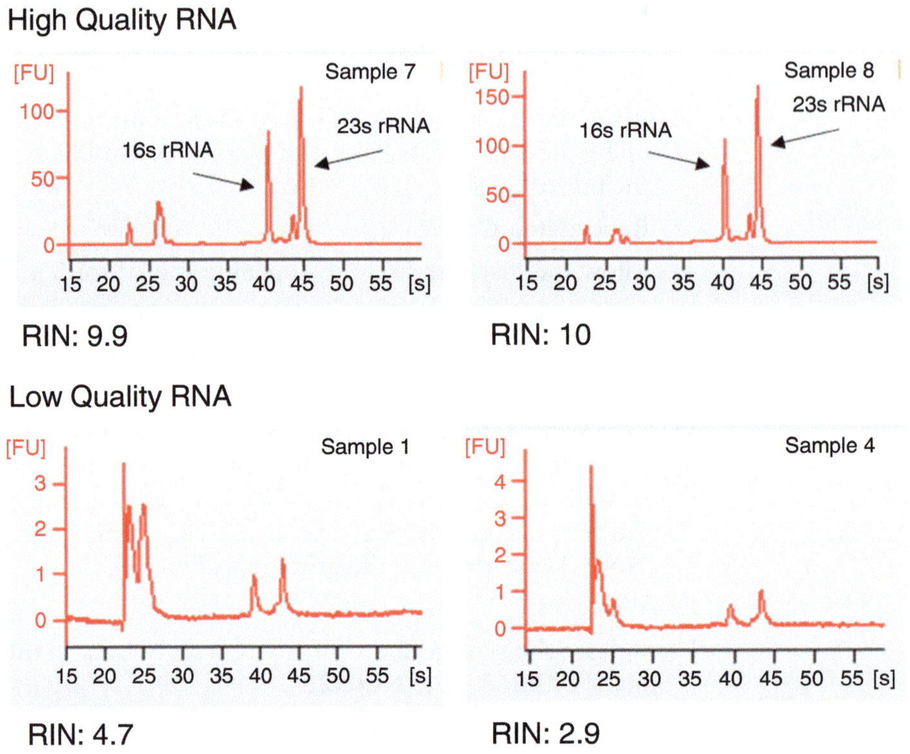

Fig. 1 Bioanalyzer traces of 16s/23s rRNA. Both high quality RNA (top panel) and low quality RNA (bottom panel) is shown. RINs are shown below each trace and in the case of high quality RNA the sharp defined peaks of 16s and 23s rRNA can be seen and are indicated with arrows

22. Rinse the machine by using a cleaning chip with 300 μL of RNase-free H_2O. Place this chip in the machine and close the lid for 30 s. Remove the chip and let the machine dry with the lid open for 1 min.

23. The output from the machine is a series of spectra that show the 16s/23s peaks. The peaks for each RNA species should be sharp, and there should be no smaller peaks (Fig. 1). Smaller peaks can indicate degraded RNA or DNA contamination. The machine also outputs an RNA Integrity Number (RIN) for each sample. Generally, RINs higher than 7.5 indicate high quality RNA and RINs from 5 to 7.5 indicate partially degraded RNA. Both the ladder and control RNA (if used) can be an indicator of whether the machine ran properly. If either of these RNA spectra look degraded then that indicates a problem with the run and the results should not be trusted. Clean the machine thoroughly as described in **steps 5** and **6** above as well as the chip priming station (*see* **Note 24**) and repeat the run.

24. After high quality RNA has been collected, prokaryotic and, depending on the experimental conditions used, eukaryotic

rRNA, must be depleted (*see* **Note 26**). Use the MicrobeEXPRESS Kit to deplete *N. gonorrhoeae* rRNA and the RiboMinus Eukaryotic Kit to deplete eukaryotic rRNA, following the manufacturer's kit instructions. Use a magnetic stand, since these kits use magnetic beads to capture and isolate rRNA strands. cDNA preparation kits generally use 10–150 ng of input RNA and since rRNA is ~90% of total rRNA, treat at least 1 μg of total RNA for rRNA depletion.

3.4 Preparation of cDNA Libraries from RNA

Once high-quality, rRNA-depleted RNA is collected, the next step is the construction of cDNA libraries for sequencing. There are a number of different sequencing technologies available and for each sequencing technology there are a variety of kits to prepare corresponding cDNA libraries (*see* **Note 6**). Illumina is the most popular sequencing platform currently in use and we have had significant success with it (*see* **Note 27** for a brief overview of library preparation and Illumina sequencing technology). Aside from sequencing technology, the read length of the sequences (the number of sequenced nucleotides in each read), the read depth (the number of reads obtained) and whether the reads will be single-end or paired-end, must all be decided. Generally, longer read lengths lead to more accurate alignment, but are more expensive and the quality of the nucleotides can drop dramatically near the end of the read. We have had success with read lengths of 75–100 bp. Read depth needs will vary between experiments, but to obtain high quality coverage of *N. gonorrhoeae*, at least 3–5 million reads should be obtained if possible. Single-end reads are sequenced from the 5′ end of the cDNA transcript only. Paired-end reads are sequenced from both the 5′ and 3′ end. This can lead to more accurate calls for each nucleotide, as several nucleotides are sequenced twice, but it is more expensive. Many reference genomes for *N. gonorrhoeae* exist (https://www.ncbi.nlm.nih.gov/genome/genomes/864), so single-end reads are generally sufficient for robust alignment of reads to genes. The approach described here uses the RapidSeqTM High Yield Directional mRNA Sample Prep Kit, which will generate single-end reads of a read length of 100 that are designed to be sequenced using an Illumina platform. The kit uses a relatively small amount of RNA (100 ng), and the protocol can be completed in a day for up to 12 libraries. The kit also comes with 12 unique aligners to allow for multiplexing (*see* **Note 28**).

1. Follow the manufacturer's instructions for the RapidSeqTM High Yield Directional mRNA Sample Prep Kit but with the protocol amendments described in **steps 2–6** below. Before the final PCR amplification of the library do a test qPCR using one-tenth the volume of the final reaction (*see* **Note 29**).
2. Prepare the PCR mix in accordance with the manufacturer's instructions. Remove 9.5 μL from each sample and place in a

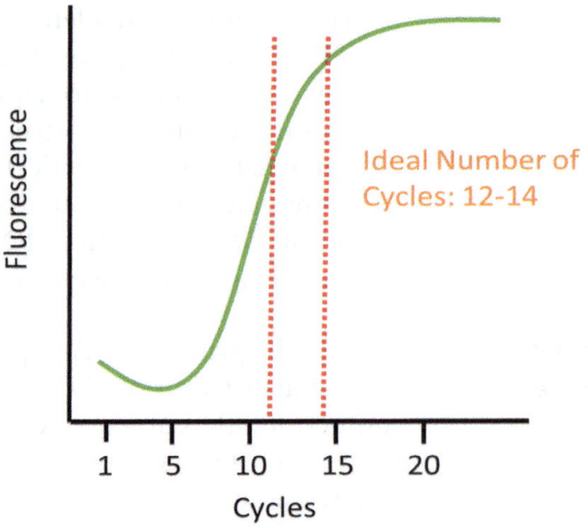

Fig. 2 Schematic of Test qPCR Analysis of cDNA Library. An example fluorescent trace against the number of cycles is shown. The ideal number of cycles for PCR amplification of the library is indicated (between the red dashed lines)

96-well plate well for use in an Applied Biosystems Real-Time PCR.

3. Add 0.5 μL of 20× EVAGREEN dye to each well.
4. Place the 96-well plate in the Real-Time PCR and run the following program: (a) 98 °C for 30 s; (b) 30 cycles of 98 °C for 10 s, 60 °C for 30 s, 72 °C for 15 s (collect fluorescent data on this step); (c) 72 °C for 10 min; and (d) hold at 4 °C.
5. Examine the exponential amplification curve of the one-tenth volume test qPCR and choose a number of cycles for the PCR amplification of the full library that is near the upper third of the curve (Fig. 2).
6. Check the concentrations after the final PCR of the cDNA libraries on a Qubit using the Qubit™ dsDNA HS Assay Kit. The protocol is similar to that described in Subheading 3.3.1 for the RNA HS Assay kit.
7. Check the quality of the cDNA library using the Agilent 2100 Bioanalyzer and the Agilent High Sensitivity DNA Kit, this protocol is similar to that described in Subheading 3.3.3 above for the RNA Pico Assay kit. Follow this protocol with one exception, which is to load the cDNA library on the Bioanalyzer at a concentration of 1–5 ng/μL (*see* **Note 30**).
8. The output of the Bioanalyzer should be a smooth wide peak of DNA (Fig. 3) and the cDNA libraries are now ready for sequencing using an Illumina platform.

   The output of the Bioanalyzer should be a smooth wide peak of DNA (Fig. 3). If this is the case the cDNA libraries are now ready for sequencing using an Illumina platform.

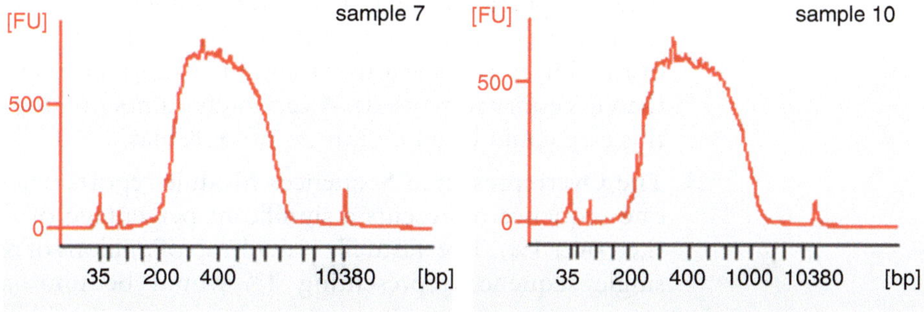

Fig. 3 Bioanalyzer traces of cDNA libraries. Two example traces of high quality cDNA libraries are shown. Both have broad peaks centering on cDNA strands of 300–500 nucleotides in length. The presence of spikes in the peaks, indicating PCR bias, is minimal

3.5 Bioinformatics

3.5.1 Sequencing of cDNA Libraries and Quality Check

Most Illumina machines are prohibitively expensive for a single laboratory and the majority of the actual RNA sequencing is done by a third party. The output of the sequencing is a set of .fastq files, one for each sample, which can then be analyzed further. Each .fastq file contains the collection of reads for a particular experiment. The sequencing data should be checked before analysis, to ensure that it is of high quality. A common, simple program to do this is FastQC (http://www.bioinformatics.babraham.ac.uk/projects/.fastqc/). This program can be run on a standard desktop or laptop and takes the .fastq file as input and runs it through a series of modules that report on the quality of the data. FastQC is designed to be run on a number of different kinds of sequencing data and as such the results of certain modules must be interpreted correctly. Generally, with RNA-seq data four Modules should be looked at closely:

1. The Per Base Sequence Quality Module reports the range and overall Phred (quality) score for the nucleotides at each position within the read. Generally, the quality drops near the end of the read but in most cases the mean Phred score for a nucleotide position should never drop below 20. Phred scores are logarithmically related to the base-calling error probabilities, so a score of 20 indicates that the read has a 99% chance of being called properly. If the scores near the end of the read dip below 20 they can be trimmed with a program such as Trimmomatic [11].
2. The Duplicate Sequences Module reports metrics on how many duplicate sequences are found in the .fastq file. Because of the small genome size of bacteria and the presence of rRNA sequences (even after depletion), it is likely that there will be several thousand duplicate sequences. Therefore, this module will often be triggered even in a .fastq file representing high quality data. However, it can be informative when PCR bias

may be a possibility (e.g., if a larger number of the PCR cycles than necessary were done on the cDNA library, or the quantity of the library is suspiciously high, indicating more PCR cycles than necessary were used). A very high number of duplicates in this case could be an indication of PCR bias.

3. The Overrepresented Sequences Module reports whether any one sequence represents a significant percentage of the reads (i.e., over 1%). The .fastq file is made up of millions of reads so a single sequence representing 1% would be unusual. Such sequences may represent contamination of the library with some other DNA (possible a PCR product) or may be rRNA/tRNA reads if they have not been properly depleted.
4. The Adapter Content Module reports whether any of several common adaptors (including those used in Illumina platforms) are found at the ends of reads. If this is the case, it suggests that the read insert between the adaptors is too short and the sequencer is sequencing all the way through the read and into the adaptor. If adaptor sequences are found, they should be removed via Trimmomatic before alignment.

3.5.2 Alignment of Reads to N. gonorrhoeae *Genome*

There are a number of programs available that align reads from . fastq files to reference genomes. Other programs determine counts of particular genes or normalize data across sequencing runs and identify differentially expressed genes. While several programs exist to do each individual step in analysis, other programs present a pipeline approach to analysis whereby all steps are automatically done starting with a .fastq file. One such program that is specifically designed for bacterial RNA-seq data and has been used with *N. gonorrhoeae* in the past is Rockhopper [4] (*see* **Note 31** for a discussion of other programs). This program can be run on a standard desktop or laptop and presents a simple user interface and, for organisms whose genomes are already uploaded into National Center for Biotechnology Information (NCBI) website, (https://www.ncbi.nlm.nih.gov/), only a .fastq file is required for analysis. Rockhopper is simple to use and has been applied with success to a number of bacterial species aside from *N. gonorrhoeae* by our group and by others [1, 5, 6, 10].

1. Load the .fastq file(s), choose the genome of interest from a drop down menu and assign the .fastq file(s) as replicates or separate experimental samples.
2. Use Rockhopper for automated alignment, counting, normalization and differential expression analysis (including statistical analysis and correction for multiple hypothesis testing). Rockhopper will output text files containing gene expression levels across all conditions. Rockhopper will also report expression

from intergenic sites in the genome, highlighting potential noncoding RNAs such as sRNAs.

3. If the organism under analysis does not have a genome in NCBI, load separately the .fasta file (file containing the genome of the organism) and .gff file (file containing information on nucleotide start and stop sites of genes) into Rockhopper for analysis.
4. Rockhopper reports *q*-values (*p*-values modified for multiple hypothesis testing using Benjamini–Hochberg) [12] for each pairwise comparison of conditions. Generally, a *q*-value of less than 0.05 indicates a statistically significant change in expression for a gene from one condition to another (a differentially expressed gene) though, since *q*-values can be more stringent than *p*-values, genes with a *q*-value of up to 0.1 may also be considered differentially regulated.
5. Once a list of differentially regulated genes is obtained, begin analysis for pathways or processes that may be affected by the change in condition.

4 Notes

1. The majority of the protocols described here will work for *N. gonorrhoeae* under almost any culture condition. In certain cases, such as collecting *N. gonorrhoeae* from human patients or possibly when *N. gonorrhoeae* has formed a biofilm, we have found that processing the samples through a QIAshredder (Qiagen) to break apart the cells can increase RNA isolation. It is critical that when cell pellets are collected they are either immediately processed for RNA extraction or stored at −80 °C.
2. TRIzol is described here as it is a common phenol–chloroform method, but other methods have also been used with success. The mirVana (ThermoFisher) and RNAPRO SOLUTION (MP Biomedicals) kits have both been used with *N. gonorrhoeae* with success [2, 9] In certain cases, a silica-membrane spin column with the capacity to bind RNA, such as the RNeasy Kit (Qiagen), can also be used. After binding to the membrane, RNA can be eluted with H_2O and stored. These kits are generally faster than phenol–chloroform extraction and also avoid the use of toxic organic chemicals. However, care must be taken when using these kits as they do not always collect RNA uniformly across sizes. Certain silica membranes preferentially bind RNA that is above 200 nucleotides, making their use inadvisable for studies that want to include sRNA transcripts in the analysis. If sRNAs are the specific focus

of the experiment, then steps can be taken to enrich the RNA sample for these transcripts. This can include electrophoresis of RNA through a gel and extracting RNA of a certain size [6], or using a kit specifically designed to capture these sRNAs, such as the mirVana kit [9].

3. Large amounts of RNase-free H_2O can also be made using diethyl pyrocarbonate (DEPC). Mix a 1 L solution of 0.1% (v/v) DEPC in H_2O. Incubate at RT overnight, followed by autoclaving at 121 °C and 2.68 kg/cm^2 for 15 min. to inactivate the DEPC.

4. We have found readings from Qubit to be more accurate than NanoDrop readings as they use a RNA-specific dye rather than a spectrophotometer approach, which can inadvertently capture both RNA and DNA.

5. The protocol described here is for an experimental approach that includes *N. gonorrhoeae* bacteria incubated with host cells and thus suggests depleting rRNA/tRNA from both prokaryotic and eukaryotic sources. If the experimental protocol includes only *N. gonorrhoeae* cultured alone for all samples, then only prokaryotic rRNA/tRNA needs to be removed.

6. We have had success with the RapidSeq High Yield Directional mRNA Sample Prep Kit (BioChain), but other kits are available as well. The following kits have been used successfully with *N. gonorrhoeae*: the SOLiD™ Total RNA-seq Kit (ThermoFisher), the Illumina TruSeq Small RNA kit (Illumina), the Small RNA v1.5 Sample Preparation Kit (Illumina), and the Ion Total RNA-Seq Kit (ThermoFisher). All of these kits produce cDNA libraries of RNA sequences that can then be sequenced on the corresponding machine, SOLiD, Illumina, Ion Torrent, 454 pyrosequencing, etc. Each of these sequencing technologies has a number of advantages and disadvantages and the correct choice for a certain experiment will depend on a number of factors including sequencing depth, presence of a genome for alignment and cost. At this point, Illumina sequencing is probably the most commonplace and popular method and many of the downstream computational approaches for alignment and analysis are designed to use output from Illumina machines. Several studies have compared and contrasted the different methods [13–15] and new methods are currently being developed that may offer faster, more accurate and less expensive sequencing.

7. This is needed only if qPCR will be used to determine the ideal number of PCR cycles for library amplification.

8. RNA is more labile than DNA and should be worked with carefully. If possible, a dedicated area of the lab should be set aside for RNA work along with a set of pipettors and other

equipment. Care should be taken that all work is done in an RNase-free environment and with RNase-free equipment. RNaseZap wipes (ThermoFisher) are available that can be used to wipe down surfaces and equipment. ELIMINase (ThermoFisher) can also be used to remove RNase, DNase and DNA contamination. Always wear gloves, as human skin is a common source of RNases. RNA should also be kept on ice whenever possible to avoid degradation. Heating of RNA, except to denature, should be avoided but if necessary any metal ions should be removed by chelation with ethylenediaminetetraacetic acid (EDTA) to avoid metal ion scission of RNA. Add EDTA to a final concentration of 0.1 M.

9. Larger volumes will be needed with lower cell densities (e.g., 0.1). If the O.D. approaches 1.0 then only 1 mL should be used. Overloading the TRIzol chemical can lead to lower quality and quantity of output RNA.
10. Alternatively, cell pellets can be flash frozen in liquid N_2 and then stored at −80 °C, stored in a propriety RNA storage solution such as RNAlater (ThermoFisher) or stored directly in 1 mL of TRIzol at −80 °C. While these approaches may be necessary for certain experiments we have had good results with simply moving the pellets immediately to −80 °C. If RNA isolation is not possible at that point, their small size leads to rapid freezing.
11. Freeze-thawing cells will release intracellular RNases that can degrade RNA; adding TRIzol immediately will inhibit these RNases.
12. One advantage of TRIzol is the ability to collect RNA, DNA and proteins from the same sample. The lower phenol layer and interstitial DNA layer can be set aside for later isolation of proteins and DNA respectively if needed.
13. It is extremely important that only the aqueous layer is collected and none of the DNA or phenol layer. If any of the DNA layer or phenol layer is accidently drawn, up then do not combine it with the aqueous layer already collected. If the phenol/DNA and aqueous layers are accidently mixed, the samples can be centrifuged again to separate them. In general, it is best to collect slightly less RNA and be sure it is pure, rather than trying to collect all RNA possible and risk contaminating the whole sample with DNA or protein.
14. The TRIzol protocol recommended by the manufacturer suggests the addition of glycogen at this step, but we have found that it is usually not required and RNA can be precipitated with it later if necessary.
15. At this point the manufacturer of TRIzol indicates that RNA should be visible as a gel-like white pellet on the bottom of the

tube. However, we have found that this is not the case with *N. gonorrhoeae*, with the RNA often not visible.

16. Certain mutant strains that we have generated do not show a pellet but the RNA is still present.
17. The manufacturer of TRIzol does not suggest a second wash step, but we have found that it is a good way to ensure quality of the RNA.
18. It is critical here to remove all of the ethanol, either by pipetting or drying. Do not allowing drying to proceed for longer than 7 min, as RNA that is too dry can become difficult to dissolve.
19. The manufacturer of TRIzol suggests heating the RNA to aid its solubility. However, heating can also damage RNA and we generally do not do this.
20. RNA can be stored in H_2O or in RNA storage solutions at −80 °C for up to 6 months or at −20 °C for up to a month, with no loss of quality. It is advisable to avoid multiple freeze–thaw cycles for RNA and always to store it on ice when it is out of the freezer.
21. We use the DNA-free kit from ThermoFisher, but other kits are available. One advantage of this kit is that the DNase Inactivation Reagent can be used to remove DNase, avoiding the necessity of recleaning the RNA.
22. It is critical that this is done: if two much DNA/RNA is present in the sample, then the efficiency of the DNase is reduced.
23. All RNA-seq library preparation protocols include a step where RNA is degraded or fragmented to force the RNA samples into a uniform size distribution. Therefore, if RNA is partially degraded (RIN between 5.0 and 7.5) it can probably still be used for RNA-seq analysis. RNA samples with a RIN below 5.0 are probably unusable.
24. The gasket of the RNA priming station can sometimes become obstructed with gel–dye mix. If this happens, the best option is to either replace the gasket or use a 22 G needle to remove the gel–dye mix that may be clogging the opening.
25. If it does not, then there may be a leak between the chip and chip priming station. Retry this step, but if the leak persists, it may be necessary to replace the gasket.
26. Under no circumstances should any rRNA depletion method be used that depends on the presence of polyA tails. Most bacterial mRNAs do not contain polyA tails but certain species, including *N. gonorrhoeae*, encode polyadenylases that may form polyA tails on mRNA. Since the exact circumstances of this addition are unknown, polyA-dependent kits should not be used.

27. Most RNA-seq experiments follow the same general process. RNA is collected from a cell population after an experimental perturbation of interest or from a population of cells in a natural setting such as during in vivo infection. After an initial quality check, the RNA is converted to cDNA via reverse transcription and is fragmented into portions of approximately the same length. Adaptors are ligated sequentially to the 3′ and 5′ ends of the cDNA strands and a PCR step is used to amplify the resulting library. This is followed by annealing to a chip containing DNA strands bound to the surface that are homologous to the adaptors ligated to the cDNA fragment. Single strand cDNA fragments from the experimental output, sandwiched between adaptors and bound to the chip, are then sequenced. Each cDNA fragment bound to the chip, usually numbering in the millions-billions, is sequenced simultaneously. Fluorescent nucleotides of all four types are sequentially added, one at a time, and if they are homologous to the cDNA fragment, they are added to the growing chain of the opposite strand, emitting a fluorescent signal as they do so. This signal is read by the sequencer and the corresponding nucleotide is added to the growing sequence of cDNA in silico. This process is repeated for a set number of reactions, termed the read length, which is usually between 50 and 250 cycles. The output of this process is a collection of millions of reads, short sequences of cDNA, each resulting from the sequencing of a single cDNA insert, which can then be aligned to a genome of choice. After normalizing for gene size, the expression level of a particular gene is directly related to the number of reads that align to it, and by comparing the read alignment of an experimental condition to a control condition, changes in gene expression levels resulting from the perturbation can be determined.
28. Multiplexing refers to loading multiple cDNA libraries onto a single flow cell of the sequencer. This is often done to reduce the cost of sequencing. Typically, Illumina sequencing machines can easily provide 50 million reads which is far more than is needed for *N. gonorrhoeae* or other bacterial samples. Usually, ~5 million reads are sufficient for robust genome coverage. This means that up to ten samples can be loaded onto the same flow cell and sequenced simultaneously. Multiplexing in this way is done by including unique barcodes of six nucleotides on the adaptors. A known barcode is used for each sample and when these barcodes are sequenced along with the actual cDNA insert, the machine can determine which sequences belong to which experiment. A demultiplexing step is then done to separate reads into .fastq files for each experimental condition. The Biochain kit described here comes with 12 adaptors, and it is critical that different adaptors are used for each sample being run on the same flow cell.

29. cDNA libraries must be amplified by PCR prior to sequencing but the number of amplification cycles to be used is often hard to determine. Generally, between 12 and 18 cycles are carried out, but if too few are done there is insufficient cDNA for sequencing. Conversely, if too many are done, then PCR biases can become amplified in the library and affect downstream analysis. An initial qPCR run can reveal how many rounds of amplification are optimal for each sample.
30. While the High Sensitivity DNA Kit has a range of input concentrations between 0.005–0.5 ng/μL, this is for samples where a majority of the DNA will be one size (e.g., a PCR product of a single gene). The cDNA library is a collection of cDNA sequences across a broad range of sizes and as such no one cDNA sequence will be within the concentration that is detectable by the High Sensitivity DNA Kit.
31. Rockhopper provides an easy to use pipeline that handles all steps of analysis of RNA-seq data: alignment to reference genome, determining normalized gene expression levels and identifying differentially expressed genes. However, in certain cases it may not be the best option. There is a lack of control for certain steps, as the program is a more automatized approach compared to doing each step separately. Also, Rockhopper is not well suited to analyzing genomes that are still parts of a large number of different scaffolds or contigs, as each of these must be loaded into Rockhopper as individual files for alignment of reads. This may be an issue when using gonococcal genomes that are in an intermediate step of assembly. Several other programs exist to carry out each of the steps that Rockhopper does and are able to work with genomes on multiple contigs. A number of alignment programs are available and have been used with *N. gonorrhoeae* including BWA, SHRiMP2, Bowtie2 and certain proprietary programs that are meant to be used with certain sequencing technologies such as SOLiD or Ion Torrent [16–18]. The HTSeq program is fast, accurate and widely used for counting the assignment of reads to each gene [19]. Normalization of counts can be done with DESeq2 [20], which has been identified as a top performer through comparison studies [21, 22], or with edgeR [23]. DESeq2 can also identify differentially expressed genes between samples.

Acknowledgments

The authors wish to thank Drs. Ana Paula Lourenco and Phillip Balzano for their helpful contributions to the manuscript.

References

1. Daou N, Yu C, McClure R et al (2013) *Neisseria* prophage repressor implicated in gonococcal pathogenesis. Infect Immun 81 (10):3652–3661. https://doi.org/10.1128/iai.00298-13
2. Isabella VM, Clark VL (2011) Deep sequencing-based analysis of the anaerobic stimulon in *Neisseria gonorrhoeae*. BMC Genomics 12:51. https://doi.org/10.1186/1471-2164-12-51
3. Jackson LA, Day M, Allen J et al (2017) Iron-regulated small RNA expression as *Neisseria gonorrhoeae* FA 1090 transitions into stationary phase growth. BMC Genomics 18(1):317. https://doi.org/10.1186/s12864-017-3684-8.
4. McClure R, Balasubramanian D, Sun Y et al (2013) Computational analysis of bacterial RNA-Seq data. Nucleic Acids Res 41(14): e140. https://doi.org/10.1093/nar/gkt444
5. McClure R, Nudel K, Massari P et al (2015) The gonococcal transcriptome during infection of the lower genital tract in women. PLoS One 10(8):e0133982. https://doi.org/10.1371/journal.pone.0133982
6. McClure R, Tjaden B, Genco C (2014) Identification of sRNAs expressed by the human pathogen *Neisseria gonorrhoeae* under disparate growth conditions. Front Microbiol 5:456. https://doi.org/10.3389/fmicb.2014.00456
7. Remmele CW, Xian Y, Albrecht M et al (2014) Transcriptional landscape and essential genes of *Neisseria gonorrhoeae*. Nucleic Acids Res 42 (16):10579–10595. https://doi.org/10.1093/nar/gku762
8. Roberts SB, Spencer-Smith R, Shah M et al (2016) Correia repeat enclosed elements and non-coding RNAs in the *Neisseria* species. Microorganisms 4(3). https://doi.org/10.3390/microorganisms4030031
9. Wachter J, Hill SA (2015) Small transcriptome analysis indicates that the enzyme RppH influences both the quality and quantity of sRNAs in *Neisseria gonorrhoeae*. FEMS Microbiol Lett 362(4). https://doi.org/10.1093/femsle/fnu059
10. Yu C, McClure R, Nudel K et al (2016) Characterization of the *Neisseria gonorrhoeae* iron and fur regulatory network. J Bacteriol 198 (16):2180–2191. https://doi.org/10.1128/jb.00166-16
11. Bolger AM, Lohse M, Usadel B (2014) Trimmomatic: a flexible trimmer for Illumina sequence data. Bioinformatics 30 (15):2114–2120. https://doi.org/10.1093/bioinformatics/btu170
12. Benjamini Y, Hochberg Y (1995) Controlling the false discovery rate: a practical and powerful approach to multiple testing. J R Stat Soc B 57:289–300
13. Liu L, Li Y, Li S et al (2012) Comparison of next-generation sequencing systems. J Biomed Biotechnol 2012:251364. https://doi.org/10.1155/2012/251364
14. Quail MA, Smith M, Coupland P et al (2012) A tale of three next generation sequencing platforms: comparison of Ion Torrent, Pacific Biosciences and Illumina MiSeq sequencers. BMC Genomics 13:341. https://doi.org/10.1186/1471-2164-13-341
15. Salipante SJ, Kawashima T, Rosenthal C et al (2014) Performance comparison of Illumina and ion torrent next-generation sequencing platforms for 16S rRNA-based bacterial community profiling. Appl Environ Microbiol 80 (24):7583–7591. https://doi.org/10.1128/aem.02206-14
16. David M, Dzamba M, Lister D et al (2011) SHRiMP2: sensitive yet practical SHort Read Mapping. Bioinformatics 27(7):1011–1012. https://doi.org/10.1093/bioinformatics/btr046
17. Langmead B, Salzberg SL (2012) Fast gapped-read alignment with Bowtie 2. Nat Methods 9 (4):357–359. https://doi.org/10.1038/nmeth.1923
18. Li H, Durbin R (2010) Fast and accurate long-read alignment with Burrows-Wheeler transform. Bioinformatics 26(5):589–595. https://doi.org/10.1093/bioinformatics/btp698
19. Anders S, Pyl PT, Huber W (2015) HTSeq—a Python framework to work with high-throughput sequencing data. Bioinformatics 31(2):166–169. https://doi.org/10.1093/bioinformatics/btu638
20. Love MI, Huber W, Anders S (2014) Moderated estimation of fold change and dispersion for RNA-seq data with DESeq2. Genome Biol 15(12):550. https://doi.org/10.1186/s13059-014-0550-8.
21. Li P, Piao Y, Shon HS et al (2015) Comparing the normalization methods for the differential analysis of Illumina high-throughput RNA-Seq data. BMC Bioinformatics 16:347. https://doi.org/10.1186/s12859-015-0778-7
22. Zyprych-Walczak J, Szabelska A, Handschuh L et al (2015) The impact of normalization methods on RNA-seq data analysis. Biomed Res Int 2015:621690. https://doi.org/10.1155/2015/621690
23. Robinson MD, McCarthy DJ, Smyth GK (2010) edgeR: a bioconductor package for differential expression analysis of digital gene expression data. Bioinformatics 26 (1):139–140. https://doi.org/10.1093/bioinformatics/btp616

Chapter 12

Bioinformatics Workflow for Gonococcal Proteomics

Fadi E. El-Rami and Aleksandra E. Sikora

Abstract

High-throughput quantitative proteomics unravels secrets of *Neisseria gonorrhoeae* biology by profiling proteome responses to environmental and endogenous cues and opens translational research paths through identification of vaccine candidates, drug targets/virulence factors, and biomarkers. Bioinformatics tools and databases are indispensable for downstream analysis of proteomic datasets to generate biologically meaningful outcomes. In this chapter, we present a workflow for proteomic data analysis with emphasis on publicly available resources, software systems, and tools that predict protein subcellular localization (CELLO, PSORTb v3.0, SOSUI-GramN, SignalP 4.1, LipoP 1.0, TMHMM 2.0) and functional annotation (EggNOG-mapper 4.5.1., DAVID v6.8, and KEGG) of *N. gonorrhoeae* proteins. This computational step-by-step procedure may help to foster new hypotheses and to provide insights into the structure–function relationship of proteins.

Key words Bioinformatics, *Neisseria gonorrhoeae*, Quantitative proteomics, Functional enrichment, Pathway mapping, Data mining, Subcellular localization, CELLO, PSORTb, SOSUI-GramN, SignalP, LipoP, TMHMM, EggNOG-mapper, DAVID, KEGG

1 Introduction

Mass spectrometry-based proteomics has become a mainstream tool for global protein profiling, assessment of posttranslational modifications, and elucidation of interactions. Stable Isotope Labeling of Amino acids in Cell culture (SILAC) and isobaric peptide tagging (isobaric Tagging for Relative and Absolute Quantitation; iTRAQ, and Tandem Mass Tagging; TMT) represent the quantitative, reliable, and commonly used global proteomic approaches [1, 2]. In SILAC, "heavy" 13C or 15N-labeled-(L-arginine, L-lysine, L-leucine, and L-proline) and "light" (nonlabeled) amino acids incorporated into proteins from the culture medium are used for in vivo proteome assessment [3]. In contrast, TMT and iTRAQ capture a specific status of proteome by using amine-specific, stable-isotope reporters that label all peptides after protein extraction from individual biological samples and protease digestion in multiplexed experimental design [1, 2, 4,

Myron Christodoulides (ed.), *Neisseria gonorrhoeae: Methods and Protocols*, Methods in Molecular Biology, vol. 1997, https://doi.org/10.1007/978-1-4939-9496-0_12,

5]. SILAC [6, 7], iTRAQ [5, 8] and TMT [9] have been successfully used to study the *Neisseria gonorrhoeae* proteome and protein–protein interactions.

Data generated using global proteomic approaches necessitate sensitive and accurate algorithms for ion peak alignment, peptide/protein identification, and integrative in-depth bioinformatic analyses [10–12]. Proteomic databases such as Uniprot [13], GenBank (www.ncbi.nlm.nih.gov/genbank/), KEGG [14, 15], and DAVID [14, 16] provide a wealth of information about the proteome of a reference *N. gonorrhoeae* strain, FA1090, and assist with the interpretation of experimental findings. Another fundamental resource for proteomics coupled to bioinformatics investigations are the publicly available, fully characterized, and annotated genomes of the 2016 World Health Organization gonococcal reference strains [17]. This panel of 14 clinical isolates represents all available phenotypes, susceptible and resistant, to antimicrobials used in the past, currently recommended, or considered for future use in gonorrhea treatment.

1.1 Prediction Platforms for Subcellular Localization of Proteins

Proteomic studies often reveal new and hypothetical proteins with unknown functions and localization. Information about where proteins reside in a cell can aid in understanding the function and organization of bacterial cells in a holistic manner. Gonococcal proteins are distributed into five subcellular compartments: cytoplasm, inner membrane, periplasm, outer membrane, and extracellular. Several predictive software systems assess protein subcellular localization, including CELLO [18], PSORTb v3.0 [19], SOSUI-GramN [20], SignalP 4.1 [21], LipoP 1.0 [22], and TMHMM 2.0 [23] (Fig. 1). Given that these software programs rely primarily on different computational algorithms to predict protein localization, one strategy for an inclusive protein location assignment could be based on the "majority vote approach" [2, 5, 8].

1.1.1 CELLO

The CELLO platform refers to the physicochemical properties of amino acids in a given protein to infer its subcellular localization, and excels in the prediction of periplasmic and cytoplasmic proteins [18]. An updated version of the program, CELLO2GO, correlates one protein with another through the same functional annotation. With a query protein sequence, this web-based system utilizes BLAST to search for homologous sequences and to obtain Gene Ontology (GO)-type categories including subcellular compartment, biological process, and molecular function [24]. The CELLO2GO is a straightforward approach with the highest (95.9%) prediction accuracy of cytoplasmic proteins in gram-negative bacteria [25].

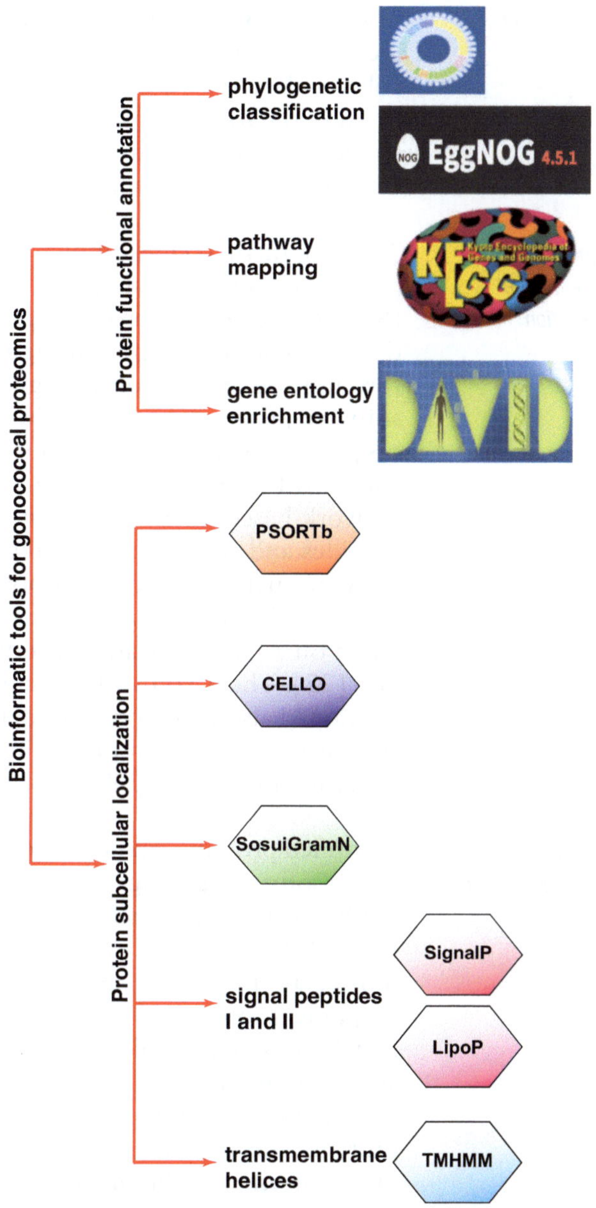

Fig. 1 The scheme outlining bioinformatic tools that can be applied in proteomic data analysis

1.1.2 PSORTb

The PSORTb program combines information accumulated from amino acid sequence comparison, presence of predicted transmembrane alpha helix segments, and domains, and is highly accurate in the prediction of outer membrane, inner membrane, periplasmic, and cytoplasmic proteins. The improved prediction of PSORTb is attributed to its high proteome-scale coverage and complementation with the Proteome Analyst predictions [19].

1.1.3 SOSUI-GramN

The SOSUI-GramN program was developed to assess the subcellular localization of proteins in gram-negative bacteria. The system collects data regarding the N- and C-terminal signal sequences, in addition to the physicochemical properties of the constituent amino acids, leading to high accuracy in predicting the extracellular, the outer membrane, and the inner membrane proteins [20].

1.1.4 Signal Peptide and Transmembrane Domain Prediction

To gain further clues about protein topology and localization, prediction of a classical signal peptide cleaved by signal peptidase I (SPaseI) and the lipoprotein signal peptide cleaved by signal peptidase II (SPaseII, lipoprotein signal peptidase Lsp) can be achieved through the application of SignalP 4.1 [21] and LipoP 1.0 [22] software programs, respectively. In addition, transmembrane helices present in integral membrane proteins can be predicted with the TMHMM 2.0 software (Fig. 1). The SignalP4.1 is a neural network-based method designed to distinguish between signal peptides and transmembrane regions, and it is a reliable predictor for gram-negative and gram-positive prokaryotes and eukaryotes [21]. The LipoP 1.0 predicts lipoproteins in gram-negative bacteria based on a Hidden Markov Model (HMM) achieving 96.8% accuracy [22]. The TMHMM protein topology prediction method was developed also based on a Hidden Markov Model and it has a high success rate (>99%) in discriminating membrane and soluble proteins, and in predicting 97–98% of the transmembrane helices [23].

1.2 Functional Annotation of Proteins

Identified proteins can be clustered into functionally related groups using software packages such as Cluster of Orthologous Genes (COG) [26, 27], which annotate protein functional classes, and Gene Ontology (GO) terms [24], that designate biological process, molecular function, and subcellular localization of proteins. Mapping proteins to diverse metabolic pathways, reactions, and metabolites is possible using the Kyoto Encyclopedia of Genes and Genomes (KEGG) software [14, 15]. KEGG provides graphic illustration of over 80 metabolic and structural pathways that link proteins, metabolites, and cell structures. EggNOG-mapper 4.5.1. [16] is another useful platform that combines COG and GO terms while the Database for Annotation, Visualization and Integrated Discovery (DAVID) v6.8 [28] offers functional clustering of proteins based on GO-type categories with references to GenBank, InterPro, and KEGG.

These aforementioned software packages make data analysis manageable and biologically meaningful, while gonococcal proteomics datasets [5, 8, 9] that are publicly available through ProteomeXchange Consortium [29, 30] foster data dissemination, standardization of working protocols and advance knowledge about *N. gonorrhoeae* biology.

2 Materials

2.1 Subcellular Localization Prediction

1. CELLO program (http://cello.life.nctu.edu.tw/) [18].
2. PSORTb v3.0 program (https://psort.hgc.jp/) [19].
3. SOSUI-GramN program (http://harrier.nagahama-i-bio.ac.jp/sosui/sosuigramn/sosuigramn_submit.html) [20].
4. SignalP 4.1 program (http://www.cbs.dtu.dk/services/SignalP/) [21].
5. LipoP 1.0 program (http://www.cbs.dtu.dk/services/LipoP/) [22].
6. TMHMM 2.0 Server v. 2.0 program (http://www.cbs.dtu.dk/services/TMHMM/) [23].

2.2 Functional Annotation of Identified Proteins

1. Mass spectrometry data that is acquired through the application of workflows coupling peptide/protein identification and quantification software programs (e.g., ProteinPilot, Proteome Discoverer, or ProgenesisQ) with multiple database search algorithms (SEQUEST, Z-Core, Mascot, Scaffold, and Byonic) for clustering and validation.
2. Computer with a processor of minimal speed 2.5 GHz; Operating System: Windows 7 or Mac OS 9 and above; memory capacity of 2 GB or more.

2.2.1 Rapid Orthology Assignments

1. EggNOG-mapper 4.5.1. program (http://eggnogdb.embl.de/#/app/home) [16].

2.2.2 GO-Term Enrichment Analysis

1. BlastP program (through PubMed: https://blast.ncbi.nlm.nih.gov/Blast.cgi).
2. Database for Annotation, Visualization and Integrated Discovery (DAVID) v6.8 program (https://david.ncifcrf.gov/) [28].

2.2.3 Pathway Analysis

1. Kyoto Encyclopedia of Genes and Genomes (KEGG) program (http://www.genome.jp/kegg/) [14, 15].

3 Methods

3.1 Subcellular Localization Prediction

3.1.1 CELLO Program

1. Visit the **CELLO** website at http://cello.life.nctu.edu.tw/.
2. Manually enter protein sequences in the FASTA format in the designated box, or upload the list of queries as a single file (Fig. 2).
3. Press ***Submit***.
4. **RESULTS** will be displayed on the "CELLO RESULTS" with reliability scores listed in descending order for each of the subcellular localizations: outer membrane, extracellular, inner membrane, periplasmic, or cytoplasmic.

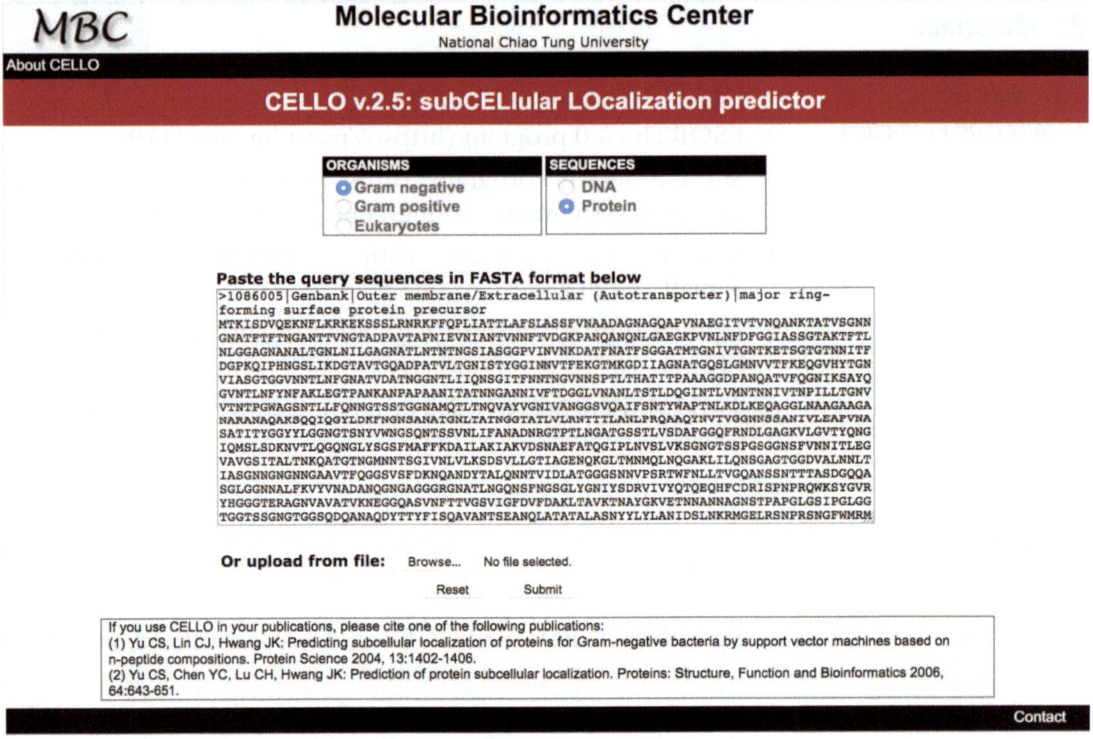

Fig. 2 The CELLO website, showing the input options

3.1.2 PSORTb Program

1. Visit the **PSORTb v3.0** website at https://psort.hgc.jp/.
2. Select the **PSORTb** prediction for gram-negative bacteria (Fig. 3).
3. Select ***Bacteria*** for the "organism type."
4. Select ***Negative*** for the "Gram stain."
5. Select the "output format" as ***Normal*** for single sequence or ***tab delimited*** in either ***Short*** or ***Long*** format for multiple sequences.
6. Select ***Send by email*** to have results sent to your selected email address (*see* **Note 1**).
7. Manually enter the protein sequences in the FASTA format in the designated box or upload the list of sequences as a single file.
8. Press ***Submit***.
9. The results will be displayed in the ***tab delimited Long*** format with the parameters shown in Table 1 (*see* **Notes 2** and **3**).

3.1.3 SOSUI-GramN Program

1. Visit the **SOSUI-GramN** website at http://harrier.nagahama-i-bio.ac.jp/sosui/sosuigramn/sosuigramn_submit.html.
2. Select a title for the search.

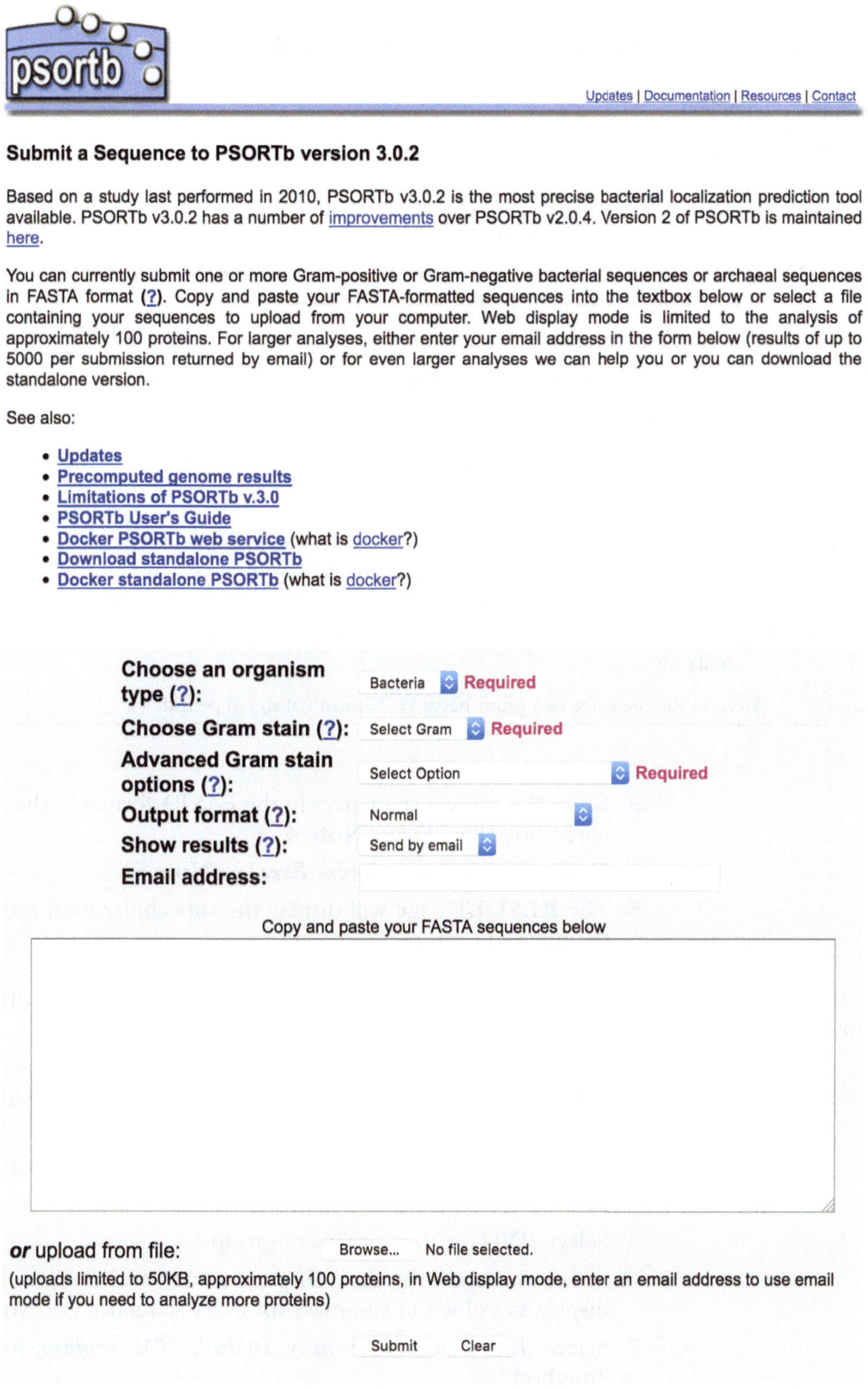

Fig. 3 The PSORTb v3.0.2 website, displaying the input options

Table 1
Parameters for the PSORTb program (*see* step 9 of Subheading 3.1.2)

Parameter	Function
CMSVM	Detects cytoplasmic membrane proteins
CytoSVM	Detects cytoplasmic proteins
ECSVM	Detects extracellular proteins
ModHMM	Detects transmembrane helices within the sequence and identifies cytoplasmic membrane protein if three or more transmembrane helices are predicted
Motif	Detects the localization site of a motif
OMPMotif	Detects outer membrane protein motifs
OMSVM	Detects outer membrane proteins
Profile	Identifies the localization site of a protein as provided in the PROSITE database
PPSVM	Detects periplasmic proteins
SCL-BLAST	Relevant to proteins of gram-positive bacteria deposited in the PSORTb database. Results reflect a match from a protein blast along with a link to the protein's record in the National Center for Biotechnology Information (NCBI) database
SCL-BLASTe	More specific than SCL-BLAST+, showing proteins that match the query with 100% similarity
Signal	Detects the presence of a gram-negative N-terminal signal peptide

3. Enter the protein sequences in the FASTA format in the designated box (Fig. 4) (*see* **Note 4**).
4. To execute the query, press ***Exec*** (*see* **Note 5**).
5. The **RESULT** page will display the subcellular localization of the queried proteins.

3.1.4 SignalP 4.1 Program

1. Visit the **SignalP 4.1** website at http://www.cbs.dtu.dk/services/SignalP/.
2. Enter the protein sequences in the FASTA format in the designated box, or upload the list of sequences as a single file (Fig. 5).
3. Select ***Gram-negative bacteria*** for the "organism group."
4. Select ***Default*** for the "D-cutoff values."
5. Select ***PNG*** for the "organism group."
6. Select ***Standard*** for the "output format" and the results will display as a plot and summary for every sequence (*see* **Note 6**).
7. Select ***Input sequences may include TM regions*** for the "method."
8. Keep the options provided for the "Positional limits" unselected.

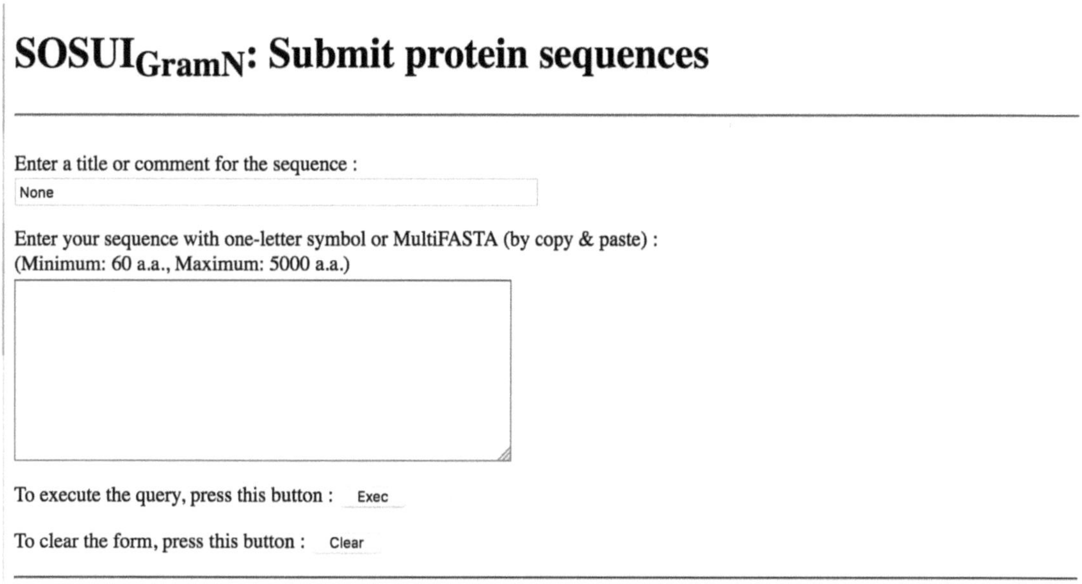

Fig. 4 The SOSUIGramN website, showing available input options

9. Press ***Submit***.
10. Results displayed on the result page will show a plot of the predicted signal peptide with the "SP" parameter indicating if a signal peptide was predicted (output: YES) or not (output: NO).

3.1.5 LipoP 1.0 Program

1. Visit the **LipoP 1.0** website at http://www.cbs.dtu.dk/services/LipoP/.
2. Enter the queries in the FASTA format in the designated box, or upload the list of protein sequences as a single file (Fig. 6).
3. Select ***Extensive, with graphics*** for the "output format" and the results will display as a plot and summary for every sequence (*see* **Note 6**).
4. Press ***Submit***.
5. The **Results** page will show the output with highlighted locations of predicted lipoprotein signal peptide "SPaseII" and a signal peptidase II cleavage site "CleavII."

3.1.6 TMHMM Server v. 2.0 Program

1. Visit the **TMHMM Server v. 2.0** website at http://www.cbs.dtu.dk/services/TMHMM/.
2. Enter the protein sequences in FASTA format (HTML 3.0 or higher) in the designated box or upload a file containing the list of sequences in FASTA format (Fig. 7).
3. Select output format ***Extensive, with graphics*** and the results will display as a plot and summary for every sequence. For multiple sequences, select ***Short button*** as an output format. This option produces one line of results per sequence query.

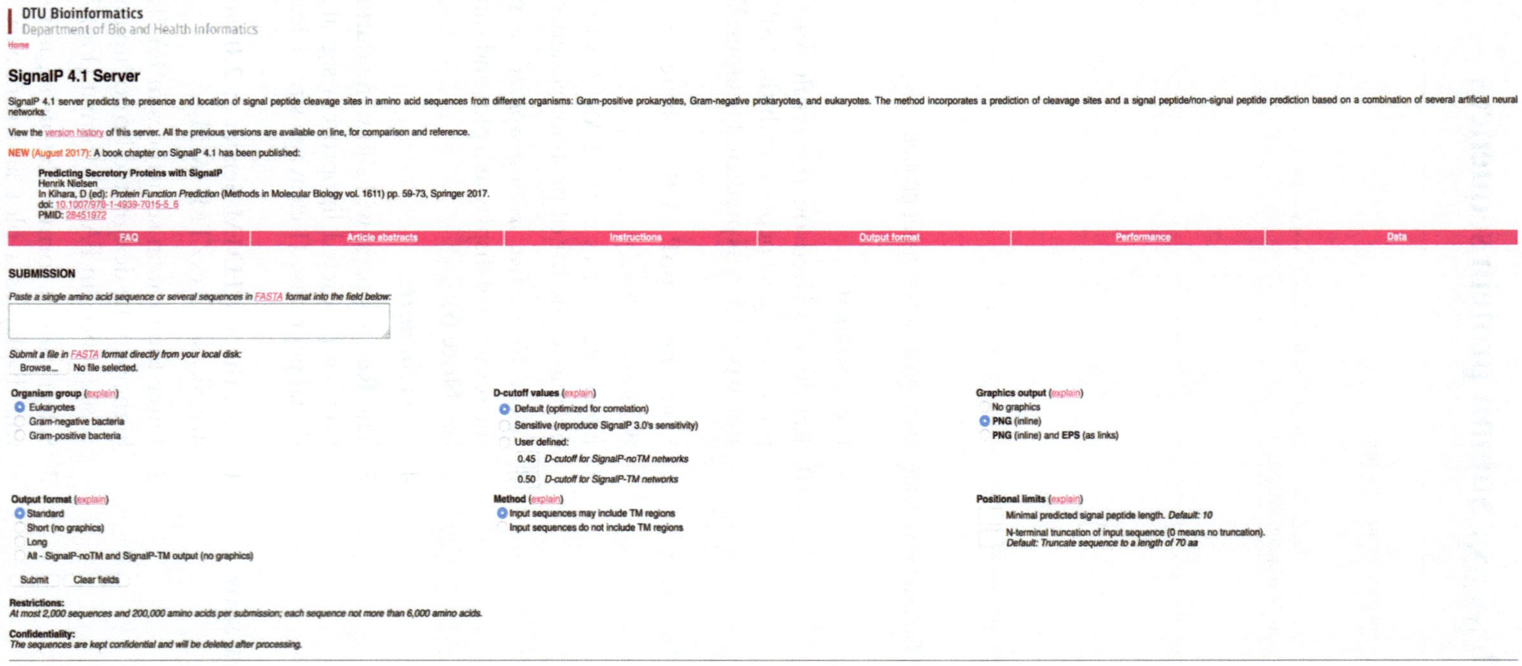

Fig. 5 The SignalP 4.1 website, featuring the input options

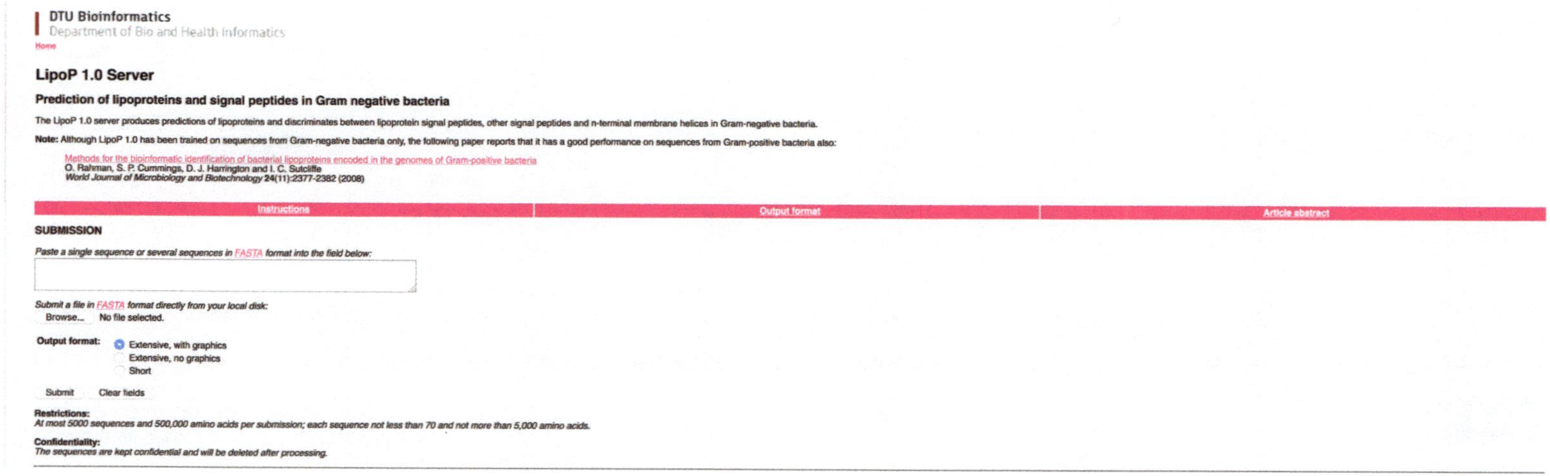

Fig. 6 The LipoP 1.0 website for prediction of signal peptide II sequence, showing different input options

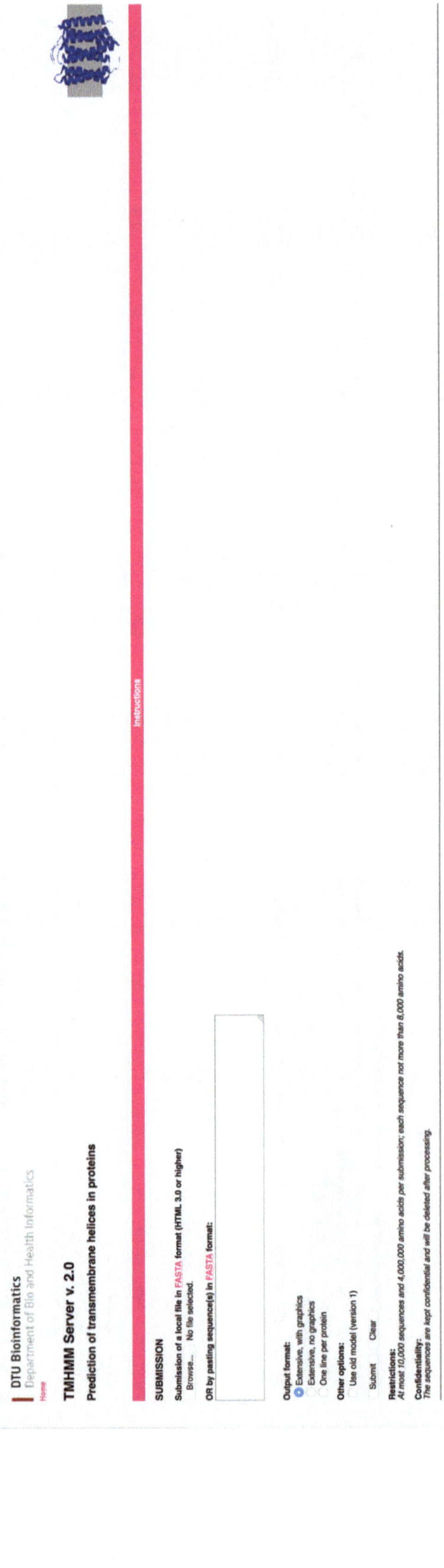

Fig. 7 The TMHMM 2.0 website for the prediction of protein transmembrane helices, displaying different input options

4. Press ***Submit***.
5. The **RESULTS** page will show output illustrated as a plot with identified transmembrane, extracellular (outside) and intracellular (inside) protein regions. In addition, the number of transmembrane fragments and the length of each hydrophobic region (the transmembrane region) will be shown with amino acid residues for each "TMhelix."

3.2 Functional Annotation of Identified Proteins

3.2.1 Rapid Orthology Assignments

Functional analyses of proteomic data sets start with the categorization of proteins into functional clusters, as designated by Cluster of Orthologous Genes (COG) groups and Gene Ontology (GO) terms (*see* **Note 7**). A COG group defines the general function of the orthologous cluster, while the GO term describes the "biological process," "molecular function," and "cellular component." The software EggNOG-mapper is a particularly useful tool for functional annotation of large sets of queries [15].

1. Enter the amino acid sequences in the FASTA format for all proteins of interest (*see* **Note 8**).
2. Visit the **EggNOG** 4.5.1 webpage (http://eggnogdb.embl.de/#/app/home).
3. Select the search engine ***eggNOG-mapper*** for analyzing the functional annotations of amino acid sequences (Fig. 8).
4. Insert the amino acid sequences and enter an email address to receive the results.
5. Two mapping modes are provided: ***DIAMOND*** and ***HMMer***. ***DIAMOND*** provides increased sensitivity for a greater number of sequences (100–1000 sequences), whereas ***HMMer*** provides better coverage for distant proteins at the expense of precision in detecting seed orthologs.
6. Selecting ***HMMer*** provides an additional option of using the ***HMM database*** which contains different clades as search pools.
7. ***Taxonomic Scope*** offers a plethora of clades to select from the ***HMM database***. Select ***Adjust automatically*** and the software will adjust the search to the "all clades deposited in database."
8. Choose ***Use all orthologs*** for better coverage or ***Restrict to one-to-one*** for better precision.
9. Finally, for the ***Gene Ontology evidence***, select ***Use non-electronic terms*** for better coverage or ***Use experimental only terms*** for better quality.
10. Press ***Run***. The results will be sent by email and will also be displayed on the webpage when the analysis is completed. For every protein query, the results will include seed ortholog, *E*-value, Score, GO terms (*see* **Note 9**), KEGG KO, COG category, and other functional descriptions.

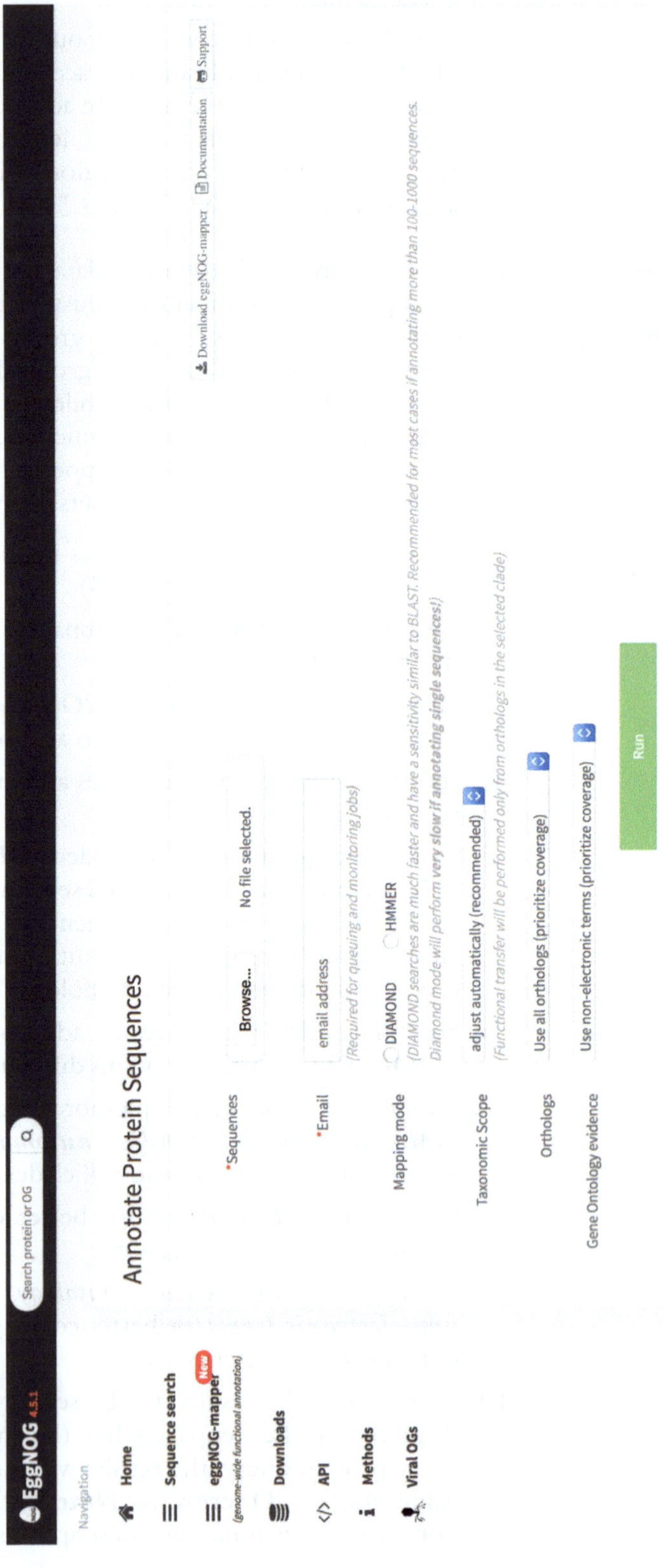

Fig. 8 The EggNOG 4.5.1. website, showing the eggNOG-mapper link on an upper left

3.2.2 GO-Term Enrichment Analysis

The GO-term enrichment analysis follows the COG and GO annotations. The aim of this step is to functionally cluster differentially regulated proteins under diverse conditions and to link observed phenotype(s) to their respective "prototype(s)." The *p*-value is the main statistical parameter in determining the over or under representation of a specific GO group among a set of proteins. The smaller the *p*-value, the more "enriched" a GO value is in one sample versus another. For large protein datasets, the DAVID software provides GO enrichment as well as pathway analysis (*see* **Note 10**).

1. Extract the list of all proteins of interest from the proteomics study. For *N. gonorrhoeae* FA1090, the nomenclature format should be "NGO" preceding directly the protein number, as follows: NGO#### (*see* **Note 11**).
2. If the query derives from gonococcal isolates other than FA1090, first use a BlastP (through PubMed: https://blast.ncbi.nlm.nih.gov/Blast.cgi) to find matching proteins and use the latter's ortholog in the format described above (*see* **Note 12**).
3. Visit the DAVID database webpage (https://david.ncifcrf.gov/).
4. Select the search engine ***Start Analysis*** using the upper left tool bar (Fig. 9).
5. Insert the protein names or upload the list file on the "Upload" page, at "Step 1: Enter Gene List".
6. Select ***LOCUS_TAG*** at "Step 2: Select Identifier".
7. Select ***Gene List*** at "Step 3: List Type".
8. Press ***Submit List*** at "Step 4: Submit List".
9. Select ***N. gonorrhoeae*** **FA1090** on the "List" page in "General List Manager."
10. Click on your selected species in the **Select Species box**.
11. A list of "Functional Annotation Tools" should appear on the right side. Select ***Functional Annotation Clustering***.
12. Functional annotation clustering based on enrichment scores, GO terms (*see* **Note 13**), count of proteins per clustering, and *p*-values will be listed in a table format.
13. For a simplified version of the findings, return to the "Annotation Summary Results" and select the "Functional Annotation Chart."
14. To obtain GO terms, Interpro annotation, KEGG pathway, and COG grouping for each protein query, return to the "Annotation Summary Results" and select the "Functional Annotation Table."

Fig. 9 The DAVID v6.8 website used for protein functional enrichment, featuring different input options. The names of genes were added in the space annotated on the left. Input options were included underneath

3.2.3 Pathway Analysis

The KEGG database is a great resource for deciphering the involvement of identified proteins in metabolic pathways (*see* **Note 14**).

1. Visit the KEGG webpage (http://www.genome.jp/kegg/).
2. In the empty box specified for code of entry KEGG organisms, type ***ngo*** or ***ngk*** to be directed to the pathways of *N. gonorrhoeae* FA1090 or *N. gonorrhoeae* NCCP11945, respectively (*see* **Note 15**; Fig. 10).
3. Select "Pathway Map" at the top left of the toolbar to view a list of KEGG pathways.
4. Choose any of the pathway(s) of interest for visualization. Every annotated protein associated with this pathway(s) will appear in a green box.
5. Select the green box for a protein of interest and assess the pathway(s) in which the protein is involved.

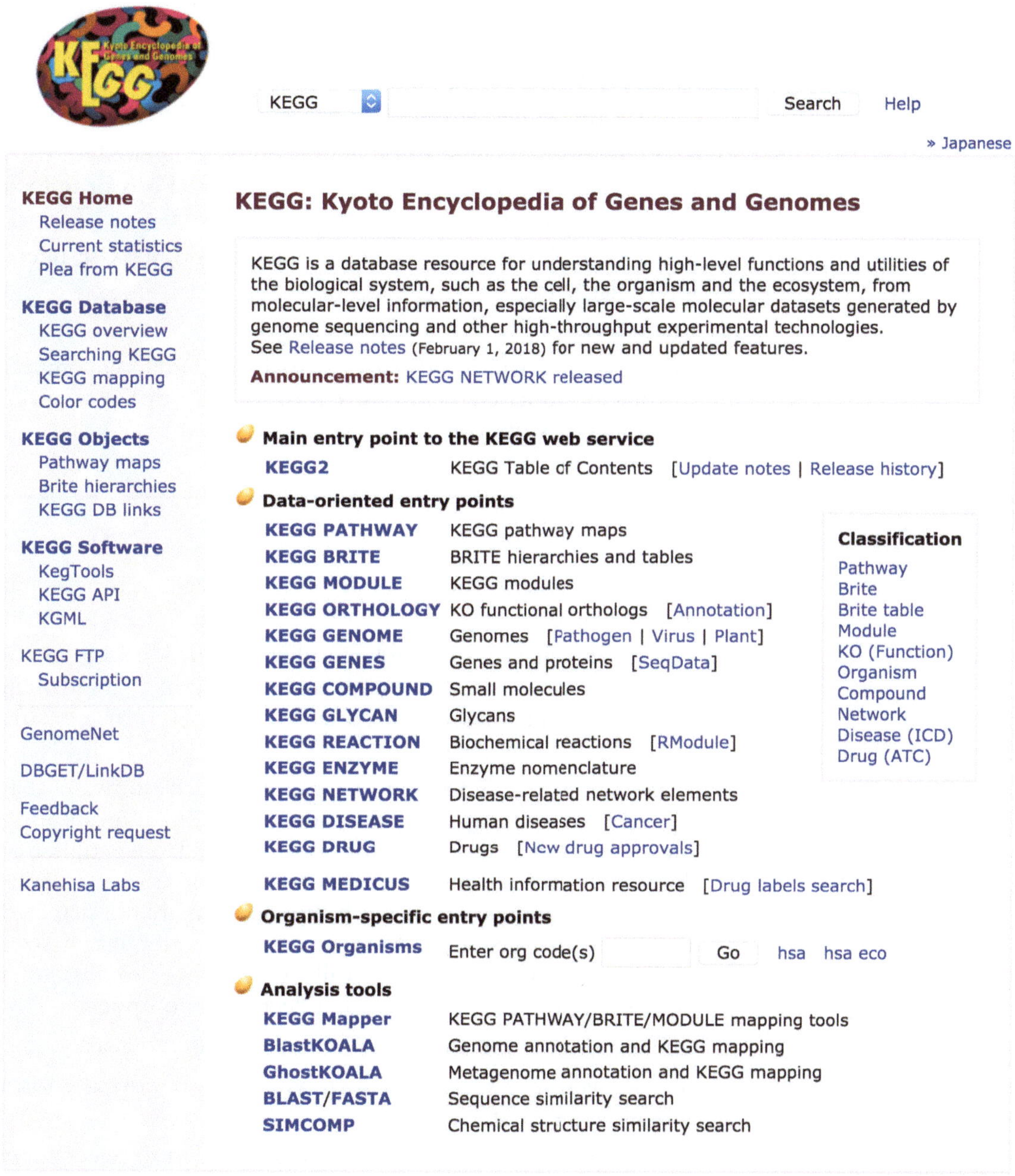

Fig. 10 The KEGG website for genome annotation and pathway mapping. The codes for KEGG organisms were included in the middle of the annotated space

6. Use the **KEGG** Mapper (http://www.genome.jp/kegg/mapper.html) for simultaneous pathway visualization of multiple proteins.
7. In the section "Pathway mapping tools," select the ***Search &Color Pathway.***

8. In the first box labeled "Search against," type ***ngo*** (for *N. gonorrhoeae* FA1090) or ***ngk*** (for *N. gonorrhoeae* NCCP11945).
9. Select ***KEGG identifiers*** for the parameter labeled ***Primary ID***.
10. Type the protein name and the color code designated for each protein per line. Type the list in the designated box or upload the file containing the query data (*see* **Notes 16** and **17**).
11. For a noncolored background, select the option "Use uncolored diagrams."
12. Press ***Exec***.
13. The next page will display links to pathway(s) that encompass the proteins queried. Select the pathway(s) of interest to check the colors' distribution (*see* **Note 18**).

4 Notes

1. It is safe to keep the results stored for further review and processing.
2. The cutoff value for subcellular localization site is a score of 7.5 or greater. If two sites have high scores, a flag of "This protein may have multiple localization sites" is also returned in the "Final Prediction" field.
3. There are some limitations to the PSORTb software that should be noted: (a) proteins residing at multiple localization sites, such as integral membrane proteins with large periplasmic domains can have multiple annotations and these proteins should be examined carefully; (b) lipoproteins are not detected in the current version; and (c) the current version emphasizes precision (specificity) over recall (sensitivity), leading to a higher confidence in results, but a decrease in output.
4. The minimum input sequence length is 60 amino acids.
5. Alternatively, press the tab ***Clear***, which will remove the form and allow a new start.
6. If using multiple sequences, it is recommended to select ***Short*** as output sequence. This will show one line of results per sequence.
7. It is noteworthy that not all proteins are fully annotated with their relevant COG groups. Therefore, we recommend searching for COG groups in orthologous proteins using Protein Identifier Cross-Reference: PICR (http://www.ebi.ac.uk/Tools/picr/).
8. FASTA sequences of *N. gonorrhoeae* FA1090 proteins can be downloaded from Genbank (www.ncbi.nlm.nih.gov/genbank/) or Uniprot databases [13].

9. A major drawback in using the GO terms for functional annotations is that 95% of the annotations are assigned in silico; only a minor portion are manually curated based on experimental findings [31].
10. In addition to GO enrichment for a protein set, DAVID offers additional features such as conversion of protein identifiers (e.g., GenBank accession numbers, locus tag, protein GI accession, Uniprot accession), mapping to diseases, and links to publications that experimentally verified the clustered function(s).
11. Although the GenBank and the Uniprot databases recognize *N. gonorrhoeae* FA1090 proteins with different nomenclatures, such as NGO#### and NGO_####, the DAVID database only recognizes *N. gonorrhoeae* FA1090 proteins in the NGO#### format.
12. Currently, the DAVID database recognizes the *N. gonorrhoeae* FA1090 proteome only. Other gonococcal strains are not identified.
13. The GO values are important for functional protein annotations but the calculated *p*-values may vary in comparison to the GO findings in other databases depending on type of the algorithm used [32].
14. Pathway analysis can help clarify the functional roles of some proteins with general GO terms such as "binding" or "cytoplasmic."
15. The KEGG database recognizes solely the proteomes of *N. gonorrhoeae* FA1090 and *N. gonorrhoeae* NCCP11945.
16. Use the NGO#### format, similar to the DAVID database. Choose the KEGG ID converter (http://www.kegg.jp/kegg/tool/conv_id.html) if you are using a different format than the NCBI or Uniprot Protein ID.
17. For different color designations, type the chosen color by the gene name in the same line.
18. KEGG annotates different colors to different pathways for easier visualization. Clicking at any part of a specific pathway will highlight the color of the whole pathway.

Acknowledgments

Funding was provided to A.E.S. by grant R01-AI117235 from the National Institute of Allergy and Infectious Diseases, National Institutes of Health.

References

1. Chahrour O, Cobice D, Malone J (2015) Stable isotope labelling methods in mass spectrometry-based quantitative proteomics. J Pharm Biomed Anal 113:2–20. https://doi.org/10.1016/j.jpba.2015.04.013
2. Zielke RA, Gafken PR, Sikora AE (2014) Quantitative proteomic analysis of the cell envelopes and native membrane vesicles derived from gram-negative bacteria. Curr Protoc Microbiol 34:1F 3 1–1F 316. https://doi.org/10.1002/9780471729259.mc01f03s34
3. Ong SE, Blagoev B, Kratchmarova I et al (2002) Stable isotope labeling by amino acids in cell culture, SILAC, as a simple and accurate approach to expression proteomics. Mol Cell Proteomics 1(5):376–386
4. Rauniyar N, Yates JR III (2014) Isobaric labeling-based relative quantification in shotgun proteomics. J Proteome Res 13 (12):5293–5309. https://doi.org/10.1021/pr500880b
5. Zielke RA, Wierzbicki IH, Baarda BI et al (2016) Proteomics-driven antigen discovery for development of vaccines against gonorrhea. Mol Cell Proteomics 15(7):2338–2355. https://doi.org/10.1074/mcp.M116.058800
6. Phillips NJ, Steichen CT, Schilling B et al (2012) Proteomic analysis of *Neisseria gonorrhoeae* biofilms shows shift to anaerobic respiration and changes in nutrient transport and outer membrane proteins. PLoS One 7(6): e38303
7. Solis N, Cordwell SJ (2011) Current methodologies for proteomics of bacterial surface-exposed and cell envelope proteins. Proteomics 11(15):3169–3189. https://doi.org/10.1002/pmic.201000808
8. Zielke RA, Wierzbicki IH, Weber JV et al (2014) Quantitative proteomics of the *Neisseria gonorrhoeae* cell envelope and membrane vesicles for the discovery of potential therapeutic targets. Mol Cell Proteomics 13 (5):1299–1317. https://doi.org/10.1074/mcp.M113.029538
9. El-Rami FE, Zielke RA, Wi T, Sikora AE, Unemo M (2019) Quantitative proteomics of the 2016 WHO *Neisseria gonorrhoeae* reference strains surveys vaccine candidates and antimicrobial resistance determinants. Mol Cell Proteomics 18(1):127–150. https://doi.org/10.1074/mcp.RA118.001125. Epub 2018 Oct 23. PMID: 30352803
10. Evanko D (2006) Systems biology for beginners. Nat Methods 3(12):964–965
11. Christodoulides M (2014) *Neisseria* proteomics for antigen discovery and vaccine development. Exp Rev Proteomics 11(5):573–591. https://doi.org/10.1586/14789450.2014.938640
12. Otto A, Becher D, Schmidt F (2014) Quantitative proteomics in the field of microbiology. Proteomics 14(4–5):547–565. https://doi.org/10.1002/pmic.201300403
13. UniProt C (2015) UniProt: a hub for protein information. Nucleic Acids Res 43(Database issue):D204–D212. https://doi.org/10.1093/nar/gku989
14. Kanehisa M, Sato Y, Kawashima M et al (2016) KEGG as a reference resource for gene and protein annotation. Nucleic Acids Res 44 (D1):D457–D462. https://doi.org/10.1093/nar/gkv1070
15. Kanehisa M, Furumichi M, Tanabe M et al (2017) KEGG: new perspectives on genomes, pathways, diseases and drugs. Nucleic Acids Res 45(D1):D353–D361. https://doi.org/10.1093/nar/gkw1092
16. Huerta-Cepas J, Szklarczyk D, Forslund K et al (2016) eggNOG 4.5: a hierarchical orthology framework with improved functional annotations for eukaryotic, prokaryotic and viral sequences. Nucleic Acids Res 44(D1): D286–D293. https://doi.org/10.1093/nar/gkv1248
17. Unemo M, Golparian D, Sanchez-Buso L et al (2016) The novel 2016 WHO *Neisseria gonorrhoeae* reference strains for global quality assurance of laboratory investigations: phenotypic, genetic and reference genome characterization. J Antimicrob Chemother 71(11):3096–3108. https://doi.org/10.1093/jac/dkw288
18. Yu CS, Chen YC, Lu CH et al (2006) Prediction of protein subcellular localization. Proteins 64(3):643–651. https://doi.org/10.1002/prot.21018
19. Yu NY, Wagner JR, Laird MR et al (2010) PSORTb 3.0: improved protein subcellular localization prediction with refined localization subcategories and predictive capabilities for all prokaryotes. Bioinformatics 26 (13):1608–1615. https://doi.org/10.1093/bioinformatics/btq249
20. Imai K, Asakawa N, Tsuji T et al (2008) SOSUI-GramN: high performance prediction for sub-cellular localization of proteins in gram-negative bacteria. Bioinformation 2 (9):417–421
21. Petersen TN, Brunak S, von Heijne G et al (2011) SignalP 4.0: discriminating signal

peptides from transmembrane regions. Nat Methods 8(10):785–786. https://doi.org/10.1038/nmeth.1701

22. Juncker AS, Willenbrock H, Von Heijne G et al (2003) Prediction of lipoprotein signal peptides in Gram-negative bacteria. Protein Sci 12(8):1652–1662. https://doi.org/10.1110/ps.0303703
23. Krogh A, Larsson B, von Heijne G et al (2001) Predicting transmembrane protein topology with a hidden Markov model: application to complete genomes. J Mol Biol 305 (3):567–580. https://doi.org/10.1006/jmbi.2000.4315
24. Harris MA, Clark J, Ireland A et al (2004) The gene ontology (GO) database and informatics resource. Nucleic Acids Res 32(Database issue):D258–D261. https://doi.org/10.1093/nar/gkh036
25. Yu CS, Cheng CW, Su WC et al (2014) CELLO2GO: a web server for protein subCELlular LOcalization prediction with functional gene ontology annotation. PLoS One 9(6):e99368. https://doi.org/10.1371/journal.pone.0099368
26. Tatusov RL, Galperin MY, Natale DA et al (2000) The COG database: a tool for genome-scale analysis of protein functions and evolution. Nucleic Acids Res 28(1):33–36
27. Guo FB, Xiong L, Teng JL et al (2013) Re-annotation of protein-coding genes in 10 complete genomes of *Neisseriaceae* family by combining similarity-based and composition-based methods. DNA Res 20 (3):273–286. https://doi.org/10.1093/dnares/dst009
28. Huang da W, Sherman BT, Lempicki RA (2009) Bioinformatics enrichment tools: paths toward the comprehensive functional analysis of large gene lists. Nucleic Acids Res 37(1):1–13. https://doi.org/10.1093/nar/gkn923
29. Vizcaino JA, Deutsch EW, Wang R et al (2014) ProteomeXchange provides globally coordinated proteomics data submission and dissemination. Nat Biotechnol 32(3):223–226. https://doi.org/10.1038/nbt.2839
30. Vizcaino JA, Csordas A, del-Toro N et al (2016) 2016 Update of the PRIDE database and its related tools. Nucleic Acids Res 44(D1): D447–D456. https://doi.org/10.1093/nar/gkv1145
31. Rhee SY, Wood V, Dolinski K et al (2008) Use and misuse of the gene ontology annotations. Nat Rev Genet 9(7):509–515. https://doi.org/10.1038/nrg2363
32. Khatri P, Draghici S (2005) Ontological analysis of gene expression data: current tools, limitations, and open problems. Bioinformatics 21 (18):3587–3595. https://doi.org/10.1093/bioinformatics/bti565

Chapter 13

Phenotypic MicroArray Screening of *Neisseria gonorrhoeae* in Chemically Defined Liquid Medium

Benjamin I. Baarda and Aleksandra E. Sikora

Abstract

Phenotype MicroArrays (PMs) provide a considerable benefit to the evaluation of potential vaccine/drug targets and the assessment of hypothetical protein function. Nearly 2000 conditions can be screened relatively quickly either to search for phenotypes associated with the loss of a protein or to understand metabolic differences between closely related bacterial isolates. The fastidious organism *Neisseria gonorrhoeae* presents an experimental challenge for phenotypic screening due to its nutrient restrictions and its autolytic activity upon reaching the stationary phase of growth. These limitations can be mitigated by modulating screening parameters. In this chapter, we describe a technique optimized for the phenotypic screening of *N. gonorrhoeae* FA1090 and isogenic mutant strains. Inoculum size and culturing times have been adjusted for growth in chemically defined, protein-free Graver-Wade liquid medium in the 96-well microtiter plate format employed by the PMs. With the conditions presented, highly reproducible gonococcal growth is achieved, and autolysis prior to the experimental endpoint is minimized.

Key words *Neisseria gonorrhoeae*, Phenotypic microarrays, Antibiotics, Biolog, Cell envelope, Vaccine antigens, Drug targets, Hypothetical proteins

1 Introduction

The application of standardized phenotypic testing can provide insights into *Neisseria gonorrhoeae* physiology, the diversity between clinical isolates, and the function of hypothetical gonococcal proteins. Hypothetical proteins present an intriguing experimental challenge because of the lack of information available and the potentially prohibitive number of conditions to test with a knockout mutant prior to the discovery of a phenotype for follow-up assessment. Biolog Phenotype MicroArray (PM) panels are useful tools for these studies. Each panel is designed to test different aspects of bacterial physiology, from nutrient utilization to responses to antibiotics, osmolytes, and various chemical probes [1]. With these tools, bacteria can be exposed to nearly 2000 conditions relatively quickly in a convenient 96-well microtiter

Myron Christodoulides (ed.), *Neisseria gonorrhoeae: Methods and Protocols*, Methods in Molecular Biology, vol. 1997,
https://doi.org/10.1007/978-1-4939-9496-0_13,

plate format. The growth of mutant bacteria can then be compared to wild type (WT) during exposure to each condition. PMs have been used to address various research questions, including (1) to examine whether metabolic changes were associated with the enhanced virulence of extraintestinal pathogenic *Escherichia coli* ST131 isolates [2]; (2) to assess the fitness costs associated with antimicrobial resistance acquisition [3]; (3) to elucidate the roles of *E. coli* two-component systems [4], as well as *xylA*, *phoP*, *malF*, *cyaA*, *astC*, and *fruR* gene products [5]; and (4) to explore the substrate specificity of the *Pseudomonas aeruginosa* ABC-type transporter NppA1A2BCD [6]. We recently applied PMs in the first such study in *N. gonorrhoeae* to gain insights into the functions of seven proteome-derived vaccine candidate proteins [7, 8] within the gonococcal cell envelope [9].

1.1 Important Considerations in PM Screening: Choice of Compounds

The choice of PM compounds is an important consideration during experimental design. The most appropriate libraries for the phenotypic analysis of *N. gonorrhoeae* are PM9, which contains compounds to probe bacterial responses to osmolytes, and PM11–PM20, which evaluate cell sensitivity to compounds that disrupt membrane integrity, including metals, antibiotics, small hydrophobic molecules, antimicrobial peptides, and dyes [1, 9]. The sole energy sources that *N. gonorrhoeae* employ—glucose, pyruvate, and lactate [10]—as well as the relatively narrow pH range of 5.8–8.3 within which the gonococcus is viable [11], preclude usage of the other available PM panels (various carbon, nitrogen, phosphorous, sulfur, and dipeptide nutrient sources, different nutrient supplements, or pH).

1.2 Important Considerations in PM Screening: Choice of Growth Medium

A second important consideration is the growth medium in which the phenotypic screening will be performed. The medium should support robust, reproducible growth and should not interfere with monitoring for changes induced by exposure to the tested compounds. We initially considered Graver-Wade liquid medium (GW) for our screening because of its successful use in a standardized in vitro time-kill curve assay with five World Health Organization *N. gonorrhoeae* reference strains exposed to nine different antibiotics [12]. An advantage of GW over gonococcal base liquid medium (GCBL) is that GW is a chemically defined, clear, colorless liquid medium that supports dense cultures from low starting inocula [13]. In addition, GW is protein-free, which enables simpler culture supernatant protein profiling [9]. In our study, gonococcal growth in GW resulted in consistent, reproducible growth kinetics for WT bacteria and seven mutant strains, in contrast to the widely variable growth rate observed for some of the mutants when cultured in GCBL [9]. These characteristics confirmed GW as the most useful medium for PM screening.

To facilitate the evaluation of vaccine/drug candidates, the study of gonococcal physiology, and the elucidation of the functions of hypothetical proteins, we describe here an optimized screening method intended for use with Biolog PM screening plates. The technique presented allows for robust, reproducible gonococcal growth and minimizes autolysis during stationary phase [14–18].

2 Materials

2.1 N. gonorrhoeae *Liquid Cultures and Phenotypic Screening*

1. M199 cell culture medium with Earle's salts, without L-glutamine, phenol red, or sodium bicarbonate. Single-use bottles for the formulation of 1.5 L of Graver-Wade (GW) Liquid Medium.
2. 1 M sodium hydroxide.
3. 1 M Hydrochloric acid.
4. pH meter (Mettler Toledo FiveEasy or equivalent).
5. 0.2 μm PES membrane vacuum filtration units, 250 or 500 mL.
6. Graver-Wade (GW) Liquid Medium: add 1 L of distilled H_2O to a clean beaker with a magnetic stir bar on a magnetic stirring plate. Add the contents of one single-use bottle of M199 cell culture medium with Earle's salts and without L-glutamine, sodium bicarbonate, or phenol red. Add additional 500 mL of distilled H_2O (*see* **Note 1**). Add the following medium components: 37 mM glucose (10.0 g), 17 mM ammonium bicarbonate (2.0 g), 4.9 mM sodium acetate trihydrate (1.0 g) (*see* **Note 2**), 3.4 mM L-glutamine (0.75 g), 919 μM spermidine (0.2 g) (*see* **Note 3**), 383 μM L-arginine (0.1 g), 245 μM hypoxanthine (dissolve 0.05 g in at least 2 mL of 1 M NaOH), 298 μM uracil (dissolve 0.05 g in at least 2 mL of 1 M NaOH), 252 μM oxaloacetate (0.05 g), 99 μM thiamine hydrochloride (0.05 g), 39 μM L-ornithine (0.01 g), 10 μM nicotinamide adenine dinucleotide (NAD, 0.01 g), 13 mM sodium DL-lactate (2.5 mL of 60% (w/w) syrup). Adjust pH to 6.8 with HCl (*see* **Note 4**) and sterilize by filtration through a 0.2 μm filter (*see* **Note 5**). Store the medium at 4 °C. Medium must be used within 3 weeks of formulation. Before use, prewarm medium at 37 °C.
7. GCBL: 15 g of BD Bacto Proteose Peptone #3 (VWR), 4 g of K_2HPO_4, 1 g of KH_2PO_4, 5 g of NaCl per liter, autoclaved to sterilize at 2.68 kg/cm^2 and 121 °C.
8. GCB plate: Difco Gonococcal medium base, BD (VWR): 18 g per 500 mL, autoclaved to sterilize as above.

9. Kellogg's supplement I: 40 g of glucose, 1 g of glutamine, 2 mg of cocarboxylase (thiamine pyrophosphate) per 100 mL, filtered to sterilize (*see* **Note 6**).
10. Kellogg's supplement II: 50 mg of $Fe(NO_3)_3{\cdot}9H_2O$ per 100 mL, filtered to sterilize.
11. *N. gonorrhoeae* strains frozen in 25% (v/v) glycerol and GCBL and stored at −80 °C.
12. Disposable sterile serological pipettes (5, 10, and 25 mL).
13. Sterile 15 and 50 mL conical tubes.
14. 1 mL disposable cuvettes.
15. 9 cm diameter petri dishes.
16. Sterile swabs (Puritan or equivalent).
17. Wooden applicators (Puritan or equivalent), autoclaved to sterilize as above.
18. Disposable pipette tips (1000, 100, and 10 μL).
19. Sterile 96 well flat bottom microtiter plates.
20. Biolog Phenotype MicroArray plates PM9, PM11C, PM12B, PM13B, PM14A, PM15B, PM16A, PM17A, PM18C, PM19, and PM20B (http://www.biolog.com/products-static/phenotype_microbial_cells_use.php).
21. Multichannel pipettors (at least two 12 channel pipettors, one capable of dispensing between 20 and 200 μL and the other between 1 and 10 μL).
22. Sterile reagent reservoirs.
23. Humidified CO_2 incubator (5% (v/v) CO_2 added to ambient air, temperature-controlled (37 °C)).
24. Orbital shaker (VWR Advanced Dura-shaker or equivalent).
25. Plate reader (BioTek Synergy HT or equivalent, capable of measuring optical density (OD) at $\lambda_{600\ nm}$ (OD_{600}) in a 96-well plate format).

2.2 Data Analysis

1. Excel computer program.
2. The R program for statistical computing and graphics if clustering analysis is a preferred method. It is a free software environment (https://www.r-project.org/).

3 Methods

3.1 N. gonorrhoeae Liquid Cultures and Phenotypic Screening

The procedure presented here was optimized for phenotypic screening of *N. gonorrhoeae* FA1090 and knockout strains derived from this background [9]. However, this technique could be adapted to assess the activity of novel antimicrobial compounds

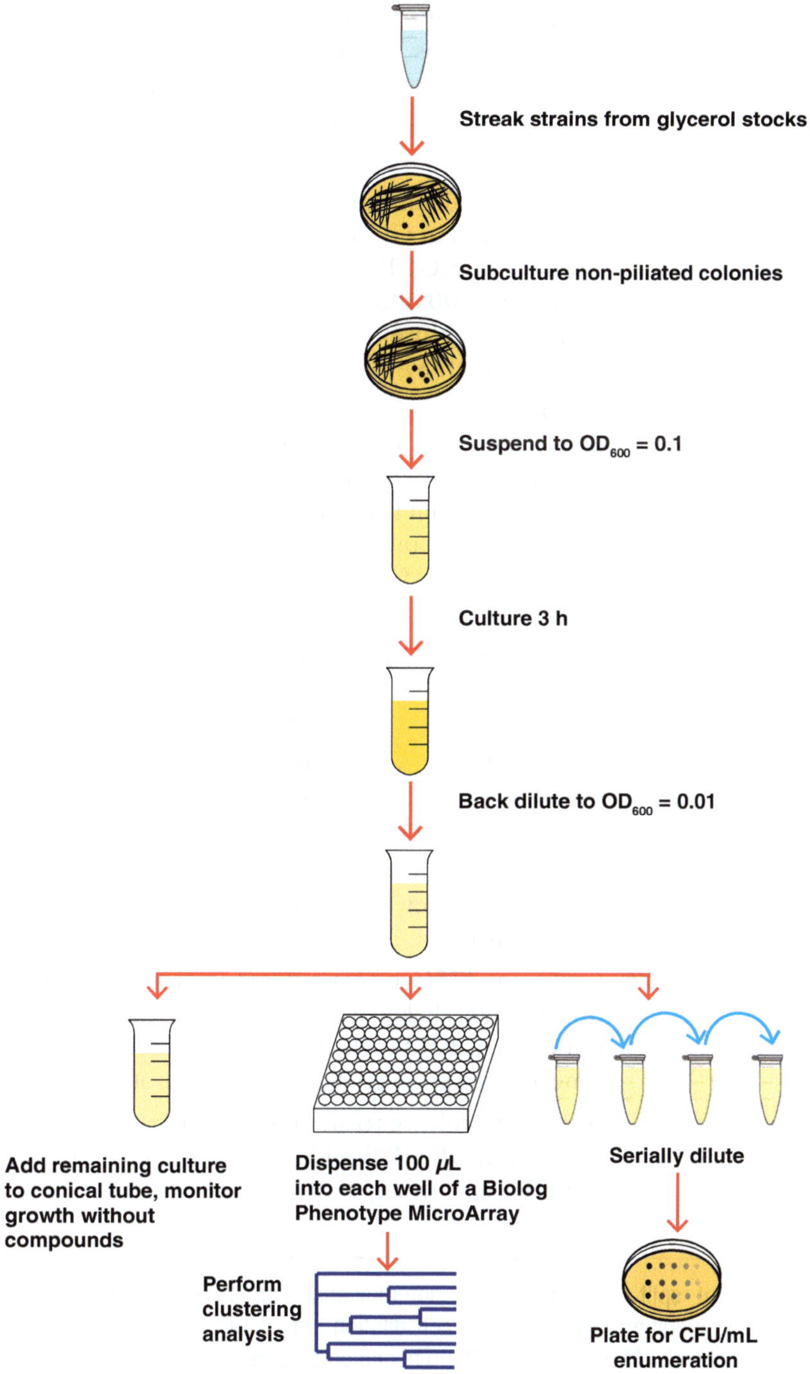

Fig. 1 Flowchart of optimized phenotypic microarray screening protocol for *Neisseria gonorrhoeae*. Streak gonococcal strains onto GCB plates from glycerol stocks stored at −80 °C. Incubate plates at 37 °C in a humidified environment with 5% (v/v) CO_2 for 18–20 h. Subculture nonpiliated colonies onto fresh GCB plates, and after incubation as described in **step 2** of Subheading 3.1, suspend nonpiliated colonies in GW liquid medium to an OD_{600} of 0.1. Culture the suspensions at 37 °C in a humidified environment with 5% (v/v) CO_2 with shaking at 220 rpm for 3 h. Dilute cultures to an OD_{600} of 0.01 in 20 mL of GW liquid medium. At this

against *N. gonorrhoeae* or for similar screening in heterologous strains. Use of this procedure with strains other than FA1090 may require some optimization (e.g., inoculum size and incubation times). A schematic of the screening procedure is presented in Fig. 1.

1. Scrape a small amount of frozen stock of each *N. gonorrhoeae* strain, stored in GCBL with 25% (v/v) glycerol at −80 °C, and streak onto a GCB plate supplemented with Kellogg's supplement I (1/100 dilution) and Kellogg's supplement II (1/1000 dilution; *see* **Note 7**). Incubate at 37 °C in a humidified environment with 5% (v/v) CO_2 for 18–20 h.
2. Using a stereomicroscope, pick single nonpiliated colonies of each strain with a sterile wooden applicator. Nonpiliated colonies appear flatter and more diffuse than piliated colonies, which are smaller and have a raised appearance (*see* **Note 8**). Subculture nonpiliated colonies onto two new GCB plates and incubate at 37 °C in a humidified environment with 5% (v/v) CO_2 for 18–20 h.
3. Prewarm GW liquid medium at 37 °C before initiating liquid cultures.
4. Using a sterile swab, collect nonpiliated bacteria from both plates and suspend in 3 mL of prewarmed GW liquid medium. Measure the OD_{600} of the suspension with a spectrophotometer.
5. Adjust the suspension to an OD_{600} of 0.1 at a final volume of 10 mL in prewarmed GW liquid medium, in a 50 mL conical tube.
6. Culture at 37 °C in a humidified environment with 5% (v/v) CO_2 on an orbital shaker at 220 rpm for 3 h (*see* **Notes 9** and **10**). The optical density of each culture should at least double during this time; do not proceed if cultures have not doubled.
7. Adjust the culture to an OD_{600} of 0.1 at a final volume of 5 mL in prewarmed GW liquid medium. Verify dilution by measuring the OD_{600} with a spectrophotometer.

Fig. 1 (continued) point, split cultures for three different analyses: (1) Serially dilute cultures (20 μL into 180 μL of prewarmed medium), spot 5 μL of each dilution onto a GCB plate, culture plate at 37 °C in a humidified environment with 5% (v/v) CO_2 for 18–20 h, and count colonies for CFU/mL enumeration of starting inoculum. (2) Dispense 100 μL of culture to each well of the PM plate; measure OD_{600} of plate; culture plate at 37 °C in a humidified environment with 5% (v/v) CO_2 with shaking at 220 rpm for 6 h; and measure OD_{600} every 2 h. Data from culture density readings can be used for clustering analysis of the phenotypic relatedness. (3) Return remaining culture (~10 mL) to a 50 mL conical tube and incubate alongside the screening plates. Measure OD_{600} of cultures every 2 h to monitor growth of strains without compounds. *GCB* gonococcal base medium, *GW* Graver-Wade liquid medium, *OD_{600}* optical density at $\lambda_{600\ nm}$, *CFU* colony forming unit

8. Dilute the culture 1/10 by combining 18 mL of prewarmed GW liquid medium with 2 mL of adjusted culture, for a final OD_{600} of 0.01 (*see* **Note 11**).
9. Using a multichannel pipettor, dispense 100 μL of $OD_{600} = 0.01$ culture into each well of the PM plate (*see* **Notes 12** and **13**). Pipette up and down to mix and to ensure compounds are uniformly suspended.
10. Serially dilute samples of each culture tenfold from 10^{-1} to 10^{-4} and spot 5 μL of each dilution onto a GCB plate and incubate at 37° in 5% CO_2 for 18–20 h to enumerate the colony forming units (CFU)/mL of the starting inoculum.
11. Culture PM plates at 37 °C in a humidified environment with 5% (v/v) CO_2 on an orbital shaker at 220 rpm for 6 h. Measure OD_{600} of plate at the initiation of the experiment and every 2 h using a plate reader (BioTek Synergy HT or equivalent) and continue incubation (*see* **Note 14**).
12. Export OD_{600} readings to Excel or a similar spreadsheet program.

3.2 Data Analysis

1. Using Excel, divide the OD_{600} value of each mutant strain for each well at the experimental endpoint by the OD_{600} value of the WT strain for the corresponding well at the experimental endpoint.
2. Perform a $\log_2$ transformation of the mutant/WT ratio by entering the following formula in the target cell: = LOG([reference of cell to be transformed], 2).
3. Select an appropriate cutoff value to distinguish "hits" from normal variation (*see* **Notes 15** and **16**).
4. Manually curate data to remove hits for which any of the following are true:
   (a) Reference strain and mutant OD_{600} values decrease over the course of the experiment.
   (b) The tested strain exhibited no or inconsistent growth for "beneficial" hits.
   (c) The reference strain exhibited no or inconsistent growth for "detrimental" hits.
5. To perform clustering analysis, compile the data based on those compounds that have similar effects on multiple tested strains (e.g., mutants). For this analysis, the magnitude of the effect and the concentration of the compound should be ignored.
6. Tested strains can be clustered based on their phenotypic similarity by hierarchical clustering, created on binary matrices that take into account (a) which of the tested strains were benefitted by each compound and (b) which of the tested strains were

attenuated by each compound. This analysis will determine how functionally similar the tested strains are, based on their responses to common compounds. Hamming distance, an informatics technique to quantify the number of differences between two pieces of data (in this case, the number of compounds that have different effects on two mutant strains) can then be used to determine the distance between each strain under investigation. A larger Hamming distance indicates that the two strains were affected by a more diverse set of compounds (*see* **Note 17**).

4 Notes

1. A small amount of water can be added to the M199 bottle and shaken to ensure that all powder is dissolved and combined with the medium preparation.
2. Different hydrates of sodium acetate can be used. Ensure that the molarity is correct for the hydrate being used.
3. Spermidine should be stored tightly closed at 4 °C, preferably with a desiccant pack. Parafilm can be used to ensure that the bottle is sealed.
4. Ensure that the pH meter is calibrated prior to adjusting the pH of the GW medium.
5. Unless GW medium will be used all at the same time, it can be filtered into several containers to minimize potential contamination.
6. Slowly add glucose to Kellogg's supplement I, allowing one portion to dissolve before the next addition.
7. Gonococci (especially strain FA1090) do not grow well on old GCB plates (3 or more days after pouring). This may be due to evaporation (i.e., the plates are too dry) or breakdown of the supplements. The iron in Kellogg's supplement II may oxidize and no longer be available for gonococcal growth. Fresh plates should be prepared 1–2 days prior to inoculation from freezer stocks or subculturing.
8. Piliated bacteria tend to clump in suspensions. Using nonpiliated bacteria in liquid culture will ensure that suspensions are homogenous.
9. For gonococcal culture in GW with shaking in a 5% (v/v) CO_2 environment, either a standalone CO_2 shaking incubator can be used, or an orbital shaker platform, such as the VWR Dura Shaker, can be placed inside a stationary CO_2 incubator chamber.

10. The initial liquid growth period may need to be adjusted for strains or mutants with altered growth rates. OD_{600} values should at least double before proceeding to the next step.
11. This second dilution will standardize the cultures at an OD_{600} of 0.01. The low starting density prevents cultures from reaching stationary phase and undergoing autolysis before the experimental endpoint (i.e., 6 h; *see* Subheading 3.2, **step 11**).
12. Biolog Phenotype MicroArray plates are 96-well half-volume microtiter plates. Volumes of 100 μL must be added to each well to ensure that compounds are at the correct concentration. Each plate contains compounds at four or more different concentrations in a dilution series; therefore, one plate must be used for each strain.
13. Biolog guidelines suggest the addition of their Redox dye, a tetrazolium-based dye that is reduced by actively respiring bacteria and generates a purple formazan by-product. However, we found addition of the dye was detrimental to gonococcal growth in GW medium. We therefore omitted the dye during screening.
14. After addition to Phenotype MicroArray plates, cultures can be returned to a 50 mL conical tube and cultured alongside the plates to monitor growth in the absence of compounds and to ensure that strains grow reproducibly. Measure OD_{600} of cultures when plates are removed for OD_{600} measurement.
15. We selected ≥ 0.5 and ≤ -0.5 to differentiate "beneficial" and "detrimental" hits, respectively. These values are equivalent to a 1.4-fold increase or decrease in mutant OD_{600} values compared to WT.
16. Repeating these screenings can be cost prohibitive. Therefore, follow-up testing is performed with "hits" identified during PM screening to repeat specific experiments. For this reason, we do not apply statistics during our data analyses.
17. After using Hamming distance to count the number of elements that differed between two mutant strains (Hamming distance is equivalent to Manhattan distance for binary data), average linkage was used to compute distances between clusters of mutant strains. Other linkage methods (single and complete) could also be explored, although all three methods resulted in similar clusters in our investigation.

Acknowledgments

Funding was provided to A.E.S. by grant R01-AI117235 from the National Institute of Allergy and Infectious Diseases, National Institutes of Health.

References

1. Shea A, Wolcott M, Daefler S et al (2012) Biolog phenotype microarrays. Methods Mol Biol 881:331–373. https://doi.org/10.1007/978-1-61779-827-6_12
2. Alqasim A, Emes R, Clark G et al (2014) Phenotypic microarrays suggest *Escherichia coli* ST131 is not a metabolically distinct lineage of extra-intestinal pathogenic *E. coli*. PLoS One 9(2):e88374. https://doi.org/10.1371/journal.pone.0088374
3. Reales-Calderon JA, Blanco P, Alcalde-Rico M et al (2016) Use of phenotype microarrays to study the effect of acquisition of resistance to antimicrobials in bacterial physiology. Res Microbiol 167(9-10):723–730. https://doi.org/10.1016/j.resmic.2016.04.004
4. Zhou L, Lei XH, Bochner BR et al (2003) Phenotype microarray analysis of *Escherichia coli* K-12 mutants with deletions of all two-component systems. J Bacteriol 185 (16):4956–4972
5. Bochner BR, Gadzinski P, Panomitros E (2001) Phenotype microarrays for high-throughput phenotypic testing and assay of gene function. Genome Res 11 (7):1246–1255. https://doi.org/10.1101/gr.186501
6. Pletzer D, Braun Y, Dubiley S et al (2015) The *Pseudomonas aeruginosa* PA14 ABC transporter NppA1A2BCD is required for uptake of peptidyl nucleoside antibiotics. J Bacteriol 197(13):2217–2228. https://doi.org/10.1128/JB.00234-15
7. Zielke RA, Wierzbicki IH, Weber JV et al (2014) Quantitative proteomics of the *Neisseria gonorrhoeae* cell envelope and membrane vesicles for the discovery of potential therapeutic targets. Mol Cell Proteomics 13 (5):1299–1317. https://doi.org/10.1074/mcp.M113.029538
8. Zielke RA, Wierzbicki IH, Baarda BI et al (2016) Proteomics-driven antigen discovery for development of vaccines against gonorrhea. Mol Cell Proteomics 15(7):2338–2355. https://doi.org/10.1074/mcp.M116.058800
9. Baarda BI, Emerson S, Proteau PJ et al (2017) Deciphering function of new gonococcal vaccine antigens using phenotypic microarrays. J Bacteriol 199:e00037. https://doi.org/10.1128/JB.00037-17
10. Morse SA (1978) The biology of the gonococcus. CRC Crit Rev Microbiol 7(2):93–189
11. Morse SA, Hebeler BH (1978) Effect of pH on the growth and glucose metabolism of *Neisseria gonorrhoeae*. Infect Immun 21(1):87–95
12. Foerster S, Golparian D, Jacobsson S et al (2015) Genetic resistance determinants, *in vitro* time-kill curve analysis and pharmacodynamic functions for the novel topoisomerase II inhibitor ETX0914 (AZD0914) in *Neisseria gonorrhoeae*. Front Microbiol 6:1377. https://doi.org/10.3389/fmicb.2015.01377
13. Wade JJ, Graver MA (2007) A fully defined, clear and protein-free liquid medium permitting dense growth of *Neisseria gonorrhoeae* from very low inocula. FEMS Microbiol Lett 273(1):35–37. https://doi.org/10.1111/j.1574-6968.2007.00776.x
14. Elmros T, Burman LG, Bloom GD (1976) Autolysis of *Neisseria gonorrhoeae*. J Bacteriol 126(2):969–976
15. Elmros T, Sandstrom G, Burman L (1976) Autolysis of *Neisseria gonorrhoeae*. Relation between mechanical stability and viability. Br J Vener Dis 52(4):246–249
16. Hebeler BH, Young FE (1976) Mechanism of autolysis of *Neisseria gonorrhoeae*. J Bacteriol 126(3):1186–1193
17. Hebeler BH, Young FE (1975) Autolysis of *Neisseria gonorrhoeae*. J Bacteriol 122 (2):385–392
18. Morse SA, Bartenstein L (1974) Factors affecting autolysis of *Neisseria gonorrhoeae*. Proc Soc Exp Biol Med 145(4):1418–1421

Chapter 14

Generation of Metal-Depleted Conditions for In Vitro Growth of *Neisseria gonorrhoeae*

Cynthia Nau Cornelissen

Abstract

Neisseria gonorrhoeae employs high-affinity metal acquisition systems to obtain necessary nutrients, such as iron (Fe) and zinc (Zn) from the environment. Because growth and replication depend upon successful metal acquisition, these high-affinity uptake systems are important virulence factors. Expression of metal acquisition systems is tightly controlled and preferentially expressed under low-metal conditions. Therefore, in order to optimally produce these transport proteins and study them in vitro, growth media must be deployed that mimic low-metal conditions. This chapter describes the chelators, media, and culturing conditions that can generate low-metal in vitro growth conditions.

Key words Iron, Zinc, Transferrin, Calprotectin, *Neisseria gonorrhoeae*, Chelator

1 Introduction

Neisseria gonorrhoeae is a human specific pathogen that does not naturally infect any other animal [1]. This selectivity is, in large part, due to the specificity that this pathogen demonstrates for many nutrients and receptors during the pathogenesis process [2–5]. In an effort to thwart bacterial growth and virulence, human hosts deploy a strategy known as "nutritional immunity" [6–8], a term used to describe the phenomenon of metal sequestration by the host in order to inhibit microbial replication. To overcome nutritional immunity, pathogens, including *N. gonorrhoeae*, produce high-affinity metal acquisition systems to compete with the sequestration proteins for the necessary metals (for review, *see* [9]). *N. gonorrhoeae*, as a human-adapted pathogen, has evolved an even more simple, yet elegant, strategy than that deployed by other bacteria. *N. gonorrhoeae* directly binds to and extracts the metals from human nutritional immunity proteins (for review, *see* [9]). Nutritional immunity proteins that sequester iron (Fe) include transferrin and lactoferrin. Proteins that prevent microbial access to zinc (Zn) include calprotectin and other S100

Myron Christodoulides (ed.), *Neisseria gonorrhoeae: Methods and Protocols*, Methods in Molecular Biology, vol. 1997, https://doi.org/10.1007/978-1-4939-9496-0_14,

proteins [10]. Receptor/transporters that enable *N. gonorrhoeae* to use these human proteins as metal sources are bona fide virulence factors as they directly contribute to the virulence of the pathogen [11, 12]. The metal uptake systems are only produced when metal concentrations are low, as in the host [13, 14]. Thus, to study these virulence factors in vitro, we must mimic these low-metal conditions to maximize production. Conditions for growing metal-deprived *N. gonorrhoeae* can be generated by adding specific chelators to complex medium or by growing in a defined medium reconstituted with water lacking contaminating metals. This chapter discusses the chelators, media and culturing conditions that can mimic a low-metal growth environment.

2 Materials

2.1 Acid-Washed Glassware

1. 1% (v/v final concentration) Aquet detergent.
2. Concentrated nitric acid (*see* **Note 1**).
3. Aluminum foil or metal caps for glassware.
4. Hot air oven (temperature controlled up to 220 °C).

2.2 Growing N. gonorrhoeae Under Low Fe or Low Zn Conditions in GC Broth

1. GC broth: dissolve 15 g of proteose peptone #3, 4 g of K_2HPO_4, 1 g of KH_2PO_4 and 5 g of NaCl in deionized H_2O, up to a final volume of 1 L.
2. 50 mM Desferal (deferoxamine mesylate) stock solution in deionized H_2O (*see* **Note 2**). Desferal is a ferric (Fe) chelator.
3. 40 mM TPEN (*N,N,N′,N′*-tetrakis-(2-pyridylmethyl)ethylenendiamine) stock solution in 100% ethanol. TPEN is a zinc (Zn) chelator.
4. Supplement I stock solution [15]: add 600 g of glucose, 10 g of L-glutamine, 20 mg of cocarboxylase dissolved in deionized H_2O, up to a final volume of 1 L.
5. 12 mM Supplement II stock solution: dissolve 1.25 g of Fe(III) nitrate [$Fe(NO_3)_3$] nonahydrate in deionized H_2O, up to a final volume of 250 mL.
6. *N. gonorrhoeae* stored in freezing medium (*see* **Note 3**) at −80 °C.
7. GC medium base (Difco) prepared as indicated by the manufacturer. Dilute Supplement I 1/100 and Supplement II 1/1000 into molten agar medium before pouring plates.
8. 50 mM zinc sulfate ($ZnSO_4$) stock solution in deionized H_2O.
9. 125 mL side-arm flasks with baffled bottoms; flasks should be acid-washed and sterilized in a hot air oven before use.
10. Hot air oven (temperature controlled up to 220 °C).

11. Klett Summerson Colorimeter (Thomas Scientific).
12. Autoclave capable of sterilization at 122 °C, 3 kg/cm^2, and slow exhaust.
13. Stir platform and stir bar.
14. Sterile petri plates (90 mm).
15. Class II Biosafety cabinet.
16. Benchtop microcentrifuge (e.g. Eppendorf).
17. Shaking incubator capable of reaching a temperature of 35 °C. Atmosphere must be supplemented with CO_2 through a port into the incubator. A gas regulator between the CO_2 tank and the port regulates the flow into the incubator.
18. Humidified CO_2 incubator for incubation of GC medium plates.

2.3 Preparation of Liquid CDM (Chelexed-Defined Medium)

1. Ingredients for liquid CDM [16] are listed in Table 1.
2. 0.45 μm filter sterilization units accommodating up to 500 mL of liquid.
3. Chelex-100 (Bio-Rad).
4. pH meter.
5. 1 N HCl.
6. Sterile plastic containers.

2.4 Growing N. gonorrhoeae Under Low Fe or Low Zn Conditions in CDM in Flasks

1. *N. gonorrhoeae* stored in freezing medium (*see* **Note 3**) stored at −80 °C.
2. GC medium base (Difco) prepared as indicated by the manufacturer and autoclaved. Dilute Supplement I 1/100 and Supplement II 1/1000 into molten agar medium before pouring plates.
3. Liquid CDM at room temperature.
4. 50 mM $ZnSO_4$ stock solution in deionized H_2O.
5. 12 mM Supplement II: $Fe(NO_3)_3{\cdot}9H_2O$ (final concentration of stock).
6. 125 mL side-arm flasks with baffled bottoms; flasks should be acid-washed and sterilized in a hot air oven before use.
7. Hot air oven (temperature controlled up to 220 °C).
8. Klett Summerson Colorimeter (Thomas Scientific).
9. Shaking incubator capable of reaching a temperature of 35 °C. Atmosphere must be supplemented with CO_2 through a port into the incubator. A gas regulator between the CO_2 tank and the port regulates the flow into the incubator.
10. Humidified CO_2 incubator.
11. Class II Biosafety cabinet.

Table 1
Composition of chelexed defined medium (CDM)

Component	Conc. (g/L) of stock solution		Vol. of stock/ 500 mL of 4× concentrate	Conc. (mM) in 1× medium
Stock solution 1			50 mL	
NaCl	233.8			100.0
K_2SO_4	40.0			5.7
NH_4Cl	8.8			4.1
K_2HPO_4	13.9			2.0
KH_2PO_4	10.9			2.0
Stock solution 2[a]			20 mL	
Thiamine HCl	0.2			0.006
Thiamine pyrophosphate-Cl (=cocarboxylase)	0.05			0.001
Calcium pantothenate	0.19			0.004
Biotin	0.3			0.012
Stock solution 3	Per 1 L	Per 4 L	250 mL	
L-Aspartic acid	4.0	16.0		3.76
L-Glutamic acid	10.4	41.6		8.84
L-Arginine	1.2	4.8		0.71
Glycine	0.2	0.8		0.33
L-Serine	0.4	1.6		0.48
L-Leucine	0.72	2.88		0.69
L-Isoleucine	0.24	0.96		0.23
L-Valine	0.48	1.92		0.51
L-Tyrosine	0.56	2.24		0.39
L-Cysteine HCl[b]	0.44	1.76		0.35
L-Cystine[b]	0.28	1.12		0.15
L-Proline	0.4	1.6		0.43
L-Tryptophan	0.64	2.56		0.39
L-Threonine	0.4	1.6		0.42
L-Phenylalanine	0.2	0.8		0.15
L-Asparagine· H_2O	0.2	0.8		0.17
L-Glutamine	0.4	1.6		0.34

(continued)

Table 1 (continued)

Component	Conc. (g/L) of stock solution	Vol. of stock/ 500 mL of 4× concentrate	Conc. (mM) in 1× medium
L-Histidine HCl	0.2 0.8		0.13
L-Methionine	0.12 0.48		0.10
L-Alanine	0.8 3.2		1.12
L-Lysine	0.4 1.6		0.29
Glutathione (reduced)	0.36 1.44		0.15
Stock solution 4		50 mL	
Glucose	200		27.8
Stock solution 5		20 mL	
Hypoxanthine	5.0		0.37
Uracil	5.0		0.45
NaOH	4.0		
N-2-Hydoxyethyl-piperazine-*N'*-2-ethane sulfonic acid (HEPES)		20.0 g	21.0
Stock solution 6			
1 M $CaCl_2 \cdot 2H_2O$	147.0		0.25 (add 125 μL per 500 mL 1× solution)
Stock solution 7			
1 M $MgSO_4$ (anhydrous)	120.4		1.07 (add 535 μL per 500 mL 1× solution)
Stock solution 8			
1 M $NaHCO_3$	84.0		20.0 (add 10 mL per 500 mL 1× solution)

[a]Stock solution 2 is prepared in 50% (v/v) ethanol; all others are prepared in deionized H_2O

[b]L-Cysteine and L-cystine are dissolved in a small amount of 1 N HCl before addition to stock solution 3

2.5 Growing N. gonorrhoeae *Under Low Fe or Low Zn Conditions in Microtiter Plates*

1. *N. gonorrhoeae* stored in freezing medium (*see* **Note 3**) stored at −80 °C.
2. GC medium plates.
3. Liquid CDM at room temperature.
4. 50 mM $ZnSO_4$ stock solution in deionized H_2O.

5. 12 mM Supplement II: $Fe(NO_3)_3 \cdot 9H_2O$ (final concentration of stock).
6. Phosphate buffered saline (PBS), pH 7.4.
7. 125 mL side-arm flasks with baffled bottoms; flasks should be acid-washed and sterilized in a hot air oven before use.
8. Hot air oven (temperature controlled up to 220 °C).
9. Sterile, flat-bottomed 96-well microtiter plates.
10. Microtiter plate reader capable of reading absorbance (optical density, OD) at $\lambda_{600\ nm}$.
11. Humidifed CO_2 incubator, with a shaking platform.
12. Class II biosafety cabinet.

2.6 Washed Agarose

1. Ultra-pure agarose (Invitrogen).
2. 100% Ethanol.
3. 100% Methanol.
4. Centrifuge capable of attaining a speed of at least 3000 × *g*.
5. Centrifuge bottles of composition able to withstand exposure to 100% ethanol and 100% methanol.
6. Metal free, acid-washed glass tray and bottle for storage.
7. Stirring platform and stir-bar.
8. Fume hood.

2.7 Solid, Low Fe or Low Zn Medium Using GC Medium Base

1. GC medium base (Difco).
2. Supplement I.
3. 12 mM Supplement II: $Fe(NO_3)_3 \cdot 9H_2O$ (final concentration of stock).
4. 50 mM Desferal stock solution.
5. 40 mM TPEN stock solution.
6. 50 mM $ZnSO_4$ stock solution in deionized H_2O.
7. Autoclave.
8. Acid-washed flasks.
9. Stirring platform and stir-bar.
10. Water bath.

2.8 Solid, Low Fe or Low Zn Medium Using CDM

1. 4× stock solution of liquid CDM (Table 1).
2. Stock solutions 6, 7 and 8 (Table 1).
3. Washed agarose (*see* Subheading 2.6).
4. Potato starch.
5. 12 mM Supplement II: $Fe(NO_3)_3 \cdot 9H_2O$ and/or 50 mM $ZnSO_4$ stock solution for Fe- and Zn-replete conditions, respectively.

6. Other possible Fe metal sources: partially saturated human transferrin (Calbiochem or Sigma-Aldrich) [2, 17], partially saturated human lactoferrin (Calbiochem or Sigma-Aldrich) [18, 19], hemin (Sigma-Aldrich) [20], hemoglobin (Sigma-Aldrich) [21, 22].
7. Other possible Zn source: calprotectin [10, 23].
8. Stirring platform and stir-bar.
9. Autoclave.
10. Water bath.
11. Sterile petri plates (90 mm).
12. Class II Biosafety cabinet or other laminar flow hood to ensure sterility.

3 Methods

3.1 Acid-Washed Glassware

1. Dilute concentrated nitric acid with four volumes of deionized H_2O. For example, in a large tub, add 1 gallon of concentrated nitric acid to 4 gallons of deionized H_2O (*see* **Notes 1** and **4**).
2. Wash glassware (*see* **Note 5**) with Aquet detergent diluted to 1% (v/v) in deionized H_2O.
3. Rinse 4–5 times with tap H_2O.
4. Rinse 4–5 times with deionized H_2O.
5. Drain in dish drying rack.
6. Place clean glassware in nitric acid bath at least overnight. Make sure the entire inside surface of the glassware is coated with acid.
7. Carefully remove glassware from acid bath and rinse five times with deionized H_2O.
8. Drain glassware until thoroughly dry.
9. Cap glassware or cover with aluminum foil.
10. Place glassware upside down in rack and bake overnight in a hot air oven set to 220 °C (*see* **Notes 6** and 7).

3.2 Growing N. gonorrhoeae *Under Low Fe or Low Zn Conditions in GC Broth*

1. Pour GC medium plates. Dissolve GC medium base (Difco) in 1 L of deionized H_2O. Add a stir bar to the flask. Autoclave the medium for 20 min on slow exhaust. Cool the medium to 56 °C in a water bath; place flask on stirring plate and add Supplements I and II. Per 1 L of medium, use 10 mL of Supplement I (1/100) and 1 mL of Supplement II (1/1000). After Supplements have been added, pour medium into petri dishes and allow to become solid. Dry the plates briefly in laminar flow hood or class II Biosafety cabinet before use.

2. Two days prior to the experiment, inoculate *N. gonorrhoeae* strains from freezer stock (*see* **Note 3**) stored at −80 °C, onto GC medium plates (*see* **Note 8**).
3. One day prior to the experiment, pass a few colonies of *N. gonorrhoeae* from GC medium plates to fresh GC medium plates. Streak for isolated, single colonies.
4. On the day of the experiment, add the desired volume of GC medium to acid-washed flasks. Add Supplement I to all flasks. Add Supplement II to all flasks *except* those that will be Fe-depleted.
5. Inoculate the flasks with *N. gonorrhoeae* from 24-h old GC medium plates. A 125 mL side-arm flask with baffles on the bottom, containing no more than 25 mL of GC medium, works best for aerated growth of *N. gonorrhoeae*. Use a cotton swab to transfer growth from plate to liquid medium. Inoculate to a density of approximately 20 Klett Units (KU), quantified using a Klett-Summerson Colorimeter (*see* **Note 9**).
6. Grow cultures in a gyratory, shaking incubator with vigorous shaking (225 rpm) at 35 °C in an atmosphere of 10% (v/v) CO_2.
7. When the culture density doubles (20 KU increases to 40 KU in 1–2 h), add the appropriate metal chelators. To initiate Fe stress, add Desferal to a final concentration of 50 μM. To initiate Zn stress, add TPEN to a final concentration of 10 μM.
8. Add Supplement II for Fe-replete cultures and no Desferal.
9. To generate Zn-replete conditions, add $ZnSO_4$ to a final concentration of 15–20 μM. No TPEN should be added to the Zn-replete cultures.
10. Grow all cultures with shaking (225 rpm) for 3–4 h.
11. At the end of the incubation period, harvest the cells by centrifugation (*see* **Note 10**) for the desired assay (Reverse Transcriptase-Polymerase Chain Reaction (RT-PCR), Western blot, binding assay, etc.). For most applications, collection of cultures in sterile 1.5 mL Eppendorf tubes followed by microcentrifugation at 16,000 × *g* for 5 min yields sufficient cell mass. Standardize the individual cultures by density reading (KU) or by protein assay.

3.3 Preparation of Liquid CDM (Chelexed Defined Medium)

1. Prepare stock solutions 1, 4 and 5 as described in Table 1. Ingredients should be added in the masses listed and dissolved in deionized H_2O (*see* **Note 2**).
2. Prepare stock solution 2 by adding the appropriate masses of the ingredients listed in Table 1 to a solution of 50% (v/v) ethanol.

3. Prepare stock solution 3 by adding the amino acids listed in Table 1 (*see* **Note 11**) in indicated masses for 1 or 4 L of stock solution. Dissolve L-cysteine and L-cystine in a small amount of 1 N HCL before addition to solution 3. After ingredients are nearly dissolved in deionized H_2O, pH the solution to above 10. At this pH, all constituents should fully dissolve.
4. Sterilize solutions 1–5 in disposable 0.45 μm filter units. Store individually in plastic tubes or bottles in the indicated volumes to be used for preparing the 4× medium (column 2, Table 1).
5. Store solutions 1–5 long-term at −20 °C.
6. Prepare solutions 6, 7 and 8 by adding the indicated mass of each ingredient (Table 1) to deionized H_2O. Solutions 6, 7 and 8 should be sterilized by filtration and stored at 4 °C until used to prepare the 1× medium.
7. To prepare a 4× concentrated solution, mix filter-sterilized solutions 1–5 with 20 g of solid HEPES (Table 1). The volumes of solutions 1–5 are as follows: 50 mL of solution 1, 20 mL of solution 2, 250 mL of solution 3, 50 mL of solution 4 and 20 mL of solution 5. Mix these premeasured solutions with solid HEPES and stir until dissolved. Adjust the pH of this solution to pH 7.4 and bring up the final volume to 500 mL with deionized H_2O (*see* **Note 2**).
8. In a separate container, add 50 g (per 500 mL of 4× concentrate to be prepared) of Chelex-100 to 1 L of deionized H_2O. Stir to wash the Chelex-100 to remove residual preservatives.
9. Filter out washed Chelex-100 into the top chamber of a 0.45 μm filter unit.
10. Add the washed Chelex-100 from **step 9** above to the 4× concentrated solution from **step** 7 above. Stir slowly at room temperature for 90 min (*see* **Note 12**).
11. Remove and discard the used Chelex-100 by filtration through a 0.45 μm filter unit.
12. Store the deferrated 4× concentrated stock at 4 °C in a sterile plastic container.
13. Immediately prior to use, dilute the 4× concentrated stock 1:4 with deionized H_2O. Before the final volume is reached, add appropriate volumes (column 3 in Table 1) of solutions 6, 7 and 8. Bring the solution up to the final volume with deionized H_2O (*see* **Note 2**) and sterilize by filtration.

3.4 Growing N. gonorrhoeae Under Low Fe or Low Zn Conditions in CDM in Flasks

1. Two days before the experiment, inoculate GC medium plates (containing both Supplements I and II) with *N. gonorrhoeae* strains from freezer stocks stored at −80 °C (*see* **Note 8**).

2. One day before the experiment, pick a few colonies from GC medium plates and streak for isolation onto fresh GC medium plates.
3. On the day of the experiment, dilute the 4× CDM down to 1×, if necessary, or warm 1× medium to room temperature, if it has been stored at 4 °C.
4. Add desired volume of CDM to acid-washed side-arm flasks. 5 mL of CDM for initial inoculation is recommended.
5. Inoculate the medium with *N. gonorrhoeae* colonies from 24-h old plates, to a density of 20 KU, quantified using a Klett Summerson Colorimeter.
6. Grow for 1–2 h at 35 °C with 10% (v/v) CO_2 with vigorous shaking (225 rpm).
7. When cultures have doubled in density (from 20 to 40 KU), back-dilute cultures by adding 3 volumes of fresh CDM (pre-warmed). This should result in a culture at a density of approximately 10 KU. For example, if the starting volume was 5 mL, add 15 mL of fresh CDM at room temperature.
8. At the time of back-dilution, for Fe-replete controls, add Supplement II to a final concentration of 10–20 μM.
9. For Zn-replete controls, add $ZnSO_4$ stock solution to a final concentration of 15–20 μM.
10. Fe-depleted and Zn-depleted cultures require no further additions.
11. Grow the bacteria for 3–4 h in a shaking incubator at 35 °C with 10% (v/v) CO_2 with vigorous shaking (225 rpm). The final density should be 60–80 KU for metal-depleted cultures. Metal-replete cultures can reach as high as 120 KU.
12. At the end of the incubation period, harvest the cells (as described in Subheading 3.2, **step 11**) for the desired assay (RT-PCR, Western blot, binding assay, etc.). Standardize the cultures by density reading (KU) or by protein assay.

3.5 Growing N. gonorrhoeae Under Low Fe or Low Zn Conditions in Microtiter Plates

1. Two days before the experiment, inoculate GC medium plates (containing both Supplements I and II) with *N. gonorrhoeae* strains from freezer stocks stored at −80 °C (*see* **Note 8**).
2. One day before the experiment, pick a few colonies from GC medium plates and streak for isolation onto fresh GC medium plates.
3. On the day of the experiment, dilute the 4× CDM down to 1×, if necessary, or warm 1× medium to room temperature, if it has been stored at 4 °C.
4. Add 10 mL of CDM to acid-washed side-arm flasks.

5. Inoculate medium with *N. gonorrhoeae* from 24-h old plates, to a density of 20 KU, quantified using a Klett-Summerson Colorimeter.
6. Incubate the flasks at 35 °C with 10% (v/v) CO_2 with vigorous shaking (225 rpm).
7. Cultures should double from 20 to 40 KU in approximately 1 h. Measure the OD of the culture by placing 100 μL of each culture into wells of the microtiter dish. Measure the OD $\lambda_{600\ nm}$ using a microtiter plate reader.
8. Dilute each culture with fresh, warm 1× CDM to a standardized density of OD $\lambda_{600\ nm} = 0.02$ (*see* **Note 13**).
9. Add 90 μL of each diluted culture to a microtiter well. Dilute desired Fe or Zn sources in filter-sterilized PBS and add a volume of these supplements so that the final culture volume in the microtiter well is 100 μL.
10. Swirl the microtiter plate to mix the culture with the supplements. Take an initial OD $\lambda_{600\ nm}$ reading before incubating the plate. Affix the plate onto a rotating platform in a stationary incubator. Incubate at 35 °C with 10% (v/v) CO_2 with the platform rotating at 200 rpm. Remove the plate at 1–2 h intervals to monitor growth via OD $\lambda_{600\ nm}$ readings taken with a microtiter plate reader. Monitor growth over 6–8 h.

3.6 Washed Agarose

1. Weigh out 50 g of ultra-pure agarose in a plastic weigh boat.
2. Add agarose to 1 L of deionized H_2O. Add a stir bar and stir for 1 h. Transfer to a centrifuge bottle and centrifuge for 15 min at 300 × *g* at room temperature. Discard H_2O by decanting carefully.
3. Add enough deionized H_2O to the agarose pellet to transfer all of the agarose to a 1 L flask. Bring volume up to 1 L with deionized H_2O. Stir for 1 h at room temperature. Transfer to centrifuge bottles and centrifuge to obtain a pellet as in **step 2**. Discard H_2O by decanting carefully.
4. Add 100% ethanol to the agarose pellet in the centrifuge bottle. Transfer to a 1 L flask. Bring up the volume to 1 L with 100% ethanol. Wash for 1 h at room temperature. Transfer to centrifuge bottles and centrifuge to obtain a pellet as in **step 2**. Discard ethanol by decanting carefully.
5. Repeat **step 4** above.
6. Add 100% methanol to each agarose pellet. Transfer to 1 L flask. Bring up the volume to 1 L with 100% methanol. Wash for 1 h at room temperature. Transfer to centrifuge bottles and centrifuge to obtain a pellet as in **step 2**. Discard methanol by decanting carefully.
7. Repeat **step 6** above.

8. Prepare a glass metal-free tray by acid-washing as described in Subheading 3.1. Transfer the washed agarose to the metal-free tray. Dry in a fume hood (*see* **Note 14**).
9. Transfer the completely dried agarose to a sterile, acid-washed glass bottle for long-term storage.

3.7 Solid, Low Fe or Low Zn Medium Using GC Medium Base

1. Weigh out 36 g of GC medium base per 1 L of medium in a plastic weigh boat. Add the powdered medium base to a 2 L flask. This flask need not be acid-washed since chelators will be added to remove residual metals.
2. Bring volume up to 1 L with deionized H_2O.
3. Add a stir bar to the flask before covering with foil. Autoclave for 20 min with slow exhaust.
4. Cool the sterilized medium to 56 °C in a water bath.
5. Remove the flask from the water bath when cool and place on stirring plate.
6. Using sterile technique, add Supplement I to all flasks: 10 mL per 1 L of medium.
7. Add Supplement II to any plates that are to be Fe-replete: 1 mL per 1 L of medium.
8. Add Zn stock solution to any plates that are to be Zn-replete: 375 μL of 50 mM stock solution per 1 L of medium.
9. Add Desferal to a final concentration of 5 μM to any plates that are to be Fe-depleted: 100 μL of 50 mM stock solution per 1 L medium.
10. Add TPEN to a final concentration of 10 μM to any plates that are to be Zn-depleted. 250 μL of 40 mM stock solution per 1 L medium.
11. Using sterile technique, pour approximately 25 mL molten medium into each sterile petri dish.
12. Allow agar to set at room temperature.
13. For long-term storage, store at 4 °C.

3.8 Solid, Low Fe or Zn Medium Using CDM

1. Per 100 mL of media for plates, mix 1 g of washed agarose with 0.5 g of potato starch in 73 mL of deionized H_2O. Add a stir bar to the flask. Cover with foil and autoclave the solution for 20 min with slow exhaust.
2. Cool the sterilized agarose to 56 °C in a water bath.
3. To the cooled, molten agarose add 25 mL of 4× CDM, 25 μL of solution 6, 107 μL of solution 7 and 2 mL of solution 8.
4. For Fe/Zn-replete conditions add 10–20 μM of ferric nitrate (Supplement II), 2.5 μM of transferrin or lactoferrin, 15–20 μM of $ZnSO_4$ or 5–10 μM of calprotectin.

5. Distribute approximately 20 mL of molten agarose medium into each petri dish: 100 mL of medium results in approximately five plates.
6. Allow the agar to set at room temperature and store long-term at 4 °C.
7. Dry plates briefly in a laminar flow hood or class II Biosafety cabinet before inoculating with *N. gonorrhoeae* (*see* **Note 15**).

4 Notes

1. Always add nitric acid to H_2O, not H_2O to concentrated nitric acid. Use appropriate personal protective equipment (PPE) including long lab coat, goggles and neoprene (acid-resistant) gloves.
2. Use deionized H_2O for all steps rather than distilled or ultrapure H_2O. In-house stills contain metal pipes and therefore reintroduce metals into the H_2O. Ultrapure H_2O systems (e.g., Milli-Q water systems) restrict metal levels too severely and *N. gonorrhoeae* will not grow in media made with this H_2O source.
3. Freezing medium for *N. gonorrhoeae* contains 3 g of trypicase soy broth, 25 mL of glycerol and make up to 100 mL with deionized H_2O. Distribute 500 μL per 1.5 mL freezer vial and autoclave to sterilize.
4. Generating the nitric acid solution as described results in a solution of greater than 3 N. However, H_2O from the glassware will dilute the acid bath solution over time. The nitric acid will take on a yellow color as it ages and accumulates Fe.
5. Plastic storage containers and disposable tubes, etc. cannot be acid-washed. These do not carry contaminating metals but virtually all glassware harbors additional metals that must be removed by acid washing.
6. Some laboratory glassware dishwashers have a final acid wash cycle. After acid washing and a final rinse, glassware should be washed additionally a minimum of three times with deionized H_2O. Glassware can then be air dried and sterilized in a hot air oven.
7. Except for media, avoid autoclaving for sterilization of any materials used for metal-depleted growth. Steam is generated from in-house H_2O systems, which contain adventitious metals. These metals will be deposited upon glassware that is autoclaved. Therefore sterilization is instead accomplished by heating in a hot air oven at 220 °C.

8. When handling *N. gonorrhoeae* cultures, use appropriate PPE including gloves and lab coat. If there is a risk of aerosolization, as in opening tubes, do so inside a Class II Biosafety cabinet.
9. 20 Klett Units corresponds to approximately 1.8×10^8 cells per mL of culture.
10. Centrifugation of *N. gonorrhoeae* cultures risks creating an aerosol; therefore, open centrifuged tubes inside a Class II Biosafety cabinet.
11. Reduced glutathione and cystine should be stored under desiccation as they are both easily oxidized in ambient air.
12. Do not incubate the medium with Chelex-100 for more than 90 min, as metals will become too limited and *N. gonorrhoeae* will not grow, even if sufficient Zn and Fe are supplied.
13. The microtiter format enables testing of multiple strains and growth conditions simultaneously. Manipulation and dilution of *N. gonorrhoeae* liquid cultures must be accomplished quickly as the pathogen is fragile and dies quickly at room temperature without CO_2.
14. If the agarose is added to the tray in a thin layer, incubation in the fume hood to remove the residual alcohols may take as little as 24 h. If the agarose layer is thick, complete evaporation of the alcohols may take more than 24 h. Make sure that all residual alcohols are evaporated from the washed agarose before adding to medium to be poured into plates. Any contaminating alcohol will inhibit the growth of *N. gonorrhoeae*.
15. When inoculating CDM plates, collect *N. gonorrhoeae* cells from GC medium base plates containing both Supplements I and II. Strains will not grow reproducibly on this chelexed, defined medium if inoculated directly from freezer stocks.

Acknowledgments

This work was supported by the National Institute of Allergy and Infectious Diseases grants R01 AI 125421 and AI 127793. The author also thanks the members of her lab for critical feedback on the manuscript.

References

1. Weyand NJ (2017) *Neisseria* models of infection and persistence in the upper respiratory tract. Pathog Dis 75(3). https://doi.org/10.1093/femspd/ftx031
2. Cornelissen CN, Sparling PF (1996) Binding and surface exposure characteristics of the gonococcal transferrin receptor are dependent on both transferrin-binding proteins. J Bacteriol 178(5):1437–1444
3. Ngampasutadol J, Ram S, Blom AM et al (2005) Human C4b-binding protein selectively interacts with *Neisseria gonorrhoeae* and results in species-specific infection. Proc Natl Acad Sci U S A 102(47):17142–17147

4. Ngampasutadol J, Ram S, Gulati S et al (2008) Human factor H interacts selectively with *Neisseria gonorrhoeae* and results in species-specific complement evasion. J Immunol 180 (5):3426–3435. https://doi.org/10.4049/jimmunol.180.5.3426
5. Sintsova A, Wong H, MacDonald KS et al (2015) Selection for a CEACAM receptor-specific binding phenotype during *Neisseria gonorrhoeae* infection of the human genital tract. Infect Immun 83(4):1372–1383. https://doi.org/10.1128/IAI.03123-14. IAI.03123-14 [pii]
6. Kochan I (1973) The role of iron in bacterial infections, with special consideration of host-tubercle bacillus interaction. Curr Top Microbiol Immunol 60:1–30
7. Weinberg ED (1977) Infection and iron metabolism. Am J Clin Nutr 30(9):1485–1490
8. Weinberg ED (1978) Iron and infection. Microbiol Rev 42(1):45–66
9. Cornelissen CN (2018) Subversion of nutritional immunity by the pathogenic *Neisseriae*. Pathog Dis 76(1). https://doi.org/10.1093/femspd/ftx112. 4553517 [pii]
10. Zackular JP, Chazin WJ, Skaar EP (2015) Nutritional immunity: S100 proteins at the host-pathogen Interface. J Biol Chem 290 (31):18991–18998. https://doi.org/10.1074/jbc.R115.645085. R115.645085 [pii]
11. Anderson JE, Hobbs MM, Biswas GD et al (2003) Opposing selective forces for expression of the gonococcal lactoferrin receptor. Mol Microbiol 48(5):1325–1337
12. Cornelissen CN, Kelley M, Hobbs MM et al (1998) The transferrin receptor expressed by gonococcal strain FA1090 is required for the experimental infection of human male volunteers. Mol Microbiol 27(3):611–616
13. Ronpirin C, Jerse AE, Cornelissen CN (2001) The gonococcal genes encoding transferrin binding proteins (Tbp) A and B are arranged in a bicistronic operon but are subject to differential expression. Infect Immun 69 (10):6336–6347
14. Yu C, McClure R, Nudel K et al (2016) Characterization of the *Neisseria gonorrhoeae* iron and fur regulatory network. J Bacteriol 198 (16):2180–2191. https://doi.org/10.1128/JB.00166-16. JB.00166-16 [pii]
15. Kellogg DS Jr, Peacock WL Jr, Deacon WE et al (1963) *Neisseria gonorrhoeae*. I. Virulence genetically linked to clonal variation. J Bacteriol 85:1274–1279
16. West SEH, Sparling PF (1987) Aerobactin utilization by *Neisseria gonorrhoeae* and cloning of a genomic DNA fragment that complements *Escherichia coli fhuB* mutations. J Bacteriol 169(8):3414–3421
17. Cornelissen CN, Biswas GD, Tsai J et al (1992) Gonococcal transferrin-binding protein 1 is required for transferrin utilization and is homologous to TonB-dependent outer membrane receptors. J Bacteriol 174 (18):5788–5797
18. Biswas GD, Anderson JE, Chen C-J et al (1999) Identification and functional characterization of the *Neisseria gonorrhoeae lbpB* gene product. Infect Immun 67(1):455–459
19. Biswas GD, Sparling PF (1995) Characterization of *lbpA*, the structural gene for a lactoferrin receptor in *Neisseria gonorrhoeae*. Infect Immun 63(8):2958–2967
20. Strange HR, Zola TA, Cornelissen CN (2011) The *fbpABC* operon is required for Ton-independent utilization of xenosiderophores by *Neisseria gonorrhoeae* strain FA19. Infect Immun 79(7):267–678
21. Chen CJ, Elkins C, Sparling PF (1998) Phase variation of hemoglobin utilization in *Neisseria gonorrhoeae*. Infect Immun 66:987–993
22. Chen C-J, Sparling PF, Lewis LA et al (1996) Identification and purification of a hemoglobin-binding outer membrane protein from *Neisseria gonorrhoeae*. Infect Immun 64 (12):5008–5014
23. Gilston BA, Skaar EP, Chazin WJ (2016) Binding of transition metals to S100 proteins. Sci China Life Sci 59(8):792–801. https://doi.org/10.1007/s11427-016-5088-4

Chapter 15

Cloning, Expression, and Purification of Recombinant *Neisseria gonorrhoeae* Proteins

María Victoria Humbert

Abstract

Modern DNA recombinant techniques and major advances in genetic engineering have resulted in the development of bacterial expression systems that guarantee an unlimited supply of valuable proteins that have potential clinical or industrial use, but which are often limited by their low natural availability. This chapter provides the reader with a general scheme to clone, express, and purify native histidine (His)-tagged proteins in the desired quantity and quality required for its intended use, and reviews the most important factors affecting the production of recombinant proteins in a soluble form. Alternative methods for purification of insoluble recombinant proteins under denaturing conditions are also discussed. An optimized protocol to successfully purify native *Neisseria gonorrhoeae* Adhesin Complex Protein (Ng-ACP; NGO1981) is used as a technical example for the processes, which could potentially be applied to any gonococcal recombinant protein of interest.

Key words *Neisseria gonorrhoeae*, Ng-ACP (NGO1981), Cloning, Recombinant protein purification, His-tag, Affinity chromatography

1 Introduction

Virtually any protein can be obtained through classical purification protocols, such as preliminary fractionation procedures in combination with chromatographic and electrophoretic methods. However, a wide variety of heterologous protein expression systems have been developed and are currently applied to produce recombinant proteins fused to a tag that facilitates their purification and detection [1, 2] and may also improve their solubility and stability [3]. Briefly, any recombinant protein production process requires (1) selection of an appropriate vector (tag), cloning (nonexpressing) host and expression host; (2) cloning: generation of the vector–insert construct and transformation of the cloning host; (3) recombinant plasmid purification, restriction, and sequencing analyses for transformation of the expression host; (4) protein pilot expression: determination of target protein solubility; and

Myron Christodoulides (ed.), *Neisseria gonorrhoeae: Methods and Protocols*, Methods in Molecular Biology, vol. 1997, https://doi.org/10.1007/978-1-4939-9496-0_15,

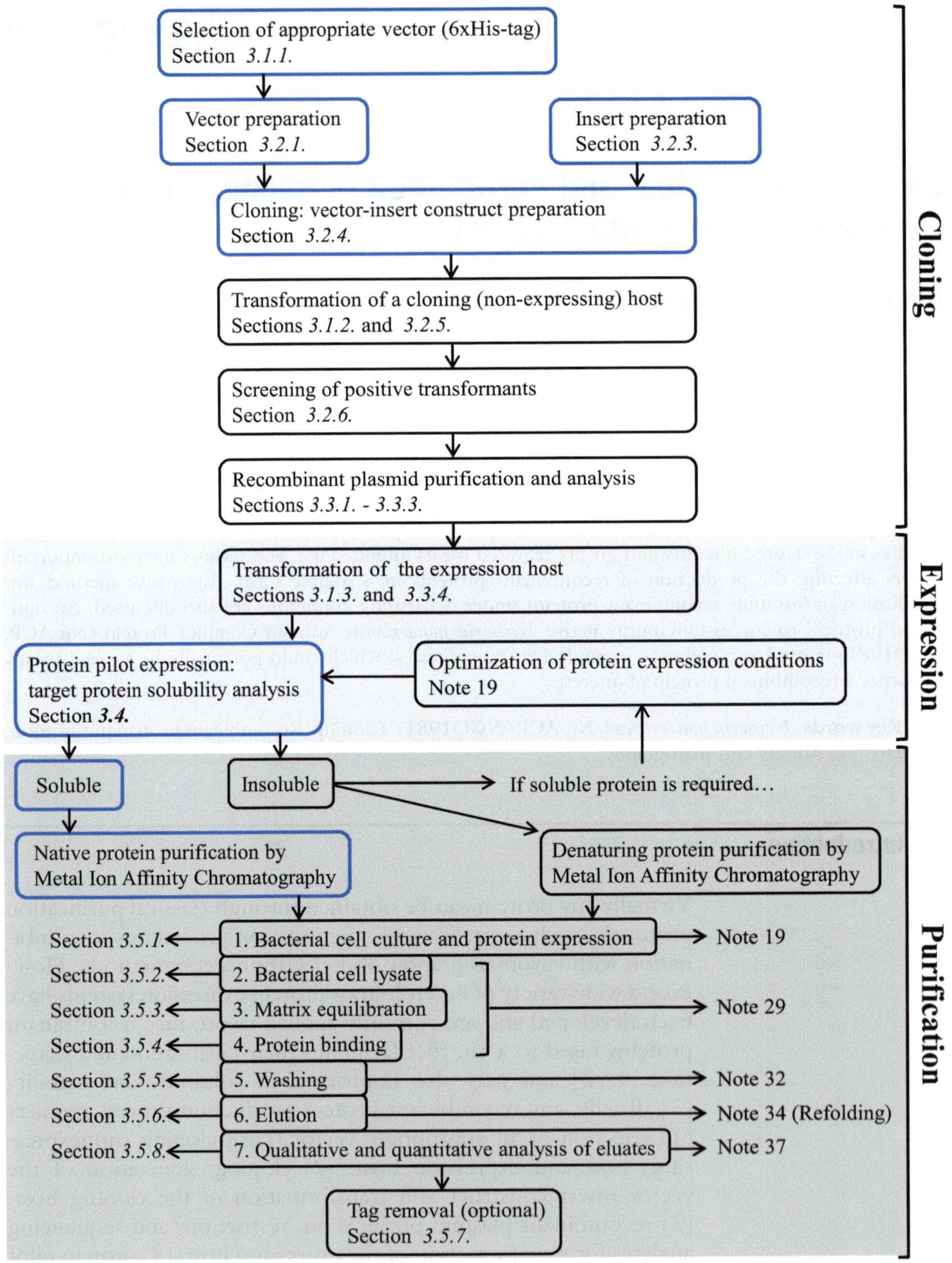

Fig. 1 Workflow diagram for cloning, expression and purification of 6×His-tagged proteins. A general strategy for His-tagged recombinant proteins production is outlined, with key steps highlighted in blue. The corresponding Subheading/Note with full detailed information for each method is also specified

(5) protein purification under native or denaturing conditions. When necessary, further handling steps can also be considered, such as dialysis for buffer exchange, refolding of insoluble proteins purified under denaturing conditions and tag removal, among others (Fig. 1). In this chapter, the process is described for producing recombinant (r)Ng-ACP protein (NGO1981) as an exemplar.

2 Materials

2.1 Vector (Tag), Cloning (Nonexpressing) and Expression Host

1. pET22b+ plasmid or any other expression vector of choice.
2. *E. coli* DH5α strain or any other suitable nonexpressing host.
3. *E. coli* BL21 (DE3) pLysS strain or any other suitable host expressing T7 RNA polymerase (if the vector of choice provides a T7 promoter for cloning).

2.2 Cloning

2.2.1 Vector Preparation

1. Luria–Berani (LB) broth: add 10 g/L of tryptone, 5 g/L of yeast extract, and 10 g/L of NaCl to distilled H_2O, adjust pH to 7 and autoclave at 121 °C, 2.68 kg/cm^2 for 15 min.
2. Sterile petri plates (90 mm).
3. Selective LB agar plates: add 12 g/L of agar to LB broth and autoclave at 121 °C, 2.68 kg/cm^2 for 15 min. Add 100 μg/ml ampicillin to cooled medium and pour into petri plates.
4. 100 mg/ml ampicillin stock solution in distilled H_2O (filter-sterilized using a 0.22 μM filter and syringe).
5. Wizard® Plus SV Minipreps DNA Purification System or any other Miniprep Kit of preference.
6. NdeI and XhoI restriction enzymes and compatible Buffer D (Promega) or other restriction enzymes selected for the cloning construct design, and their corresponding digestion buffer.
7. Alkaline phosphatase and corresponding Dephosphorylation Buffer (Promega) or any other commercial phosphatase of choice.
8. Equipment: Platform shaker; humidified CO_2 (5% v/v) incubator; thermal block or water bath.

2.2.2 Agarose Gel Electrophoresis

1. Ultrapure agarose, 1× TAE buffer: 40 mM Tris, 20 mM acetate and 1 mM Ethylenediaminetetraacetic acid (EDTA) in distilled H_2O, pH ~8.6.
2. SYBR™ Safe™ DNA Gel Stain.
3. DNA Loading buffer.
4. Appropriate DNA molecular weight marker.
5. Wizard® SV Gel and PCR Clean-Up System.

6. Equipment: microwave; agarose gel electrophoresis unit; UV- or blue light-transilluminator; NanoDrop™ UV-Vis spectrophotometer.

2.2.3 Insert Preparation

1. A plasmid clone with the target gene and appropriate restriction enzymes and restriction buffer for insert release; or appropriate primers, template, and reagents for PCR amplification of the gene of interest. Alternatively, the target insert can be synthesized in vitro.

2.2.4 Cloning: Vector–Insert Construct Preparation

1. Linearized, purified cloning vector.
2. Prepared insert.
3. T4 DNA Ligase and 1× Ligase Buffer.

2.2.5 Transformation of a Cloning (Nonexpressing) Host

1. LB broth and selective LB agar plates with 100 μg/ml of ampicillin.
2. Sterile 250 ml Erlenmeyer flask.
3. 0.1 M $MgCl_2$ solution in distilled H_2O (sterilized by autoclaving at 121 °C, 2.68 kg/cm^2 for 15 min).
4. 0.1 M $CaCl_2$ solution in distilled water (sterilized by autoclaving at 121 °C, 2.68 kg/cm^2 for 15 min).
5. 100% glycerol (sterilized by autoclaving at 121 °C, 2.68 kg/cm^2 for 15 min).
6. Cryotubes.
7. DNA ligation mixture.
8. Equipment: platform shaker; spectrophotometer and cuvettes for visible spectroscopy; thermal block or water bath; centrifuge (standard, low speed); dry incubator.

2.2.6 Direct Colony PCR

1. T7 promoter (forward) and T7 terminator (reverse) primers (suitable for most pET vectors); or any other pair of primers appropriate for the cloning vector of choice.
2. LB broth and selective LB agar plates supplemented with 100 μg/ml of ampicillin.
3. Polymerase chain reaction (PCR) master mix.
4. All agarose gel electrophoresis reagents (*see* Subheading 2.2.2).
5. 100% glycerol (sterile).
6. Cryovials.
7. Equipment: thermal block or water bath; centrifuge (standard, low speed); thermal cycler; platform shaker.

2.3 Transformation of the Expression Host

2.3.1 Plasmid Purification

1. Wizard® Plus SV Minipreps DNA Purification System.

2.3.2 Restriction Analysis

1. Purified recombinant plasmid.
2. NdeI and XhoI restriction enzymes and compatible Buffer D (Promega) or same pair of restriction enzymes used for cloning.
3. All agarose gel electrophoresis reagents (*see* Subheading 2.2.2).

2.3.3 Sequencing Analysis

1. Purified recombinant plasmid.
2. T7 promoter (forward) and T7 terminator (reverse) primers (suitable for most pET vectors); or any other pair of primers appropriate for the cloning vector of choice.

2.3.4 Transformation of the Expression Host

1. All reagents for preparation and transformation of competent cells (*see* Subheading 2.2.5).
2. LB agar plates supplemented with 100 μg/ml of ampicillin and 34 μg/ml of chloramphenicol (if using *E. coli* BL21 (DE3) pLysS as the expression host).

2.4 Protein Pilot Expression

1. LB broth.
2. 100 mg/ml ampicillin stock solution in distilled H_2O (filter-sterilized using a 0.22 μM filter and syringe).
3. 34 mg/ml chloramphenicol stock solution in 95% (v/v) EtOH.
4. Sterile 250 ml Erlenmeyer flasks.
5. 100 mM isopropyl-β-D-thiogalactopyranoside (IPTG) solution in distilled H_2O (filter-sterilized using a 0.22 μM filter and syringe).
6. Phosphate Buffered Saline (PBS), pH 7.4.
7. 0.5% (w/v) Sodium Dodecyl Sulfate (SDS) in PBS.
8. Appropriate protein molecular weight marker.
9. 14% Separating gel for SDS–polyacrylamide gel electrophoresis (PAGE): 2.33 ml of 30% (w/v) acrylamide, 1.9 ml of Separating Buffer, 0.7 ml of distilled H_2O, 50 μl of 10% (w/v) SDS, 20 μl of 10% (w/v) ammonium persulfate (APS), and 5 μl tetramethylethylenediamine (TEMED).
10. Stacking gel for SDS-PAGE: 0.3 ml of 30% (w/v) acrylamide, 1 ml of Stacking Buffer, 0.6 ml of distilled H_2O, 20 μl of 10% (w/v) SDS, 10 μl of 10% (w/v) APS, and 2 μl of TEMED.
11. Separating Buffer for SDS-PAGE: 1.5 M Tris–HCl, pH 8.8.

12. Stacking Buffer for SDS-PAGE: 0.5 M Tris–HCl, pH 6.8.
13. 4× SDS-PAGE Loading Buffer for SDS-PAGE: 4.8 ml of 0.5 M Tris–HCl pH 6.8, 0.7 ml of distilled H_2O, 4 ml of 100% glycerol, 0.8 g of SDS, and 4 mg of Bromophenol Blue. Make 950 μl aliquots and add 50 μl of β-mercaptoethanol per aliquot (and store at −20 °C).
14. 10× Running Buffer for SDS-PAGE: 30.3 g of Tris, 144 g of glycine, and 10 g of SDS in distilled H_2O. Adjust to pH 8.3 with HCl and make to a final volume of 1 L with H_2O.
15. Staining Solution for SDS-PAGE: 10% (v/v) acetic acid, 20% (v/v) isopropanol, 0.5 mg/ml Coomassie Blue in distilled H_2O.
16. Destaining Solution for SDS-PAGE: 10% (v/v) acetic acid, 10% (v/v) isopropanol in distilled H_2O.
17. Equipment: platform shaker; spectrophotometer and cuvettes for visible spectroscopy; centrifuge (standard, low speed); probe sonicator (MSE Soniprep 150); SDS-PAGE unit.

2.5 Protein Purification

2.5.1 Bacterial Cell Culture and Protein Expression

1. LB broth.
2. 100 mg/ml ampicillin stock solution in distilled H_2O (filter-sterilized using a 0.22 μM filter and syringe).
3. 34 mg/ml chloramphenicol stock solution in 95% (v/v) EtOH.
4. Sterile 250 ml Erlenmeyer flask.
5. 100 mM IPTG solution (filter-sterilized using a 0.22 μM filter and syringe).
6. Equipment: platform shaker; spectrophotometer and cuvettes for visible spectroscopy; centrifuge (standard, low speed).

2.5.2 Bacterial Cell Lysate

1. Native Binding Buffer: 50 mM NaH_2PO_4, 300 mM NaCl, and 10 mM imidazole in distilled H_2O. Adjust pH to 8.0 with NaOH.
2. Equipment: probe sonicator (MSE Soniprep 150); centrifuge (standard, low speed).

2.5.3 Matrix Equilibration

1. Nickel-charged nitrilotriacetic acid (Ni-NTA) agarose.
2. Native Binding Buffer (*see* **step 1** of Subheading 2.5.2).

2.5.4 Protein Binding in Batch Suspension

1. Equilibrated Ni-NTA agarose.
2. Cleared lysate.
3. Equipment: rotating shaker; glass chromatography column (1 cm wide × 20 cm long).

2.5.5 Wash

1. Native Washing Buffer: 50 mM NaH_2PO_4, 300 mM NaCl, and 20 mM imidazole in distilled H_2O. Adjust pH to 8.0 with NaOH.

2.5.6 Elution and Ni-NTA Resin Regeneration

1. Native Elution Buffer: 50 mM NaH_2PO_4, 300 mM NaCl, and 250 mM imidazole in distilled H_2O. Adjust pH to 8.0 with NaOH.
2. Regeneration Buffer: 6 M guanidine hydrochloride (GuHCl) and 0.2 M acetic acid in distilled H_2O.
3. 2% (w/v) SDS in distilled H_2O.
4. 25% (v/v), 30% (v/v), 50% (v/v), 75% (v/v), and 100% EtOH.
5. 100 mM EDTA solution in distilled H_2O, pH 8.0.
6. 100 mM nickel sulfate ($NiSO_4$) solution in distilled H_2O.
7. Native Binding Buffer (*see* **step 1** of Subheading 2.5.2).

2.5.7 His-Tag Removal

1. Appropriate protease or reagents suitable for the cloning vector of choice.

2.5.8 Analyses of the Eluates

1. Fast protein liquid chromatography (FPLC) equipment (optional).
2. All reagents for SDS-PAGE analysis (*see* **steps 8–17** in Subheading 2.4).
3. Blotting Buffer: 200 ml of 1× Running Buffer for SDS-PAGE (*see* **step 14** of Subheading 2.4) and 50 ml methanol.
4. Tween–Tris buffered saline (TTBS): 2.42 g of Tris base and 29.24 g of NaCl. Adjust to pH 7.5 with HCl and to a final volume of 1 L with distilled H_2O. Add 0.05% (v/v) of Tween 20 (500 μl).
5. Blocking Buffer: 5% (w/v) skimmed milk powder (or bovine serum albumin, BSA) dissolved in TTBS.
6. Substrate Buffer (for colorimetric method): 1.21 g of Tris base, 0.58 g of NaCl, and 0.04 g of $MgCl_2$. Adjust to pH 9.5 with HCl and to a final volume of 100 ml with distilled H_2O. Immediately before use add substrates: 200 μl of nitroblue tetrazolium (NBT) solution and 200 μl of 5-bromo-4-chloro-3-indolyl-phosphate (BCIP) solution to 20 ml of Substrate Buffer. Substrates (prepare right before use): 0.06 g of NBT in 2 ml of 70% (v/v) *N,N*-dimethyl formaldehyde (DMF) and 0.03 g of BCIP in 2 ml of 100% DMF. Prepare both reagents in glass containers only and cover with foil to avoid exposure to light.
7. Nitrocellulose or polyvinylidene difluoride (PVDF) membrane.
8. Whatman filter paper.
9. Ponceau S staining solution.

10. Anti-His antibodies or antibody that is specific for the target protein.
11. Alkaline phosphatase (AP)-conjugated secondary antibody compatible with the first antibody of choice; or any other convenient secondary antibody (e.g., horseradish peroxidase (HRP)-conjugated).
12. Equipment: western blotting equipment; spectrophotometer and cuvettes for visible spectroscopy.
13. Protein assay kit (bicinchoninic acid, Lowry or Bradford reagents).

3 Methods

The following methods are a description of optimized protocols for cloning, expression and affinity chromatography purification of recombinant (r)Ng-ACP fused to a His tag using a pET system. They can be considered as general guidelines, but optimization of these techniques might be required for different protein candidates. In fact, other expression systems and different purification methods might be more appropriate for other target proteins.

3.1 Vector (Tag), Cloning (Nonexpressing) Host, and Expression Host

3.1.1 Vector (Tag)

The choice of a suitable vector and the design of the fusion protein are dependent on the intended use of the purified recombinant protein [4, 5]. Well characterized expression vectors are commercially available and generally consist of (1) an origin of replication (ORI), (2) promoter and regulatory sequences to suit a specific expression host, (3) a selection marker for a straightforward screening of transformants, (4) an N-terminal or C-terminal tag with or without a cleavage site for optional tag removal, and (5) a multiple cloning site for insertion of the target gene. The pET system, for example, has been developed for cloning and expression of recombinant proteins in *E. coli* as protein fusions to a wide variety of tags, where the target gene is under transcriptional control of the strong bacteriophage T7 promoter sequence and expression is induced by an inducible T7 RNA polymerase provided by the expression host [6]. Hence, an advantage of this system is its ability to maintain the target gene transcriptionally silent under noninduced conditions, which could be of prime importance when working with proteins potentially toxic to the host.

The pET22b+ vector was chosen as the expression system for Ng-ACP. This plasmid carries a pBR322 ORI, a T7 promoter and transcription start site, an N-terminal *pelB* signal sequence for potential periplasmic localization, an optional C-terminal 6×His tag, a T7 terminator sequence, a *lacI* coding sequence and a *lac* operator sequence for transcriptional repression under noninduced conditions and a *bla* coding sequence for positive selection of transformants.

3.1.2 Cloning (Nonexpressing) Host

Suitable bacterial hosts for cloning and plasmid storage include the *E. coli* K12 NovaBlue, JM109, and DH5α strains. These strains are convenient nonexpressing hosts for initial cloning of target DNA into pET vectors and for maintaining plasmids because they are *recA*$^{-}$ *endA*$^{-}$ and give high transformation efficiencies and good plasmid yields. Furthermore, background expression is minimal in the absence of T7 RNA polymerase because the host RNA polymerases do not initiate from T7 promoters. This is particularly advantageous when working with recombinant proteins that might be toxic for the host and therefore inhibit or limit transformation efficiency and endanger bacterial storage. *E. coli* DH5α was used in our exemplar for recombinant DNA manipulation and plasmid storage.

3.1.3 Expression Host

Selection of the expression host affects not only the expression of the protein but also the way in which the product can be subsequently purified. Therefore, the amount and the degree of purity of the product, as well as its biological integrity and potential toxicity should be considered before deciding which host is most suitable. For example, bacterial expression systems are not suitable if post-translational modification is required to produce a fully functional recombinant product. When no posttranslational modification is needed, however, *E. coli* is often used as an expression host, since it facilitates protein expression by its relatively simple, inexpensive and fast high-density cultivation, its well-known genetics and the large number of compatible molecular tools available [7].

For expression of rNg-ACP, we selected BL21 (DE3) pLysS strain. This strain contains a chromosomal copy of the gene encoding for T7 RNA polymerase under transcriptional regulation of a *lacUV5* promoter, which is inducible by IPTG. In addition, BL21 is deficient in the *lon* protease and lacks the *ompT* outer membrane protease that can degrade proteins during purification, thus increasing recombinant protein stability. This variant of the λDE3 lysogen of strain BL21 also provides additional stability to target genes as it contains a compatible plasmid (that confers chloramphenicol resistance) encoding for T7 lysozyme, and it also binds to T7 RNA polymerase, inhibiting transcription. This facilitates the disruption of the bacterial inner membrane (lysis), enhancing production yields and at the same time conferring an additional mechanism of transcriptional control. The presence of pLysS increases the tolerance of λDE3 lysogens for plasmids with toxic inserts, so the combination of a strong T7 *lac* promoter-containing vector and pLysS is preferable.

3.2 Cloning

Many different strategies can be used for cloning a protein-coding fragment of DNA (insert) into a pET vector, but using two different (unique) restriction sites found in the multiple cloning region of the vector is the most convenient way to guarantee defined gene orientation relative to the plasmid promoter sequence and minimize plasmid religation. When using C-terminal tags (as with pET22b+), the insert must be cloned in frame with both the ATG start codon and the 3′-6×His coding sequence in the vector.

3.2.1 Vector Preparation

1. Acquire the selected vector from a commercial source: they are supplied generally as purified plasmids in a lyophilized form. Conversely, use a glycerol stock of a transformed nonexpressing *E. coli* strain (e.g., DH5α) with the plasmid.
2. For bacteria transformed with the desired vector, inoculate 3–10 ml of LB broth supplemented with the corresponding selective marker (100 μg/ml ampicillin for DH5α/pET22b+) with two to three individual *E. coli* colonies from overnight (ON) growth on an LB-agar (100 μ/ml ampicillin) plate. Incubate ON at 37 °C with shaking at 200 rpm.
3. Purify the plasmid from the bacterial culture using a Wizard® Plus SV Minipreps DNA Purification System following the manufacturer's recommendations (*see* **Note 1**). A final plasmid preparation of at least 200–300 ng/μl will be sufficient for cloning purposes, although higher concentrations (300–2000 ng/μl) are preferred.
4. Digest ≥200 ng of the purified vector at two different (unique) restriction sites of the multiple cloning region of the vector by following the manufacturer's recommendations for restriction buffer and incubation conditions for the selected enzymes (*see* **Note 2**). Many combinations of enzymes are compatible in the same buffer. In our example of Ng-ACP cloning, we double-digest 1 μg of pET22b+ with 1 U of both NdeI and XhoI in 1× Buffer D in a final reaction volume of 50 μl, for 3 h at 37 °C (Fig. 2a) (*see* **Note 3**).
5. Dephosphorylate the vector to decrease the background of non-recombinants due to self-ligationof the vector (*see* **Note 4**). For pET22b+ dephosphorylation, add 0.1 U/μl alkaline phosphatase in 1× Dephosphorylation Buffer (80 μl final volume), incubate at 37 °C for 30 min, and finally inactivate the enzyme at 65 °C in a thermal block for 15 min. This dephosphorylation reaction can be performed right away after digestion (**step 4**), without any previous plasmid purification.
6. Gel-purify the vector to remove residual nicked and supercoiled plasmid, which transform very efficiently relative to the desired ligation product. Extract the linearized plasmid from the corresponding agarose gel band using the Wizard® SV Gel and PCR Clean-Up System, following the manufacturer's instructions. This membrane-based system guarantees up to 95% recovery depending upon the DNA fragment size (100 bp to 10 kbp), and purified DNA can be eluted in as little as 15 μl, allowing an increased final concentration of the sample (*see* **Note 5**).

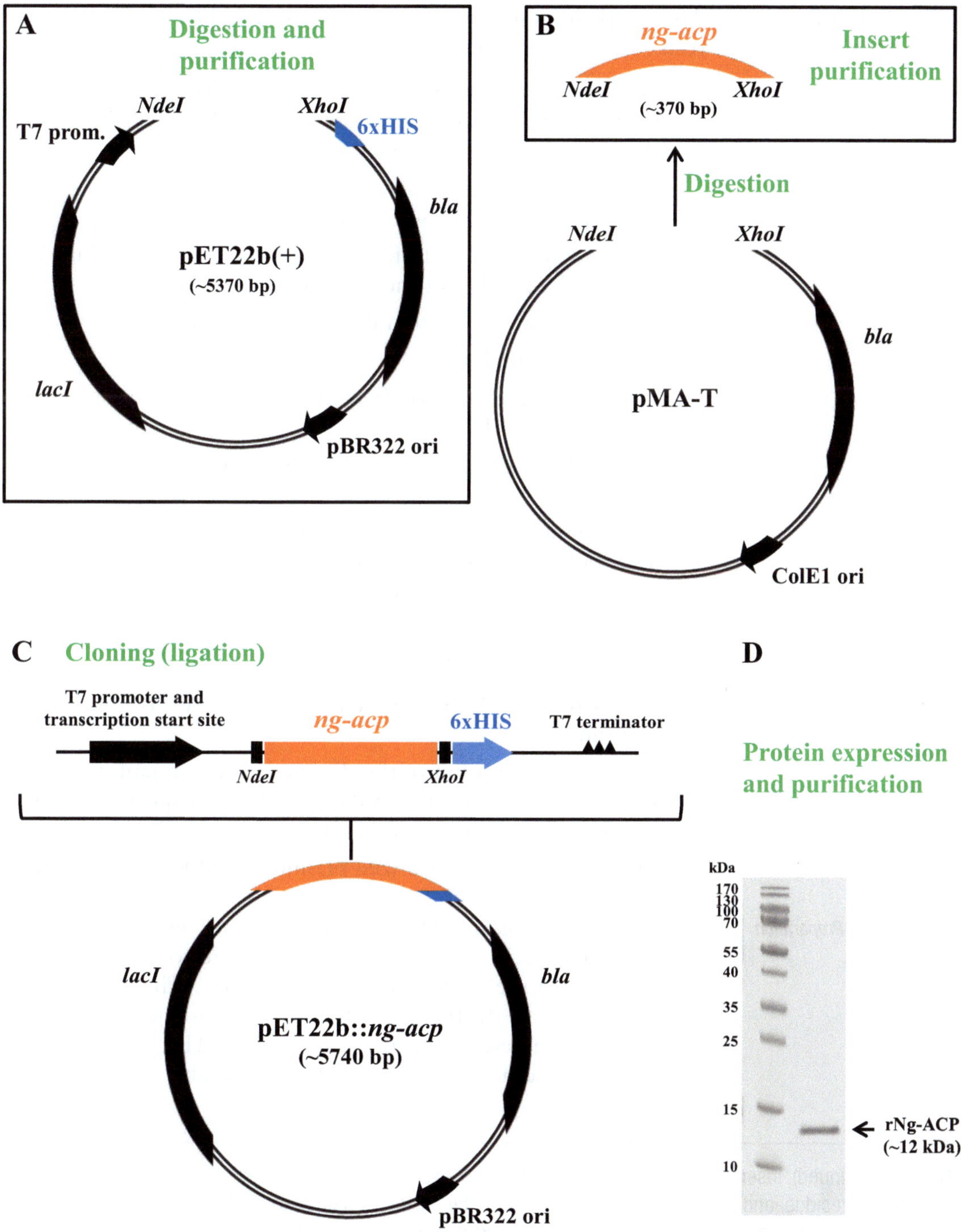

Fig. 2 Construction of pET22b::*ng-acp* for recombinant protein expression and purification. (**a**) Vector preparation. pET22b+ vector was linearized by double digestion with restriction enzymes NdeI (CATATG) and XhoI (CTCGAG), and purified by extraction of the band from an agarose gel. *Bla* and *lacI* sequences encode for β-lactamase (ampicillin resistance) and Lactose Operon Repressor, respectively. (**b**) Insert preparation. The *ng-acp* gene sequence, optimized for *E. coli* expression, was synthesized in vitro (GeneArt, Invitrogen). The

3.2.2 Agarose Gel Electrophoresis

Analyze the digested vector with agarose gel (1% w/v) electrophoresis to confirm efficient digestion prior to the ligation step.

1. Dissolve 1 g of agarose in 100 ml of 1× TAE buffer. Cover the flask and microwave for 2 min or until boiling (the agarose should be completely dissolved).
2. Cool down the agarose solution for ~5 min.
3. Add 10 μl of SYBR™ Safe™ DNA Gel Stain or any other appropriate DNA Gel Stain (e.g., 5 μg/ml of ethidium bromide). Mix gently but thoroughly.
4. Pour the agarose solution into the casting tray.
5. Place the comb into the casting tray.
6. Let the agarose solution become solid (~10–15 min at room temperature (RT)).
7. Add DNA loading buffer to the sample at a final concentration of 1×.
8. Prepare the gel electrophoresis unit and fill the gel box with 1× TAE buffer until the gel is completely covered.
9. Carefully load the samples (and an appropriate MW marker) into the gel.
10. Run the gel at 70–80 V for ~1 h.
11. Visualize the DNA fragments with a UV- or blue light-transilluminator.
12. After gel-purification of the digested vector, quantify the final product (e.g., using a NanoDrop™ UV-Vis spectrophotometer, for which only 1–2 μl of sample is required).
13. Store the purified, linearized plasmid at −20 °C until needed.

3.2.3 Insert Preparation

1. Preparation of the DNA target sequence encoding for the desired protein (insert) is essential for it to be successfully placed into the backbone vector of choice. The quickest and easiest way to do this is to design the construct in silico and have it commercially synthesized in vitro. The product can be purchased as a purified piece of linear double-stranded DNA

Fig. 2 (continued) insert was specifically designed to exclude the start codon (ATG) for the N-terminal methionine residue and the stop codon (TAA) from the *ng-acp* open reading frame. Insert release was achieved by double digestion of commercial pMA-T::*ng-acp* vector with NdeI and XhoI, followed by purification from an agarose gel band. (**c**) Cloning. The purified *ng-acp* insert was cloned into linearized/purified pET22b+ vector by ligation ON at 4 °C and subsequently transformed into *E. coli* DH5α. Positive transformants were screened by Colony PCR. (**d**) Recombinant protein expression and purification. Recombinant (r)Ng-ACP protein expression was induced from a log-phase culture of BL21 (DE3) pLysS/pET22b::*ng-acp* and then purified by Ni^{2+} metal ion affinity chromatography under native conditions. Quality of purified rNg-ACP (~12 kDa) was assessed by SDS-PAGE

with the sequence of interest, or cloned into a standard vector from which it will need to be released by digestion for further subcloning into the vector of choice. In our cloning example, the *ng-acp* gene sequence, optimized for *E. coli* expression and encoding the entire coding sequence for Ng-ACP protein, was synthesized in vitro (GeneArt, Invitrogen). The insert was specifically designed to exclude the start codon (ATG) for the N-terminal methionine residue and the stop codon (TAA) from the *ng-acp* open reading frame (ORF). The start codon was meant to be replaced by the ATG codon within the 5′-end NdeI (CATATG) restriction site, so no removal of the ATG start codon from the *ng-acp* ORF would have eventually resulted in codon duplication after insertion into the cloning vector. Exclusion of the stop codon, on the other hand, allows C-terminal translational fusion of Ng-ACP to the C-terminal His tag (the termination regulatory element is being provided by the vector downstream of the tag sequence). An advantage of this method over a classical PCR based cloning technique (*see* **Note 6**) is that no further sequencing of the insert is required, as quality control of the insert sequence is routinely performed after its synthesis.

2. Double-digest with NdeI and XhoI enzymes (*see* **step 4** of Subheading 3.2.1) to release the *ng-acp* insert from the commercial vector pMA-T (GeneArt, Invitrogen).
3. Purify the released *ng-acp* insert from an agarose gel (*see* **step 6** of Subheading 3.2.1) as described for the cloning vector preparation (Fig. 2b).
4. Quantify the digested, purified insert by absorbance and store at −20 °C until needed (*see* **step 12** of Subheading 3.2.2).

3.2.4 Cloning: Vector–Insert Construct Preparation

The following technique applies to any insert cloning, regardless of the method used to generate it (amplified by PCR or released by digestion from a different recombinant construct).

1. Ligate the prepared insert (*see* Subheading 3.2.3) to the linearized, purified cloning vector (*see* Subheading 3.2.1). When cloning a DNA fragment into a plasmid vector, a molar ratio of vector and insert of 1:3 or 1:5 is generally recommended. Use the following equation to convert molar ratios to mass ratios:

$$\frac{(\text{ng of vector} \times \text{kbp size of insert})}{\text{kbp size of vector}} \times \text{molar ratio of insert/vector} = \text{ng of insert}$$

The following example illustrates the conversion of molar ratios to mass ratios for 100 ng of a 5.5 kbp plasmid (pET) and a 0.4 kbp DNA fragment insert (*ng-acp*), with a desired vector–insert ratio of 1:5:

$$\frac{(100\,\text{ng pET} \times 0.4\ \text{kbp}\ \textit{ng-acp})}{5.5\,\text{kbp pET}} \times (5/1) = 36.4\,\text{ng}\ \textit{ng-acp}$$

2. To generate the pET22b::*ng-acp* recombinant construct, mix 100 ng of digested pET22b+ with 36.4 ng of *ng-acp* insert in a ligation mixture together with 1 U of T4 DNA Ligase and 1× Ligase buffer, in a final volume of 20 μl (Fig. 2a–c). To confirm no circularization of the vector, prepare a negative control in the same conditions but without insert. Do the ligation reaction ON at 4 °C, although different protocols might be recommended for other commercial DNA ligases. Store the ligation mixture at −20 °C until required, or directly transform the mixture into a competent cloning host with no need to inactivate the ligase prior to transformation.

3.2.5 Transformation of a Cloning (Nonexpressing) Host

Initial cloning should be done in a *recA*$^{-}$ cloning strain, such as DH5α, or other similar host that lacks the gene for T7 RNA polymerase (refer to Subheading 3.1.2 for further information).

1. Inoculate 3 ml of LB broth with a host *E. coli* DH5α colony and incubate ON at 37 °C with shaking (200 rpm). Seed 25 ml of LB in a sterile 250 ml Erlenmeyer flask with 1 ml of the ON culture, and grow at 37 °C with shaking (200 rpm) to an optical density (OD) of 0.4–0.5 at $\lambda_{600\ \text{nm}}$.
2. Stop bacterial growth by chilling the culture on ice for 20 min.
3. Centrifuge the cells at 3000 × *g*, at 4 °C, for 10 min and discard the supernatant (SN).
4. Suspend the cell pellet in 5 ml of ice-cold, sterile 0.1 M $MgCl_2$ buffer.
5. Centrifuge as described in **step 3**, discard the SN and suspend the pellet in 1 ml of ice-cold, sterile 0.1 M $CaCl_2$ buffer.
6. Leave on ice for at least 1 h.
7. To store competent cells, add 200 μl of sterile 100% glycerol, distribute ~50–150 μl aliquots into cryotubes and store at −80 °C until needed (*see* **Note** 7).
8. To chemically transform the cells, add 10 μl of ice-cold pET22b::*ng-acp* DNA ligation mixture DNA (ligation) to 50 μl of competent cells and leave on ice for at least 30 min. Shake occasionally by hand (*see* **Note 8**).
9. Heat-shock the cells at 42 °C for 45 s and immediately place them on ice and leave for 5 min.
10. Add 1 ml of LB to each tube and incubate at 37 °C for 1 h.

11. Centrifuge at 5000 × *g* for 5 min and remove ~1 ml of the SN. Suspend the bacterial pellet in the remaining SN (*see* **Note 9**).
12. Plate on selective medium (e.g., LB agar supplemented with the corresponding antibiotic to which resistance is conferred by the cloning vector; 100 μg/ml ampicillin for pET22b) and incubate ON at 37 °C (*see* **Note 10**).

3.2.6 Direct Colony PCR

There are several methods for identifying the positive clones, such as Colony Polymerase Reaction (PCR) or plasmid miniprep and restriction analysis. The quickest and simplest way to simultaneously screen several transformant candidates is by direct Colony PCR using primers binding in the vector sequence, upstream and downstream of the target insert. For most pET vectors, the T7 promoter (forward) and the T7 terminator (reverse) primers, both commercially available, are useful for determining the presence (or absence) of the target insert as well as its correct orientation relative to the vector promoter sequence.

1. Pick a positive transformant colony candidate using a 200 μl pipette tip or a sterile toothpick. Touch the tip to a new selective LB agar plate (for generating a copy of the desired colony) and suspend the remaining bulk of the colony in 50 μl of sterile H_2O in a 0.5 ml tube. Vortex thoroughly to disperse the bacterial cells. Repeat this procedure with as many colonies as desired.
2. Place the tubes in boiling water or a heat block at 100 °C for 5 min to lyse the cells and denature DNases.
3. Centrifuge at maximum speed (e.g., 13,000 × *g*) for 1 min to remove cell debris and leave on ice.
4. Mix the components shown in Table 1 (except for the template) together in a single Eppendorf tube for a master PCR reaction mix using reagent volumes corresponding to the number of reactions desired (*see* **Note 11**). Place 19 μl aliquots into 0.5 ml PCR tubes and add 1 μl of the corresponding template (from **steps 1–3**). Keep the tubes on ice.
5. Run the thermal cycler for 30–35 cycles as shown in Table 2.
6. To analyze the PCR reactions, add DNA loading dye buffer to a final concentration of 1× to 3–10 μl of each sample and load them onto a 1% (w/v) agarose gel (*see* Subheading 3.2.2) together with an appropriate MW marker (*see* **Note 13**). Run the gel at 80 A for about 1 h and visualize the gel in a UV- or blue light trans-illuminator.
7. To make glycerol stocks of the positive transformants (PCR positives with the correct band size and insert orientation), inoculate 3–5 ml of LB broth (supplemented with 100 μg/ml

Table 1
Components of the colony PCR master mix

Reagent	Volume (μl)
Master mix (2×)	10
Forward primer (10 μM)	1
Reverse primer (10 μM)	1
DNA Template (suspended colony)	1
H_2O	7
Total volume	20

Table 2
Thermal cycler parameters for the colony PCR protocol (*see* Note 12)

Cycle step	Process stage	Temperature (°C)	Time (min)
1	Lid	105	–
2	Initial DNA denaturation	95	2
3	DNA denaturation	95	0.5
4	Primers annealing	60–65	0.5
5	Extension	72	1
(Go back to **steps 2–5** 30–35 times)			
6	Final extension	72	5
7	Pause	4	∞

ampicillin) with the corresponding colonies previously grown on a selective LB-agar plate (*see* **step 1** of Subheading 3.2.6). Incubate ON at 37 °C with shaking at 200 rpm. Dispense 1 ml aliquots into cryovials and add 10–20% (v/v) sterile glycerol. Mix and store at −80 °C.

3.3 Transformation of the Expression Host

3.3.1 Plasmid Purification

1. Isolate the plasmids from the identified positive clones (*see* Subheading 3.2.6) using standard miniprep kits (*see* **step 3** of Subheading 3.2.1) following the manufacturer's recommendations. Purified plasmids can then be used for restriction, sequencing analyses, and transformation into the expression host.

3.3.2 Restriction Analysis

1. Digest the recombinant vectors with the same pair of restriction enzymes used for cloning (*see* **step 4** of Subheading 3.2.1).
2. Analyze the digested samples by agarose gel electrophoresis (*see* Subheading 3.2.2). Positive clones should release an insert of

the correct molecular weight, thus confirming the colony PCR results (*see* Subheading 3.2.6).

3.3.3 Sequencing Analysis

1. Vector-specific T7 promoter (forward) and T7 terminator (reverse) primers are generally used for sequencing from pET vectors. Sequencing is not necessary, however, when the insert has been released by enzymatic digestion from a different recombinant vector (which has been previously sequence-verified), or when the insert has been synthesized in vitro (as a sequence quality control is always guaranteed) (*see* **Note 14**).

3.3.4 Transformation of the Expression Host

For protein expression, the recombinant vector first needs to be transformed into an appropriate expression host (e.g., BL21 (DE3) pLysS—*see* Subheading 3.1.3).

1. Prepare BL21 (DE3) pLysS competent cells as described in **steps 1**–7 of Subheading 3.2.5 or obtain them from a commercial source (*see* **Note** 7). The transformation protocol with pET22b::*ng-acp* described for DH5α can also be used for transforming the expression host (*see* **steps 8–12** of Subheading 3.2.5), except that the selective plates used in **step 12** should also be supplemented with 34 μg/ml chloramphenicol (antibiotic resistance provided by the pLysS vector).
2. Incubate the plates ON at 37 °C, repeat-streak transformants for single colonies and prepare glycerol stocks as described in **steps 1** and 7 of Subheading 3.2.6 (*see* **Note 15**).

3.4 Protein Pilot Expression

A transformed λDE3 lysogen containing the recombinant construct can be induced to express the target protein (under transcriptional control of a T7 *lac* promoter encoded in the pET vector) by the addition of IPTG to a growing culture (*see* Subheading 3.1.3).

1. Pick a single colony of an ON agar plate culture of BL21 (DE3) pLysS/pET22b::*ng-acp* and inoculate 3–5 ml of LB broth supplemented with 100 μg/ml ampicillin and 34 μg/ml chloramphenicol. Incubate with shaking (200 rpm) ON at 37 °C.
2. Inoculate 50 ml of LB broth (supplemented with 100 μg/ml ampicillin and 34 μg/ml chloramphenicol) in a sterile 250 ml Erlenmeyer flask with 1 ml of this ON culture. Incubate with shaking (200 rpm) at 37 °C until mid-log phase is reached (OD_{600} ~0.6) (*see* **Note 16**). Prepare a second flask in parallel as a negative control of the induction process.
3. Add 500 μl of 100 mM IPTG to one of the flasks (IPTG final concentration of 1 mM). Leave the negative control flask untreated.

4. Incubate both flasks with shaking (200 rpm) at 37 °C for a further 4 h.
5. Harvest the cells (1 ml aliquots) by centrifugation (6000 × *g* for 5 min), remove the SN and store the cell pellets at −80 °C until needed (or continue with **step 6**) (*see* **Note 17**).
6. Thaw individual cell pellets (if frozen) and suspend in 100 μl of PBS. Prepare lysates by sonication with a probe sonicator (applying 2–5 pulses of 15 s, with 30 s intervals, at ~40 MHz) and centrifuge the samples for 5 min at maximum speed (13,000 × *g*).
7. Transfer the SN (soluble proteins) to new Eppendorf tubes and suspend the pellets (insoluble proteins and cell debris) in 100 μl of PBS containing 0.5% (w/v) SDS.
8. Dispense 3–5 μl volumes of each sample (soluble and insoluble for both induced and noninduced flasks) into new Eppendorf tubes, make up the volume to 9 μl with H_2O and add 3 μl of 4× SDS-PAGE loading buffer. Boil the samples for 5 min at 100 °C.
9. Load the samples and an appropriate MW Marker onto an SDS-PAGE gel and run the gel in 1× Running Buffer for ~1 h at 200 V.
10. Stain the gel by soaking it in Staining Solution (with gentle agitation) at RT for at least 1 h.
11. Destain the gel by soaking it in Destaining Solution (with gentle agitation) at RT for at least 1 h. Several changes of the Destaining Solution are advisable for clear gels.

The presence of the target recombinant protein of correct molecular weight in either the soluble or the insoluble fraction of the IPTG-induced culture will determine solubility (*see* **Note 18**).

3.5 Protein Purification

The method chosen for protein purification depends on a number of factors, including the properties of the protein of interest, its cellular localization, the nature of its tag, the expression strain background, the protein expression level and solubility, the amount of protein needed and the intended application for the expressed protein [2]. One advantage of the pET system is that the target recombinant protein is highly expressed, and therefore constitutes a high percentage of the total cell mass, making its purification process relatively easy and straightforward. Proteins can be purified under native or denaturing conditions, which mainly depend on their solubility (*see* Subheading 3.4 and **Note 19**).

rNg-ACP is a soluble (active), mature protein (~12.5 kDa) with the leader peptide (LP) cleaved by the *E. coli* Sec system when the protein is translocated into the periplasm after translation. Presence of the C-terminal 6×His tag allows for purification under native conditions (*see* **Note 20**). His-tagged proteins have a high selective affinity for Ni^{2+} and several other metal ions that can be immobilized on chromatography media using chelating ligands. Consequently, a His-tagged protein will selectively bind to a metal-ion-charged substrate, such as a nickel-nitrilotriacetic acid (Ni-NTA) matrix, while other cellular proteins will not bind or will bind weakly. This chromatographic technique is often termed immobilized metal ion affinity chromatography (IMAC) [8] (*see* **Note 21**).

An overview of a typical purification workflow for His-tagged proteins under native conditions is detailed below, and considerations for insoluble proteins are discussed in the corresponding Notes for each purification step.

3.5.1 Bacterial Cell Culture and Protein Expression

Small-scale pilot experiments are always recommended to establish empirically the optimal conditions of culture and induction of expression (*see* Subheading 3.4).

1. Picking a single colony of an ON agar plate culture of BL21 (DE3) pLysS/pET22b::*ng-acp* and inoculate 15 ml of LB broth supplemented with 100 μg/ml ampicillin and 34 μg/ml chloramphenicol. Incubate with shaking (200 rpm) ON at 37 °C (*see* **Note 22**).
2. Inoculate 500 ml of LB broth (supplemented with 100 μg/ml ampicillin and 34 μg/ml chloramphenicol) in a sterile 2 L Erlenmeyer flask with 10 ml of the ON culture. Incubate with shaking (200 rpm) at 37 °C until mid-log phase is reached (OD_{600} ~0.6) (*see* **Note 23**).
3. Add 5 ml of sterile 100 mM IPTG to reach a final concentration of 1 mM and incubate with shaking (200 rpm) at 37 °C for further 4 h.
4. Harvest the cells by centrifugation (6000 × *g* for 30 min), remove the SN and store the cell pellets at −80 °C until needed (or continue with Subheading 3.5.2) (*see* **Note 24**).

3.5.2 Bacterial Cell Lysate

1. Suspend the bacterial cell pellet in 30 ml of ice-cold native binding buffer (*see* **Notes 25–27**).
2. Sonicate the bacterial suspension using a probe sonicator, until a clear lysate is achieved. Keep the cells and protein solutions at 0–4 °C at all times to prevent protein degradation; the addition of protease inhibitors may also be necessary (*see* **Note 28**).
3. Clarify the lysate by centrifugation (10,000 × *g* for 30 min at 4 °C) to remove all cell debris particles and insoluble material that may interfere with drainage of the column during purification. Alternatively, filter the unclarified lysate using a 0.45 μm filter with syringe. Keep the lysate at 0–4 °C at all times. (For cell lysate preparation in denaturing conditions, *see* **Note 29**.)

3.5.3 Matrix Equilibration

The Ni-NTA matrix can be purchased prepacked inside a column (column purification) or as a reagent that can be added and suspended straight into the lysate (batch purification) and then packed manually into an empty column for further purification steps (washes and elution). Once the resin has been packed into the column, protein purification can be done using either gravity-flow or a syringe for manual use, a laboratory pump or a chromatography system. Both column and batch-mode purifications allow efficient binding of the target protein to the resin [9]. However, the latter is recommended when the His-tag is not fully accessible or when the target protein in the lysate is at very low concentrations. An equally effective alternative to improve column-mode purifications yield is to decrease the flow rate or pass the sample through the column several times.

Equilibrate the matrix with the same binding buffer used to prepare the lysate (*see* Subheading 3.5.2) prior to the protein binding step (*see* Subheading 3.5.4) for maximum target protein capture efficiency.

1. When prepacked inside a column, pass binding buffer at ~10–20 times the volume of Ni-NTA resin, and preferably at a slow flow-rate, to equilibrate the matrix.
2. For batch purification procedures, suspended the Ni-NTA resin first in binding buffer to equilibrate the matrix and centrifuge at 3000 × *g* for 5–10 min to produce a pellet.

3.5.4 Protein Binding in Batch Suspension

1. Suspend 2 ml of the Ni-NTA resin pellet (*see* **step 2** of Subheading 3.5.3) into 30 ml of cleared cell lysate (*see* Subheading 3.5.2) and maintain at 4 °C for 1 h with constant, slow rotation (*see* **Note 30**).
2. Pack the protein-resin complex into an empty chromatography column by gravity-flow (*see* **Note 31**).

3.5.5 Wash

Since the tagged protein binds specifically to the resin ligand, host cell proteins can be washed away by passing through the column, a washing buffer containing slightly higher concentrations of imidazole than the binding buffer. Stringent washes (>20 mM imidazole) may increase purity but at the cost of reducing yield. Consequently, the imidazole concentration used for washing will then depend of the intended use of the purified recombinant protein.

1. Pass ~20 resin-equivalent volumes of ice-cold washing buffer through the column, preferably at a slow flow-rate.
2. Collect an aliquot of the wash flow-through to determine if there is possible removal of bound target protein during the washing step. For washing in denaturing conditions, *see* **Note 32**.

3.5.6 Elution and Ni-NTA Resin Regeneration

During the elution step, the imidazole concentration is increased to promote dissociation of the 6×His-tagged protein from the resin ligand, as it can no longer compete for binding sites on the Ni-NTA resin. Imidazole concentrations of 250–300 mM will efficiently remove all His-tagged proteins from the column within a small volume of elution buffer.

1. Pass ice-cold elution buffer through the column, preferably at a slow flow-rate. The volume of elution buffer needed to recover most target protein will depend on the amount of protein bound, on the volume of resin and on the flow-rate (*see* **Note 33**).
2. Collect the eluates in 0.5–1.0 ml fractions for further qualitative and quantitative analyses (*see* Subheading 3.5.8). For protein elution in denaturing conditions, *see* **Note 34**.
3. Follow the regeneration protocol shown in Fig. 3 in order to reuse the Ni-NTA resins (*see* **Note 35**).

3.5.7 His-Tag Removal

The 6×His-tags generally do not affect biological activity of native recombinant proteins and their removal may be unnecessary for some applications (e.g., antibody generation). However, His-tag removal may be desirable for some applications such as X-ray crystallography or NMR, in which case, a protease cleavage site should be present between the vector-encoded tag sequence and the coding sequence of the target gene. The pET system offers a vast collection of plasmids encoding for thrombin, enterokinase, or factor Xa protease cleavage sites, but the pET22b+ used for cloning Ng-ACP does not contain one and consequently the C-terminal 6×His tag cannot be removed from this recombinant protein [6]. Other commercial expression vectors may offer the possibility of chemical instead of enzymatic methods to cleave fusion proteins at designed sites, but only the natural proteolytic enzymes have the requisite specificity to be broadly useful reagents for this purpose (*see* **Note 36**).

3.5.8 Analyses of the Eluates

When columns are coupled to Fast Protein Liquid Chromatography (FPLC) systems, the recombinant protein of interest can easily be identified by monitoring the absorbance ($\lambda_{280\ \mathrm{nm}}$) of the elution fractions. Protein purity can be assessed qualitatively by analysing samples of each protein-containing elution fractions by SDS-PAGE (Fig. 2d). For native purifications, the eluate samples can be loaded onto the SDS-PAGE gel without further treatment (*see* **steps 8–11** of Subheading 3.4). For SDS-PAGE analysis of elution samples in denaturing conditions, *see* **Note 37**.

Western Blot can also be used as a specific and sensitive method for target protein detection and estimation of expression levels, and the protocol can be conveniently done with commercial anti-His

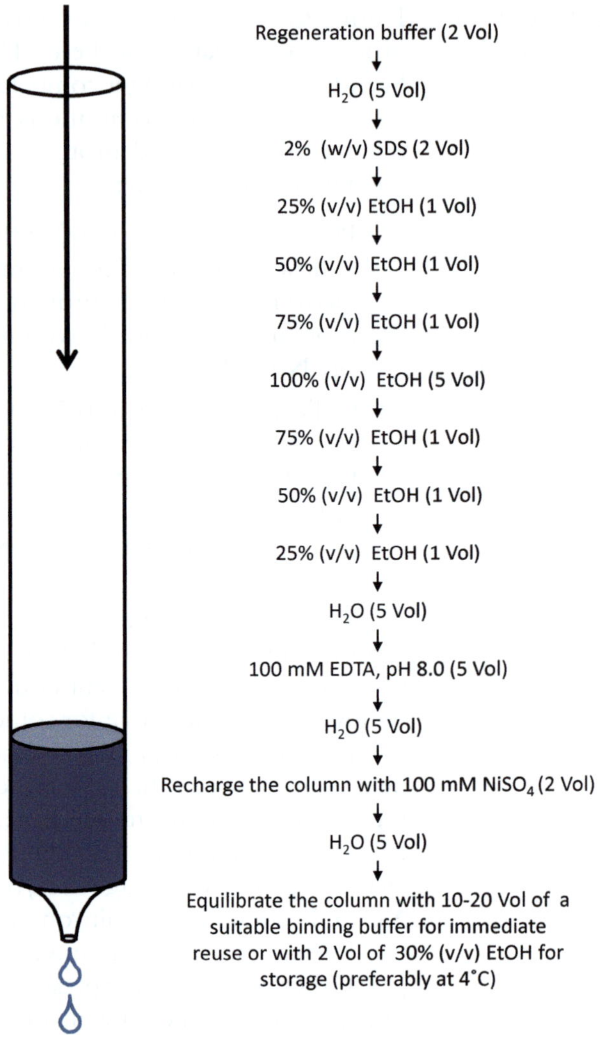

Fig. 3 Flow diagram showing sequential steps to regenerating Ni-NTA affinity chromatography resins

tag specific antibodies or target protein-specific antibodies. The Western Blot protocol described below may need optimization for each particular recombinant protein.

1. Add the appropriate amount of 4× SDS-PAGE loading buffer to the (native) eluate samples and boil them for 5 min. For denaturing eluate samples *see* **Note 37**.
2. Load and run the SDS-PAGE gel (*see* **step 9** of Subheading 3.4). Do not Coomassie-stain the gel after electrophoresis.
3. Transfer the proteins from the gel to the membrane, which can be either a nitrocellulose or PVDF membrane. If using PVDF, activate the membranes with methanol for 1 min and then rinse them with transfer buffer before preparing the stack.

4. Assemble the transfer stack as follows: over the positive electrode, stack two pieces (of the same size of the gel) of Whatman filter paper, then the membrane, then the gel, and finally two other pieces of Whatman filter paper. Soak all components in blotting buffer and ensure that there are no bubbles between the layers as these would prevent the electric current going through the stack evenly. Place the negative electrode on top.
5. Transfer for 30 min at 10 V (*see* **Note 38**).
6. Disassemble the stack and move the membrane to a dish, add blocking buffer for 1 h at RT or ON at 4 °C.
7. Remove the blocking buffer and incubate the membrane with appropriate dilutions of primary antibody (e.g., anti-His antibodies) prepared in blocking buffer. Keep the blot at 4 °C ON to allow binding to low concentrations of the target protein in the samples and/or with low affinity antibodies. Other conditions may need to be optimized.
8. Wash the membrane three times with TTBS, 5 min each.
9. Incubate the membrane with the recommended dilution of conjugated secondary antibody in blocking buffer at RT for 1 h.
10. Repeat **step 8**.
11. For signal development using a colorimetric method, incubate the membrane with substrate buffer (for alkaline phosphatase (AP)-conjugated secondary antibodies) and stop the reaction by rinsing the membrane with H_2O when the desired band intensity has been achieved. For signal development by chemiluminescence, follow the kit manufacturer's recommendations (*see* **Note 39**).

Finally, protein precipitation and solubilization or dialysis are useful techniques that allow changing the elution buffer with a more appropriate one depending on the intended use of the recombinant protein produced (e.g., removal of the imidazole may be necessary for certain assays in which it might interfere).

4 Notes

1. This kit provides a simple and reliable method for rapid isolation of plasmid DNA within 30 min or less, depending on the number of samples processed. This system can be used to isolate any plasmid from *E. coli* hosts, but works most efficiently when the plasmid is less than 20 kbp in size. Plasmid yield will vary depending on a number of factors, including culture volume, plasmid copy number, type of culture medium and the bacterial strain used.

2. It is essential that no restriction sites for the selected restriction enzymes are present within the target gene, as this will result in undesired cleavage of the insert. Digestion with two different enzymes is always recommended over digestion with only one, as having two incompatible ends will significantly reduce the probability of the vector religating without incorporating the insert, and will warrant the correct orientation of the gene once inserted into the construct.
3. Nonoptimal reaction conditions (e.g., high concentration of glycerol (>5% v/v) and high ratio of enzyme units to μg DNA) may cause the restriction enzymes to display "star activity," that is, the enzyme cuts a sequence that is an imperfect match to its known target, thus preventing cloning success.
4. This step is highly recommended when cloning into a single site, but it is also useful to dephosphorylate vectors cut with two enzymes, especially when the sites are close together or if one of the enzymes is a poor cutter (incomplete digestion with one of the enzymes is undetectable by agarose gel analysis and it will prevent the target DNA from inserting into the vector).
5. Gel-purification is optional but highly recommended to reduce the effort required to screen for the correct construction.
6. PCR-based cloning allows amplification of the desired piece of DNA, while adding specific restriction sites to its ends so that it can be easily cloned into the plasmid of interest. Genomic (g) DNA (or a different recombinant plasmid harboring the insert) could be used as templates for the PCR reaction. This technique is particularly useful when the insert is cloned into a different vector that does not have the restriction sites of choice; therefore, the insert cannot be released using the same pair of digestion enzymes used in **step 4** of Subheading 3.2.1. However, amplifying the insert by PCR carries the risk of incorporating mutations into the gene of interest, and it is therefore important to use a high fidelity Taq polymerase in order to minimize the probability of generating such mutations.

   In general, successful PCR reactions require important features for primer design to be taken into account, such as the following:

   (a) The 3′-end of the primer molecule is critical for the specificity and sensitivity of the PCR. One or two G or C nucleotide bases are highly recommended at this position as these bases will provide a much stronger binding affinity than a T or an A base. However, three or more G or C bases at this position may stabilize nonspecific annealing of the primer. A 3′-end T should also be avoided, since it is more prone to mispriming than the other three nucleotides (G, C, or A). Complementarity at the 3′-end of the primer pair should be circumvented as this may lead to primer-dimer formation, resulting in lower PCR yields or even in complete inhibition of the PCR reaction.

(b) Bases at the 5′-end of the primer are less critical for primer annealing, allowing for the addition of restriction sites without putting at risk the efficient annealing of primers to the template. Extra base pairs on the 5′ end of the primer assist with restriction enzyme digestion (usually 3–6 bp right next to the restriction site), but are not essential.

(c) A start codon (usually ATG) should be included in the 5′ (forward) primer when the gene of interest is not expressed together with an N-terminal tag or fusion partner, in which case the ATG is provided by the cloning vector.

(d) A stop codon should be included when no C-terminal tag is used. In this case, TAA is preferred over TAG and TGA because it is less prone to read-through. Furthermore, using two or three stop codons in series increases the efficiency of translational termination.

(e) Usually a primer length of 18–30 bases is optimal for most PCR applications and the region of the primer complementary to the sequence to be amplified is usually 15–20 bp long. Shorter primers could lead to amplification of nonspecific PCR products.

(f) Good results are also strongly dependent on the melting temperature of the primers (T_m, temperature at which half of the primer has annealed to the template). Standard recommendations for successful PCR amplification are that both primers have similar T_m values (within 2–4 °C) and preferably above 60 °C (for which the GC content should be between 40% and 60%). The T_m can be estimated using the following formula:

$$T_m = 2^\circ C \times (A + T) + 4^\circ C \times (C + G).$$

Alternative cloning strategies to PCR are routine; for example, the insert could be released from a different recombinant vector by double digestion with the same pair of restriction enzymes used to linearize the plasmid of choice (*see* **step 4** of Subheading 3.2.1) for further subcloning.

7. For convenience and consistent performance, relevant host strains are commercially available as competent cells, ready for high-efficiency chemical transformation. Alternatively, competent cells could also be transformed by electroporation

(or electropermeabilization) with similar transformation yields, but an electroporator is required [10].

8. Handle only the very top of the tube and the tube cap to prevent the cells from warming. Keep the cells on ice whenever possible. It is highly recommended to transform also with a ligation control (no insert), and to keep **50** μl of the competent cells untransformed as a negative control of the selection screening of transformants.
9. The smaller the final volume of the suspended bacteria, the easier it will be to plate.
10. A successful cloning implies, usually, many more colonies for the transformation with the ligation in the presence of the insert than with the negative control ligation (no insert). However, cloning could also be successful even though the number of colonies on the two plates is roughly equivalent. Colonies on the negative control ligation plate suggest possible religation of the vector due to, for example, low efficiency digestion with one of the two restriction enzymes (*see* **step 4** of Subheading 3.2.1) or incomplete dephosphorylation after digestion (*see* **step 5** of Subheading 3.2.1). Conversely, there should be no colonies on the negative control of untransformed bacteria, as the cloning host should not carry the resistance cassette necessary for growing in the selective media.
11. Commercial PCR master mixes generally provide the DNA polymerase enzyme and its appropriate buffer. However, they may vary in stock concentration, depending on the brand. To account for pipetting losses, increase the volumes of reagents (e.g., prepare master mix for 11–12 colonies when testing 10).
12. The PCR thermal cycler settings described have been used successfully to amplify *ng-acp* insert. However, for other PCR reactions, the duration of the extension stage will depend on the size of DNA to be amplified. Normally, DNA polymerases can process 1 kbp/min, so smaller or bigger inserts will require a shorter or longer extension stage, respectively. Furthermore, the primers' annealing temperatures should be $<T_m$ calculated for the particular pair of primers being used (*see* **Note 6(f)**).
13. Some commercial master mixes are already supplemented with loading buffer, in which case you can load the PCR samples directly onto the gels.
14. In the case of PCR-based cloning, for which the insert is generated by PCR, it is highly advisable that any identified positive clone is checked by sequencing, to verify no errors in the target sequence due to mismatches during the PCR synthesis of the insert. A single sequence reaction (forward or reverse) should be enough for sequencing the small inserts (approximately ≤700 bp). However, double sequencing from

both ends of the insert (forward and reverse) is required for longer inserts due to processing limitations of the sequencing technique.

15. In this case, transformants do not need to be screened as previously explained for the cloning host (*see* Subheading 3.2.6) because a pure, closed (sequence-checked) plasmid is being used for transformation instead of a ligation mixture.
16. For an initial 1/50 dilution of the ON culture, it takes ~2 h to reach OD_{600} ~0.6 in the conditions detailed.
17. Preparing 1 ml aliquots from both flasks and harvesting the cells in 1.5 ml Eppendorf tubes may facilitate further analysis of the samples.
18. Coomassie Blue staining of whole cell lysates will detect all protein bands in the gel, but the band corresponding to the target protein should be much thicker (and present only in the IPTG-induced sample), as it is being transcribed from a strong T7 *lac* promoter. No band should appear in the noninduced culture (negative control).
19. Solubility partly depends on the expression vector used (construct design) and it will also vary for each protein candidate. Moreover, growth conditions during the induction process, the nature and position of the tag and/or selected lysis buffer for preparing the whole cell extracts for protein purification may strongly influence target protein solubility [11, 12]. Growth at 37 °C, for example, may cause some proteins to accumulate as inclusion bodies, while incubation at 30 °C may favor the production of soluble proteins [13]. When the target protein is meant to be exported using the signal sequences present in a number of pET vectors, then growth and induction at 25–30 °C may be more convenient. In some other cases, an ON induction at low temperatures (15–20 °C) may prove optimal for higher soluble protein yields. Solubility does not necessarily indicate that a protein is folded properly; some proteins form soluble species that are inactive [14].

    Insoluble target proteins accumulated as inclusion bodies could be an advantage because these protect the protein from proteolysis and are easily isolated by centrifugation, yielding highly concentrated and relatively pure recombinant protein. In addition, if the target protein was toxic for the expression host, it may not inhibit cell growth when present in inactive form as inclusion bodies, allowing higher purification yields [11]. Several protein folding methods are available for recombinant proteins that are needed in their active form, but the optimal protocol will vary for each particular target protein and must therefore be determined empirically.

Proteins containing hydrophobic or highly charged domains may associate with bacterial lipids or other cellular components, what may severely affect their partitioning into the soluble or insoluble fraction. In this case, proteins in the insoluble fraction may not actually be present in inclusion bodies. Addition of nonionic or zwitterionic detergents in the lysis buffer may turn lipid or membrane-associated insoluble proteins into soluble ones, but the addition of detergent may affect downstream purification procedures. Addition of salt to the lysis buffer may favor highly charged proteins to dissociate from other cellular components (e.g., DNA, cell debris) and consequently partition into the soluble fraction.

20. Even when most of the protein is present in inclusion bodies, there is generally a soluble fraction that can be purified in its native form. In this situation, however, since larger culture volumes might be necessary to compensate for the lower yield of soluble material, denaturing purification and subsequent refolding may be considered if the recombinant is needed in a native and biologically active form. However, efficient renaturing procedures are not always available.

21. Binding of the 6×His tag does not depend on the three-dimensional structure of the protein and is not affected by most detergents and denaturants such as 8 M urea or 6 M GuHCl, so insoluble His-tagged proteins can also be purified using the same chromatography matrix.

22. A small flask is preferred over a Falcon tube for better aeration of the bacterial culture.

23. For an initial 1/50 dilution of the ON culture, it takes ~2 h to reach OD_{600} ~0.6 in the conditions detailed.

24. The duration of the centrifugation will mainly depend on the volume of the centrifuge bottles used. For this protocol, centrifugation at 6000 × *g* for 30 min successfully produced a cell pellet when splitting the induced culture (2 × 250 ml) into two bottles of 300 ml each.

25. For other proteins not expressed highly, bacteria from a large culture volume should be lysed in the smallest possible volume of binding buffer so as to increase the target protein concentration in the lysate.

26. For native purification, handling the sample on ice is important to preserve the target protein native structure.

27. Under native conditions, there is a higher potential for binding background contaminants than under denaturing conditions. Imidazole competes with proteins for binding to metal-ion-charged media and at low concentrations it reduces nonspecific binding of host cell proteins. Therefore, adding low

concentrations of imidazole (10–20 mM) to the binding buffer leads to greater purity in fewer steps. Higher concentrations of imidazole will also decrease binding of the His-tagged protein, so the imidazole concentration used in each step of the purification process should be optimized for each particular target protein to ensure the best balance between high purity (low binding of host cell proteins) and high yield (strong binding of His-tagged target protein). For some proteins, 10–20 mM imidazole might drastically affect the final purification yield, in which case the amount of imidazole should be reduced to 1–5 mM.

High concentrations of strong reducing (e.g., 1,4-dithiothreitol, (DTT)) or chelating (e.g., EDTA or EGTA (ethylene glycol-bis(β-aminoethyl ether)-*N*,*N*,*N'*,*N'*-tetraacetic acid)) agents in the binding buffer should be avoided as they reduce the available nickel ions in the Ni-NTA matrix and thereby prevent binding of His-tagged proteins. Strong electron-donating groups or amino acids such as arginine, glutamine, glycine, or histidine should also be avoided [15].

28. Freeze and thaw cycles of the bacterial suspension prior to sonication might improve cell lysis, but may not be practical (time-consuming) when handling large volumes of samples. If the sample is too viscous, dilute it with binding buffer to prevent it from clogging the column; increase lysis treatment (sonication, homogenization); or add DNase/RNase to reduce the size of nucleic acid fragments. Moreover, if the sonicated cell lysate is frozen before use, precipitation and aggregation may increase on thawing. A second sonication of the lysate is then advisable.

29. For denaturing conditions, suspend the bacterial pellet in 30 ml of denaturing binding buffer (10 mM Tris–HCl buffer containing 100 mM NaH_2PO_4, 6 M GuHCl, and 10 mM imidazole; adjust pH to 8.0 with NaOH). 8 M urea can also be used instead of 6 M GuHCl. Strong denaturants such as 6 M GuHCl or 8 M urea completely solubilize inclusion bodies, allowing insoluble 6×His-tagged proteins to remain in solution. Thorough cell disruption is efficiently achieved with either chaotropic agent, so sonication is not required, although centrifugation or filtration to clear the cell lysate is advisable before purification. Under denaturing conditions, the 6×His tag on a protein will be fully exposed so that binding to the Ni-NTA matrix will increase, and the efficiency of the purification procedure will be maximized by reducing the potential for nonspecific binding. However, adding low concentrations of imidazole (10–20 mM) to the denaturing binding buffer is also recommended. Samples can be handled at RT.

30. The amount of resin used will depend on the binding capacity of the resin and on the concentration of the target tagged-protein in the lysate. Excessive amounts of resin may increase nonspecific binding, while insufficient amounts of resin will reduce the final target protein yield. The ideal amount of resin should be determined empirically in small pilot purifications and scaled up accordingly for bigger lysate preparations.
31. Collecting the lysate flow-through is advisable for testing for possible unbound target protein in the supernatant. If the sample was purified (bound) in-column, equilibrate the resin (*see* Subheading 3.5.3) and slowly apply the cell lysate onto the column. Lower flow-rates (or reloading the flow-through into the column) will increase the time of protein–resin interaction and the amount of target protein binding.
32. For denaturing conditions, washes (10 mM Tris–HCl buffer containing 100 mM NaH_2PO_4, 6 M GuHCl, and 10 mM imidazole; adjust pH to 8.0 with NaOH) can be done at RT.
33. His-tagged proteins can be eluted by other methods than using imidazole. For example, lowering the pH within the range of 2.5–6.5 will disassemble the tagged protein-resin complex, allowing elution and recovery of the target protein. This can be done most effectively by gradually decreasing the pH of the wash and elution buffers. The histidine residues in the 6×His-tag have a pK_a of ~6.0 and will become protonated if the pH is reduced (pH 4.5–5.3). Under these conditions, the 6×His-tagged protein can no longer bind to the Ni^{2+} ions and will dissociate from the Ni-NTA resin. Below pH 4, metal ions will be stripped from the chromatography medium (turning the resin from blue to white/grey) and the column will need to be recharged with metal ions if it is to be reused.

    It is not always possible, however, to elute with lower pH when using a metal ion other than Ni^{2+}. This is protein and metal ion dependent. Strong chelating compounds (e.g., EGTA and EDTA (up to 1 mM)) can also be used for protein elution as they will strip the metal ions from the resin NTA groups. In this case, the chelated ions can be removed easily from the target protein solution with a desalting column. When using EGTA or EDTA at pH $\leq$4, the column will need to be recharged with metal ions before the next purification.

    Whereas all elution methods are equally effective, imidazole is the mildest and is recommended under native conditions. Furthermore, high-purity imidazole will give very low or no background absorbance at $\lambda_{280\ nm}$, allowing target protein concentration in the eluate to be quantified without removing the imidazole from the sample. However, imidazole can be removed easily from the purified protein with a desalting column if required.

The optimal conditions for protein binding (*see* Subheading 3.5.4), washing (*see* Subheading 3.5.5) and elution may vary for each target protein and must be determined empirically. Step gradients for lowering the pH or increasing the imidazole concentrations are preferable because they are much more effective than linear gradients to separate unbound and nonspecifically bound material from the target protein, allowing higher purity to be achieved. However, an initial linear gradient is recommended to determine the most convenient concentrations to be used in further step-gradient elutions. Coupling the column to a laboratory pump or a chromatography system is advisable for more precise and reproducible gradient conditions.

34. For denaturing protein purifications, elution (10 mM Tris–HCl buffer containing 100 mM NaH_2PO_4, 6 M GuHCl, and 250 mM imidazole; adjust pH to 8.0 with NaOH) of the protein can be done at RT. Solubilization and refolding are necessary to restore biological function of proteins expressed as inclusion bodies [16–19]. Most protocols suggest dialyzing or slowly diluting the insoluble recombinant protein into a nondenaturing buffer where refolding occurs. Moreover, many proteins that are insoluble when refolded in solution can be successfully refolded while immobilized on the Ni-NTA matrix, which appears to prevent intermolecular interactions that lead to aggregate formation. Regardless of the method chosen, refolding conditions must be determined empirically for each individual protein.

35. Ni-NTA resins can be used several times before disposal, preferably with the same recombinant protein to prevent cross-contamination with other tagged-protein purifications. Ion metal stripping can occur at very low pH, in the presence of chelating agents (*see* **Note 33**) and/or as a consequence of several purification rounds, even when eluting with imidazole only. To guarantee optimal binding capacity, a maximum of 5 runs per column and washing the resin for 30 min with 0.5 M NaOH after each use is recommended. Ni-NTA matrices should be stored in 30% (v/v) ethanol (preferably at 4 °C) to inhibit microbial growth, and can also be stored for up to 1 week in any of the denaturing buffers.

36. After protease digestion, the protein of interest should be repurified in order to remove cleaved tags and uncleaved material. The optimal method to do this will depend mainly on the tag and the proteolytic digestion method used, and these are reviewed elsewhere [20, 21].

37. For denaturing purifications, GuHCl must be removed from the elution fractions as this chemical causes SDS precipitation,

consequently interfering with the PAGE electrophoresis of proteins. Conventional protocols for GuHCl removal require protein precipitation and solubilization in an appropriate buffer compatible with SDS. Protein precipitation using trichloroacetic acid (TCA) is detailed below:

(a) Add 1 volume of the elution sample to 5 volumes of H_2O (e.g., 20 μl of sample in 100 μl of H_2O).

(b) Add 1 volume of 10% (w/v) of TCA (e.g., add 120 μl of TCA into 120 μl of diluted sample).

(c) Keep on ice for 20 min.

(d) Centrifuge at top speed ($\geq$10,000 × *g*) for 15 min (preferably at 4 °C, but it can be done at RT).

(e) Remove the SN and wash the pellet with ice-cold 100% EtOH.

(f) Repeat **step (d)** for 5 min.

(g) Discard the SN and allow the protein pellets to dry (the pellet may be difficult to see).

(h) Leaving the Eppendorf tubes open and upside-down on tissue paper inside an incubator at 37 °C will accelerate the drying process. Do not overdry the pellets as this may hinder further solubilization of the samples.

(i) Suspend the pellets in 15–20 μl of 1× SDS-PAGE loading buffer.

(j) Boil the samples for 5 min and run them on an SDS-PAGE gel (with an appropriate MW marker).

Alternatively, protein precipitation can be achieved with acetone, chloroform–methanol, or EtOH instead of TCA. As a second example, a protein precipitation method using EtOH is described as follows:

(a) Add 9 volumes of cold 100% EtOH to 1 volume of protein solution.

(b) Mix thoroughly and keep at −20 °C for at least 60 min. Leave the samples at −20 °C ON for optimal protein precipitation or when handling low concentrated samples.

(c) Centrifuge at top speed ($\geq$10,000 × *g*) for 10 min (preferably at 4 °C, but it can also be done at RT).

(d) Carefully remove the SN and retain the pellet (pellet may be difficult to see).

(e) Wash pellet with cold 100% EtOH (preferably kept at −20 °C).

(f) Repeat **step (c)** for 5 min.

(g) Dry the protein pellets by inversion of the open tubes on tissue paper at 37 °C. Do not overdry.

(h) Suspend the pellets and analyze them by SDS-PAGE. When purification in denaturing conditions is critical, for example, because the recombinant protein is insoluble in native buffers, then using 8 M urea is preferable to 6 M GuHCl, because this milder denaturant is compatible with SDS. Therefore, elution samples can be analyzed directly by SDS-PAGE without further treatment, as described for native purification samples.

38. The time and voltage of transfer may require optimization. We recommend following the manufacturer's instructions. Transfer of proteins to the membrane can be checked using Ponceau S staining before the blocking step (which can be washed away by simply rinsing the membrane with H_2O before blocking).

39. SDS-PAGE and Western blot can be used for quantitation, if the samples are processed together with a suitable standard curve. Specific detection of the target protein can also be done by functional assays (relying on the tag or on the protein itself, if native and biologically active). Furthermore, the relative yield of the target protein can easily be determined by measuring the absorbance at $\lambda_{280\ nm}$, for which the extinction coefficient of the target protein will be needed. A good estimation may be obtained by theoretical calculation from the amino acid composition of the protein [22]. Alternatively, the protein yield may also be determined by standard chromogenic methods (e.g., Lowry, BCA protein assay, Bradford).

Acknowledgments

I thank Professor Myron Christodoulides for valuable experimental advice and critical reading of the manuscript. This work was funded by the Medical Research Council, UK (grant number MR/K027131/1).

References

1. Malhotra A (2009) Chapter 16: Tagging for protein expression. In: Burgess RR, Deutscher MP (eds) Methods in enzymology, vol 463. Academic Press, San Diego, CA, pp 239–258. https://doi.org/10.1016/S0076-6879(09)63016-0
2. Wingfield PT (2015) Overview of the purification of recombinant proteins. Curr Protoc Protein Sci 80:6.1.1–6.135. https://doi.org/10.1002/0471140864.ps0601s80
3. Esposito D, Chatterjee DK (2006) Enhancement of soluble protein expression through the use of fusion tags. Curr Opin Biotechnol 17(4):353–358. https://doi.org/10.1016/j.copbio.2006.06.003
4. Young CL, Britton ZT, Robinson AS (2012) Recombinant protein expression and purification: a comprehensive review of affinity tags and microbial applications. Biotechnol J 7(5):620–634. https://doi.org/10.1002/biot.201100155
5. Waugh DS (2005) Making the most of affinity tags. Trends Biotechnol 23(6):316–320. https://doi.org/10.1016/j.tibtech.2005.03.012

6. Mierendorf RC, Morris BB, Hammer B et al (1998) Expression and purification of recombinant proteins using the pET system. Methods Mol Med 13:257–292. https://doi.org/10.1385/0-89603-485-2:257
7. Vincentelli R, Cimino A, Geerlof A et al (2011) High-throughput protein expression screening and purification in *Escherichia coli*. Methods 55 (1):65–72. https://doi.org/10.1016/j.ymeth.2011.08.010
8. Porath J (1992) Immobilized metal ion affinity chromatography. Protein Expr Purif 3 (4):263–281. https://doi.org/10.1016/1046-5928(92)90001-D
9. Spriestersbach A, Kubicek J, Schäfer F et al (2015) Chapter 1: Purification of His-tagged proteins. In: Lorsch JR (ed) Methods in enzymology, vol 559. Academic Press, San Diego, CA, pp 1–15. https://doi.org/10.1016/bs.mie.2014.11.003
10. Lessard JC (2013) Transformation of *E. coli* via electroporation. Methods Enzymol 529:321–327. https://doi.org/10.1016/B978-0-12-418687-3.00027-6
11. Kane JF, Hartley DL (1988) Formation of recombinant protein inclusion bodies in *Escherichia coli*. Trends Biotechnol 6(5):95–101. https://doi.org/10.1016/0167-7799(88)90065-0
12. Leibly DJ, Nguyen TN, Kao LT et al (2012) Stabilizing additives added during cell lysis aid in the solubilization of recombinant proteins. PLoS One 7(12):e52482. https://doi.org/10.1371/journal.pone.0052482
13. Schein CH, Noteborn MHM (1988) Formation of soluble recombinant proteins in *Escherichia coli* is favored by lower growth temperature. Biotechnology 6:291. https://doi.org/10.1038/nbt0388-291
14. Sorensen HP, Mortensen KK (2005) Soluble expression of recombinant proteins in the cytoplasm of *Escherichia coli*. Microb Cell Fact 4 (1):1. https://doi.org/10.1186/1475-2859-4-1
15. Ishibashi M, Tsumoto K, Tokunaga M et al (2005) Is arginine a protein-denaturant? Protein Expr Purif 42(1):1–6. https://doi.org/10.1016/j.pep.2005.03.028
16. Clark EDB (2001) Protein refolding for industrial processes. Curr Opin Biotechnol 12 (2):202–207. https://doi.org/10.1016/S0958-1669(00)00200-7
17. Mukhopadhyay A (1997) Inclusion bodies and purification of proteins in biologically active forms. Adv Biochem Eng Biotechnol 56:61–109
18. Rudolph R, Lilie H (1996) *In vitro* folding of inclusion body proteins. FASEB J 10(1):49–56
19. Marston FA, Hartley DL (1990) Solubilization of protein aggregates. Methods Enzymol 182:264–276
20. Arnau J, Lauritzen C, Petersen GE et al (2006) Current strategies for the use of affinity tags and tag removal for the purification of recombinant proteins. Protein Expr Purif 48 (1):1–13. https://doi.org/10.1016/j.pep.2005.12.002
21. Waugh DS (2011) An overview of enzymatic reagents for the removal of affinity tags. Protein Expr Purif 80(2):283–293. https://doi.org/10.1016/j.pep.2011.08.005
22. Gill SC, von Hippel PH (1989) Calculation of protein extinction coefficients from amino acid sequence data. Anal Biochem 182 (2):319–326. https://doi.org/10.1016/0003-2697(89)90602-7

Chapter 16

Complement-Dependent Serum Bactericidal Assays for *Neisseria gonorrhoeae*

Sunita Gulati, Peter A. Rice, and Sanjay Ram

Abstract

The global spread of multidrug-resistant gonorrhea has spurred efforts to develop a safe and effective vaccine against gonorrhea. Complement plays an important role in host defenses against *Neisseria* infections. Complement-dependent bactericidal activity of antibodies (either natural antibodies or those elicited by immunization) is a well-established correlate of protection against meningococcal infections. Although correlates of protection against gonococcal infection have not been defined, there is evidence to suggest that complement-mediated killing may also predict vaccine efficacy against this disease. This chapter describes methods to prepare human complement sources and perform bactericidal assays against *Neisseria gonorrhoeae*.

Key words Complement, *Neisseria gonorrhoeae*, Bactericidal assay

1 Introduction

1.1 Complement and Neisseria Infections

The complement system comprises several fluid-phase and membrane-associated proteins [1, 2]. Complement activation in the fluid phase occurs through three pathways, which are called the classical, lectin, and alternative pathways. All pathways of complement converge at the level of C3, and C3 fragments deposited on microorganisms serve as opsonins. Further activation of complement recruits complement components C5 through C9, which together form the membrane attack complex pore and can lyse gram-negative bacteria. Defects in alternative and terminal pathway complement proteins are associated with recurrent invasive meningococcal infections [3, 4]. Several cases of disseminated gonococcal infection (DGI) have also been reported in persons with congenital or acquired complement defects [5–9], which highlights a role for the bactericidal activity of complement in protection against gonococcal infections.

Myron Christodoulides (ed.), *Neisseria gonorrhoeae: Methods and Protocols*, Methods in Molecular Biology, vol. 1997, https://doi.org/10.1007/978-1-4939-9496-0_16,

1.2 Complement-Dependent Bactericidal Activity and Protection Against Neisseria Infections

Measurement of serum bactericidal activity provides an appraisal of the function of the humoral immune response to bacterial pathogens or vaccines and correlates well with protection from certain infections. Complement-dependent bactericidal activity is a well-established correlate of protection against invasive meningococcal disease [10]. Whether complement-dependent bactericidal activity also protects against gonorrhea remains to be determined. While it is reasonable to assume that antibodies that mediate complement-dependent bactericidal activity may be protective, antibodies that target and block important bacterial functions (independent of complement activation) may also serve as effective vaccine candidates.

1.3 Complement Resistance in *Neisseria gonorrhoeae*

An understanding of gonococcal complement resistance is important for choosing test strains and a source of complement for serum bactericidal assays. All strains of *N. gonorrhoeae* recovered directly from human urethral secretions, that is, without passage on routine culture media such as Thayer-Martin agar, and incubated with homologous normal human serum (NHS) are fully resistant to complement-dependent killing [11]. The ability of gonococci to resist killing by NHS is termed serum resistance. Some strains of *N. gonorrhoeae* remain serum-resistant even after passage onto culture media; this is termed stable serum resistance. On the other hand, strains that are serum-resistant ex vivo but become sensitive to killing by NHS after passage on routine culture media are termed "unstably" serum-resistant. Unstable serum resistance in vivo results from the addition of *N*-acetylneuraminic acid to lipooligosaccharide (LOS), also referred to as sialylation. Neu5Ac is a "sialic acid," a family of 9-carbon sugars (nonoses, or nonulosonates (NulOs) [12–14]. Routine gonococcal growth media such as Thayer-Martin media, does not contain 5′-cytidinemonophospho-*N*-acetylneuraminic acid (CMP-Neu5Ac), the sialic acid substrate that is added to gonococcal LOS by the enzyme, sialyltransferase (Lst); Lst is provided by the gonococcal organism itself.

Intact NHS, although it contains naturally occurring (endogenous) antibodies, may be used a source of complement provided the test strain(s) are serum-resistant—that is, stably serum-resistant strains, or gonococci are grown in media supplemented with CMP-Neu5Ac to induce complement-resistance. However, it may be desirable to use human complement that lacks naturally occurring anti-gonococcal IgG or IgM that may themselves activate complement and synergize with immune antibodies, otherwise being tested for bactericidal activity in the assay. In some instances, "blocking antibodies" (endogenous or elicited by immunization or infection) may antagonize bactericidal antibodies. Anti-reduction modifiable protein (Rmp) antibodies are an example of blocking antibodies against *N. gonorrhoeae* [15].

2 Materials

2.1 Complement

2.1.1 Normal Human Serum (NHS)

1. Needle for phlebotomy (19 or 21 gauge).
2. Container to clot blood and separate clot from plasma (e.g., BD Medical Vacutainer® Glass Blood Collection Tube) or 15 ml conical centrifuge tube with sterile glass beads.
3. Refrigerated centrifuge that can accommodate BD Vacutainers or 15 ml centrifuge tubes.
4. Dry-ice/100% ethanol bath.
5. Microcentrifuge tubes.
6. Sterile Pasteur pipette.
7. Glass storage vials.

2.1.2 NHS Complement, Depleted of IgG and IgM Antibodies (see **Note 1**)

1. Protein A/G Plus agarose, of binding capacity >50 mg human IgG/ml settled resin (*see* **Note 2**).
2. Anti-human IgM agarose (Sigma), binding capacity 2–3 mg human IgM/ml settled resin (*see* **Note 2**).
3. 5 ml centrifuge column (total column capacity 8 ml) for Protein A/G Plus agarose (*see* **Note 3**).
4. 10 ml centrifuge column (total column capacity 23 ml) for anti-human IgM agarose (*see* **Note 3**).
5. 15 and 50 ml conical polypropylene centrifuge tubes.
6. EMD Millipore™ Amicon™ Ultra-15 Centrifugal filter Unit (10,000 NMWL) or Thermo Scientific™ Slide-A-Lyzer™ 10 K MWCO Dialysis Cassette or dialysis unit (10 kDa molecular mass cutoff) suitable for dialysis of 5 ml sample.
7. 0.22 μm syringe filter.
8. Refrigerated centrifuge that can accommodate 15 and 50 ml centrifuge tubes.
9. 0.5 M ethylenediaminetetraacetic acid (EDTA, sterile, ice-cold) (*see* **Note 4**).
10. 5 M NaCl (sterile, ice-cold) (*see* **Note 4**).
11. Dulbecco's phosphate buffered saline (without Ca or Mg), pH 7.4 (PBS, sterile, ice-cold).
12. Reagents for hemolytic CH50 assay, or equivalent.

2.1.3 Adsorbing NHS Complement Against Test Strain

1. *Neisseria gonorrhoeae* strains/isolates stored frozen at −80 °C in trypticase soy broth (TSB) containing 20% (v/v) glycerol. To make TSB/20% (v/v) glycerol, add 2.4 g of TSB powder to 80 ml of double-distilled H_2O and 20 ml of glycerol, mixing thoroughly and autoclaving at 121 °C and 2.68 kg/cm^2 for 20 min. Skimmed milk powder (10% w/v) may also be used to maintain freezer stocks of *N. gonorrhoeae* (*see* **Note 5**).

2. 90 mm sterile plastic petri plates.
3. Chocolate agar plates may be purchased commercially. Alternatively, add 37 g of GC Agar Base to 500 ml of distilled H_2O in a 1 L bottle at room temperature and stir vigorously until thoroughly dissolved. In a separate bottle, add 11 g of dried bovine hemoglobin to 500 ml of distilled H_2O at room temperature and stir vigorously until thoroughly dissolved. Place a magnetic stir bar to the bottle containing hemoglobin and sterilize both solutions by autoclaving at 121 °C and 2.68 kg/cm^2 for 30 min. Cool the solutions to 56 °C in a water bath, and add 500 ml of GC Agar solution to the 500 ml of hemoglobin solution (1000 ml total). Add one tube of 10 ml of Isovitalex or Isovitalex equivalent to the 1000 ml mixture and stir for 5 min. Pour the mixture into sterile 90 mm plastic petri dishes (approximately 20 ml per plate) using sterile technique and allow the agar to set and cool prior to use. Store the plates at 4 °C until use.
4. IsoVitaleX may be purchased (BD BBL™ IsoVitaleX Enrichment). Alternatively, prepare an Isovitalex equivalent by first dissolving 1.1 g of L-cystine, 25.9 g of L-cysteine hydrochloride, 1.0 g of adenine, and 10.0 g of L-glutamine, 0.1 g of cocarboxylase, 0.25 g of diphosphopyridine nucleotide (oxidized), 0.03 g of guanine hydrochloride, 0.02 g of $FeNO_3 \cdot 9H_2O$, 0.003 g of thiamine hydrochloride, and 0.013 g of p-aminobenzoic acid into 200 ml of distilled H_2O. Add 6 N HCl dropwise until the solution is clear. Next, add 0.01 g of vitamin B12 to the solution. In a separate beaker, dissolve 100 g of dextrose in 300 ml of distilled H_2O. Mix the two solutions and add water to bring the volume to 1.0 L. Sterilize the Isovitalex equivalent through a 0.22 μm filter and store as aliquots of 10 ml at −20 °C until use.
5. Cotton or Dacron-tipped swabs.
6. Freshly prepared NHS.
7. Dulbecco's PBS (without Ca or Mg).
8. 25% (v/v) glutaraldehyde stock solution.
9. 1 mg/ml of L-lysine in PBS.
10. 15 ml conical centrifuge tubes.
11. 0.22 μm syringe filter.
12. Refrigerated centrifuge for 15 ml polypropylene centrifuge tubes.
13. Humidified incubator set at 37 °C with 5% (v/v) CO_2 or a candle jar to grow gonococci.
14. Reagents for hemolytic CH50 assay, or equivalent.

2.2 Serum Bactericidal Assay (SBA)

2.2.1 Growth of Bacteria in Liquid Media

1. *Neisseria gonorrhoeae* strains/isolates stored frozen at −80 °C in TSB/20% (v/v) glycerol.
2. Chocolate agar plates.
3. 10 μl sterile inoculation loops.
4. Solution A (Morse A): dissolve 1.5 g of proteose peptone No. 3, 0.4 g of potassium phosphate (dibasic, K_2HPO_4), 0.1 g of potassium phosphate (monobasic, KH_2PO_4), 0.5 g of sodium chloride (NaCl) and 0.1 g of soluble starch in 100 ml of deionized/distilled H_2O. Sterilize by autoclave at 121 °C and 2.68 kg/cm^2 for 20 min, cool to room temperature and store at 4 °C until use. Solution A has a shelf life of 2 weeks.
5. Solution B (Morse B): dissolve 0.042 g of sodium bicarbonate (Na_2CO_3) and 4.0 g of glucose in 90 ml of deionized/distilled H_2O. Sterilize by filtration through a 0.22 μm filter and store the solution at 4 °C until use.
6. IsoVitaleX (BD BBL™ IsoVitaleX Enrichment) or Isovitalex equivalent.
7. Gonococcal (GC) liquid growth media: mix 9 ml of Solution A with 1 ml of Solution B and 0.1 ml of IsoVitaleX (or equivalent).
8. 50 mg/ml of Cytidine-5′-monophospho-*N*-acetylneuraminic acid (CMP-Neu5Ac) stock solution in sterile PBS. Store in aliquots at −70 °C.
9. Rotary shaker.
10. Water bath.
11. Spectrophotometer.
12. Cotton or Dacron-tipped swabs.
13. 12 mm × 75 mm polystyrene tubes.
14. Humidified incubator set at 37 °C with 5% (v/v) CO_2.
15. Vortex mixer.

2.2.2 Growth of Bacteria Harvested from Chocolate Agar Plates

1. *Neisseria gonorrhoeae* strains/isolates stored frozen at −80 °C in TSB/20% (v/v) glycerol.
2. Chocolate agar plates.
3. Spectrophotometer.
4. Cotton or Dacron-tipped swabs.
5. 12 mm × 75 mm polystyrene tubes.
6. Humidified incubator set at 37 °C with 5% (v/v) CO_2.

2.2.3 SBA Procedure

1. Complement source.
2. Immune serum (*see* **Note 6**).

3. Hanks Balanced Salt Solution (HBSS) containing 0.15 mM $CaCl_2$ and 1 mM $MgCl_2$ ($HBSS^{++}$).
4. Chocolate agar plates.
5. Water bath at 37 °C with a shaking platform for incubating bactericidal reaction tubes.
6. Humidified incubator set at 37 °C with 5% (v/v) CO_2 or a candle jar to grow gonococci.
7. 12 mm × 75 mm polystyrene tubes.
8. Vortex mixer.
9. Cotton or Dacron-tipped swabs.
10. Spreaders (either plastic or glass) to evenly spread out bacteria.
11. Culture plate rotator.
12. Spectrophotometer.

3 Methods

3.1 Complement

3.1.1 Normal Human Serum (NHS)

1. Draw 12 ml of blood into BD Medical Vacutainer® Glass Blood Collection Tubes or into a plastic 15 ml conical polypropylene centrifuge tube containing sterile glass beads (*see* **Note 7**).
2. Allow blood to clot at room temperature (~22 °C) for 20–30 min (no more than 60 min; *see* **Note 8**).
3. Once the clot has fully retracted, centrifuge at 1000–3000 × *g* at 22 °C for 10 min.
4. Carefully remove the serum from above the clot with a pipette and transfer to a sterile 15 ml conical tube on ice.
5. Remove the cellular fragments contained within this serum with a second high speed centrifugation step (10,000–20,000 × *g* for 2–5 min at 4 °C) either in microcentrifuge tubes or conventional centrifuge tubes.
6. Remove the supernatants with a sterile Pasteur pipette and chill on ice for IgG and IgM depletion or prepare aliquots preferably into glass vials, which have high thermal conductance and then flash freeze in a dry-ice/100% ethanol bath and store in single-use aliquots at −80 °C (*see* **Note 9**).

3.1.2 NHS Complement, Depleted of IgG and IgM Antibodies

Unless otherwise specified, all buffers should be chilled to 4 °C and all steps carried out at 4 °C to minimize complement activation.

1. Equilibrate the Protein A/G Plus agarose and anti-human IgM agarose in PBS containing 0.1 mM EDTA (prepared from the stock solution of 500 mM EDTA). Place the Protein A/G Plus agarose in a 5 ml centrifuge column (total column capacity 8 ml) and the anti-human IgM agarose in a 10 ml centrifuge column (total column capacity 23 ml) (*see* **Notes 2** and **3**).

2. Remove the top screw cap and bottom plug from the column containing Protein A/G Plus, place the 5 ml column in a 15 ml conical centrifuge tube, and spin at 500–1000 × *g* for 1 min to remove excess buffer. Replace the bottom plug.
3. Similarly, place the 10 ml column containing the anti-human IgM agarose in a 50 ml conical centrifuge tube and remove excess buffer by centrifugation at 1000 × *g* for 1 min.
4. Put 200 μl of NHS aside in a sterile tube on ice for a hemolytic assay.
5. Add NaCl (from the sterile 5 M stock solution) to a final concentration of 1 M and EDTA (from a sterile stock solution of 0.5 M) to a final concentration of 10 mM to freshly prepared NHS (*see* **Note 10**).
6. Add 5 ml of NHS containing 1 M NaCl and 10 mM EDTA to the Protein A/G Plus agarose in the 5 ml column, and suspend the agarose in NHS by very gentle pipetting and/or tapping the column (*see* **Note 11**).
7. Once the agarose is suspended in NHS, allow the resin to settle, remove the bottom plug and place in a 15 ml conical tube on ice and allow NHS to drain by gravity.
8. When NHS stops draining, remove the remaining NHS by centrifugation at 500–1000 × *g* for 1 min.
9. Transfer the IgG-depleted serum to the 10 ml column containing anti-human IgM agarose.
10. Gently suspend the anti-human IgM agarose in NHS as described in **step 5** above (*see* **Note 11**).
11. When the agarose has settled, remove the bottom plug and allow NHS to drain by gravity into a 50 ml conical centrifuge tube placed on ice.
12. Remove residual NHS from the column by centrifugation at 500–1000 × *g* for 1 min.
13. Regenerate and store Protein A/G and anti-human IgM gels according to the manufacturer's instructions (*see* **Note 12**).
14. Remove EDTA and excess NaCl in the IgG/IgM depleted serum (human complement) either by spin-concentration dialysis or dialysis against ice-cold PBS. Use an EMD Millipore™ Amicon™ Ultra-15 Centrifugal filter Unit (10,000 NMWL) or equivalent to dialyze by spin concentration. Add 10 ml of cold PBS to 5 ml of complement in the filtration unit. Centrifuge at 4000 × *g* if a swinging bucket rotor is used (or at 5000 × *g* with the membrane panel facing up if a fixed angle rotor is used) for 20–30 min at 4 °C. The volume at the end of the spin should be 4–5 ml. Add 10 ml of PBS and repeat the centrifugation step. Repeat the process twice, that is, to a total

of four centrifugation steps. Alternatively, dialyze the complement against ice-cold PBS (500–1000 ml, two exchanges for 2 h each) using a Thermo Scientific™ Slide-A-Lyzer™ 10 K MWCO Dialysis Cassette (or another dialysis method with similar characteristics).

15. Filter the IgG and IgM depleted serum using a 0.22 μm syringe filter and store serum, preferably in glass vials, at −70 °C in single-use aliquots.
16. Determine the hemolytic activity of complement relative to the hemolytic activity of NHS (set aside in **step 4**) by using a CH50 assay that employs antibody-sensitized sheep erythrocytes, such as from Complement Technologies, Inc. or Diamedix 24EZ Complement Cells. A detailed description of the CH50 assay can be obtained from the company's website, (https://www.complementtech.com/product-description/Cells/EA.htm). Alternatively, ELISA assays such as the Classical Complement Pathway CH50 Eq EIA kit (The Binding Site, Inc.) may be used to estimate total complement activity of IgG and IgM depleted human serum. The hemolytic activity of complement should be at least 50% that of the original (unmanipulated) NHS. The complement source contains ~0.1 mM EDTA. Therefore, $CaCl_2$ and $MgCl_2$, each at a final concentration of 1 mM must be added to complement prepared by this method to restore hemolytic activity for functional assays (*see* **Note 13**).

3.1.3 Adsorbing NHS Complement Against Test Strain

This method removes antibodies directed against a specific test strain. Serum should be absorbed against each individual strain, or a mixture of strains if more than one strain is to be tested.

1. Prepare NHS as described in Subheading 3.1.1.
2. Inoculate *Neisseria gonorrhoeae* test strain from a frozen culture stock in TSB/20% (v/v) glycerol (stored at −70 °C) onto a chocolate agar plate using a 10 μl inoculating loop, spread the inoculum evenly over the plate using a sterile cotton or Dacron swab and incubate overnight in an humidified incubator at 37 °C with 5% (v/v) CO_2 or in a candle jar.
3. Transfer the test bacterial strain, after overnight growth on the heavily streaked culture plate, into a tube containing 1.0 ml of Dulbecco's PBS.
4. Wash the bacteria three times in 1.0 ml of 0.1 M PBS, with centrifugation at 5000 × *g* for 5 min between each wash.
5. Incubate the bacteria in 1.0 ml of 0.25% (v/v) final concentration of glutaraldehyde in PBS, for 30 min at 25 °C.
6. Repeat **step 4**.
7. Incubate the bacteria with 1.0 ml of 1 mg/ml of L-lysine in PBS for 20 min at 25 °C to quench residual glutaraldehyde.

8. Repeat **step 4**.
9. Incubate the bacteria in 5 ml of serum in an ice bath for 1 h, with frequent stirring of the suspension. We recommend about 1–2 ml of packed organisms to absorb the antibodies present in 5 ml of serum.
10. Centrifuge the suspension at 5000 × *g* at 4 °C for 10 min to remove the bacteria.
11. Sterilize the supernatant through a 0.22 μm filter and determine the level of complement activity after absorption by routine CH50 assays as described in **step 16** of Subheading 3.1.2 and it should be at least 50% that of nonmanipulated NHS.
12. Store the serum in aliquots at −70 °C until use as the complement source for a bactericidal assay involving the homologous bacterial strain against which it had been absorbed.

3.2 Serum Bactericidal Assay

3.2.1 Growth of Bacteria in Liquid Media

This method more reliably yields organisms in the mid-log phase of growth. It also is the preferred method to more uniformly and reproducibly sialylate gonococci.

1. Inoculate the test gonococcal strain from frozen culture stock (−70 °C) onto a chocolate agar plate as described in **step 2** of Subheading 3.1.3. Grow overnight in an humidified incubator at 37 °C with 5% (v/v) CO_2 or in a candle jar.
2. After overnight growth, inoculate a sterile swab full of bacteria from the plate into 1 ml of Morse (Solution A) in a 12 mm × 75 mm sterile polystyrene tube. Vortex to evenly suspend the bacteria.
3. Inoculate the bacterial suspension, a few drops at a time, into a sterile tube containing GC liquid growth medium until an Optical Density (OD) at $\lambda_{600\ nm}$ of ~0.1 is reached. If desired, add CMP-Neu5Ac to the medium to a final concentration of 2 μg/ml (*see* **Note 14**).
4. Grow the culture at 37 °C with vigorous shaking (200–250 rpm), to a mid-log phase concentration of approximately 10^8 colony-forming units (CFU)/ml (OD $\lambda_{600\ nm} = 0.2$).
5. Make four, tenfold dilutions and another twofold dilution of the liquid culture with Solution A to obtain an inoculum with a concentration of approximately 5 × 10^4 CFU/ml. Incubate in a 37 °C water bath until use (maximum of 10 min) (*see* **Note 15**).

3.2.2 Growth of Bacteria Harvested from Chocolate Agar Plates

1. Inoculate the test strain from frozen culture stock (−70 °C) onto a chocolate agar plate as described in **step 2** of Subheading 3.1.3 and incubate overnight in an humidified incubator at 37 °C with 5% (v/v) CO_2.

2. After overnight growth, inoculate a sterile swab full of bacteria from the plate onto a fresh chocolate agar plate that has been prewarmed to 37 °C. If needed, sialylation of gonococcal LOS (*see* **Note 14**) can be achieved on bacteria grown on chocolate agar plates as follows: prior to inoculating bacteria, flood the plate with 500 μl of CMP-Neu5Ac (25 μg/ml in sterile H_2O). Spread the CMP-Neu5Ac solution evenly over the plate using a sterile plastic or glass spreader and allow the CMP-Neu5Ac to absorb into the agar in a laminar flow hood (keep the plate horizontal and the lid off). This process usually takes 15–30 min.
3. Grow the bacteria in a humidified incubator at 37 °C with 5% (v/v) CO_2 for 4–5 h.
4. Harvest the bacteria from the plate using a sterile cotton- or Dacron-tipped swab and suspend in Solution A to an OD $\lambda_{600\ nm}$ of 0.2.
5. Dilute the bacteria as described in **step 5** of Subheading 3.2.1.

3.2.3 Assay Procedure

1. Add Hank's balanced salt solution (HBSS) containing 0.15 mM $CaCl_2$ and 1 mM $MgCl_2$ ($HBSS^{++}$) to a sterile 12 mm × 75 mm sterile polystyrene capped tube. Adjust the amount such that the final volume of the bactericidal reaction mixture is 150 μl (*see* **Note 16**).
2. Add 25 μl of bacterial suspension prepared in **step 5** of Subheading 3.2.1 or **step 5** of Subheading 3.2.2 (*see* **Note 15**).
3. Add test (immune) serum that has been heat-inactivated by heating at 56 °C for 30 min in a water bath to inactivate complement (*see* **Note 6**). The maximum amount is usually 50 μl, that is, a final serum dilution of 1/3. Adjust the volume to obtain dose-responses for killing.
4. Thaw the NHS/complement quickly at 37 °C and placed on ice and use within 60 min.
5. Add NHS or human complement (IgG/IgM depleted human serum, or NHS absorbed against the test strain). The final concentration of human complement used will depend on its hemolytic activity relative to NHS. In general, fully active NHS is used at a final concentration of 10–20% (v/v) in the bactericidal reaction mixture. Adjust the volume of complement accordingly to simulate the activity provided by 10–20% (v/v) NHS. Add the complement source last to the mixture.
6. Include the following control reactions:
   (a) An organism growth control, that is, bacteria in $HBSS^{++}$.
   (b) Controls for both active and heat-inactivated (56 °C for 30 min) human complement (no killing expected with active or inactivated complement sources).

(c) A killing control, that is, an antibody source with known bactericidal activity against the test strain.

7. Vortex to suspend the bacteria thoroughly. Inoculate duplicate chocolate agar plates with 25 μl of the reaction mixture immediately (within 15 s) after complement addition to serve as baseline (0-min incubation) growth control. Place the reaction tubes in a 37 °C water bath with a shaking platform (100 rpm).
8. Spread the 0-min incubation 25 μl reaction mixture inocula onto chocolate agar plates thoroughly with a sterile glass rod (heated then cooled) or individual sterile plastic spreader, while rotating the inoculated plates on a culture plate rotator. Avoiding spreading out to the edges of the plate where accumulation of colonies is hard to count.
9. Incubate the reaction mixture for 30 min at 37 °C (30 min incubation point) and then sequentially inoculate duplicate chocolate agar plates with 25 μl of the reaction mixture (repeating **step 8**). Maintain a constant 30 min interval of incubation of each of the tubes. The maximal number of tubes per assay should be about 30, allowing for a minute per tube at the outset to process.
10. Incubate the plates overnight in a humidified incubator at 37 °C with 5% (v/v) CO_2.
11. Count viable colonies on all plates after incubation.
12. Bactericidal activity is expressed as the percentage of bacteria surviving in the 30 min reaction mixture divided by the percentage surviving in the time zero control. The following criteria for killing should be met to validate results obtained in an individual experiment:
    (a) Less than 50% survival in the 30 min sample compared with survival in the time zero sample (a killing end point between 50% and 90% may be selected).
    (b) Percentage survival in the test reaction mixture divided by percentage survival in the active complement control of less than 50%.
    (c) 90% or greater survival at 30 min in the active and heat-inactivated complement control mixtures.

    These criteria ensure that agglutination is not a significant factor in the reduction of colony counts in the 30 min samples. Agglutination of organisms (e.g., with piliated gonococci) may yield erratic assay results. Duplicate plating of 25 μl samples should not result in colony counts that vary by more than 10–15%.

4 Notes

1. NHS Complement, Depleted of IgG and IgM antibodies is now commercially available from Pel-Freez Biologicals (http://www.pelfreez-bio.com/products/complement-and-ancillary-products/depleted-human-complement/). The product, called IgG/IgM Depleted Human Complement, has been derived either from NHS from a single donor or from pooled NHS (NHS combined from several donors).
2. Other sources of protein G agarose may be used; the volume of gel used should be adjusted such that the binding capacity of the gel exceeds the estimated amount of IgG in NHS by about 25–50%. For depleting IgG and IgM from 5 ml of NHS, we recommend 2 ml of Pierce Protein A/G Plus Agarose and 5 ml of anti-human IgM agarose. To ensure maximal depletion, be certain that the IgG and IgM binding capacities of the columns exceed the expected amount of IgG and IgM in NHS, to be absorbed, by at least 25%. On average, 5 ml of NHS contains about 50–75 mg of IgG and about 5–7 mg of IgM. Thus, the IgG and IgM binding capacities of the resin used should be at least 100 and 10 mg, respectively.
3. The volumes of the centrifuge columns should be sufficient to accommodate the settled gel plus serum. The columns should fit into 15 or 50 ml centrifuge tubes and should withstand at least 1000 × *g*. We recommend Pierce Centrifuge Columns, but other columns that handle equivalent volumes can be substituted.
4. Stock solutions of EDTA (500 mM) and NaCl (5 M) should be sterilized either by autoclaving or by passage through a 0.22 μm filter.
5. Skimmed milk (10% w/v) is prepared by adding 10 g of dry skimmed milk powder to 100 ml of ddH_2O, mixing thoroughly and autoclaving at 121 °C and 2.68 kg/cm^2 for 20 min.
6. Immune serum refers to sera that may be obtained following immunization of rodents such as mice or rabbits, nonhuman primates, or humans. Immune serum should be heat-inactivated (56 °C for 30 min in a water bath) to destroy endogenous complement activity.
7. For functional studies involving human complement, it is preferable to prepare serum separated from clotted whole blood rather than serum prepared by clotting plasma after removal of the cellular components of blood. Clotting and clot retraction should take place at room temperature (around 22 °C) in a container that allows blood to "wet" the surface (for example, glass surfaces). If plastic tubes are used, they should contain glass beads to provide a wettable surface.

8. Blood should not be chilled before coagulation and before clot retraction has occurred. Serum should not be separated from the clot until full clot retraction has occurred. For blood volumes up to 50 ml, this usually takes about 20 min and no more than 1 h. Cell lysis and fibrinolysis could compromise complement activity [16–18].
9. Storage of NHS or human complement at −20 °C is not suitable; they must be stored at −70 °C. In general, aliquots of NHS or complement should not be thawed more than once.
10. NaCl prevents loss of C1q that occurs when serum is incubated with Protein G or anti-human IgM sepharose; EDTA minimizes complement activation during serum manipulation.
11. Avoid excessive and/or vigorous pipetting when suspending agarose in serum.
12. In general, Protein A/G and anti-human IgM agarose can be used five times without significant loss in binding capacity.
13. The classical and lectin pathways of complement require Ca^{2+}, while the alternative pathway requires Mg^{2+} for activity.
14. The addition of CMP-Neu5Ac to growth media renders gonococci that are otherwise sensitive to killing by normal human serum (NHS; contains "natural" IgG and IgM), serum-resistant.
15. Spectrophotometers differ and the described dilution series are only guidelines and will have to be modified by each laboratory to ensure about 100–300 CFUs per plate at 0 min. Avoid prolonged incubation in buffer as viability of bacteria may decrease over time.
16. In some instances, the amount of immune (e.g., vaccine) sera available may be limited. In this instance, the final reaction volume may be reduced to 75 μl. The volumes of reagents should be reduced in a corresponding fashion and plates should be inoculated with 12.5 μl of bacterial suspension (instead of 25 μl). The number of bacteria used should be adjusted such that plating 12.5 μl at t_0 yields about 200 colonies. The bactericidal procedure described here may be adapted to microtiter wells as has been done for *N. meningitidis* [19]. Several other serum bactericidal assays methods have been described previously [20], where the pros and cons of each method have been discussed. The method described here uses relatively small amounts of immune serum and complement, and is most suitable for evaluating vaccine sera.

References

1. Ricklin D, Hajishengallis G, Yang K et al (2010) Complement: a key system for immune surveillance and homeostasis. Nat Immunol 11 (9):785–797
2. Walport MJ (2001) Complement. First of two parts. N Engl J Med 344(14):1058–1066
3. Figueroa JE, Densen P (1991) Infectious diseases associated with complement deficiencies. Clin Microbiol Rev 4(3):359–395
4. Ram S, Lewis LA, Rice PA (2010) Infections of people with complement deficiencies and patients who have undergone splenectomy. Clin Microbiol Rev 23(4):740–780
5. Keiser HD (1997) Recurrent disseminated gonococcal infection in a patient with hypocomplementemia and membranoproliferative glomerulonephritis. J Clin Rheumatol 3 (5):286–289
6. McWhinney PH, Langhorne P, Love WC et al (1991) Disseminated gonococcal infection associated with deficiency of the second component of complement. Postgrad Med J 67 (785):297–298
7. Snyderman R, Durack DT, McCarty GA et al (1979) Deficiency of the fifth component of complement in human subjects. Clinical, genetic and immunologic studies in a large kindred. Am J Med 67(4):638–645
8. Hublikar S, Maher WE, Bazan JA (2014) Disseminated gonococcal infection and eculizumab—a "high risk" connection? Sex Transm Dis 41(12):747–748
9. Gleesing J, Chiwane S, Rongkavilit C (2012) Gonococcal septic shock associated with eculizumab treatment. Pediatr Infect Dis J 31 (5):543
10. Goldschneider I, Gotschlich EC, Artenstein MS (1969) Human immunity to the meningococcus. I. The role of humoral antibodies. J Exp Med 129(6):1307–1326
11. Ward ME, Watt PJ, Glynn AA (1970) Gonococci in urethral exudates possess a virulence factor lost on subculture. Nature 227 (256):382–384
12. Nairn CA, Cole JA, Patel PV et al (1988) Cytidine 5′-monophospho-N-acetylneuraminic acid or a related compound is the low Mr factor from human red blood cells which induces gonococcal resistance to killing by human serum. J Gen Microbiol 134 (Pt 12):3295–3306
13. Parsons NJ, Ashton PR, Constantinidou C et al (1993) Identification by mass spectrometry of CMP-NANA in diffusible material released from high M(r) blood cell fractions that confers serum resistance on gonococci. Microb Pathog 14(4):329–335
14. Smith H, Parsons NJ, Cole JA (1995) Sialylation of neisserial lipopolysaccharide: a major influence on pathogenicity. Microb Pathog 19 (6):365–377
15. Joiner KA, Scales R, Warren KA et al (1985) Mechanism of action of blocking immunoglobulin G for *Neisseria gonorrhoeae*. J Clin Invest 76(5):1765–1772
16. Amara U, Rittirsch D, Flierl M et al (2008) Interaction between the coagulation and complement system. Adv Exp Med Biol 632:71–79
17. Foley JH, Peterson EA, Lei V et al (2015) Interplay between fibrinolysis and complement: plasmin cleavage of iC3b modulates immune responses. J Thromb Haemost 13 (4):610–618
18. Seya T, Nagasawa S, Matsukura M et al (1985) Generation of C3d,g and C3d by urokinase-treated plasma in association with fibrinolysis. Complement 2(2–3):165–174
19. Maslanka SE, Gheesling LL, Libutti DE et al (1997) Standardization and a multilaboratory comparison of *Neisseria meningitidis* serogroup A and C serum bactericidal assays. The Multilaboratory Study Group. Clin Diagn Lab Immunol 4(2):156–167
20. McQuillen DP, Gulati S, Rice PA (1994) Complement-mediated bacterial killing assays. Methods Enzymol 236:137–147

Chapter 17

Basic Methods for Examining *Neisseria gonorrhoeae* Interactions with Host Cells In Vitro

Hannia Liliana Almonacid-Mendoza and Myron Christodoulides

Abstract

The obligate human pathogen *Neisseria gonorrhoeae* colonizes primarily the mucosal columnar epithelium of the male urethra and the female endocervix. In addition, gonococci can infect the anorectal, pharyngeal, and gingival mucosae and epithelial cells of the conjunctiva. More rarely, the organism can disseminate through the bloodstream, which can involve interactions with other host cell types, including blood vessel endothelial cells and innate immune cells such as dendritic cells, macrophages, and neutrophils. "Disseminated gonococcal infection" is a serious condition with various manifestations resulting from the seeding of organs and tissues with the pathogen. The host response to gonococcal infection is inflammatory. Knowledge of the biology of gonococcal interactions has been served well through the use of a wide variety of ex vivo models using host tissues and eukaryotic cell monocultures. These models have helped identify bacterial surface adhesins and invasins and the corresponding cell surface receptors that play roles in gonococcal pathogenesis. Furthermore, they have been useful for understanding virulence mechanisms as well as innate and adaptive immune responses. In this chapter, readers are provided with protocols for examining the basic interactions between gonococci and a representative human cell line.

Key words Cell culture, Association, Invasion, Cell death, Microscopy, Cytokine production

1 Introduction

Members of the genus *Neisseria* colonize exposed mucosal epithelial surfaces of mammalian species, and the ability to disseminate from sites of colonization also provides the opportunity for bacterial interactions with a wide variety of host cell types and organ systems. *Neisseria gonorrhoeae* (the gonococcus) is the causative agent of gonorrhea, and depending on the anatomical site of exposure, clinical infection can be urogenital, anorectal or pharyngeal. The mucosal columnar epithelium of the urethra and the endocervix are the primary site of male and female infection, respectively. The anorectal mucosa is also lined with simple columnar epithelium. The mucosa of the oropharynx and pharynx proper are lined by largely nonkeratinizing stratified squamous epithelium and the

Myron Christodoulides (ed.), *Neisseria gonorrhoeae: Methods and Protocols*, Methods in Molecular Biology, vol. 1997,
https://doi.org/10.1007/978-1-4939-9496-0_17, © Springer Science+Business Media, LLC, part of Springer Nature 2019

nasopharynx is mainly lined by ciliated columnar epithelium. Gonococci can also interact with gingival mucosa and nonkeratinized epithelial cells of the conjunctiva.

The host response to mucosal infection is inflammatory. In men, gonococcal urogenital infection is characterized by dysuria and the production of a urethral discharge of inflammatory exudate, mainly neutrophils. Inflammatory complications can include acute epididymitis, penile lymphangitis, periurethral abscess, acute prostatitis, and seminal vesiculitis. In women, the syndrome of pelvic inflammatory disease encompasses conditions such as cervicitis, endometritis or Fallopian tube salpingitis [1]. Local inflammatory responses to gonococcal infection can include pharyngitis, conjunctivitis in adults and infants (*ophthalmia neonatorum*), gingivitis, intraoral and cutaneous abscess formation. Although an uncommon occurrence, gonococci can disseminate through the blood stream to cause "disseminated gonococcal infection" (DGI), with arthritis-dermatitis syndrome the most common presentation. Characteristic dermatitis may present as papules and pustules, often with a hemorrhagic component. Occasionally, endocarditis, meningitis, osteomyelitis, septic shock, Acute Respiratory Distress Syndrome and acute perihepatitis (Fitz-Hugh–Curtis syndrome) are complications of DGI. The ability to disseminate suggests that gonococci interact with many other host cells, including innate immune cells such as dendritic cells, macrophages and neutrophils, and blood vessel endothelial cells.

There is a substantial primary article and review literature on the biology of gonococcal interactions with a wide variety of ex vivo host cells and tissues [2–19]. These studies are important for identifying bacterial surface adhesins and invasins and the corresponding cell surface receptors that play roles in gonococcal pathogenesis. They are also useful for understanding virulence mechanisms as well as innate and adaptive immune responses. The criticism of monocultures is that they do not reflect the multicellular nature of pathogen interactions, the contribution of the microbiome and the plethora of in situ mechanisms involved in pathogen detection and host defense. Nevertheless, they have been critical to our understanding of how the gonococcus colonizes exposed mucosal epithelia and interacts with other cellular barriers and cells of the innate immune system.

This chapter provides the reader with protocols for examining the basic interactions between gonococci and a representative human epithelial cell line. The protocols describe how to measure gonococcal association and internalization and bacteria-induced cell death as the basis for more detailed investigations, using advanced microscopy techniques and the variety of "-omics" platforms that are increasingly available.

2 Materials

2.1 Cell Culture

1. Human Chang Conjunctival Epithelial cells, purchased from the European Type Culture Collection or equivalent (*see* **Note 1**).
2. Phosphate buffered saline (PBS), pH 7.4 (*see* **Note 2**).
3. Dulbecco's Modified Eagle's Medium (DMEM) with Ultra-Glutamine 1 and 4.5 g/L of glucose.
4. Decomplemented fetal calf serum (dFCS): heat FCS in a water bath at 56 °C for 1 h. Usually, we heat-inactivate 500 mL of FCS and store in sterile aliquots of 20 mL at −20 °C.
5. 5000 U penicillin and 5000 U streptomycin mixture. Add 1 mL of the antibiotic mixture per 200 mL of medium.
6. Growth Medium: DMEM containing 10% (v/v) dFCS and antibiotics.
7. Infection Medium: DMEM containing 1% (v/v) dFCS and no antibiotics (*see* **Note 3**).
8. 25 and 75 cm^2 cell culture flasks.
9. Sterile 24- and 96-well cell culture plates.
10. Sterile 10 μL inoculating loops.
11. Trypsin–ethylenediaminetetraacetic acid (EDTA) solution: Hank's balanced salt solution (HBSS) containing 0.5 g/L of trypsin and 0.2 g/L of EDTA.
12. Cryopreservation Medium: DMEM containing 10% (v/v) dFCS and 10% (v/v) dimethyl sulfoxide (DMSO).
13. Sterile cryopreservation vials.
14. Sterile plastic Universal bottles and Bijou bottles.
15. Class II Biosafety cabinet.
16. Humidified incubator set to 37 °C and 5% (v/v) CO_2.
17. Benchtop centrifuge.

2.2 Gonococcal Inocula

1. GC medium agar for growth of *N. gonorrhoeae*: dissolve the following chemicals in 500 mL of distilled H_2O—5 g of Bacto-agar (no. 1, Oxoid), 0.5 g of KH_2PO_4, 2.62 g of $K_2HPO_4 \cdot 3H_2O$, 5 g of proteose peptone, 2.5 g of NaCl, 0.5 g of starch.
2. Supplement A: dissolve the following in 800 mL of distilled H_2O—100 mg of cocarboxylase, 10 mg of cyanocobalamin, 20 mg of ferric nitrate, 100 mg of glucose, 10 g of L-glutamine, 13 mg of para-amino-benzoic acid, 3 mg of thiamine hydrochloride and 250 mg of β-nicotinamide adenine dinucleotide.

Sterilize the solution through a 0.2 μM filter and store in aliquots at −20 °C (*see* **Note 4**).

3. Supplement B: dissolve the following in 100 mL of boiling 0.1 M hydrochloric acid—1 g of adenine, 30 mg of guanidine hydrochloride, 320 mg of hypoxanthine and 800 mg of uracil. Allow the solution to cool to room temperature and then add 26 g of cysteine and add distilled H_2O to make a final volume of 200 mL. Sterilize the solution through a 0.2 μM filter and store in aliquots at −20 °C (*see* **Note 4**).
4. GC agar plates: sterilize the GC medium agar by autoclaving for 15 min at 120 °C with 2.68 kg/cm^2 pressure. Allow sterile medium to cool to ~50 °C, and under aseptic conditions, add 4 mL of Supplement A and 1 mL of Supplement B per 500 mL and pour into sterile petri dishes and allow setting at room temperature.
5. GC broth: GC medium with Supplements A and B, but omit the agar and starch (*see* **Note 5**).
6. *N. gonorrhoeae* strains stored at −80 °C or under liquid nitrogen in sterile 10% (w/v) proteose peptone broth containing 10% (v/v) glycerol.
7. Growth Medium.
8. Infection Medium.
9. Sterile 24- and/or 96-well cell culture plates containing cell monolayers.
10. Sterile plastic petri dishes (20 mL volume, 90 mm dish).
11. Sterile 96-well microtiter plates.
12. Sterile plastic Pasteur pipettes (fine-tip).
13. Trypsin–ethylenediaminetetraacetic acid (EDTA) solution: Hank's balanced salt solution (HBSS) containing 0.5 g/L of trypsin and 0.2 g/L of EDTA.
14. DMEM base medium.
15. PBS, pH 7.4.
16. Dulbecco's B salts: 0.5 mM $MgCl_2$ and 0.9 mM $CaCl_2$ prepared in sterile distilled H_2O.
17. PBSB: sterile PBS, pH 7.4 containing Dulbecco's B salts. Add 0.5 mL of sterile B salts to 100 mL of PBS.
18. Lysis buffer: 0.1% (w/v) sodium hydroxide and 1.0% (w/v) sodium dodecyl sulfate (SDS) in distilled H_2O.
19. 0.4% (v/v) trypan blue solution.
20. Glass hemocytometer.
21. Light microscope.
22. UV Spectrophotometer and UV cuvettes.

23. Humidified incubator set to 37 °C with 5% (v/v) CO_2.
24. Class II Biosafety cabinet.

2.3 Gonococcal Association

1. PBS, pH 7.4.
2. Infection Medium.
3. Cell lysis solution: 1% (w/v) saponin from *Quillaja* Bark prepared in 10 mL of sterile PBS containing 1% (v/v) dFCS. Sterilise by filtration using a polysulfone membrane filter (0.2 μM pore size) and syringe.
4. Sterile 24- and 96-well cell culture plates.
5. Sterile plastic Pasteur pipettes (fine tip).
6. Sterile mini-Eppendorf tubes.
7. GC agar plates.
8. Light microscope.
9. Humidified incubator set to 37 °C and 5% (v/v) CO_2.
10. Class II Biosafety cabinet.

2.4 Gonococcal Internalization

1. Cytochalasin D (CD) solution: Infection Medium containing 2 μg/mL of CD.
2. Infection Medium.
3. Infection Medium containing 200 μg/mL of gentamicin.
4. Sterile 24- and 96-well cell culture plates.
5. GC agar plates.
6. Humidified incubator set to 37 °C and 5% (v/v) CO_2.
7. Class II Biosafety cabinet.

2.5 Host Cell Viability

2.5.1 Lactate Dehydrogenase (LDH) Release

1. Sterile 96-well cell culture plates.
2. Infection Medium.
3. PBS, pH 7.4.
4. CytoTox 96® Non-radioactive Cytotoxicity Assay kit (Promega, or equivalent).
5. Nonsterile ELISA plates.
6. Eppendorf tubes.
7. Humidified incubator set to 37 °C and 5% (v/v) CO_2.
8. ELISA spectrophotometer.
9. Class II biosafety cabinet.

2.5.2 LIVE/DEAD® Assays

1. Transparent Transwell ThinCerts™, 0.4 μM pore size with appropriate cell culture plate (12, 24, or 48 well) for holding inserts.
2. Infection Medium.

3. PBS, pH 7.4.
4. LIVE/DEAD® Viability/Cytotoxicity Assay (Molecular Probes, or equivalent).
5. Scalpel blade, fine-tipped forceps, glass microscope slides and coverslips, finger nail polish.
6. Humidified incubator set to 37 °C and 5% (v/v) CO_2.
7. Class II biosafety cabinet.
8. Laser-scanning confocal microscope (LSCM).

2.5.3 Cell Proliferation Assay

1. Sterile 96-well cell culture plates.
2. Infection Medium.
3. PBS, pH 7.4.
4. Colorimetric Cell Proliferation Assay (e.g., WST-1).
5. Humidified incubator set to 37 °C and 5% (v/v) CO_2.
6. Class II biosafety cabinet.
7. ELISA spectrophotometer.

2.5.4 Annexin V Assay

1. Sterile 8-chamber tissue culture-treated slides.
2. Infection Medium.
3. Apoptosis Detection Kit containing Annexin V–Fluorescein isothiocyanate (FITC) conjugate and propidium iodide (PI).
4. 1 μM Staurosporine.
5. Scalpel blade, fine-tipped forceps, glass microscope slides and coverslips, finger nail polish.
6. Humidified incubator set to 37 °C and 5% (v/v) CO_2.
7. Class II biosafety cabinet.
8. Laser-scanning confocal microscope (LSCM).

2.5.5 DNA Fragmentation Assay

1. Sterile 96-well cell culture plates.
2. Infection Medium.
3. Trypsin–ethylenediaminetetraacetic acid (EDTA) solution: Hank's balanced salt solution (HBSS) containing 0.5 g/L trypsin and 0.2 g/L EDTA.
4. Wizard Genomic DNA Purification Kit (Promega, or equivalent).
5. Source of UV radiation (e.g., UV cross-linker).
6. 1.5% (w/v) agarose gel electrophoresis system with 1× Tris–acetate–EDTA (TAE) buffer and 0.5 μg/mL ethidium bromide (EtBr).

3 Methods

3.1 Cell Culture

1. Remove the cryovial of human cells from storage under liquid nitrogen, thaw quickly in a water bath at 37 °C and transfer the contents to a sterile Universal bottle. Add 8 mL of Growth Medium and centrifuge at 1000 × *g* for 3 min to pellet the cells. Remove the supernatant containing DMSO and suspend the pellet in either 7 or 14 mL of fresh Growth Medium for a T25 and T75 cm^2 cell culture flask, respectively. Incubate at 37 °C with 5% (v/v) CO_2 in an humidified atmosphere and monitor until >90% confluence is reached.
2. Remove the Growth Medium and wash the monolayers twice with prewarmed PBS. To detach cell monolayers, add 1 mL of trypsin–EDTA solution to a T25 cm^2 flask (2 mL to a T75 cm^2 flask) and incubate at 37 °C with 5% (v/v) CO_2 until the monolayer detaches visibly (usually within 5–10 min). Add fresh Growth Medium to the trypsin–EDTA treated cell monolayer, which inactivates the enzyme, and centrifuge at 1000 × *g* for 3 min to pellet the cells. Remove the supernatant and gently tap to dislodge the cell pellet, and then add fresh Growth Medium to suspend. The suspended cells can be further diluted and volumes added to 24-well and 96-well cell culture plates as desired (*see* **Note 6**). Incubate the cell culture plates until monolayer confluence is reached.
3. Expand the cell cultures or prepare cells for long-term storage in liquid nitrogen for future use. For long-term storage, grow cell cultures in T75 cm^2 flasks, detach the monolayers and suspend the pellet(s) in 1–2 mL of Cryopreservation Medium, which is then distributed in 0.6 mL volumes into cryopreservation vials for storage under liquid nitrogen (*see* **Note 6**).

3.2 Gonococcal Inocula

1. Measure the number of cells in the monolayers grown in 24- and/or 96-well cell culture plates. Detach the cells with 50–100 μL of trypsin–EDTA solution per well as described in **step 2** of Subheading 3.1 and suspend cells in 1 mL of DMEM base medium. Transfer 10 μL into a microcentrifuge tube and mix with 10 μL of trypan blue solution. Incubate for 5 min at 37 °C and then place 10 μL of the mix onto a glass hemocytometer. View under a light microscope and count the cells.
2. Remove *N. gonorrhoeae* bacteria from an overnight growth on GC agar using a 10 μL sterile loop and suspend in 2 mL of PBSB. Use a wide-bore pipette tip to gently produce an homogenous suspension and if necessary, centrifuge the bacterial suspension at 200 × *g* for 1 min to remove any large clumps (*see* **Note 7**). Remove the supernatant to a fresh sterile Bijou.

3. Calculate the number of colony forming units (CFU) units/mL in the suspension, by adding a 25 μL aliquot to 975 μL of lysis buffer and measuring the optical density (OD) on a spectrophotometer at $\lambda_{260\ \mathrm{nm}}$ in UV cuvettes using lysis buffer (975 μL) with PBS (25 μL) alone as the blank. With our P9 isolates, a suspension with an OD $\lambda_{260\ \mathrm{nm}}$ of 1.0 absorbance units is equivalent to ~1–2 × 10^9 CFU/mL (*see* **Note 8**).
4. Prepare bacterial suspensions of different multiplicities of infection (MOI) in Infection Medium. The MOI refers to the number of bacteria per eukaryotic cell and needs to be calculated after estimation of cell density within the attached monolayers (**step 1**). Prepare different volumes for each MOI according to the size of the well to be infected: for 96-well cell culture plates, add a volume of 200 μL containing bacteria of a given MOI per monolayer and for a 24-well cell culture plate, add a volume of 1000 μL per monolayer.
5. Remove the Growth Medium from the wells with a sterile plastic Pasteur pipette and wash gently the cell monolayers a minimum of three times with sterile PBS, prewarmed to 37 °C, in order to remove antibiotics and any cell debris (*see* **Note 9**). Next, add the appropriate volume of bacterial suspension of different MOI. Add Infection Medium alone to control wells. In general, triplicate wells are infected with each MOI per time point.
6. Place cell culture plates in an humidified incubator at 37 °C with 5% (v/v) CO_2.
7. Quantify the actual initial MOI by serial tenfold dilution of the bacterial suspension in PBS (e.g., use a microtiter plate with wells containing 90 μL of PBS per well, into which 10 μL volumes are serially diluted). Plate out in aliquots of 15 μL onto triplicate GC agar plates. Grow cultures overnight at 37 °C with 5% (v/v) CO_2 and count colonies to calculate the MOI.

3.3 Gonococcal Association

1. Infect the cell cultures as described in **step 5** of Subheading 3.2 and incubate the cell culture plates for 3 h (*see* **Note 10**).
2. Remove the bacterial suspensions with a sterile Pasteur pipette and wash the infected monolayers gently four times with prewarmed PBS (*see* **Note 11**).
3. After washing, add 250 μL/well of the saponin lysis solution for 24-well cell cultures or 50 μL/well for 96-well cell cultures, and incubate the plate at 37 °C with 5% (v/v) CO_2 for 15 min. Examine the monolayers under a microscope and incubate for longer until the monolayers are detached and lysed. Scrape the wells with a sterile pipette tip and pool the volumes for each triplicate set of wells to give a final volume of 750 μL (24-well

cell cultures) or 150 μL (96-well cell cultures). Make serial dilutions of the lysates in PBS in a separate sterile 96-well plate or in sterile mini-Eppendorf tubes and plate out each dilution onto GC agar plates in triplicate (15 μL per plate). Incubate overnight at 37 °C with 5% (v/v) CO_2 and count the colonies on the following day.

4. Use the equation, CFU/monolayer = [(750 μL/15 μL) × dilution factor]/3 to calculate total association per monolayer (*see* **Note 12**).

3.4 Gonococcal Internalization

The level of bacterial invasion can be estimated using a standard gentamicin assay. Gentamicin is an antibiotic that poorly penetrates eukaryotic cells in the short term. The assay also introduces Cytochalasin D (CD), a cell-permeable reagent and a potent inhibitor of actin polymerization. These experiments can be done in both 24-well or 96-well formats and the protocol for the former is provided.

1. Grow human cells in 24-well cell culture plates following the cell culture protocol as described in Subheading 3.1 and the layout of the experimental plate shown in Fig. 1. In the same experiment, measurements of total bacterial association, in the presence and absence of CD are also made, and two strains can be tested on a single plate.
2. Remove the existing Growth Medium from the monolayers and wash the cells with prewarmed PBS, gently four times.
3. Add 500 μL/well of Cytochalasin D solution to 6 wells and 500 μL/well of the same medium without CD to another six wells, per strain to be tested. Incubate for 30 min at 37 °C with 5% (v/v) CO_2.
4. Prepare gonococcal suspensions in Infection Medium as described in **steps 2–4** of Subheading 3.2 and add 500 μL/well of 2× concentrated MOI to each well, to give a final dose of ~2×10^7 CFU/mL (i.e., ~2×10^7 CFU/monolayer) (*see* **Note 10**). Incubate the plate(s) for 3 h at 37 °C and 5% (v/v) CO_2.
5. After 3 h incubation, remove the medium from the wells for quantifying internalization and add 1 mL of Infection Medium containing 200 μg/mL of gentamicin and incubate for 90 min at 37 °C and 5% (v/v) CO_2. To the remaining wells, remove the culture fluid and quantify bacterial association as described in **steps 2–4** of Subheading 3.3 (*see* **Note 13**).
6. Remove the gentamicin solution after 90 min, wash the monolayers gently with prewarmed (37 °C) PBS and then add lysis solution (250 μL/well). Count the recovered bacteria by serial

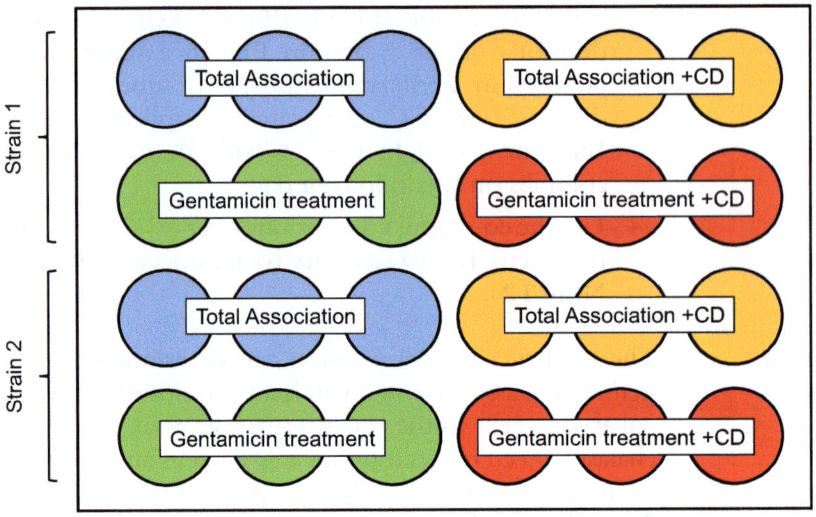

Fig. 1 Schematic for a 24-well cell culture plate infection experiment to measure total gonococcal association and internalization. The plate layout can accommodate testing of two different strains

dilution of pooled lysate solutions from the triplicate wells as described in **steps 3** and **4** of Subheading 3.3.

7. Use the equation, CFU/monolayer = [(750 μL/15 μL) × dilution factor]/3 to calculate bacterial internalization (*see* **Note 14**).
8. Compare the levels of total association and internalization in the presence or absence of CD with, for example, a one-way ANOVA with Dunnett's multiple comparison test or unpaired t-tests, with P values <0.05 denoting significance (*see* **Note 15**).

3.5 Host Cell Viability

Host cell viability in response to gonococcal infection can be measured with several simple nonradioactive biological assays (*see* **Note 16**).

3.5.1 Lactate Dehydrogenase (LDH) Release

1. Grow cells to confluence in sterile 96 well cell culture plates following the cell culture protocol as described in Subheading 3.1 and wash cells three times with sterile prewarmed PBS.
2. Prepare different MOI of gonococci in Infection Medium as described in **steps 2–4** of Subheading 3.2 and add 200 μL volumes of each MOI to triplicate wells. Maintain uninfected monolayers also for measurements of spontaneous and maximum LDH release. Maintain wells with Infection Medium alone for background absorbance readings.
3. Incubate the plates at 37 °C in a humid atmosphere with 5% (v/v) CO_2 and measure LDH release over time (e.g., at 3, 6, 9, 24, and 48 h).

4. At 45 min before each time point measurement, add the proprietary lysis solution (20 μL/well) directly to the maximum LDH release monolayers.
5. Remove 50 μL from each well (infected and control) and add to a fresh ELISA plate; if monolayer detachment is evident, remove the whole cell volume and briefly centrifuge in an Eppendorf tube (3000 × *g*, 5 min) to pellet debris and bacteria, and then remove a 50 μL aliquot.
6. Add 50 μL of proprietary substrate solution to each well and incubate the plate in the dark at room temperature for 30 min.
7. Add 50 μL/well of proprietary stop solution and measure absorbance at $\lambda_{495\ \mathrm{nm}}$ on an ELISA spectrophotometer.
8. Calculate the release of LDH as a measure of cytotoxicity by subtracting initially the medium alone background control value from all readings, and then dividing the average LDH release value of the test by the average maximum LDH release value (from lysed, uninfected cells) using the equation:

   LDH release of test (%) = (LDH release of test/ maximum LDH release) × 100.

9. Analyze the data, for example, using one-way ANOVA with Dunnett's multiple comparison test or unpaired *t*-tests, with *P* values <0.05 denoting significance.

3.5.2 LIVE/DEAD® Assays

1. Grow cells to confluence on transparent Transwell inserts (*see* **Note 17**) following the cell culture protocol as described in Subheading 3.1. Use inserts that fit 24-well cell culture plates (*see* **Note 18**). Wash the confluent monolayers gently with prewarmed PBS.
2. Prepare different MOI of gonococci in Infection Medium as described in **steps 2–4** of Subheading 3.2 and add 1 mL volumes of each MOI to duplicate Transwells. Maintain negative (uninfected cell monolayers) as controls. Incubate the Transwells at 37 °C in a humid atmosphere with 5% (v/v) CO_2 and examine cell death over time (e.g., at 3, 6, 9, 24, and 48 h).
3. Prepare fresh Calcein-AM/ethidium homomdimer-1 reagent mix at each time point, following the manufacturer's instructions (*see* **Note 19**).
4. At each time point, wash the cell monolayers three times gently with prewarmed PBS, and add 300 μL of reagent mix per monolayer and incubate at room temperature for 45 min in the dark.
5. Cut the membrane at the bottom of the Transwell insert with a scalpel blade and transfer the membrane using fine-tipped

forceps to a glass microscope slide, mount a coverslip and seal the edges with finger nail polish to avoid evaporation.

6. View under a laser-scanning confocal microscope and obtain images with two-channel scanning at $\lambda_{488\ \mathrm{nm}}$ and $\lambda_{568\ \mathrm{nm}}$ to excite the green and red signals. Examine at least $n = 6$ independent fields of view and count the green (live) and red (dead) cells to determine the percentage of cell death, using the equation:

$$\%\text{Cell Death} = [\text{Mean number of red cells}\ \ (n = 6)/$$
$$\text{Total number of cells}\ \ (\text{red and green},\ \ n = 6)] \times 100.$$

7. Analyze the data, for example, using one-way ANOVA with Dunnett's multiple comparison test or unpaired t-tests, with P values <0.05 denoting significance.

3.5.3 Cell Proliferation Assay

1. Grow cells to confluence in 96-well cell culture plates following the cell culture protocol as described in Subheading 3.1 and wash monolayers gently with prewarmed PBS.
2. Prepare different MOI of gonococci in Infection Medium as described in **steps 2–4** of Subheading 3.2 and add 200 μL volumes of each MOI to triplicate wells. Maintain negative (uninfected cell monolayers) as controls. Incubate the cell culture plates at 37 °C in a humid atmosphere with 5% (v/v) CO_2 and examine cell proliferation over time (e.g., at 3, 6, 9, 24, and 48 h).
3. In this example, we use the colorimetric and nonradioactive cell proliferation assay that contains the stable tetrazolium salt WST-1 (*see* **Note 20**). Wash the cell monolayers gently three times with PBS and then once with Infection Medium. Add 100 μL of fresh Infection Medium and 10 μL/well of the proprietary WST-1 reagent. Incubate the cultures for 2 h at 37 °C and measure absorbance in an ELISA spectrophotometer at $\lambda_{450\ \mathrm{nm}}$.
4. Compare the mean levels of cell proliferation between infected and uninfected monolayers using a two-sample t-test, with P values <0.05 denoting significance.

3.5.4 Annexin V Assay

1. Grow cells to confluence in sterile 8-chamber tissue culture treated slides following the cell culture protocol as described in Subheading 3.1 (*see* **Notes 21** and **22**) and wash monolayers gently with prewarmed PBS. Each chamber holds a volume of 500 μL of medium.
2. Prepare different MOI of gonococci in Infection Medium as described in **steps 2–4** of Subheading 3.2 and add 500 μL

volumes of each MOI to duplicate chambers. Maintain negative (uninfected cell monolayers) as controls. Incubate the chamber slides at 37 °C in a humid atmosphere with 5% (v/v) CO_2.

3. Add 1 μM of staurosporine to duplicate uninfected cells and incubate for 6 h to induce apoptosis as a positive control. Add the reagent 6 h before each test time point.
4. Examine Annexin V and PI labeling of cells over time (e.g., at 3, 6, 9, 24, and 48 h) using a commercial kit following the manufacturer's instructions (*see* **Note 23**). Add 5 μL of Annexin V-FITC conjugate and 10 μL of PI per 500 μL volume in each chamber and incubate at room temperature for exactly 10 min and protect from light.
5. After 10 min, remove the upper chamber with a scalpel blade, add a glass coverslip and view immediately by LSCM with FITC excitation/emission at $\lambda_{494/528\ \text{nm}}$ and PI at $\lambda_{535/617\ \text{nm}}$, respectively. Cells in early apoptosis will stain with the Annexin V-FITC Conjugate alone. Live cells will show no staining by either PI or Annexin V-FITC Conjugate. Late apoptosis in cells is shown by Annexin-V FITC staining surrounding a PI nucleus. Necrotic cells will be stained by PI with little or no Annexin V-FITC Conjugate staining. Count ~200 cells from at least $n = 6$ independent fields of view.
6. Quantify the number of apoptotic/necrotic/live cells using the equation:

$$[\text{Mean } \% \text{ of cells/Mean total number of cells present in normal uninfected monolayers}] \times 100.$$

7. Analyze the data, for example, using one-way ANOVA with Dunnett's multiple comparison test or unpaired *t*-tests, with *P* values <0.05 denoting significance.

3.5.5 DNA Fragmentation Assay

1. Grow cells to confluence in 96-well cell culture plates following the cell culture protocol as described in Subheading 3.1.
2. Prepare different MOI of gonococci in Infection Medium as described in **steps 2–4** of Subheading 3.2 and add 200 μL volumes of each MOI to triplicate wells. Maintain negative (uninfected cell monolayers) as controls. Incubate the cell culture plates at 37 °C in a humid atmosphere with 5% (v/v) CO_2.
3. At various time points over a 48 h period, remove the Infection Medium, wash the cells gently three times with prewarmed PBS and add 50 μL/well of trypsin–EDTA solution. Incubate for ~10 min to dissociate the cells. Pool the cells from each triplicate test condition and from the control, uninfected cells.

4. Expose a sample of cells to UV radiation in a UV cross-linker for 30–60 min as a positive control for DNA fragmentation.
5. Isolate genomic DNA from each pooled cell suspension using a Wizard Genomic DNA purification kit (or equivalent), following the manufacturer's instructions.
6. Analyze samples of genomic DNA by electrophoresis in 1.5% (w/v) agarose gels in 1× TAE buffer, run for 1 h at 90 V. Stain the gel with 0.5 μg/mL of EtBr in 1× TAE for 20 min to visualize DNA fragments. Apoptotic cells show distinctive DNA fragmentation/laddering patterns that are absent in control, untreated cells.

4 Notes

1. There is a plethora of human primary and immortalised cell lines available commercially that can be used in vitro for studying host cell-pathogen interactions. Primary cell lines are limited by passage, but arguably offer a more realistic cellular and morphological profile. However, care needs to be taken that they do not become dominated by fibroblasts (unless intended). Continuous cultures will senesce with excessive cycles of passage, so it is important to prepare a series of Master and Working cell banks. Care needs to be exercised with commercial cell cultures or cultures acquired from other laboratories and the user should validate the cell line prior to use. There are cell line authentication services available using STR (short tandem repeat) DNA profiling. Cell line contamination and misidentification potentially invalidate research findings [20, 21].
2. Tablets are a convenient way to prepare small volumes of PBS. For larger volumes, use the recipe: 8 g/L of NaCl; 0.2 g/L of KCl, 1.15 g/L of Na_2HPO_4 and 0.2 g/L of KH_2PO_4 in distilled water and autoclave at 121 °C and 2.68 kg/cm^2 for 15 min.
3. All trace of antibiotics must be removed prior to the addition of bacteria.
4. Supplement A and B can be substituted with commercial enrichment reagents, (e.g., Vitox or Isovitalix) and used at concentrations recommended by the manufacturers.
5. It is recommended to omit the starch from GC broth if the organism is grown in liquid culture, as its inclusion makes the broth cloudy and difficult to read optical density for growing bacterial cultures.
6. The time needed for expanding cell cultures to reach >90% confluence depends on many factors e.g., cell growth rate,

which is often slower for primary cells than immortalized cells, and is influenced by the initial cell seeding density. It is useful to count the cells at seeding and adjust with number and medium volume to allow confluence to be reached slowly or quickly, depending on need. This comes with experience and also with the different cell lines available. For most cell lines, subculture by passage is usually best before confluence is reached in order to maintain cell characteristics.

7. If clumping is an issue, it sometimes helps to let the bacterial suspension stand for a few minutes to allow hydration of the clumps. Gentle pipetting will produce an homogeneous suspension without needing to centrifuge to remove clumps.

8. If the operator is using an optical density measurement either at $\lambda_{260\ nm}$ for lysed cultures, which pose no hazard, or at $\lambda_{600\ nm}$ for live cultures grown in GC broth, it is important that a standard CFU/mL is calculated. Normally, we make triplicate measurements of bacterial numbers at an OD of ~1.0 to provide an estimate and we do this for every strain.

9. The cell monolayers can be washed with PBS the night before and the Growth Medium replaced with Infection Medium with reduced dFCS (1% v/v) and without antibiotics. This step saves some time in washing the cells on the day of the experiment; remove this Infection Medium and replace with medium containing different MOI of gonococci or fresh medium alone for uninfected control cell monolayers.

10. For a rapid analysis of bacterial adherence and invasion, we follow the protocol from Virji et al. [22], in which cultures are infected with ~2×10^7 CFU per monolayer for 3 h, which is sufficient time to allow initial gonococcal association and enable significant invasion, if the phenotype permits. This assay is useful when comparing different wild-type, phenotypic variant and mutant strains together. This time point also reduces the extent of bacteria-induced cell death.

11. The association infection experiment is done over 3 h, but operators can leave bacterial infections on host cells for longer periods of time, and use different MOI. At each time point, the Infection Media and control uninfected media that is removed can be examined in other ways. A viable count can be made of growing, unattached bacteria and the suspensions can then be centrifuged at $3000 \times g$ for 5 min to pellet the bacteria. The supernatants can be stored at -20 °C and used for quantifying the release of cellular biomarkers such as cytokines/chemokines, reactive oxygen and nitrogen species and other molecules. There are a variety of ways to quantify released protein biomolecules, including specific single ELISA assay kits from

many different manufacturers to using multiplex platforms (e.g., Luminex).

If protein measurements are not required, and the operator is interested only in gene and mRNA expression profiles, these can be determined by removing the medium, washing the infected and uninfected control cell monolayers and then using a commercial kit to extract RNA. The RNA can be stored for future analyses by qRT-PCR for a wide variety of host cell signals and for transcriptomics analyses. It is also possible to extract RNA from the associated bacteria for dual transcriptomics analyses. For these early infection events, sample at 1, 2, 4 and 6 h to catch the up/down-regulated signals.

12. The protocol provides a measure of total bacterial association, which is the sum of gonococcal adherence and invasion.
13. Generally, the addition of cytochalasin D (CD) causing some visible cell shrinkage, but not cell death over the time course of the adherence and invasion experiments. CD does not influence gonococcal adherence to epithelial cells in our experience.
14. The levels of gonococcal internalization can be calculated as a percentage of the total associated number. It is recommended that viable counting of internalization is confirmed by additional assays (e.g., Z stack analyses from confocal microscopy or transmission electron microscopy) to visualize intracellular bacteria. Low numbers of recovered internalized bacteria may suggest an artefact. Imaging techniques can also demonstrate whether invasion of a confluent monolayer occurs via an intercellular route, which could be caused by gonococcal loosening of cellular junctions. The latter can be visualised with high-resolution microscopy techniques.
15. Generally, we do a minimum of $n = 3$ infection experiments with gonococci and calculate mean association and internalization values with standard errors of the mean. A one-way ANOVA with Dunnett's multiple comparison test or unpaired t-tests are useful for testing data, with P-values <0.05 denoting significance.
16. There is a large choice of viability and cytotoxicity assays available commercially, for measuring both prokaryotic and eukaryotic cell viability. Generally, we do at least 2/3 different assays to quantify host cell death, including release of LDH as a measure of cell membrane permeability, differential staining assays (e.g., acridine orange and ethidium bromide) through to measurements of caspase and inflammasome activation by qRT-PCR [23].
17. Transwell inserts are available with several pore sizes, the most common being 0.4, 1.0, 3.0 and 8.0 μm. For growing cells and infection, we find that the 0.4 and 1.0 μm Transwells are good

for allowing cell attachment and no passage of the bacteria through to the lower chamber. A pore size of 3.0 μm is useful if measuring bacterial penetration by intra/intercellular route and allows bacteria to accumulate in the lower chamber. Cell culture is difficult on the largest pore size. Culture medium in both the lower and upper chambers can be sampled for bacterial and host cell biomarkers of infection.

18. We usually use Transwell inserts that fit a 24-well plate format, for ease of handling and for providing a large surface area for analysis. Other sizes are available and Transwells that fit 12- and 48-well plates are also popular.
19. The kit contains the green-fluorescent dye Calcein-AM and the red-fluorescent dye ethidium homodimer-1, which allows simultaneous and rapid discrimination of live (green-stained) and dead (red-stained) cells. The Calcein-AM indicates intracellular esterase activity and ethidium homodimer-1 indicates loss of plasma membrane integrity. Several different stains are available for discriminating between live and dead cells. The ratio of Calcein-AM to ethidium homomdimer-1 that discriminates between live and dead cells with little fluorescence overlap must be determined empirically for different cell types infected with different bacteria.
20. Cell proliferation assays work on essentially the same principle, but can use different tetrazolium salts (e.g., XTT (sodium 2,3-bis(2-methoxy-4-nitro-5-sulfophenyl)-5-[(phenylamino)-carbonyl]-2H-tetrazolium inner salt), MTT (3-(4,5-Dimethylthiazol-2-yl)-2,5-diphenyltetrazolium bromide), WST-1 (4-[3-(4-Iodophenyl)-2-(4-nitrophenyl)-2H-5-tetrazolio]-1,3-benzene disulfonate)), all of which are reduced to formazan by cellular enzymes. Other cell proliferation assays can incorporate 5-bromo-2′-deoxyuridine (BrdU) in place of thymidine to monitor DNA synthesis and cell proliferation.
21. Cells grown in 8-chamber slides and on Transwell inserts are useful for visualizing host cell-gonococcal interactions by microscopy, including laser scanning confocal microscopy (LSCM) and scanning 2D and 3D and transmission electron microscopy. The format also allow for real-time imaging of bacterial attachment and invasion. Gonococci that express a green fluorescent protein (GFP) [24] or other fluorescent dye are valuable for fixed and real-time LSCM, although fixation with paraformaldehyde is recommended for the former, so as not to bleach the GFP. Counterstaining can be done for a variety of host cell surface proteins, intracellular filaments, etc. More advanced platforms such as atomic force microscopy are becoming more common for studying host cell–pathogen interactions [25].

22. We have used Nunc™ Lab-Tek™ II Chamber Slides successfully, and these consist of a removable upper polystyrene media chamber attached to a standard glass slide. The slide surface is cell-culture treated to improve cell adherence and is available in 1-, 2-, 4-, and 8-well formats. Similar produces are available from other manufacturers.
23. The kit uses recombinant Annexin V conjugated to green-fluorescein isothiocyante (FITC) to label phosphatidylserine sites on the membrane surface. The kit includes propidium iodide (PI) to label the cellular DNA in necrotic cells where the cell membrane has been totally compromised. This combination allows the differentiation among early apoptotic cells (Annexin V positive, PI negative), necrotic cells (Annexin V positive, PI positive), and viable cells (Annexin V negative, PI negative).

Acknowledgments

H.L.A.M. was a Ph.D. student funded by Colciencias–Colfuturo, Government of Colombia, Scholarship 568.

References

1. Marazzo J, Handsfield HH, Sparling PF (2010) *Neisseria gonorrhoeae*. In: Mandell GL, Douglas RG, Bennett JE et al (eds) Mandell, Douglas, and Bennett's principles and practice of infectious diseases, 7th edn. Elsevier/ Churchill Livingstone, Philadelphia, PA, pp 2753–2769
2. McGee ZA, Stephens DS, Hoffman LH et al (1983) Mechanisms of mucosal invasion by pathogenic *Neisseria*. Rev Infect Dis 5(Suppl 4):S708–S714
3. Merz AJ, So M (2000) Interactions of pathogenic *Neisseriae* with epithelial cell membranes. Annu Rev Cell Dev Biol 16:423–457. https://doi.org/10.1146/annurev.cellbio.16.1.423
4. Muenzner P, Dehio C, Fujiwara T et al (2000) Carcinoembryonic antigen family receptor specificity of *Neisseria meningitidis* Opa variants influences adherence to and invasion of proinflammatory cytokine-activated endothelial cells. Infect Immun 68(6):3601–3607
5. Dehio C, Gray-Owen SD, Meyer TF (2000) Host cell invasion by pathogenic *Neisseriae*. Subcell Biochem 33:61–96
6. Koomey M (2001) Implications of molecular contacts and signaling initiated by *Neisseria gonorrhoeae*. Curr Opin Microbiol 4(1):53–57
7. Boyle EC, Finlay BB (2003) Bacterial pathogenesis: exploiting cellular adherence. Curr Opin Cell Biol 15(5):633–639
8. Massari P, Ram S, Macleod H et al (2003) The role of porins in neisserial pathogenesis and immunity. Trends Microbiol 11(2):87–93
9. Singleton TE, Massari P, Wetzler LM (2005) Neisserial porin-induced dendritic cell activation is MyD88 and TLR2 dependent. J Immunol 174(6):3545–3550
10. Virji M (2009) Pathogenic *Neisseriae*: surface modulation, pathogenesis and infection control. Nat Rev Microbiol 7(4):274–286
11. Sadarangani M, Pollard AJ, Gray-Owen SD (2011) Opa proteins and CEACAMs: pathways of immune engagement for pathogenic *Neisseria*. FEMS Microbiol Rev 35(3):498–514. https://doi.org/10.1111/j.1574-6976.2010.00260.x
12. Criss AK, Seifert HS (2012) A bacterial siren song: intimate interactions between *Neisseria* and neutrophils. Nat Rev Microbiol 10 (3):178–190. https://doi.org/10.1038/nrmicro2713
13. Hung MC, Christodoulides M (2013) The biology of *Neisseria* adhesins. Biology (Basel) 2(3):1054–1109. https://doi.org/10.3390/biology2031054

14. Edwards JL, Jennings MP, Apicella MA et al (2016) Is gonococcal disease preventable? The importance of understanding immunity and pathogenesis in vaccine development. Crit Rev Microbiol 42(6):928–941. https://doi.org/10.3109/1040841x.2015.1105782
15. Chateau A, Seifert HS (2016) *Neisseria gonorrhoeae* survives within and modulates apoptosis and inflammatory cytokine production of human macrophages. Cell Microbiol 18 (4):546–560. https://doi.org/10.1111/cmi.12529
16. Hooda Y, Shin HE, Bateman TJ et al (2017) Neisserial surface lipoproteins: structure, function and biogenesis. Pathog Dis 75(2). https://doi.org/10.1093/femspd/ftx010
17. Mubaiwa TD, Semchenko EA, Hartley-Tassell LE et al (2017) The sweet side of the pathogenic *Neisseria*: the role of glycan interactions in colonisation and disease. Pathog Dis 75(5). https://doi.org/10.1093/femspd/ftx063
18. Weyand NJ (2017) *Neisseria* models of infection and persistence in the upper respiratory tract. Pathog Dis 75(3). https://doi.org/10.1093/femspd/ftx031
19. Quillin SJ, Seifert HS (2018) *Neisseria gonorrhoeae* host adaptation and pathogenesis. Nat Rev Microbiol 16(4):226–240. https://doi.org/10.1038/nrmicro.2017.169
20. Capes-Davis A, Theodosopoulos G, Atkin I et al (2010) Check your cultures! A list of cross-contaminated or misidentified cell lines. Int J Cancer 127(1):1–8. https://doi.org/10.1002/ijc.25242
21. Development Organization Workshop ATCC Standards (2010) Cell line misidentification: the beginning of the end. Nat Rev Cancer 10 (6):441–448. https://doi.org/10.1038/nrc2852
22. Virji M, Kayhty H, Ferguson DJP et al (1991) The role of pili in the interactions of pathogenic *Neisseria* with cultured human endothelial cells. Mol Microbiol 5:1831–1841
23. Cendra MM, Christodoulides M, Hossain P (2017) Signaling mediated by Toll-like receptor 5 sensing of *Pseudomonas aeruginosa* flagellin influences IL-1β and IL-18 production by primary fibroblasts derived from the human cornea. Front Cell Infect Microbiol 7:130. https://doi.org/10.3389/fcimb.2017.00130
24. Christodoulides M, Everson JS, Liu B et al (2000) Interaction of primary human endometrial cells with *Neisseria gonorrhoeae* expressing green fluorescent protein. Mol Microbiol 35:32–43
25. Zhou Z, Leake MC (2016) Force spectroscopy in studying infection. Adv Exp Med Biol 915:307–327. https://doi.org/10.1007/978-3-319-32189-9_19

Chapter 18

Use of Human Monocyte-Derived Macrophages to Study *Neisseria gonorrhoeae* Infection

Jimena Gatica, Paula I. Rodas, and Alejandro Escobar

Abstract

Macrophages are critical cells in the innate immune response to microorganisms sensed in the tissues. During infections, the interaction between pathogens and macrophages leads to a macrophage response that includes cytokine production, antigen processing and presentation in the context of MHC molecules, expression of T cell costimulatory molecules and recruitment of innate defense effectors, which results in clearance of infection. However, *Neisseria gonorrhoeae* can suppress the protective immune response at this level, avoiding its detection and elimination. Studies addressed to develop the interactions between macrophages and *Neisseria gonorrhoeae* allow us to find potential targets to be exploited with vaccines and therapeutic drugs. In this chapter, we describe protocols to generate human monocyte-derived macrophages and assess their response to infection with *Neisseria gonorrhoeae*.

Key words *Neisseria gonorrhoeae*, Monocyte-derived macrophages, Phagocytosis, Cytokines, T cell proliferation

1 Introduction

Effective immune response of the host to foreign organisms requires the coordinated action of the innate and adaptive immune system. The innate immune response provides the first line of defense against microorganisms but also provides the biological context and signals that will instruct the adaptive immune response [1]. Macrophages belong to the innate immune system and are present in almost all tissues, where they are responsible for early response against microbes [2]. Macrophages express pattern recognition receptors (PRRs) for many microbial pathogen-associated molecular patterns (PAMPs), allowing engulfment and digestion of the invading pathogen, followed by secretion of cytokines and chemokines to recruit more immune cells to the area of infection for restriction of pathogen invasion [3].

Macrophages can be derived from the yolk sac, fetal liver monocytes, or adult bone-marrow monocytes (BM-MOs) and are capable

Myron Christodoulides (ed.), *Neisseria gonorrhoeae: Methods and Protocols*, Methods in Molecular Biology, vol. 1997, https://doi.org/10.1007/978-1-4939-9496-0_18, © Springer Science+Business Media, LLC, part of Springer Nature 2019

of entering tissues during inflammation [2, 4]. In response to several signals, macrophages may drive to classical M1 activation or alternative M2 activation [5]. The M1 phenotype promotes a Th1 response and strong microbicidal and tumoricidal activity. In contrast, the M2 phenotype helps in parasite clearance, dampens inflammation, promotes tissue remodeling, tumor progression, and possesses immune-regulatory functions [6]. Interestingly, during microbial infection, macrophages are polarized to M1 or M2 cells in response to microbial components and host immune mediators. Depending on the bacterial species, M1 or M2 polarization can play a beneficial or detrimental role in disease outcomes [7, 8]. In humans, peripheral blood monocytes are the most commonly used precursors for generating macrophages in vitro. The culture of monocyte-derived macrophages (MDM) permits polarization of monocytes towards different macrophage phenotypes [5, 6, 9]. Macrophage-driven innate immune responses can be subverted by *Neisseria gonorrhoeae* and numerous studies have evaluated the different mechanisms used by this pathogen to dampen macrophage activity [10–17]. These studies have led to better understanding about macrophage interactions with the gonococcus. In this chapter, we describe the methodologies that can be used to investigate the interactions of human monocyte-derived macrophages and *N. gonorrhoeae* in a model of infection, and the protocols used to assess the bacteria-induced macrophage responses.

2 Materials

All the reagent solutions for human sample processing for MDM are prepared with cell culture grade consumables. A Class II biosafety (BSL-2) cabinet and a humidified cell culture incubator set to 37 °C with 5% (v/v) CO_2 are mandatory for handling both human and gonococcal samples as well as doing most of the protocols detailed in this chapter.

2.1 Generation of Human Monocyte-Derived Macrophages (MDM)

1. Human buffy coats (*see* **Note 1**).
2. Phosphate buffered saline, pH 7.4 (PBS).
3. Ammonium Chloride Potassium Carbonate EDTA (ACK) Lysis Buffer (Biolegend). Alternatively, prepare a solution containing 8.02 g of NH_4Cl, 1.0 g of $KHCO_3$ and 0.037 g of EDTA in 1 L of ultrapure H_2O. Sterilize by filtration through a 0.22 μm filter before use.
4. Histopaque-1077 (Ficoll).
5. Complete RPMI 1640 culture medium: RPMI with L-glutamine, 50 U/mL of penicillin, 50 μg/mL of streptomycin, and 10% (v/v) fetal bovine serum (FBS).
6. Trypan blue.

7. RosetteSep™ Human Monocyte Enrichment Cocktail (STEMCELL Technologies).
8. Human Macrophage colony-stimulating factor (M-CSF) (*see* **Note 2**).
9. Hemocytometer.
10. Sterile tissue culture 6 well plates (*see* **Note 3**).
11. Sterile Pasteur pipettes.
12. 10–20 mL sterile serological pipette.
13. 15 and 50 mL sterile conical centrifuge tubes.
14. Scissors.
15. Centrifuge with a swing bucket rotor for 15 and 50 mL conical tubes.
16. Inverted microscope.
17. Biosafety Class II Cabinet.
18. Humidified incubator set to 37 °C with 5% (v/v) CO_2.

2.2 Culture of Neisseria gonorrhoeae

1. *Neisseria gonorrhoeae* strain P9 variant 17 (Pil+ Opa+) [18] or other gonococcal strains (*see* **Note 4**).
2. GC agar base: add 10 g of proteose peptone, 1.01 g of corn starch, 4.03 g of K_2HPO_4, 1.01 g of KH_2PO_4, 5.03 g of NaCl, and 10.07 g of agar to 1 L of distilled H_2O and autoclave for 15 min at 121 °C and 2.68 kg/cm^2 [19].
3. BBL IsovitaleX™ Enrichment.
4. Kellogg's Supplement I (100×): add 40 g of glucose and 2 mg of thiamine pyrophosphate to 100 mL of distilled H_2O and sterilize by filtration using a 0.22 μm filter, then store at −20 °C.
5. Kellogg's Supplement II (1000×): add 50 mg of $Fe(NO_3)_3{\cdot}9H_2O$ to 100 mL of distilled H_2O and sterilize by filtration using a 0.22 μm filter. Store at −20 °C [19].
6. 90 mm × 15 mm sterile petri dishes.
7. GC agar plates: add 10 mL of Supplement I and 1 mL of Supplement II per 1 L of GC agar base cooled to ~50 °C in a water bath, prior to pouring into Petri dishes (*see* **Note 5**).
8. GC broth: add 1.5 g of proteose peptone, 0.4 g of K_2HPO_4, 0.1 g to KH_2PO_4, and 0.1 g of NaCl to distilled H_2O, adjust the pH to 7.2 with HCl and make up to a final volume of 100 mL with H_2O [19]. Autoclave for 15 min at 121 °C and 2.68 kg/cm^2 and store at room temperature.
9. GC broth supplemented with 1% (v/v) IsovitalexTM Enrichment or Kellogg's Supplements I and II and 10% (v/v) dimethyl sulphoxide (DMSO).

10. 10 μL sterile disposable inoculation loops.
11. Stereomicroscope.
12. Infection Medium: RPMI 1640 with no phenol red. Do not add antibiotics or FBS.
13. 50–2000 μL volume cuvettes.
14. Spectrophotometer.
15. Ultracentrifuge.
16. 10 mg/mL of stock gentamicin solution.

2.3 Infection of MDM with Neisseria gonorrhoeae

1. MDM cultures.
2. PBS, pH 7.4.
3. Complete RPMI medium with no antibiotics.
4. *N. gonorrhoeae* strain(s).
5. *Escherichia coli* lipopolysaccharide (LPS).
6. 100 μg/mL of gentamicin solution, prepared in complete RPMI.
7. Humidified incubator set to 37 °C with 5% (v/v) CO_2.

2.4 Gentamicin Protection Assay

1. Infection Medium.
2. 10 mg/mL of stock gentamicin solution.
3. 1% (w/v) saponin in cell culture grade H_2O or PBS.
4. 15 mL sterile centrifuge tubes.
5. 1.5 mL sterile microcentrifuge tubes.
6. GC broth.
7. GC agar plates.
8. 1.5 mL sterile microcentrifuge tubes.
9. 96-well cell culture plates.
10. Multichannel pipette.

2.5 Determining Gonococcal Internalization by Confocal Microscopy

1. MDM cell cultures.
2. *N. gonorrhoeae* rs_GFP strain(s).
3. 12 mm poly-D-lysine coated coverslips.
4. 12- or 24-well cell culture plates.
5. PBS, pH 7.4.
6. 1% (w/v) paraformaldehyde (PFA) in cold PBS.
7. Rhodamine–Phalloidin.
8. 4′,6-diamidino-2-phenylindole (DAPI).
9. 4% (w/v) bovine albumin serum (BSA) in PBS.
10. 0.2% (v/v) Triton X-100 in PBS.

11. Fluoromont® aqueous mounting medium.
12. Microcentrifuge tubes.
13. Parafilm.
14. Microscope slides.
15. Humid chamber.
16. Paper tissue.
17. Pasteur pipettes.
18. Humidified incubator set to 37 °C with 5% (v/v) CO_2.
19. Confocal microscope.

2.6 Assessing Human MDM Activation

2.6.1 Flow Cytometry

1. Cell scrapers.
2. Antibodies (Table 1).
3. 2% (v/v) FBS in PBS staining buffer/wash buffer.
4. 10% (v/v) of inactivated human AB serum in PBS (*see* **Note 6**).
5. 4% (v/v) PFA in PBS.
6. Fluorescence-activated cell sorting (FACS) buffer: PBS containing 2% (v/v) FBS, 1 mM EDTA and 0.1% (w/v) sodium azide.
7. 1.5 mL microcentrifuge tubes.
8. 96-well V-bottom plates.
9. 5 mL polystyrene round bottom FACS tubes.
10. Centrifuge with a rotor for microcentrifuge tubes and/or plate rotor.
11. FACScalibur flow cytometer using CellQuest Pro software (or equivalent, *see* **Note 7**).

2.6.2 Cytokine Production

1. Cytokine ELISA kits using matched antibody pairs.
2. 96-well plates.
3. ELISA microplate reader.

2.6.3 Real-Time Polymerase Chain Reaction (RT-PCR)

1. RLT Buffer (Qiagen).
2. RNeasy mini kit (Qiagen).
3. M-MLV RT kit (Invitrogen).
4. Fast SybrGreen Master Mix (Fermentas).
5. 0.2 mL PCR tubes.
6. NanoDrop spectrophotometer.
7. Primers (Table 2).
8. −80 °C freezer.
9. ABI PRISM 7900 Sequence Detector System (or equivalent).

Table 1
List of antibodies used for flow cytometry

Antibodies	Clone	Catalog N°	Company	Test size (μL)
CD14 FITC	61D3	11-0149-41	eBiosciences	2
CD40 FITC	HI40a	MA1-10222	ThermoScientific	0.5
CD86 PE-Cy5	IT2.2	46-0869	eBiosciences	0.5
CD163 PE	eBioGHI/61	12-1639	eBiosciences	5
CD206 PECy5	1502	321108	BioLegend	2
HLA-DR PECy5	LN3	45-9956-41	eBiosciences	5
CD4 FITC	MEM-241	MA1-19603	Thermo Scientific	1

Table 2
Forward and reverse primers used for cytokines amplification by quantitative real-time PCR

Gene	Forward (5′–3′)	Reverse (5′–3′)
IL-1β	AAACAGATGAAGTGCTCCTTCCAGG	TGGAGAACACCACTTGTTGCTCCA
IL-6	AACCTGAACCTTCCAAAGATG	TCTGGCTTGTTCCTCACTACT
IL-10	AGAACCTGAAGACCCTCAGGC	CCACGGCCTTGCTCTTGTT
IL-12	AAATATGCGGCCGCTAAGCCACCATGGG TCAC	CCGCTCGAGCGTTAGGAAGCA TTCAGATAG
IL-23	GCAGATTCCAAGCCTCAGTC	TTCAACATATGCAGGTCCCA
TNF-α	CCCAGGGACCTCTCTCTAATCA	GGCCTTGTAGACACCTTGGTC
18s	CTAAACACGGGAAACCTCAC	CGCTCCACCAACTAAGAACG

2.6.4 Alloproliferation Assay

1. MDM cell cultures.
2. *N. gonorrhoeae* strain(s).
3. Complete RPMI 1640 culture medium.
4. Cold complete RPMI 1640 culture medium containing 10% (v/v) FCS.
5. Peripheral Blood Lymphocytes (PBL).
6. T150 cm^2 cell culture flasks.
7. Trypan blue.
8. 5 μM carboxyfluorescein succinimidyl ester (CFSE).
9. Monoclonal antibodies to CD4 T cells.
10. 5 μg/mL of phytohemagglutinin (PHA).
11. 96-well flat-bottomed sterile plates.
12. 24-well sterile plates.

13. Hemocytometer.
14. 5 mL polystyrene round bottom FACs tubes.
15. 15 mL conical tubes.
16. 50 mL centrifuge tubes.
17. Humidified incubator set to 37 °C and 5% (v/v) CO_2.
18. Centrifuge.

3 Methods

3.1 Generation of Human Monocyte-Derived Macrophages (MDM)

1. Collect human buffy coats from the blood bank.
2. In a clean Biosafety Class II cabinet cut the hose of buffy coat bags with scissors and then place blood cells into a 50 mL conical tube. Each buffy coat bag contains approximately 20 mL of concentrated blood cells. Take 15 mL of blood and transfer to another 50 mL conical tube and dilute it with PBS to a final volume of 30 mL, mix gently and take off all the bubbles on the surface using a P1000 micropipette. Save 10 mL of blood for later.
3. Add 10 mL of Ficoll onto the bottom of a 50 mL conical tube, in a very slow and continuous flow avoiding the blending of blood and Ficoll. Centrifuge at 400 × *g* for 18 min at room temperature (RT, 15–20 °C), setting acceleration to low and without the brake, in order to prevent disruption of the gradient.
4. As soon as the centrifugation is over, recover the peripheral blood mononuclear cell (PBMC) fraction, which is found in the white layer between plasma and Ficoll, by gently aspiration using a P1000 micropipette. Add the PBMC fraction to a new 50 mL conical tube. Ficoll can be taken if necessary if the monocytes layer is not distinguishable.
5. Wash the PBMC by adding 20–30 mL of PBS and centrifuge at 300 × *g* for 5 min at RT (*see* **Note 8**).
6. Discard the supernatant gently, leaving approximately 50–100 μL to avoid disturbing the pellet. Repeat **step 5**.
7. Add the 10 mL of blood from **step 2** and add 50 μL/mL of the RosetteSep™ Human Monocyte Enrichment Cocktail. Mix well and incubate for 30 min at 37 °C.
8. Add 10 mL of PBS with 2% (v/v) FBS and mix well, removing all the bubbles on the surface. Add 10 mL of Ficoll as described in **step 3** and centrifuge at 400 × *g* for 18 min at RT setting acceleration and brake to low.
9. Take the layer of enriched cells between the Ficoll and the plasma. If it is difficult to see the cells at the interface, Ficoll

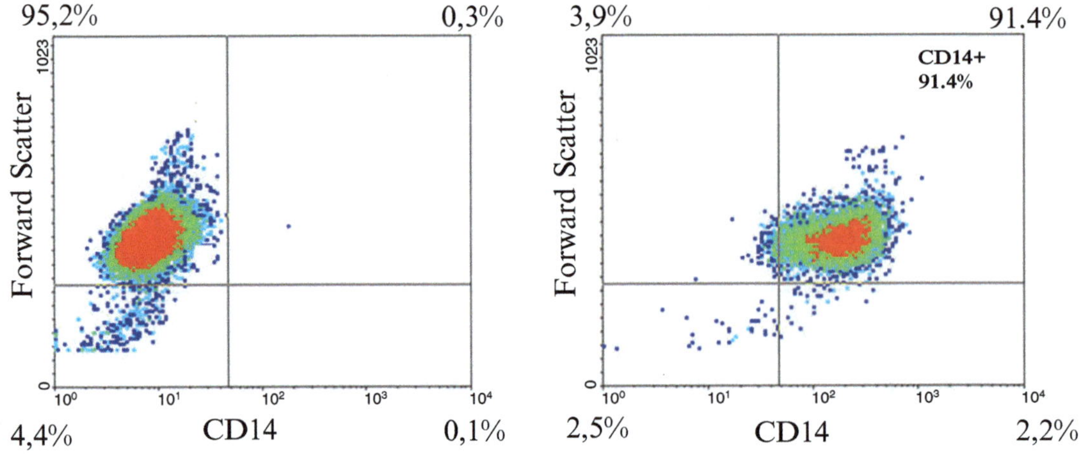

Fig. 1 Isolation of purified CD14+ monocytes. CD14+ monocytes were isolated from PBMCs by RosetteSep™ separation (negative selection) and stained using an anti-CD14 antibody. Following gating on single, live cells, the % CD14+ cells was established

can be removed along with the enriched cells to recover them all, but make sure not to aspirate blood.

10. Place the cells into a new conical tube and wash twice with PBS at 300 × *g* for 5 min at RT. If red blood cells are observed in the pellet, lyse the red cells using ACK Lysing Buffer. Remove all the supernatant and add 10 mL of ACK Lysing Buffer (1×). Incubate at RT for 3–5 min. Lysis of the red cells should be evident during this incubation. Leave the pellet with the minimum amount of PBS, because it may interfere with the lysis buffer. Centrifuge at 300 × *g* for 5 min at RT. Discard gently the supernatant leaving approximately 50 μL to avoid disturbing the pellet, and then add 5 mL of PBS. Wash the pellet twice with PBS, the first wash with centrifugation at 300 × *g* for 5 min and the second wash with centrifugation for 10 min at 100 × *g* at RT, in order to eliminate platelets that may affect culture.

11. Count purified monocytes using trypan blue, and suspends the cells in prewarmed complete RPMI medium at a density of 1×10^6 cells/mL.

12. Prepare an aliquot of cells to stain for CD14 to check for cell purity (Fig. 1), following the staining protocol used for MDMs described in Subheading 3.6.1.

13. On Day 0, seed 2×10^6 cells in 3 mL of complete RPMI supplemented with 50 ng/mL of M-CSF in 6 well plates. This can be reduced proportionally to 1.5 mL in 12 well plates. Incubate the cells in a humidified incubator at 37 °C with 5% (v/v) CO_2.

14. On Day 3, remove 1.5 mL of the culture medium from each well and discard and then add 1.5 mL of fresh medium with the exact amount of M-CSF as added on Day 0.
15. On Day 6 of differentiation, the cells are ready to be infected or polarized (*see* **Note 9**).
16. Culture cells for an additional 24 h and harvest to downstream assays.

3.2 Culture of Neisseria gonorrhoeae

1. Plate bacteria from a stock of bacterial cell suspension, frozen in GC broth supplemented with 1% (v/v) Isovitalex™ Enrichment or Kellogg's Supplements I and II and 10% (v/v) DMSO onto GC plates (*see* **Note 5**). Grow bacteria for 18–24 h in a humidified incubator at 37 °C with 5% (v/v) CO_2.
2. Check the piliated (Pil+) and opacity protein (Opa+) colony phenotype by viewing the bacterial growth on the agar plates with a stereo microscope (*see* **Note 4**). For phagocytosis experiments, the gonococci used contain the red-shift mutant Green Fluorescent Plasmid (rs_GFP).
3. Collect gonococcal colonies from the plate using a sterile 10 μL disposable inoculation loop and place the bacteria into a 1.5 mL microcentrifuge tube containing 1 mL of infection medium. Measure the optical density (O.D.) of the suspension with a spectrophotometer at an absorbance of $\lambda_{600\ \mathrm{nm}}$. An O.D. of 1 unit is equivalent to 3.2×10^9 gonococci/mL.
4. Adjust the bacterial suspension to 1×10^9 gonococci/mL ready for use in infection experiments.

3.3 Infection of MDM with Neisseria gonorrhoeae

1. Remove the culture medium from the MDM cultures prepared in Subheading 3.1 carefully followed by gentle washing with 2 mL of PBS.
2. Replace with 2 mL of complete RPMI without antibiotics.
3. Add the gonococci at an appropriate multiplicity of infection (e.g., an MOI of 10 is 10 bacteria to 1 macrophage) to the MDM cells. Also, microbial products such as purified lipopolysaccharide (LPS) can be used to stimulate MDM. Usually, *E. coli* LPS is used at a concentration of 10–100 ng/mL to control macrophage stimulation.
4. After 4 h, supplement the culture with 100 μg/mL of gentamicin to kill extracellular gonococci. Then keep the culture for an additional 20 h at 37 °C with 5% (v/v) CO_2 and harvest the cells to downstream assays (Fig. 2).

3.4 Gentamicin Protection Assay

For MDM cultures infected with *N. gonorrhoeae*, we use a protocol described previously [20] with some modifications [12].

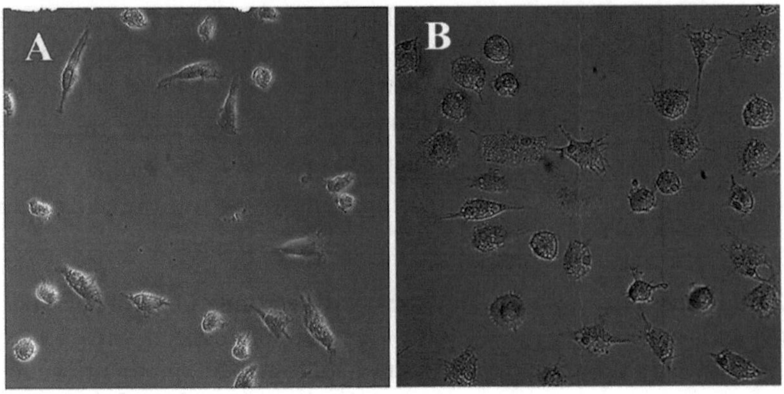

Fig. 2 Morphology of monocyte-derived macrophages produced in vitro. (**a**) Microscopic visualization (400×) optical magnification of differentiated unstimulated M0 macrophages (MDM) during 7 days using MCSF. (**b**) MDM following gonococcal infection

1. Dilute the stock solution of 10 mg/mL gentamicin in Infection Medium to a working concentration of 100 μg/mL in a sterile 15 mL conical tube. Prewarm this solution at 37 °C.
2. Wash the infected MDM cultures three times gently with prewarmed PBS and then add 2 mL (for 6-well) or 1 mL (for 12-well) of gentamicin solution. Incubate for 1 h at 37 °C with 5% (v/v) CO_2 to kill extracellular gonococci.
3. Wash the MDM cultures three times gently with prewarmed PBS and then add 150 μL of 1% (v/v) saponin lysis solution to each well. Incubate for 30 min at RT to permeabilize the cells.
4. Transfer 50 μL of the bacteria–saponin solution into a well of a 96-well cell culture plate containing 50 μL of prewarmed GC broth (*see* **Note 11**). Dilute this suspension serially over an additional 4 wells containing 50 μL of prewarmed GC broth. Plate 20 μL of each bacterial dilution, as well as the undiluted bacteria–saponin solution, in duplicate onto prewarmed GC agar plates and incubate at 37 °C with 5% (v/v) CO_2 for 24 h.
5. Count colony forming units (CFU) and determine the percentage of internalized bacteria by comparing with the initial count of gonococci (consider the initial CFU count as 100%).

3.5 Determining Gonococcal Internalization by Confocal Microscopy

1. Prepare MDM cells as described in Subheading 3.1, and seed the cells into 12-well cell culture plates containing 12 mm round poly-D-lysine-coated coverslips (*see* **Note 12**).
2. Infect the cells with *N. gonorrhoeae* rs-GFP strain at a MOI of 100 as described in Subheading 3.3.
3. Incubate for 4 h at 37 °C with 5% (v/v) CO_2. Remove the supernatant and rinse gently five times with RPMI culture medium.

4. Add 1 mL of cold 1% (w/v) PFA solution for 10 min at RT to fix the cells.
5. Remove the 1% (w/v) PFA solution gently with a Pasteur pipette and wash the fixed cells three times with 1 mL of PBS each (*see* **Note 13**).
6. Add 500 μL of 0.2% (v/v) Triton X-100 in PBS to permeabilize the cells (*see* **Note 14**). Incubate for 10 min at RT.
7. Remove the 0.2% (v/v) Triton X-100 in PBS solution and wash four times with PBS.
8. Add 500 μL of 4% (w/v) BSA in PBS to each coverslip-containing well. Incubate for 30 min at RT to reduce nonspecific background staining.
9. Dilute the Rhodamine–Phalloidin conjugate 1/200 in 4% (w/v) BSA in PBS in a microcentrifuge tube. Estimate the volume of conjugate required (25 μL per coverslip) (*see* **Note 15**).
10. Place 25 μL of Rhodamine–Phalloidin conjugate (1/200) onto a piece of parafilm. Place the coverslips cell-side down onto the dye dilution and incubate for 20 min at RT. To avoid evaporation, keep the coverslips inside a humid chamber.
11. Place the coverslips cell-side up into wells of a 12- or 24-well cell culture plate containing 500 μL of PBS, and wash four times with PBS.
12. Stain cell nuclei with DAPI at a dilution of 1/3000 for 10 min.
13. Rinse twice in PBS for 5 min each time.
14. Dry the coverslips with paper tissue.
15. Put 10 μL of the bubble-free Fluoromount® mounting medium onto a microscope slide, and then put the coverslip cell-side down onto the mounting medium. Incubate overnight at RT in a cool and dark place. When assembling coverslips over the mounting medium on a slide, leave ~3–5 mm of the slide free between them (*see* **Note 16**).
16. View the fixed sample slices using a confocal microscope, following the equipment manufacturer's instructions (Fig. 3).

3.6 Assessing Human MDM Activation

Flow cytometry is a useful technique to determine surface membrane phenotype and expression of antigen presentation and activation molecules on human MDM.

3.6.1 Flow Cytometry

1. Gently scrape the cultured MDM cells with a cell scraper. Make a soft and steady movement, just in one direction while you move the plate to reach all the spaces (*see* **Note 17**).
2. Put the cells into a microcentrifuge tube or V-bottom 96-well plate in an aliquot of $0.3–0.5 \times 10^6$ cells per test. Centrifuge at $600 \times g$ for 3 min at RT. Discard supernatant.

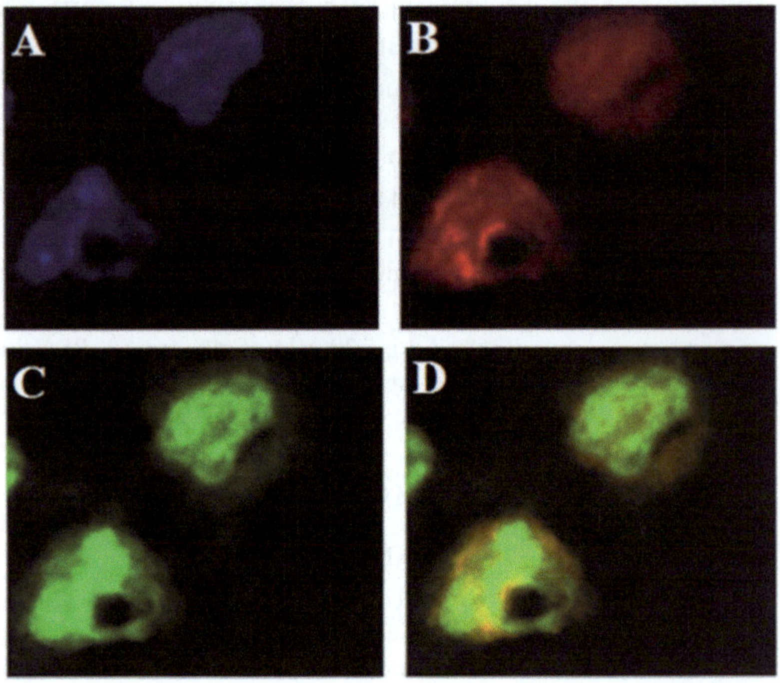

Fig. 3 Phagocytosis of *Neisseria gonorrhoeae* P9-17 by monocyte-derived macrophages. Confocal microscopy of M0 macrophages incubated with a variant of *Neisseria gonorrhoeae* expressing GFP (green) for 4 h. Optical magnification 60×, digital magnification 5×, Kalman 4 sequential mode. (**a**) DAPI; (**b**) Phalloidin; (**c**) *Neisseria gonorrhoeae* GFP; (**d**) Merge

3. Wash twice with 2% (v/v) FBS in PBS (1 mL for microcentrifuge tube or 200 μL for plate well). Centrifuge at 600 × *g* for 3 min at RT. Discard the supernatant.
4. Block the Fc γ receptors with 100 μL of 10% (v/v) of inactivated AB serum in PBS on ice for 10 min. Centrifuge at 600 × *g* for 3 min at RT. Discard supernatant (*see* **Notes 18** and **19**).
5. Leave the cells in 100 μL of 2% (v/v) FBS in PBS.
6. Stain the cells with antibodies (Table 1) for 30 min on ice and in the dark (*see* **Note 20**).
7. Wash the cells twice in 2% (v/v) FBS in PBS (1 mL for Eppendorf tube or 200 μL for plate well).
8. Suspend the cells in 200–300 μL of FACS Buffer. The samples are now ready to be analyzed on a flow cytometer (*see* **Note 21**).

3.6.2 Cytokine Production

Cytokine production can be determined at the protein level using enzyme-linked immunosorbent assay (ELISA) or intracellular cytokine staining using flow cytometry. Assessing cytokine production by ELISA is useful in the absence of an heterogeneous cell population, where is not necessary to determine which cell is producing a particular cytokine by fluorescent labeling.

1. To assess cytokine production by MDM using ELISA, following infection of MDM with gonococci as described in Subheading 3.3, harvest the cell culture medium from individual wells over different time points.
2. Centrifuge the cell culture media samples at 4000 × *g* for 5 min to remove cell debris. Store the supernatants at −80 °C until required. Stimulation for 18–24 h represents the peak of cytokine accumulation in the MDM medium, but it is essential to carry out a time course for different microbial stimuli as the kinetics may vary.
3. Quantify cytokine protein production using commercial ELISA kits according to the respective manufacturer's instructions. Examples of cytokines to study include TNF-α, IL-1β, IL-6, IL-10, IL-12, and IL-23 (*see* **Note 22**).

3.6.3 Real-Time Polymerase Chain Reaction (RT-PCR)

Cytokine production can also be assessed at the RNA transcription level using RT-PCR.

1. Remove cell culture medium and rinse MDM cells thoroughly with PBS.
2. Add 350 μL of RLT buffer in each well to lyse adherent MDM cells. Store the lysates in a −80 °C freezer until further processing.
3. Extract total RNA using a RNeasy mini kit. Store the RNA in a −80 °C freezer until new use.
4. Measure RNA concentration using a NanoDrop spectrophotometer and use 500–1000 ng RNA to synthesize cDNA using a high-capacity RNA to cDNA kit.
5. Measure cDNA concentration using a NanoDrop spectrophotometer. Dilute the cDNA with PCR-grade H_2O in work solution with known concentrations (150 ng/μL). Store the cDNA at −20 °C until PCR reaction.
6. Set up a PCR reaction using 5 μL of Fast SybrGreen Master Mix, 0.25 μL of forward and reverse primers (10 μM stocks, final primer concentration is 500 nM), and 1 μL of diluted cDNA. The final volume is 10 μL per well.
7. Run the RT-PCR reaction using the following cycle program: 95 °C for 10 min, followed by 40 cycles of 95 °C for 15 s, 60 °C for 30 s, and 72 °C for 30 s.
8. Calculate the expression of *cytokines* using the ΔΔCT method. Use average expression of the housekeeping gene (*18s*), to calculate the relative expression of MDM phenotypic markers.

3.6.4 Alloproliferation Assay

Macrophages are Antigen Presenting Cells that can stimulate T cells. Here, we describe a method to determine the proliferation of allogenic CD4+T cells upon stimulation with infected-MDM cells.

1. Infect MDM cells with *N. gonorrhoeae* as described in Subheading 3.3.
2. Harvest MDM cells into 15 mL conical tubes, and then centrifuge at 300 × *g* for 5 min at RT. Suspend the cells in complete RPMI 1640 medium and count with a hemocytometer using trypan blue.
3. Prepare a human allogenic PBMC suspension as described in **steps 1–6** of Subheading 3.1 and then obtain Peripheral Blood Lymphocytes (PBL) from these PBMC, by depleting monocytoid cells from the mononuclear cell preparation using adherence to a plastic tissue culture flask as follows: suspend the PBMC cells in complete RPMI 1640 medium to a final concentration of 1–2 × 10^6 cell/mL. Transfer the cell suspension to a tissue culture flask. An appropriate volume for a T150 tissue culture flask is 50 mL. Incubate the cells for 1 h at 37 °C in a humidified incubator with 5% (v/v) CO_2. Gently pour the supernatant containing nonadherent lymphoid cells into a 50-mL centrifuge tube. Rinse the flasks gently with 10 mL of complete RPMI 1640 medium and add to a centrifuge tube. Count and assess viability using trypan blue exclusion.
4. Label the PBL with CFSE as follows: suspend 1 × 10^7 cells/mL in complete RPMI 1640 medium in the presence of 5 μM CFSE. Incubate the cells for 10 min at 37 °C. Stop the labeling reaction by adding 20 mL of cold RPMI 1640 medium containing 10% (v/v) FCS. Centrifuge the cells at 300 × *g* for 5 min and wash the cells a further two times with RPMI 1640 medium containing 10% (v/v) FCS to remove any excess CFSE (*see* **Note 23**). Count and assess viability using trypan blue exclusion.
5. Co-culture PBL with infected MDM in 96-well U-bottomed plates in a total volume of 200 μL. Usually, 10^4 MDM are used as stimulator cells in co-culture with a range of PBL concentrations, depending on the PBL/MDM ratio. For example, a PBL/MDM ratio of 1:10 requires 10^5 PBL cells (*see* **Note 24**). Include 5 μg/mL of PHA in co-culture as a positive control for T cell proliferation.
6. Incubate the co-cultures for 7 days at 37 °C in a humidified incubator with 5% (v/v) CO_2.
7. To assess the phenotype of proliferating cells, label the cells with anti-CD4 antibodies as described in **steps 2–8** of Subheading 3.6.1.

4 Notes

1. Patient consent is mandatory and must be approved previously by the local Scientific Ethics Committee. Additionally, human samples are considered as biohazards, thus handling and disposal must be done under biohazard safety regulations.
2. It is critical to batch test M-CSF, since this cytokine is essential for inducing the correct differentiation of the MDM. In our experience, research grade cytokine has worked well. It is recommended that samples should be stored frozen in small aliquots. Avoid repeated freeze/thaw cycles.
3. Cell culture plastic flasks and plates with tissue culture (TC)-treated surfaces are suitable to grow and expand cells because their hydrophilic, negatively-charged surface favors cell attachment. We usually use Falcon® or Corning® consumables.
4. The use of *N. gonorrhoeae* variants, that is, pilus (Pil) or/and phase-variable colony opacity (Opa) phenotypes, depends on the purpose of the experiment. For example, it has been shown in monocytes that Opa-positive and piliated gonococci caused a differential oxidative response [13].
5. As an alternative method, *N. gonorrhoeae* can be grown using Thayer-Martin (TM) medium [21].
6. Human AB serum can be obtained from the blood bank. Commercial Human AB serum can also be used. This serum is not heat inactivated, and must therefore be treated at 56 °C for 30 min to inactivate it.
7. Currently there are free and paid software to analyze the results of flow cytometry. In addition, more modern flow cytometers can be used.
8. Ficoll can be noxious to the cells in prolonged exposure; therefore this step must be done quickly.
9. Monocytes differentiated in the absence of polarizing cytokines lead to the M0 phenotype. On Day 6 you also can differentiate monocytes into M1 or M2 phenotypes by adding 100 ng/mL of LPS plus 20 ng/mL of IFN-γ or 1000 U/mL of IL-4 respectively.
10. To facilitate the downstream use of MDM cells, we recommend seeding monocytes in a format that will be used. If you require replate cells after infection or differentiation, you can harvest the cells using cell scrapers or trypsin.
11. Use GC broth without Isovitalex™ Enrichment or Kellogg's supplements for serial dilution and plating of *N. gonorrhoeae* after gentamicin protection assay. Do not use PBS as a buffer for bacterial suspension.

12. Poly-D-Lysine coated coverslips can be prepared using the following recipe: coat coverslips with 1 mg/mL of poly-D-Lysine in cell culture grade H_2O for 2 h in a Class II cabinet. Wash the coverslips gently with H_2O twice and a single wash with absolute ethanol, air dry on a piece of filter paper and keep them in a dry container until use. Expose to UV-light for 5 min before cell seeding [22].
13. PFA-fixed MDM can be stored overnight at 4 °C in 2 mL of PBS in each well of the cell culture plate. The immunofluorescence staining protocol can be continued after.
14. Do not use a cold solution of 0.2% (v/v) Triton X-100/PBS, but prewarm at room temperature before use.
15. Be careful with direct light exposure of dye stock and working solutions, or use opaque microcentrifuge tubes.
16. Avoid any air bubbles in the mounting medium, since they can scatter light and interfere with the fluorescence observation.
17. It is known that scraping will involuntarily lead to a loss of cells (no more than 5–15%, depending on your way of scraping). Trypsin can be used instead to harvest the cells.
18. Commercial Fc-blocking antibody (anti-CD16/32) can also be used to Block Fc γ receptors.
19. It is important that the cells are kept on ice or at 4 °C to stop internalization of antibodies by Fc γ receptors expressed on MDM.
20. It is essential to include isotype controls of the same antibody subclass to make sure that your antibody signal is specific and is not being caused by the nonspecific uptake of the antibody by the cells.
21. Samples must be fresh from the same day. If you cannot acquire on the same day, suspend the pellet in 200 μL of 4% (w/v) PFA and fix for 15 min at RT and store for no more than 5 days until used.
22. We usually use human cytokine ELISA kits Ready-SET-Go! from eBioscience, but other manufacturer's reagents are also suitable for use in this instance.
23. It is important to confirm that the CFSE labeling has worked by checking the cells for CFSE incorporation by flow cytometry.
24. It is essential to optimize the PBL/MDM ratio for each set of experiments.

Acknowledgments

This work was supported by FONDECYT regular 1180666 (J.G., P.I.R., and A.E.). PCI-CONICYT '*Apoyo a la Formación de Redes Internacionales Para Investigadores en Etapa Inicial*' REDI170370 (P.I.R., A.E.).

References

1. Janeway CJ, Travers P, Walport M et al (2001) Immunobiology: the immune system in health and disease, 5th edn. Garland Science, New York
2. Cole J, Aberdein J, Jubrail J et al (2014) The role of macrophages in the innate immune response to *Streptococcus pneumoniae* and *Staphylococcus aureus*: mechanisms and contrasts. Adv Microb Physiol 65:125–202. https://doi.org/10.1016/bs.ampbs.2014.08.004
3. Vladimer GI, Marty-Roix R, Ghosh S et al (2013) Inflammasomes and host defenses against bacterial infections. Curr Opin Microbiol 16(1):23–31. https://doi.org/10.1016/j.mib.2012.11.008
4. van de Laar L, Saelens W, De Prijck S et al (2016) Yolk sac macrophages, fetal liver, and adult monocytes can colonize an empty niche and develop into functional tissue-resident macrophages. Immunity 44(4):755–768. https://doi.org/10.1016/j.immuni.2016.02.017
5. Biswas SK, Mantovani A (2010) Macrophage plasticity and interaction with lymphocyte subsets: cancer as a paradigm. Nat Immunol 11 (10):889–896. https://doi.org/10.1038/ni.1937
6. Sica A, Mantovani A (2012) Macrophage plasticity and polarization: in vivo veritas. J Clin Invest 122(3):787–795. https://doi.org/10.1172/JCI59643
7. Benoit M, Desnues B, Mege JL (2008) Macrophage polarization in bacterial infections. J Immunol 181(6):3733–3739
8. Mege JL, Mehraj V, Capo C (2011) Macrophage polarization and bacterial infections. Curr Opin Infect Dis 24(3):230–234. https://doi.org/10.1097/QCO.0b013e328344b73e
9. Mantovani A, Biswas SK, Galdiero MR et al (2013) Macrophage plasticity and polarization in tissue repair and remodelling. J Pathol 229 (2):176–185. https://doi.org/10.1002/path.4133
10. Chateau A, Seifert HS (2016) *Neisseria gonorrhoeae* survives within and modulates apoptosis and inflammatory cytokine production of human macrophages. Cell Microbiol 18 (4):546–560. https://doi.org/10.1111/cmi.12529
11. Ellis CD, Lindner B, Anjam Khan CM et al (2001) The *Neisseria gonorrhoeae* lpxLII gene encodes for a late-functioning lauroyl acyl transferase, and a null mutation within the gene has a significant effect on the induction of acute inflammatory responses. Mol Microbiol 42(1):167–181
12. Escobar A, Candia E, Reyes-Cerpa S et al (2013) *Neisseria gonorrhoeae* induces a tolerogenic phenotype in macrophages to modulate host immunity. Mediat Inflamm 2013:127017. https://doi.org/10.1155/2013/127017
13. Knepper B, Heuer I, Meyer TF et al (1997) Differential response of human monocytes to *Neisseria gonorrhoeae* variants expressing pili and opacity proteins. Infect Immun 65 (10):4122–4129
14. Mosleh IM, Huber LA, Steinlein P et al (1998) *Neisseria gonorrhoeae* porin modulates phagosome maturation. J Biol Chem 273:35332–35338
15. Ortiz MC, Lefimil C, Rodas PI et al (2015) *Neisseria gonorrhoeae* modulates immunity by polarizing human macrophages to a M2 profile. PLoS One 10(6):e0130713. https://doi.org/10.1371/journal.pone.0130713
16. Zhou X, Gao X, Broglie PM et al (2014) Hexa-acylated lipid A is required for host inflammatory response to *Neisseria gonorrhoeae* in experimental gonorrhea. Infect Immun 82 (1):184–192. https://doi.org/10.1128/iai.00890-13
17. Zughaier SM, Kandler JL, Balthazar JT et al (2015) Phosphoethanolamine modification of *Neisseria gonorrhoeae* lipid A reduces autophagy flux in macrophages. PLoS One 10(12): e0144347. https://doi.org/10.1371/journal.pone.0144347
18. Lambden PR, Heckels JE, James LT et al (1979) Variations in surface protein

composition associated with virulence properties in opacity types of *Neisseria gonorrhoeae*. J Gen Microbiol 114:305–312

19. Dillard JP (2011) Genetic manipulation of *Neisseria gonorrhoeae*. Curr Protoc Microbiol. Chapter 4:Unit4A.2. https://doi.org/10.1002/9780471729259.mc04a02s23
20. Gomez-Duarte OG, Dehio M, Guzman CA et al (1997) Binding of vitronectin to Opa-expressing *Neisseria gonorrhoeae* mediates invasion of HeLa cells. Infect Immun 65 (9):3857–3866
21. Thayer JD, Martin JE Jr (1964) A selective medium for the cultivation of *N. gonorrhoeae* and *N. meningitidis*. Public Health Rep 79:49–57
22. Sawasdichai A, Chen HT, Abdul Hamid N et al (2010) In situ subcellular fractionation of adherent and non-adherent mammalian cells. J Vis Exp (41). https://doi.org/10.3791/1958

Chapter 19

Protocols to Interrogate the Interactions Between *Neisseria gonorrhoeae* and Primary Human Neutrophils

Stephanie A. Ragland and Alison K. Criss

Abstract

Neisseria gonorrhoeae (Gc) infection of its obligate human host results in a robust neutrophil-driven immune response. Despite neutrophils' intrinsic ability to neutralize microbes, Gc can survive in the presence of neutrophils. To interrogate how this pathogen evades killing by neutrophils, we employ an ex vivo model of Gc infection with Interleukin-8-primed and adhered primary human neutrophils. This chapter will describe how primary human neutrophils are purified from venous blood, how Gc is prepared for infection, how to assess Gc survival in the presence of human neutrophils by enumeration of colony forming units, and how to determine Gc internalization by human neutrophils using an immunofluorescence-based approach.

Key words Neutrophils, Polymorphonuclear leukocytes, *Neisseria gonorrhoeae*, Phagocytosis, Infection

1 Introduction

Infection with the obligate-human pathogen *Neisseria gonorrhoeae* (Gc) results in a robust neutrophil-driven inflammatory response [1]. Although neutrophils employ a multitude of mechanisms to neutralize microbes, viable Gc can be cultured from patient exudates [2]. This observation suggests that Gc can withstand killing by neutrophils. Intriguingly, Gc does not possess canonical virulence factors, like the production of toxins; this implicates other bacterial factors in Gc that provide resistance to neutrophil killing. To test how various bacterial and host factors affect Gc–neutrophil interactions, we developed an ex vivo model using adherent primary human neutrophils [3–6].

Peripheral blood neutrophils are in an inactive state that serves to limit collateral damage of host tissues in the absence of infection. As a consequence, resting neutrophils are less responsive to and less effective in killing microbes [7]. During an active infection, extravasating neutrophils receive priming signals via adherence and

Myron Christodoulides (ed.), *Neisseria gonorrhoeae: Methods and Protocols*, Methods in Molecular Biology, vol. 1997,
https://doi.org/10.1007/978-1-4939-9496-0_19,

exposure to infection-related cytokines, chemokines, and bioactive lipids [7, 8]. Priming enhances the ability of neutrophils to respond to a second stimulus, making neutrophils markedly more capable in responding to and killing microbes [7]. In particular, the chemokine interleukin-8 (IL-8) is secreted in humans with gonorrhea [9, 10]. Thus, to mimic the physiological state of tissue-migrated neutrophils in vivo, we treat adherent human neutrophils with IL-8 prior to exposure to Gc.

Herein, we describe how to purify primary human neutrophils from venous blood, how to grow Gc for infection with human neutrophils, how to assess bacterial survival in the presence of human neutrophils by enumerating colony forming units, and, finally, how to assess Gc internalization by human neutrophils using an immunofluorescence-based approach.

2 Materials

2.1 Purification of Human Neutrophils

1. Fresh, human venous blood collected with an anticoagulant (*see* **Note 1**).
2. Sterile 50 mL polypropylene conical tubes with screwcaps.
3. Sterile H_2O for irrigation, *USP* grade (*see* **Note 2**). Store at room temperature.
4. 3% (w/v) dextran and 0.9% (w/v) sodium chloride in sterile H_2O for irrigation. Add 9 g of NaCl (*see* **Note 3**) and 30 g of dextran (from *Leuconostoc* spp., M_r 450,000–650,000) to 800 mL of H_2O (sterile, irrigation) in a sterile, nonpyrogenic container with lid. Heat in a 37 °C water bath with occasional, vigorous shaking until all reagents have dissolved. Bring up to 1 L with sterile H_2O for irrigation and sterilize through a 0.2 μm filter. Store at room temperature.
5. Tabletop centrifuge with refrigeration, adjustable acceleration and brake speeds, and swinging bucket rotor and adaptors to centrifuge 50 mL conical tubes.
6. 0.1% (w/v) dextrose in Dulbecco's phosphate-buffered saline (DPBS, minus calcium chloride/minus magnesium chloride) (*see* **Notes 3** and **4**). Add 0.5 g of dextrose (i.e., D-glucose) to 500 mL of sterile DPBS in its original container. Shake to mix. Store at 4 °C. Bring to room temperature just before use.
7. Sterile large-bore polypropylene transfer pipettes.
8. Ficoll-Paque™ PLUS. Store at 4 °C once opened. Bring to room temperature just before use.
9. 1.7% (w/v) NaCl in sterile H_2O for irrigation. Add 17 g of NaCl to 1 L of sterile H_2O for irrigation in its original container (*see* **Note 3**). Shake to mix. Store at room temperature.

10. Kimwipes®.
11. 2% (v/v) glacial acetic acid in H_2O. Store at room temperature.
12. Hemocytometer.
13. Waste container.

2.2 Growth of N. gonorrhoeae to Mid-Logarithmic Phase

1. *Neisseria gonorrhoeae* strains/isolates. Bacteria are stored in GCBL containing 20% glycerol in cryogenic vials at −80 °C.
2. Platinum-iridium inoculating loop. A nichrome loop is an alternative but does not cool down as rapidly. Alternatively, 10 μL sterile disposable plastic inoculating loops can be used.
3. Bunsen burner for sterilizing metal inoculating loop. This is unnecessary if disposable inoculating loops are used.
4. Gonococcal base medium (GCB): 42% (w/w) Proteose Peptone #3 (15.10 g), 3% (w/w) cornstarch (1.01 g), 11% (w/w) K_2HPO_4 (4.03 g), 3% (w/w) KH_2PO_4 (1.01 g), 14% (w/w) NaCl (5.03 g), 28% (w/w) agar (10.07 g) [11] (*see* **Note 5**). Mix dry and store at room temperature. GCB can also be purchased from Difco.
5. Kellogg's Supplement 1 [11, 12] (100×): 40% (w/v) glucose, 1% (w/v) glutamine, 0.002% (w/v) thiamine pyrophosphate. Add 40 g of glucose, 1 g of glutamine, and 2 mg of thiamine pyrophosphate to a total of 100 mL of H_2O. Sterilize through a 0.2 μm filter, dispense into aliquots, and store at −20 °C (*see* **Note 5**).
6. Kellogg's Supplement 2 [11, 12] (1000×): 0.05% (w/v) $Fe(NO_3)_3{\cdot}9H_2O$ in H_2O. Add 50 mg of $Fe(NO_3)_3{\cdot}9H_2O$ to a total of 100 mL of H_2O. Limit solution exposure to light. Sterilize through a 0.2 μm filter, dispense into aliquots, and store at −20 °C (*see* **Note 5**).
7. Modified GCB agar plates: 3.625% (w/v) GCB and 0.125% (w/v) agar. Dissolve 36.25 g of GCB and 1.25 g of agar in 1 L of H_2O and autoclave at 121 °C and 3.39 kg/cm^2 for 30 min with a stir bar added. Cool the autoclaved medium to 56 °C and add Kellogg's Supplements 1 and 2 to a final concentration of 1× each (*see* **Note 5**). Mix on a stir plate, taking care to not introduce bubbles, and pour medium into sterile petri dishes. Store solidified agar plates at 4 °C.
8. 100 mm × 15 mm sterile Petri dishes.
9. Bottom-lit binocular microscope with a "type B" Bausch and Lomb type base, and Nicholas illuminator as light source.
10. Dacron polyester swab (*see* **Note 6**).
11. Gonococcal base liquid medium (GCBL): 1.5% (w/v) Proteose Peptone #3, 0.4% (w/v) K_2HPO_4, 0.1% (w/v) KH_2PO_4, and 0.1% (w/v) NaCl. Dissolve 15 g of Proteose Peptone #3,

4 g of K_2HPO_4, 1 g of KH_2PO_4, and 1 g of NaCl in a total of 1 L of H_2O. Adjust pH to 7.2 with HCl. For storage medium, dissolve the components above in 800 mL of H_2O, then add 200 mL of glycerol. Autoclave at 121 °C and 3.39 kg/cm^2 for 30 min and store at room temperature (*see* **Note 5**).

12. Modified Kellogg's Supplement 1, 100×: 4.2% (w/v) sodium bicarbonate in 100× Kellogg's Supplement 1 (*see* **Note 5**). Add 0.42 g of $NaHCO_3$ to 10 mL of Kellogg's Supplement 1. Vortex to mix and sterilize through a 0.2 μm filter and store at room temperature. Use within 1 week.
13. Modified GCBL: Add modified Kellogg's Supplement 1 and Kellogg's Supplement 2 to a final concentration of 1× each in GCBL [11, 12] (*see* **Note 7**). Store at room temperature.
14. Humidified incubator with 5% (v/v) CO_2 set at 37 °C.
15. 30 °C incubator with a cell culture roller drum (e.g., Fisherbrand 17 mm tissue culture rotator, or equivalent) (no CO_2; *see* **Note 5**).
16. 37 °C incubator with a cell culture roller drum (e.g., Fisherbrand 17 mm tissue culture rotator, or equivalent) (no CO_2; *see* **Note 5**).
17. Microcentrifuge.
18. RPMI-10: Roswell Park Memorial Institute (RPMI) 1640 medium, no glutamine, with 10% (v/v) heat-inactivated fetal bovine serum (*see* **Note 8**). Store at 4 °C. Bring an aliquot to room temperature before use.
19. Visible light spectrophotometer for 1 cm^2 cuvettes.
20. Disposable 1 cm^2 cuvettes.
21. Autoclave for sterilizing media (e.g., Amsco 3021 gravity autoclave, or equivalent).

2.3 Measuring N. gonorrhoeae Survival

1. Highly purified neutrophils.
2. 0.1% (w/v) bovine serum albumin (BSA; low endotoxin) in DPBS (minus calcium chloride/minus magnesium chloride) (*see* **Note 4**). Sterilize through a 0.2 μm filter and store at room temperature.
3. 50 μM of recombinant Human Interleukin-8 (IL-8, CXCL8) protein, carrier free (*see* **Note 9**) in sterile DPBS containing 0.1% (w/v) BSA. Alternatively, reconstitute IL-8 according to the manufacturer's recommendations. Aliquot and store at −20 °C.
4. RPMI-10, room temperature.
5. 10 nM of IL-8 in RPMI-10. Make fresh from a stock solution of 50 μM recombinant Human IL-8 in RPMI-10.

6. 13 mm diameter, sterile, endotoxin-free, plastic tissue culture-treated coverslips.
7. 24-well polystyrene plates: sterile, flat bottom, tissue culture-treated. Add a 13 mm coverslip to appropriate wells, except for Gc alone wells (*see* **Note 10**).
8. Humidified incubator with 5% (v/v) CO_2 set at 37 °C.
9. Dried, modified GCB agar plates: Gonococcal base medium (GCB) modified with Kellogg's Supplements 1 and 2. Remove lids from GCB plates and allow GCB to dry in biosafety cabinet for 30–60 min (*see* **Note 11**). Store at room temperature. Use dried plates within 24 h.
10. Gc in mid-logarithmic phase growth.
11. Gonococcal base liquid medium (GCBL).
12. 96-well plates: round bottom, polystyrene. Fill each plate with 180 μL of GCBL (*see* **Note 12**).
13. Ice packs (from shipping containers for refrigerated items), stored at 4 °C (*see* **Note 13**).
14. Tabletop centrifuge with refrigeration, adjustable acceleration and brake speeds, and swinging bucket rotor and adaptors to centrifuge multi-well plates.
15. Phosphate buffered saline (PBS), 1×: 137 mM NaCl, 2.7 mM KCl, 10 mM Na_2HPO_4 (dibasic, anhydrous), 2 mM KH_2PO_4 (monobasic, anhydrous) in H_2O. Adjust pH to 7.4 with HCl.
16. 1% (w/v) saponin in PBS. Make a 10% (w/v) saponin solution in H_2O and store at 4 °C. Make fresh 1% (w/v) saponin in PBS from the stock on the day of the experiment and store at room temperature.
17. Waste container.

2.4 Immunofluorescence

2.4.1 Gc Infection of Neutrophils

1. 12 mm acid-washed, glass coverslips (*see* **Note 14**). To acid-wash the coverslips, submerge them in 1 N HCl and incubate at 56 °C for 4 h. Rinse coverslips with several changes of double-distilled (dd)H_2O by swirling the coverslips a few times in the liquid before pouring it off. Rinse coverslips with several changes of 70% (v/v) ethanol as done with ddH_2O. Rinse once with 95% (v/v) ethanol, pour off the liquid, aspirate the excess liquid, and dry coverslips in tissue culture hood overnight.
2. 24-well plates: sterile, polystyrene, flat-bottom, tissue culture-treated. Add a glass coverslip to each well needed (*see* **Note 15**). Use one plate per time point.
3. 0.9009 mM of calcium chloride and 0.4926 mM of magnesium chloride in DPBS, available from numerous commercial sources (also *see* **Note 4**).

4. 50% (v/v) normal human serum (*see* **Note 16**) solution in DPBS, with 0.9009 mM of calcium chloride and 0.4926 mM of magnesium chloride. Store at −20 °C.
5. Highly purified neutrophils.
6. RPMI-10, room temperature.
7. 10 nM of IL-8 in RPMI-10.
8. Humidified incubator with 5% (v/v) CO_2, set at 37 °C.
9. Gc in mid-logarithmic phase growth.
10. Gonococcal base liquid medium (GCBL).
11. 96-well plates: round bottom, polystyrene. Fill each plate with 180 μL of GCBL (*see* **Note 12**).
12. Dried, modified GCB agar plates: Gonococcal base medium (GCB) modified with Kellogg's Supplements 1 and 2. Open GCB plates and allow GCB to dry in biosafety cabinet for 30–60 min (*see* **Note 11**). Store at room temperature. Use dried plates within 24 h.
13. Tabletop centrifuge with refrigeration, adjustable acceleration and brake speeds, and swinging bucket rotor and adaptors to centrifuge 24-well plates.
14. Vacuum flask outfitted for aspiration.

2.4.2 Staining Extracellular Gc Only

1. Vacuum flask outfitted for aspiration.
2. 5 mM of magnesium sulfate ($MgSO_4$) in DPBS (minus calcium chloride/minus magnesium chloride) (*see* **Note 4**).
3. Phosphate buffered saline (PBS), 1×.
4. 4% (v/v) paraformaldehyde in PBS. Store at −20 °C.
5. 10% (v/v) normal goat serum (*see* **Note 17**) in PBS. Store at −20 °C.
6. Microforceps.
7. Humidified chamber (*see* **Note 18**).
8. Primary antibody against Gc surface antigen (*see* **Note 19**).
9. Secondary antibodies that recognize the anti-Gc primary antibody. Each secondary antibody should be coupled to its own fluorophore (*see* **Note 20**).

2.4.3 Staining Total Gc, Including Intracellular Gc

1. Vacuum flask outfitted for aspiration.
2. Phosphate buffered saline (PBS), 1×.
3. 10% (v/v) normal goat serum with 0.2% (w/v) saponin in PBS. Store at −20 °C.
4. Microforceps.
5. Humidified chamber (*see* **Note 18**).

6. Primary antibody against Gc surface antigen (*see* **Note 19**).
7. Secondary antibodies that recognize the anti-Gc primary antibody. Each secondary antibody should be coupled to its own fluorophore (*see* **Note 20**).

2.4.4 Mounting and Imaging

1. Microforceps.
2. Kimwipes®.
3. Mounting Medium (*see* **Note 21**).
4. 25 mm × 75 mm glass microscope slides.
5. Clear nail polish.
6. Holder for horizontal storage of microscope slides.
7. Fluorescence microscope, conventional or confocal laser scanning, with dedicated camera and image capture software.

3 Methods

3.1 Purification of Human Neutrophils

The following method for purifying human neutrophils from venous blood takes approximately 2 h to complete (*see* Fig. 1). Venous blood should be collected in accordance with institutional Human Subjects in Research guidelines. The yield for neutrophils using this preparation should be at least 1×10^6 cells per 1 mL of venous blood and will vary depending on the donor. The purity of neutrophils is high, with greater than 95% granulocyte content [13]. All reagents and procedures should be brought to room temperature prior to use, unless otherwise indicated. All procedures are done on the benchtop with sterile technique (*see* **Note 22**).

1. Add fresh, heparinized venous blood into a sterile 50 mL conical tube, by gently pouring the blood down the inside of the tube. Add 10–20 mL of blood per tube (*see* **Note 23**).
2. Gently pour an equivalent volume of 3% (w/v) dextran and 0.9% (w/v) NaCl solution down the inside of the 50 mL conical tube containing the heparinized blood. Mix by gently rotating end-over-end for 2–3 times. Loosen the screwcap and allow tubes to stand upright for at least 20 min (*see* **Note 24**).
3. Transfer the top, light red layer by gentle pipetting to a fresh, sterile 50 mL conical tube (*see* **Note 25**).
4. Centrifuge at 200 × *g* for 10 min, with slow acceleration and slow brake (*see* **Note 26**).
5. Pour off the supernatant into a waste container, taking care not to disrupt the loose red pellet, which contains the neutrophils and other cells (*see* **Note 27**).

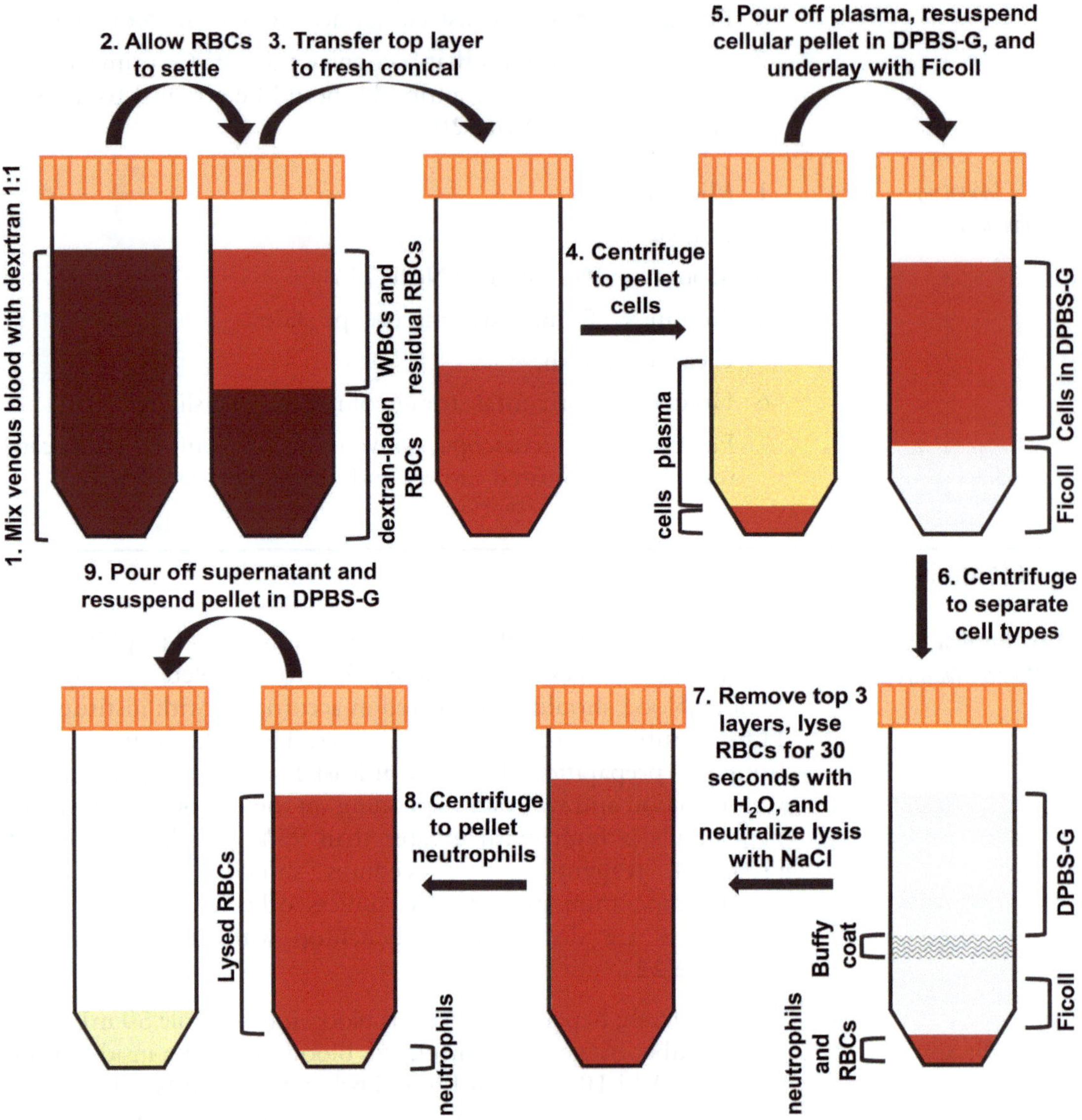

Fig. 1 Purification of neutrophils from human venous blood. Heparinized venous blood is first mixed with dextran to sediment the majority of the red blood cells (RBCs). The mostly RBC-free top layer is then transferred to a fresh conical tube and centrifuged. The plasma supernatant is removed and the pellet suspended in 0.1% (w/v) dextrose in DPBS. The suspension is then underlayed with Ficoll-Paque™ PLUS, followed by centrifugation. The 0.1% (w/v) dextrose in DPBS, mononuclear cell layer ("Buffy coat"), and Ficoll layers are carefully removed. The remaining RBCs are then lysed using a hypotonic solution. Lysis is arrested using a NaCl solution, followed by centrifugation. The neutrophil-rich pellet is then suspended in 0.1% (w/v) dextrose in DPBS and placed on ice until use

6. Suspend the pellet in 5 mL of 0.1% (w/v) dextrose-DPBS using a sterile transfer pipette (*see* **Note 28**). Gently pour 0.1% (w/v) dextrose-DPBS solution down the side of the conical tube to achieve a final volume of 35 mL. The suspension will be a light red color.

7. Underlay the suspension with 10 mL of Ficoll-Paque™ PLUS (*see* **Note 29**). To this end, fill a serological pipette with 10 mL of Ficoll. Place the tip of the pipette against the bottom of the conical tube. Slowly dispense the Ficoll, swirling the tip of the pipette against the angled sides of the conical tube to suspend any cellular material. Take care not to release any air bubbles that might disrupt the gradient.
8. Centrifuge at 200 × *g* for 30 min, with slow acceleration and slow brake (*see* **Note 30**).
9. Remove the supernatant from the loose, red pellet that contains the neutrophils as follows (*see* **Note 30** and Fig. 1): first, use a serological pipette to remove the top 0.1% (w/v) dextrose–DPBS layer and the interface with the Ficoll layer that contains the mononuclear cells. It is important to remove this entire interface so as to not contaminate the neutrophil preparation with monocytes. Then, carefully remove the Ficoll layer without disturbing the loose red pellet. These solutions should be dispensed into the waste container used in **step 5**.
10. Lyse the remaining red blood cells within the loose red pellet using a hypotonic solution. To this end, add 20 mL of sterile, irrigation H_2O to the pellet, and pipet up and down for 30 s (*see* **Note 31**). Return the solution to isotonicity by adding immediately 20 mL of 1.7% (w/v) NaCl solution (*see* **Note 32**).
11. Centrifuge at 130 × *g* for 10 min, with slow acceleration and slow brake.
12. If lysis is complete, the pellet should be off-white in color while the supernatant is light red (*see* **Notes 33** and **34**). Pour off the supernatant into a waste container. While the conical tube is still inverted, use a Kimwipes® to absorb any drops of residual liquid from the opening of the tube.
13. Suspend the neutrophil pellet with 0.1% (w/v) dextrose-DPBS solution using a transfer pipette and store on ice (*see* **Notes 35–37**).
14. Remove 10 μL of the neutrophil suspension into 190 μL of 2% (v/v) acetic acid (*see* **Note 38**). Vortex the neutrophil–acetic acid mixture and load 10 μL into the counting chamber of a hemocytometer. Determine the cell concentration according the manufacturer's instructions.

3.2 Growth of N. gonorrhoeae *to Mid-Logarithmic Phase*

Gc undergoes autolysis when it reaches stationary phase [14, 15]. For assays examining Gc–neutrophil interactions, bacteria are grown in rich liquid medium with several successive back dilutions to enrich for a population of Gc that is predominantly viable and in mid-logarithmic phase [13] (*see* Fig. 2). Gc must be handled under biosafety level 2 conditions with appropriate personal

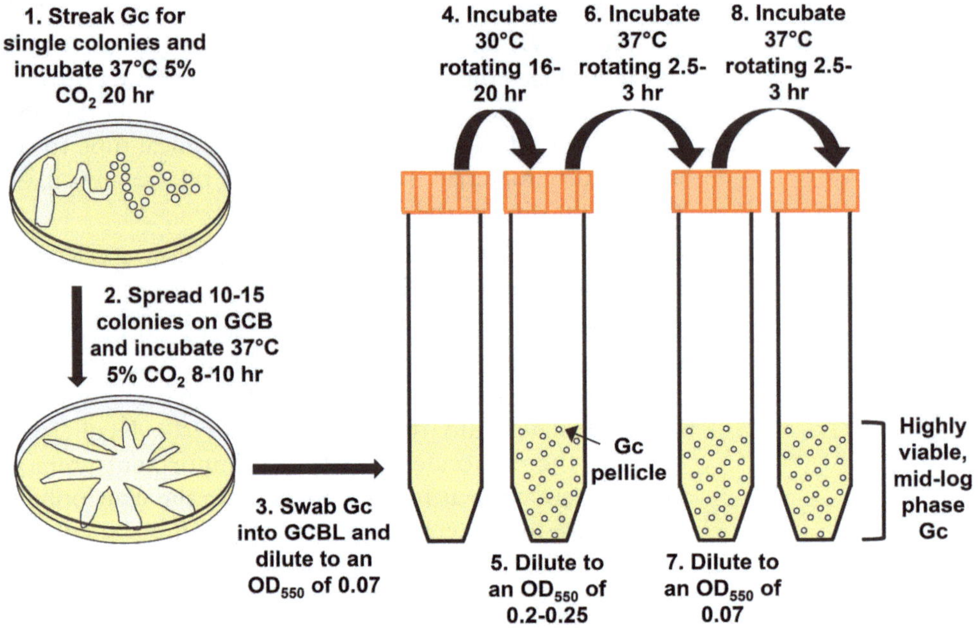

Fig. 2 Gc growth to mid-logarithmic phase. Gc are first streaked from a glycerol stock onto a GCB plate with Kellogg's supplements and incubated for 20 h at 37 °C with 5% (v/v) CO_2. Next, 10–15 colonies of desired morphology are spread onto a fresh GCB plate with Kellogg's supplements and incubated for another 8–10 h as before. Gc are subsequently swabbed into GCBL medium with Kellogg's supplements and sodium bicarbonate and diluted to an OD $\lambda_{550\ nm}$ of 0.07. Liquid cultures are then incubated for 16–20 h at 30 °C, rotating in a roller drum. Cultures are then diluted to an OD $\lambda_{550\ nm}$ of 0.2–0.25 and incubated for 2.5–3 h at 37 °C rotating in a roller drum. Finally, the cultures are diluted to an OD $\lambda_{550\ nm}$ of 0.07 and incubated for 2.5–3 h at 37 °C rotating in a roller drum. Gc are now enriched for high viability and are also in mid-logarithmic phase of growth

protective equipment, including goggles when handling liquid cultures. All steps are performed using sterile technique.

1. Remove sufficient material from a frozen (−80 °C) glycerol stock of Gc using a sterile pipette tip or a sterile inoculating loop and streak onto a modified GCB agar plate (*see* **Note 39**). Incubate for 20 h at 37 °C with 5% (v/v) CO_2.
2. Place the agar plate(s) under a bottom-lit dissecting microscope and use a sterile inoculating loop to select 10–15 colonies of desired piliation and opacity protein expression state (*see* **Note 40**). Spread colonies across the entire surface of a fresh, warm modified GCB agar plate (*see* **Note 41**). Incubate for 8–12 h at 37 °C with 5% (v/v) CO_2.
3. Use a Dacron polyester swab (*see* **Note 6**) to swab the lawn of bacteria into a 15 mL conical tube containing 6 mL of modified GCBL medium. Vortex (*see* **Note 42**) and remove 1 mL to measure the Optical Density (OD) at $\lambda_{550\ nm}$ with a spectrophotometer. Dilute the culture with modified GCBL to an OD

at $\lambda_{550\ nm}$ of 0.07 in a total volume of 5 mL. Add tubes to the center of a cell culture roller drum and rotate on low speed (setting 2 on a Fisherbrand 17 mm tissue culture rotator) at 30 °C for 12–16 h (*see* **Note 5**).

4. Dilute cultures to an OD at $\lambda_{550\ nm}$ of 0.2–0.25 in a total of 5 mL with modified GCBL medium (*see* **Note 43**). Add tubes to the edge of a cell culture roller drum and rotate on medium speed (setting 5 on a Fisherbrand 17 mm tissue culture rotator) at 37 °C for 2.5–3 h (*see* **Note 5**).
5. Dilute cultures to an OD at $\lambda_{550\ nm}$ of 0.07 in a total of 5 mL with modified GCBL medium. Add cultures to the edge of a cell culture roller drum and rotate on medium speed at 37 °C for 2.5–3 h (*see* **Note 5**).
6. Measure the OD of cultures at $\lambda_{550\ nm}$ and calculate the Colony Forming Units (CFU)/mL based on an empirically calculated ratio (*see* **Note 44**).
7. Remove the volume of culture that is equivalent to the desired number of bacteria (generally 1×10^8 CFU for the assays below) and pellet in a microfuge at 10,000 × *g* for 3 min. Remove the supernatant and suspend the pellet in room temperature RPMI-10 medium, or other applicable medium, to the appropriate concentration for addition to human neutrophils. Alternatively, remove the supernatant and suspend the pellet in a desired opsonin (e.g., antibody or serum).

3.3 Measuring N. gonorrhoeae Survival

The following protocol describes how to measure Gc survival in the presence of adherent, IL-8-primed human neutrophils (Fig. 3). This assay involves a synchronous infection with multiple time points. At each time point, nonassociated Gc are removed, neutrophils are lysed, total Gc (both adherent and intracellular) are plated, and viable CFU are enumerated. This assay is conducive for multiple conditions and time points. However, this assay is less sensitive than other approaches and cannot distinguish viability of intracellular from extracellular bacteria (*see* **Note 45**).

1. Add purified human neutrophils produced as described in Subheading 3.1 to 10 of nM IL-8 in RPMI-10 medium, where the final concentration of neutrophils is ~2.5×10^6 neutrophils/mL. Gently invert the conical tube to mix.
2. Place one 13 mm coverslip in each well of a 24-well plate (*see* **Note 46**). Add 400 μL of neutrophil–IL-8 mixture to each well (~1×10^6 neutrophils/well). Prepare 3–5 wells per strain and/or condition as technical replicates for each time point (*see* **Notes 10** and **47**). Incubate for 30–60 min at 37 °C with 5% (v/v) CO_2. Neutrophils should be loosely attached to the coverslip when viewed by light microscopy.

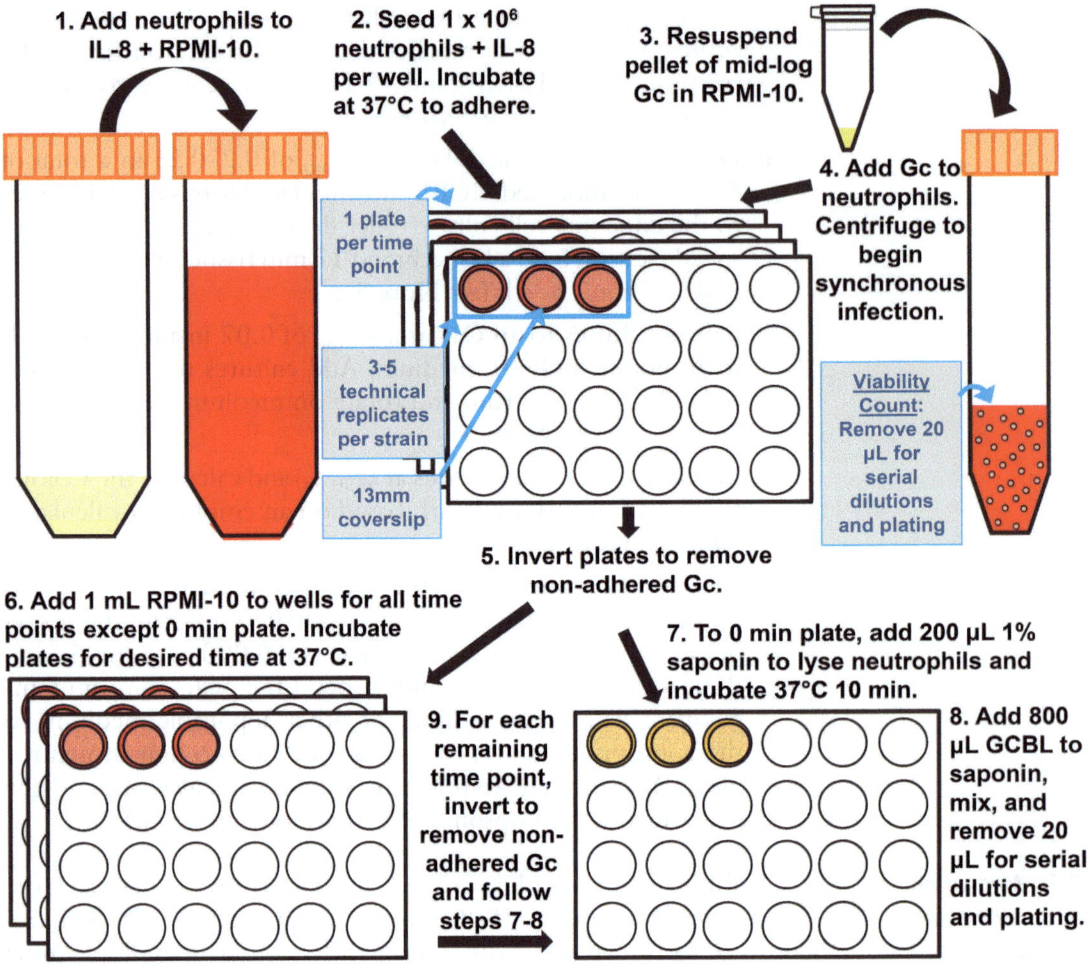

Fig. 3 Assessing Gc survival in the presence of primary human neutrophils by enumerating colony forming units. Neutrophils are treated with IL-8 and allowed to adhere to plastic coverslips in a 24-well plate. Sufficient IL-8 treated, adherent neutrophils should be prepared for one plate per time point and at least three technical replicates per condition or strain tested. At the desired MOI, highly viable, mid-logarithmic phase Gc are centrifuged onto the neutrophils to begin the synchronous infection. After nonadherent Gc are removed, all plates receive RPMI with 10% (v/v) fetal bovine serum and are incubated at 37 °C with 5% (v/v) CO_2, with exception of the 0 min plate. For the 0 min plate, add 1% (w/v) saponin in PBS to lyse the neutrophils, mix, and plate viable CFU. Continue lysis and plating procedure for each remaining time point

3. During the incubation in **step 2**, centrifuge ~1×10^8 CFU of mid-logarithmic phase Gc, grown as described in Subheading 3.2 at $10{,}000 \times g$ for 3 min. Remove the supernatant and suspend Gc in 10 mL of RPMI-10 ("Gc-RPMI-10") for a concentration of ~1×10^7 CFU/mL (*see* **Note 48**).
4. Chill the neutrophils on ice packs for 5 min (*see* **Note 49**).
5. While the neutrophils chill, vortex the "Gc-RPMI-10" suspension and transfer 20 µL to a 96-well plate, each well containing

180 μL of GCBL medium. Serially dilute tenfold by mixing and transferring 20 μL to successive wells to a final dilution of 10^{-6}. Spot 10 μL of the 10^{-1}–10^{-6} dilutions in triplicate on dried, modified GCB agar plates (*see* **Note 50**). This will determine the actual Multiplicity Of Infection (MOI) (*see* **Note 51**).

6. For a “Gc alone” control, add 100 μL of vortex-mixed “Gc-RPMI-10” to each of two replicate wells containing 900 μL of RPMI-10 medium in a separate 24-well plate. Mix well, using a single-channel pipette set on 1000 μL. Inoculate 20 μL of Gc into a 96-well plate containing 180 μL of GCBL medium for serial dilution and plating as described in **step 5**. Set “Gc alone” plate aside until **step 9** (*see* **Note 52**).
7. Add 100 μL of vortex-mixed Gc to each well of chilled neutrophils, and then centrifuge Gc onto the neutrophils at 600 × *g* for 4 min at 12 °C using a tabletop refrigerated centrifuge with microtiter plate carriers (with high acceleration and high break).
8. Gently, but quickly, pour the supernatant from each plate into a waste container to remove nonadhered Gc (*see* **Note 53**). Place the plates back on ice.
9. For all plates except the 0 min plate, pipet 1 mL of RPMI-10 medium gently down the side of each well of Gc + neutrophils. Transfer these plates, as well as “Gc alone” plate, to the incubator at 37 °C with 5% (v/v) CO_2. The time when the plates are placed in the incubator is the official start of the experiment.
10. For the 0 min plate, add 200 μL of 1% (w/v) saponin in PBS solution to each well of Gc + neutrophils to lyse the neutrophils (*see* **Note 54**). Incubate for 10 min at 37 °C with 5% (v/v) CO_2.
11. Add 800 μL of GCBL medium to each well of the 0 min plate containing 200 μL of 1% (w/v) saponin in PBS solution (*see* **Note 55**). Mix the well contents several times with a single-channel pipette set on 1000 μL, and remove 20 μL aliquots for serial dilutions (*see* **Note 56**). Spot 10 μL of the 10^{-1}–10^{-4} dilutions in triplicate on dried, modified GCB agar plates (*see* **Note 50**).
12. At each time point, remove the corresponding Gc + neutrophil plate from the incubator. Gently, but quickly, pour off the supernatant to remove nonadhered Gc as described in **step 8**. Repeat **steps 10** and **11**. Also at each time point, remove the “Gc alone” plate from the incubator, mix the contents well, remove 20 μL aliquots for serial dilutions, and return the plate to the incubator until the next time point.

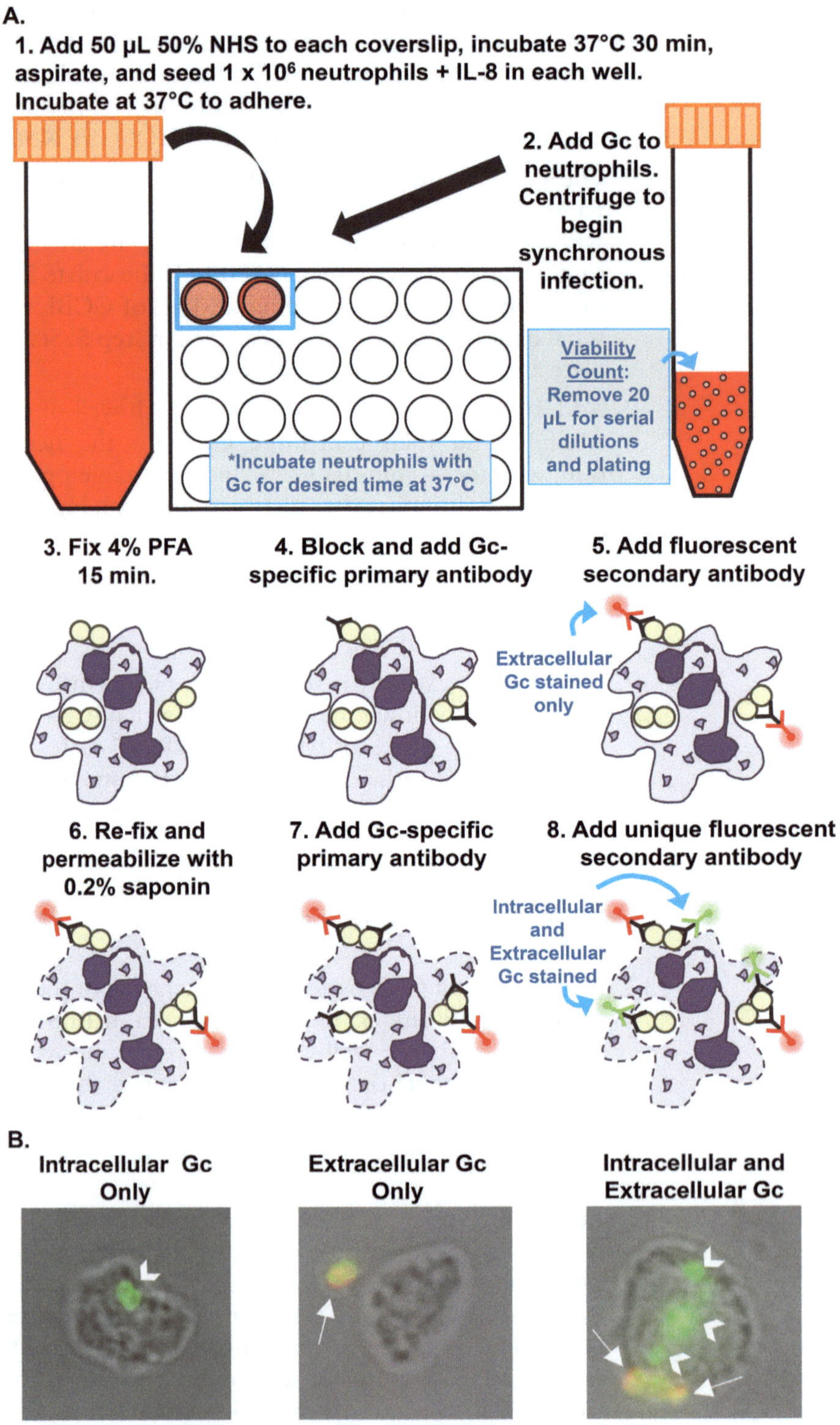

Fig. 4 Immunofluorescence-based approach for the discrimination of intracellular from extracellular Gc within primary human neutrophils. (**a**) Workflow for experimental setup and immunofluorescence staining. IL-8-

13. Incubate 10 µL spotted dilutions of Gc on GCB agar plates in a 37 °C incubator with 5% (v/v) CO_2 for 20 h. Enumerate CFU and calculate CFU/mL (*see* **Note 57**).

3.4 Immunofluorescence

The following protocol is an immunofluorescence-based approach for assessing Gc binding and internalization by human neutrophils (*see* Fig. 4). This approach is simple and relatively inexpensive, depending on the availability and cost of using a fluorescence microscope with image capturing software (*see* **Note 58**). The assay involves discriminating bound from internalized Gc by their differential accessibility to an anti-Gc antibody before and after permeabilizing neutrophils with saponin. As one alternative approach, Gc can be labeled with carboxyfluorescein succinimidyl ester (CFSE) (*see* **Note 59**), prior to neutrophil exposure; extracellular bacteria are then recognized with a Gc-specific antibody without permeabilizing neutrophils. GFP-expressing Gc have also been developed [16].

3.4.1 Gc Infection of Neutrophils

1. Add acid-washed glass coverslips to each well of a 24-well plate (*see* **Note 15**).
2. Add 50 µL of 50% (v/v) normal human serum in PBS to each coverslip (*see* **Note 60**). Incubate at 37 °C with 5% (v/v) CO_2 for 30–60 min. Aspirate the serum using a sterile pipette tip connected to a vacuum flask.
3. Add purified human neutrophils, prepared as described in Subheading 3.1, to 10 nM of IL-8 in RPMI-10 medium, where the final concentration of neutrophils is ~2.5×10^6 neutrophils/mL. Gently invert the conical tube to mix. Add 400 µL (~1×10^6 cells) of the cell suspension to each serum-coated glass coverslip (*see* **Note 46**). Incubate for 30–60 min at 37 °C with 5% (v/v) CO_2 (*see* **Note 61**). Neutrophils should appear attached to the coverslip when viewed by light microscopy.
4. While neutrophils adhere, centrifuge ~1×10^8 CFU of mid-logarithmic phase Gc, grown as described in Subheading 3.2 at 10,000 × *g* for 3 min. Remove the supernatant and suspend

Fig. 4 (continued) primed human neutrophils are adhered to normal human serum (NHS)-coated glass coverslips. Gc are centrifuged onto neutrophils to begin a synchronous infection. At the desired length of infection, fixed, infected cells are stained with a primary antibody against a Gc-specific surface antigen and subsequently stained with a fluorescent secondary antibody to label extracellular Gc. Infected cells are then fixed again, permeabilized, and stained with the same primary antibody and a new fluorescent secondary antibody to label total Gc. (**b**) Representative images of Gc associated with human neutrophils, as prepared in (**a**). Overlays are shown of phase contrast and fluorescent images. Extracellular Gc were labeled red while total Gc were labeled green. Arrows indicate extracellular Gc (green and red), and arrowheads indicate intracellular Gc (green only). Images were taken using a Nikon E800 with Hamamatsu Orca-ER digital camera using Openlab software and were processed in Adobe Photoshop CS3

Gc in 10 mL of RPMI-10 ("Gc-RPMI-10") (*see* **Note 48**). Alternatively, use CFSE-labeled Gc (*see* **Note 59**).

5. Chill the neutrophils on ice packs for 5 min (*see* **Note 49**).
6. While neutrophils chill, vortex the "Gc-RPMI-10" suspension and transfer 20 μL of vortex-mixed Gc to a 96-well plate, each well containing 180 μL of GCBL medium. Serially dilute tenfold by mixing and transferring 20 μL aliquots to successive wells to a final dilution of 10^{-6}. Spot 10 μL of the 10^{-1}–10^{-6} dilutions in triplicate on dried, modified GCB agar plates (*see* **Note 50**). This will determine the actual MOI (*see* **Note 51**). Incubate Gc plates at 37 °C with 5% (v/v) CO_2 for 20 h. Enumerate CFU and determine MOI (*see* **Note 51**).
7. Add 100 μL of vortex-mixed Gc to each well of chilled neutrophils, and then centrifuge Gc onto the neutrophils at 600 × *g* for 4 min at 12 °C using a tabletop refrigerated centrifuge with microtiter plate carriers (high acceleration and high break).
8. Without touching the coverslip, gently aspirate the media using a pipette tip connected to a vacuum flask.
9. Gently pipet 1 mL of RPMI-10 medium down the inside of each well. Incubate at 37 °C with 5% (v/v) CO_2 for the desired time (*see* **Note 62**).

3.4.2 Staining Extracellular Gc Only (See ***Note 63****)*

1. Gently aspirate media using a sterile pipette tip connected to a vacuum flask. Gently rinse the coverslips in the original 24-well plate twice in PBS containing 5 mM $MgSO_4$, removing the wash solution each time by aspiration using a sterile pipette tip connected to a vacuum flask. Add 250 μL of 4% (v/v) PFA in PBS to each well (*see* **Note 64**). Incubate at room temperature for 15 min.
2. Aspirate the 4% (v/v) PFA in PBS solution with a sterile pipette tip connected to a vacuum flask. Rinse three times with 500 μL of PBS per well. Add 250 μL of 10% (v/v) normal goat serum in PBS to each well to block nonspecific antibody binding. Incubate at room temperature for 10 min. Aspirate the blocking buffer using a sterile pipette tip connected to a vacuum flask.
3. Transfer coverslips using microforceps to a humidified chamber, neutrophil side up (*see* **Note 18**).
4. Add 50 μL of anti-Gc antibody (*see* **Note 19**), diluted in 10% (v/v) normal goat serum in PBS. Incubate at room temperature for 1 h.
5. Transfer coverslips using microforceps into the original 24-well plate (or a fresh 24-well plate), neutrophil side up, and wash three times with 500 μL of PBS for 5 min each.

6. Keep the coverslips in the 24-well plate and add 250 μL of fluorescent secondary antibody diluted in 10% (v/v) normal goat serum in PBS to each well (*see* **Note 65**). Incubate at room temperature, in the dark (*see* **Note 66**) for 1 h.
7. Aspirate the secondary antibody suspension with a sterile pipette tip connected to a vacuum flask. Wash the coverslips three times with 500 μL of PBS for 5 min each.
8. Fix the coverslips by adding 250 μL of 4% (v/v) PFA in PBS (no $MgSO_4$) to each well (*see* **Note 64**). Incubate at room temperature for 15 min.
9. Rinse the coverslips three times with 500 μL of PBS per well (*see* **Note 67**).

3.4.3 Staining Total Gc, Including Intracellular Gc

1. Add 250 μL of 10% (v/v) normal goat serum and 0.2% (w/v) saponin in PBS to each well in order to permeabilize neutrophils and block nonspecific binding (*see* **Note 68**). Incubate at room temperature for 10 min in the dark. Aspirate the medium using a sterile pipette tip connected to a vacuum flask.
2. Transfer coverslips with microforceps to a humidified chamber, neutrophil side up (*see* **Note 18**).
3. Add 50 μL of primary antibody diluted in 10% (v/v) normal goat serum with 0.2% (w/v) saponin in PBS solution to each coverslip (*see* **Note 69**). Incubate at room temperature in the dark for 1 h.
4. Transfer the coverslips with microforceps into the original 24-well plate (or a fresh 24-well plate), neutrophil side up, and wash three times with 500 μL of PBS for 5 min each in the dark.
5. Keep the coverslips in the wells and add 250 μL of fluorescent secondary antibody, diluted in 10% (v/v) normal goat serum with 0.2% (w/v) saponin in PBS to each well. The antibody chosen must be conjugated to a different fluorophore than the secondary antibody used in **step 6** of Subheading 3.4.2. Incubate at room temperature in the dark for 1 h.
6. Aspirate the secondary antibody suspension with a sterile pipette tip connected to a vacuum flask. Wash three times with 500 μL of PBS for 5 min each in the dark.
7. Aspirate the last PBS wash with a sterile pipette tip connected to a vacuum flask and add ddH_2O to each well.

3.4.4 Mounting and Imaging

1. Remove coverslips from wells with microforceps. Touch the edge of the coverslip to a piece of filter paper or a Kimwipes® to wick away excess liquid. Gently place coverslips, neutrophil side down, into a 5 μL drop of mounting medium on a 25 mm × 75 mm glass microscope slide (*see* **Note 70**).

2. Seal the edges of the coverslip with clear nail polish and dry horizontally at room temperature in the dark before viewing on a fluorescence microscope.

4 Notes

1. We use a BD Vacutainer™ that is spray-coated with sodium heparin to collect venous blood (catalog #BD 367874). Other anticoagulants are available.
2. This water is *United States Pharmacopeia* (USP) grade, non-pyrogenic, and hemolytic (not isotonic). It is important to use water lacking any microbial traces to prevent adverse activation of neutrophils.
3. We advise keeping all reagents to be used for neutrophil preparation solutions, such as NaCl, dextrose, and dextran, separate from common reagents. This will limit the potential for contamination.
4. Using a buffer lacking calcium or magnesium is critical for limiting neutrophil activation during purification. DPBS (minus calcium chloride/minus magnesium chloride) can be made in-house, taking care to limit contamination with endotoxin. Add KCl (2.7 mM), KH_2PO_4 (1.5 mM), NaCl (136.9 mM), and $Na_2HPO_4 \cdot 7H_2O$ (8.9 mM) to sterile H_2O for irrigation in a sterile, nonpyrogenic container with lid. Shake to dissolve. Filter sterilize (0.2 μm) and store at room temperature.
5. Gc are fastidious bacteria, and rich media like GCB/GCBL is required for growth [11]. The addition of glucose via Kellogg's Supplement 1 and iron via Kellogg's Supplement 2 to GCB and GCBL is also required [12]. In addition, for growth on GCB plates, bacteria are grown in the presence of 5% (v/v) CO_2 in a 37 °C incubator. For liquid growth in GCBL, bacteria are grown in the presence of $NaHCO_3$ as CO_2 source, in an incubator lacking CO_2, set at 37 °C or 30 °C as needed.
6. Use a Dacron polyester swab, as opposed to a cotton swab, because cotton swabs may contain residual fatty acids that inhibit growth of some strains of Gc [11].
7. We recommend using modified GCBL medium within 48 h after addition of Kellogg's Supplements 1 and 2 and $NaHCO_3$.
8. We heat inactivate fetal bovine serum (Hyclone; characterized, triple-filtered) in a 56 °C water bath for 1 h, with occasional swirling to ensure consistent heating throughout. Bottles of fetal bovine serum are typically shipped at −80 °C and must be completely thawed prior to heat inactivation, for instance by

storage at 4 °C for 18–24 h. Heat-inactivated fetal bovine serum is then dispensed into aliquots and stored at −20 °C.

9. We use human CXCL-8 from R&D Systems (208-IL-050/CF).
10. Prepare one 24-well plate per time point. Prepare an additional 24-well plate for the "Gc alone" control that does not receive plastic coverslips.
11. We dry the modified GCB agar plates for 30–60 min in a biosafety cabinet with laminar flow. This allows the plates to quickly absorb the 10 μL drops of Gc serial dilutions. It takes about 10 min for drops of Gc to be absorbed on the benchtop at room temperature. After the plate absorbs the spots, it can be placed upside down in a 37 °C incubator with 5% (v/v) CO_2 for 20 h.
12. We recommend doing serial dilutions in a volume of 200 μL total (180 μL GCBL medium plus 20 μL Gc), which yields more than enough volume for dilutions and plating replicate spots per well. Unmodified GCBL medium is used for serial dilutions.
13. Ice buckets are an alternative, but become unwieldy with the numbers of plates used in the assay.
14. Acid-washing coverslips thoroughly cleans the glass to allow for better cellular adherence.
15. Prepare at least two coverslips (i.e., two wells) per strain or condition in the event of one breaking.
16. We source our normal human serum from Sigma (#S7023) and do not heat-inactivate the serum prior to use. However, we have verified that this serum source does not contain active complement components. Verification should be done if other sources are used.
17. We source our normal goat serum from Thermo-Fisher Scientific (#16210064), but other sources are available.
18. To make a humidified chamber, place a piece of Parafilm on the bottom of a flat-bottom rectangular container with lid. Roll up 1–2 small square Kimwipes®, wet them so they are soaked through, and place along the edges of the chamber. The coverslips are placed on the Parafilm, neutrophil-side up.
19. We use the Meridian Life Science (#B65111R) rabbit polyclonal anti-*N. gonorrhoeae* antibody at a final concentration of 20 μg/mL, but any immunofluorescence-compatible antibody raised against an abundant surface antigen should work. In our hands, the Meridian rabbit anti-*N. gonorrhoeae* antibody works well for strain FA1090 but less well for other strains such as MS11.

20. Fluorophores should be selected based on their compatibility with the microscope filters or lasers available and to achieve minimal spectral overlap. We typically use Alexa Fluor 488- and 555-conjugated secondary antibodies for immunofluorescence microscopy.
21. We use Fluoromount G (Southern Biotechnology) as a mounting medium. Before use, add 2.5 mg of propyl gallate to 1 mL of Fluoromount G. Rotate end-over-end until combined and store protected from light at 4 °C.
22. Human blood should be handled under biosafety level 2 conditions. Neutrophil purification can be done within a biosafety cabinet if there are heightened concerns about the potential for exposure to blood-borne pathogens. Our protocol specifies that donors are generally healthy and are not known to be infected with blood-borne pathogens, or have conditions that affect neutrophil numbers or function.
23. Heparinized blood from a single donor may have been collected by venipuncture into multiple tubes (such as Vacutainer® tubes). These can be combined into the same 50 mL conical tube. However, blood or neutrophils from different donors should be kept separate and considered independent biological specimens.
24. The blood–dextran mixture can be left upright up to 30 min. During this step, the majority of red blood cells will absorb dextran, thus increasing their density and causing them to sediment to the bottom of the tube. This bottom layer will be dark red in color while the top layer, which contains the white blood cells, will be light red to amber in color and more translucent. It is important not to exceed a volume of ~40 mL in the conical tube to allow for efficient sedimentation of red blood cells.
25. In order to maximize neutrophil yield, it is important to remove as much of the light red layer as possible without including any of the dark red material underneath, that is, predominantly erythrocytes.
26. Neutrophils, other white blood cells, and remaining erythrocytes will form a loose pellet in this step. The supernatant is yellow and will contain plasma components. On a Sorvall™ Legend™ XF centrifuge, acceleration and brake are set at 2.
27. We typically use empty tissue culture media bottles as waste containers. Containers should be capped when not being used. Bleach (final concentration of 10% v/v) should be added to the waste container at the end of the purification procedure to inactivate the blood products for at least 10 min, after which it can be flushed down the sink drain with water.

28. Use transfer pipettes and serological pipettes with a wide tip in order to reduce any shear stress that may activate the neutrophils during purification. When suspending any pellet containing neutrophils, do so gently and with as little introduction of air bubbles as possible. Never vortex a suspension containing neutrophils. Avoid dramatic changes in temperature by using solutions at room temperature until purification is completed. These steps help reduce unwanted neutrophil activation.
29. Ficoll is denser than 0.1% (w/v) dextrose in DPBS and separates cells based on differences in cellular density.
30. After centrifugation, the conical tube will contain four distinct layers: an uppermost, clear layer of 0.1% (w/v) dextrose in DPBS (~35 mL); a small, cloudy layer of predominantly mononuclear cells at the interface between 0.1% (w/v) dextrose in DPBS and Ficoll; the Ficoll layer (~10 mL); and a loose red pellet containing neutrophils, other granulocytes, and residual red blood cells.
31. When lysing red blood cells, pipet up and down at a slow to medium speed and take care to not introduce air bubbles. Two pellets from the same donor can be combined at this step. To do so, first add 20 mL of sterile, irrigation H_2O to one pellet. Transfer the suspended cellular material to the second pellet, and then pipette up and down for 30 s. Do not exceed 60 s of red blood cell lysis or this may adversely affect neutrophil viability.
32. Prior to the lysis step, aliquot 20 mL of 1.7% (w/v) NaCl solution into a 50 mL conical tube, and leave the cap loose. This solution can then be quickly poured directly into the red blood cell hypotonic lysis solution. Pour the 1.7% (w/v) NaCl solution down the inside of the tube containing the lysed red blood cells.
33. If the pellet is reddish in color, indicating there are still red blood cells present, the hypotonic lysis can be repeated. Lysis should not be performed more than twice.
34. If the neutrophils are intact and not activated, the pellet should appear compact at the bottom of the tube. If stringy material is observed or if the pellet is smeared up the sides of the tube, this may be an indication of unwanted neutrophil activation. Potential steps to troubleshoot are contamination of reagents with microbial components, overly aggressive handling during purification, or overly long hypotonic lysis.
35. We generally suspend the neutrophils at a final concentration of ~10–20 × 10^6 cells/mL, which are then diluted into media for subsequent experimentation.

36. For large volumes of blood from the same donor that are prepared in different conical tubes, the final neutrophil pellets can be combined into one tube.

37. Neutrophils kept on ice will last for at least 2 h, but are best used as soon as they are purified to maximize viability and responsiveness in functional assays.

38. Suspending neutrophils in 2% (v/v) acetic acid solution allows for better visualization of their lobulated nuclei, to discriminate them from mononuclear cells that lack this morphological characteristic. This step is important for verifying that the purification was successful.

39. Inoculate and spread Gc onto GCB agar plates such that there are single colonies after overnight growth.

40. We and others have shown that expression of opacity-associated proteins and type IV pili affect Gc survival in the presence of human neutrophils [17–20]. However, both features are variable in a natural population of Gc, complicating the interpretation of experimental results unless care is taken to select for particular Opa and pilus expression states when colonies are selected for lawns of Gc. Bacterial colonies can be visualized using a bottom-lit dissecting microscope with a "type B" Bausch and Lomb type base and Nicholas illuminator. Colonies with piliated bacteria appear more compact and domed in shape than colonies with nonpiliated bacteria. Colonies with Gc expressing Opa proteins tend to appear darker in color if visualized using the frosted side of the sub-base illuminator or glittery if using the mirrored side.

41. GCB plates used for making lawns should be no older than 2 weeks. Equilibrating GCB plates to 25–37 °C also helps ensure optimal growth of Gc under these conditions.

42. Pilus-expressing Gc can form pellicles in liquid culture (Fig. 2). Always pulse vortex the cultures to sufficiently disaggregate pellicles before attempting to measure optical density or dilute cultures.

43. Since bacteria that phase-vary off pilus expression tend to outgrow piliated bacteria in in vitro culture, it may be desirable to enrich for pilus-expressing Gc after 12–16 h in liquid medium. To this end, allow pellicles to settle to the bottom of the conical tube. Using a sterile pipette, remove the settled pellicles into fresh modified GCBL medium. Dilute to an appropriate OD $\lambda_{550\text{ nm}}$ in a total of 5 mL with modified GCBL medium.

44. The CFU/mL per OD $\lambda_{550\text{ nm}}$ ratio should be calculated for each strain background used in each lab; piliation and opacity protein expression state can also affect this ratio. This is accomplished by growing a liquid culture to mid-logarithmic phase,

measuring the OD $\lambda_{550\ nm}$, and enumerating CFU from serial dilutions of the culture. Enumeration should be repeated on multiple days to achieve an average CFU per OD $\lambda_{550\ nm}$.

45. For a more sensitive fluorescence-based assay that can distinguish intracellular from extracellular bacterial viability, please refer to ref. 21.

46. Add neutrophils directly to the coverslip. A volume of 400 μL will form a dome on the coverslip and then will evenly disperse throughout the well. Take care to not jostle the plate, to ensure the coverslip remains at the bottom of the well and neutrophils are on top of the coverslips.

47. It is helpful to include Gc with a known sensitivity to killing by neutrophils as a positive control [19, 22].

48. To adjust the MOI, increase or decrease the concentration of bacteria, maintaining the volume of bacterial suspension to be added at 100 μL.

49. To chill the neutrophils, we suggest placing the 24-well plates on ice packs stored at 4 °C. Chilled neutrophils are less phagocytic [23] and thus allow for better synchronization of the experimental time points.

50. We use a multichannel pipette to spot serial dilutions in triplicate. Use agar plates that are appropriately dried beforehand (*see* **Note 11**). For an MOI of 1, dilute lysates from the Gc–neutrophil wells from 10^{-1}–10^{-4}. We use one multichannel pipette set at 200 μL to mix and another set at 20 μL to transfer to the next set of wells. Changing tips between dilutions is essential for accurate CFU enumeration. CFU will likely be enumerated from the 10^{-3}–10^{-4} dilution for the MOI calculation, that is, viability count, and the 10^{-2} dilution at 0 min for Gc + neutrophils condition.

51. To calculate the actual MOI, first determine the CFU/mL of the starting culture based on the viability count. To this end, enumerate CFU from diluted spots yielding 30–150 colonies, which is the optimal counting range. From the replicate spots, determine the average CFU per 10 μL spot and then account for the dilution factor associated with the spots counted. For example, an average of 100 CFU per 10 μL spot at the 10^{-3} dilution gives a starting concentration of 1×10^7 CFU/mL. Because 100 μL of bacteria was added to the neutrophils, next determine how many CFU were contained within 100 μL. Then, divide this number by the number of neutrophils per well (1×10^6) to yield the actual MOI.

52. The Gc alone control ensures that any decrease in CFU during the assay with neutrophils is due to the presence of neutrophils rather than an inherent growth defect in the bacteria under the assay conditions.

53. When the coverslips are wet and have been centrifuged, they will remain in the well despite inversion of the plate to pour off contents. To mitigate infectious risk associated with Gc in liquid suspension, an alternative method is to gently aspirate the supernatant from each well using a pipette tip connected to a vacuum flask, taking care to not disrupt the coverslips. However, for an experiment with multiple time points and conditions, aspiration can take several minutes, which could prevent accurate synchronization of the experiment. If biosafety issues are a concern (for instance, using multidrug-resistant Gc), the plate inversion can be done in a biosafety cabinet.

54. This concentration of saponin lyses neutrophils but does not usually affect Gc viability. However, this assumption should be experimentally validated for new strains or mutants, especially those that may have reduced envelope integrity.

55. Add 800 μL of GCBL medium directly onto the coverslip. The neutrophils should immediately lift from the coverslip and disperse, indicating that they have been lysed.

56. Pipet up and down 8–10 times, keeping this number consistent between wells. Typically, we mix 3–5 wells sequentially and then remove 20 μL from each mixed well into a 96-well plate for serial dilutions. This approach is to ensure CFU are enumerated from an evenly dispersed suspension of lysed neutrophils.

57. To calculate percentage of Gc survival, first calculate the actual CFU/mL remaining in each well as described in **Note 51**. Average the CFU/mL for the 3–5 technical replicates for each time point and condition. Divide the average CFU/mL of one condition from each time point by the average CFU/mL of the same condition at 0 min. Multiply by 100%.

58. For a more sensitive and high-throughput approach to assess Gc binding and internalization by human neutrophils using imaging flow cytometry, *see* refs. 24, 25.

59. To label bacteria with CFSE, centrifuge ~1×10^8 CFU Gc, aspirate the supernatant, and suspend the pellet in 20 μg/mL of CFSE, diluted in DPBS (minus calcium chloride/minus magnesium chloride) with 5 mM $MgSO_4$. Incubate for 20 min in a 37 °C water bath. Centrifuge Gc, remove the supernatant, and wash the pellet with 1 mL of DPBS with 5 mM $MgSO_4$. Centrifuge as before, remove the supernatant, and suspend the pellet in RPMI-10 medium to the desired concentration. The Gc pellet should be an obvious yellowish-green color if CFSE labeling is successful. All steps with CFSE-labeled Gc should be done in the dark, which is sufficiently accomplished by placing a small cardboard box over the plate during room temperature incubations. Because using CFSE-labelled Gc facilitates total Gc staining, skip Subheading 3.4.3.

60. Use pooled normal human serum (*see* **Note 16**). If necessary, use a pipette tip to spread the serum over the coverslip. Serum promotes neutrophil adherence and spreading. Do not allow coverslips to dry out after serum coating.
61. We have noted human donor variability in the ability of primary neutrophils to adhere to coverslips.
62. The kinetics of internalization vary, depending on the surface components expressed by Gc [26]. Typically, a 1 h infection allows for adequate binding and internalization by most variants. Shorter times may be advantageous for assessing the contributions of opacity-associated proteins and opsonins.
63. If no antibody is available that recognizes the Gc strain of interest, lectins such as soybean lectin (SBL) can be used. It must first be empirically determined that the lectin binds to the Gc strain of interest. Wash the coverslips with 500 μL of 3-(*N*-morpholino)propanesulfonic acid (MOPS) buffer [21]. Aspirate the buffer and add 250 μL of SBL, coupled to a fluorophore of interest, diluted in MOPS to a final concentration of 5 μg/mL. Incubate at room temperature in the dark for 10 min. Aspirate and rinse once with MOPS buffer. Fix with 4% (v/v) PFA in PBS as described in **step 10** of Subheading 3.4.2.
64. A quick fixation step is important because overnight fixation can cause unwanted permeabilization in this context.
65. Use a secondary antibody raised against immunoglobulin from the animal in which the primary antibody is made. Dilute the secondary antibody according to the manufacturer's suggestion. Working antibody concentrations should be empirically optimized for each bacterial strain.
66. Once fluorescent reagents are used, all incubations are performed in the dark to prevent photobleaching. This is also true for fluorescently labeled Gc (*see* **Note 59**).
67. If desired, 24-well plates containing coverslips can be stored overnight in PBS at 4 °C in the dark.
68. 0.2% (w/v) saponin will permeabilize the neutrophils but will not permeabilize Gc.
69. The same Gc-specific primary antibody used to stain extracellular bacteria can be used again to stain total (intracellular + extracellular) bacteria. Since antibody binding is not always saturated, intracellular bacteria will appear a single color while extracellular bacteria will be two colors.
70. If necessary, gently press the coverslip down with the flat side of a pipette tip to remove any bubbles from underneath the coverslip.

References

1. Stevens JS, Criss AK (2018) Pathogenesis of *Neisseria gonorrhoeae* in the female reproductive tract: neutrophilic host response, sustained infection, and clinical sequelae. Curr Opin Hematol 25(1):13–21. https://doi.org/10.1097/MOH.0000000000000394
2. Wiesner PJ, Thompson SE 3rd (1980) Gonococcal diseases. Dis Mon 26(5):1–44
3. Stohl EA, Criss AK, Seifert HS (2005) The transcriptome response of *Neisseria gonorrhoeae* to hydrogen peroxide reveals genes with previously uncharacterized roles in oxidative damage protection. Mol Microbiol 58 (2):520–532. https://doi.org/10.1111/j.1365-2958.2005.04839.x
4. Criss AK, Katz BZ, Seifert HS (2009) Resistance of *Neisseria gonorrhoeae* to non-oxidative killing by adherent human polymorphonuclear leucocytes. Cell Microbiol 11(7):1074–1087. https://doi.org/10.1111/j.1462-5822.2009.01308.x
5. Johnson MB, Criss AK (2013) *Neisseria gonorrhoeae* phagosomes delay fusion with primary granules to enhance bacterial survival inside human neutrophils. Cell Microbiol 15 (8):1323–1340. https://doi.org/10.1111/cmi.12117
6. Ball LM, Criss AK (2013) Constitutively Opa-expressing and Opa-deficient *Neisseria gonorrhoeae* strains differentially stimulate and survive exposure to human neutrophils. J Bacteriol 195(13):2982–2990. https://doi.org/10.1128/JB.00171-13
7. Miralda I, Uriarte SM, McLeish KR (2017) Multiple phenotypic changes define neutrophil priming. Front Cell Infect Microbiol 7:217. https://doi.org/10.3389/fcimb.2017.00217
8. Nemeth T, Mocsai A (2016) Feedback amplification of neutrophil function. Trends Immunol 37(6):412–424. https://doi.org/10.1016/j.it.2016.04.002
9. Ramsey KH, Schneider H, Cross AS et al (1995) Inflammatory cytokines produced in response to experimental human gonorrhea. J Infect Dis 172(1):186–191
10. Hedges SR, Sibley DA, Mayo MS et al (1998) Cytokine and antibody responses in women infected with *Neisseria gonorrhoeae*: effects of concomitant infections. J Infect Dis 178 (3):742–751
11. Dillard JP (2011) Genetic manipulation of *Neisseria gonorrhoeae*. Curr Protoc Microbiol. Chapter 4:Unit4A.2. https://doi.org/10.1002/9780471729259.mc04a02s23
12. Kellogg DS Jr, Peacock WL Jr, Deacon WE et al (1963) *Neisseria gonorrhoeae*. I. Virulence genetically linked to clonal variation. J Bacteriol 85:1274–1279
13. Criss AK, Seifert HS (2008) *Neisseria gonorrhoeae* suppresses the oxidative burst of human polymorphonuclear leukocytes. Cell Microbiol 10(11):2257–2270. https://doi.org/10.1111/j.1462-5822.2008.01205.x
14. Morse SA, Bartenstein L (1974) Factors affecting autolysis of *Neisseria gonorrhoeae*. Proc Soc Exp Biol Med 145(4):1418–1421
15. Hebeler BH, Young FE (1975) Autolysis of *Neisseria gonorrhoeae*. J Bacteriol 122 (2):385–392
16. Zenni MK, Giardina PC, Harvey HA et al (2000) Macropinocytosis as a mechanism of entry into primary human urethral epithelial cells by *Neisseria gonorrhoeae*. Infect Immun 68(3):1696–1699
17. Johnson MB, Criss AK (2011) Resistance of *Neisseria gonorrhoeae* to neutrophils. Front Microbiol 2:77. https://doi.org/10.3389/fmicb.2011.00077
18. Sadarangani M, Pollard AJ, Gray-Owen SD (2011) Opa proteins and CEACAMs: pathways of immune engagement for pathogenic *Neisseria*. FEMS Microbiol Rev 35(3):498–514. https://doi.org/10.1111/j.1574-6976.2010.00260.x
19. Johnson MB, Ball LM, Daily KP et al (2015) Opa+ *Neisseria gonorrhoeae* exhibits reduced survival in human neutrophils via Src family kinase-mediated bacterial trafficking into mature phagolysosomes. Cell Microbiol 17 (5):648–665. https://doi.org/10.1111/cmi.12389
20. Stohl EA, Dale EM, Criss AK et al (2013) *Neisseria gonorrhoeae* metalloprotease NGO1686 is required for full piliation, and piliation is required for resistance to H_2O_2- and neutrophil-mediated killing. MBio 4(4). https://doi.org/10.1128/mBio.00399-13
21. Johnson MB, Criss AK (2013) Fluorescence microscopy methods for determining the viability of bacteria in association with mammalian cells. J Vis Exp (79). https://doi.org/10.3791/50729
22. Handing JW, Criss AK (2015) The lipooligosaccharide-modifying enzyme LptA enhances gonococcal defence against human neutrophils. Cell Microbiol 17(6):910–921. https://doi.org/10.1111/cmi.12411
23. Peterson PK, Verhoef J, Quie PG (1977) Influence of temperature on opsonization and

phagocytosis of staphylococci. Infect Immun 15(1):175–179

24. Smirnov A, Solga MD, Lannigan J et al (2015) An improved method for differentiating cell-bound from internalized particles by imaging flow cytometry. J Immunol Methods 423:60–69. https://doi.org/10.1016/j.jim.2015.04.028
25. Smirnov A, Solga MD, Lannigan J et al (2017) High-throughput particle uptake analysis by imaging flow cytometry. Curr Protoc Cytom 80:11.22.11–11.22.17. https://doi.org/10.1002/cpcy.19
26. Smirnov A, Daily KP, Criss AK (2014) Assembly of NADPH oxidase in human neutrophils is modulated by the opacity-associated protein expression state of *Neisseria gonorrhoeae*. Infect Immun 82(3):1036–1044. https://doi.org/10.1128/IAI.00881-13

Chapter 20

Analysis of Host Responses to *Neisseria gonorrhoeae* Using a Human Three-Dimensional Endometrial Epithelial Cell Model

Paweł Łaniewski and Melissa M. Herbst-Kralovetz

Abstract

Neisseria gonorrhoeae infections have been associated with complications including chronic endometritis and pelvic inflammatory disease. Robust in vitro models of the female reproductive tract are urgently needed to better understand the biological mechanisms leading to these pathophysiological changes. Our human three-dimensional (3D) endometrial epithelial cell (EEC) model, which is generated using the HEC-1A cell line and rotating wall vessel (RWV) bioreactor technology, replicates several hallmarks of endometrial tissue in vivo. Studying the interactions of *N. gonorrhoeae* with the host using this newly characterized human 3D EEC model allows for the investigation of unique mechanisms of gonococcal pathogenesis in the upper female reproductive tract. In this chapter, we describe methodologies that can be used to investigate the interactions of *N. gonorrhoeae* with the human 3D endometrial epithelium. Protocols for generating the human 3D EEC model using the RWV technology and assessing the host response (including morphological/ultrastructural changes to the epithelial cells; cytokine/chemokine secretion or gene expression changes) following infection with *N. gonorrhoeae* are presented.

Key words Female reproductive tract, *Neisseria gonorrhoeae*, Host–pathogen interactions, Human epithelial cells, Innate immunity, Organotypic mucosal model, Rotating wall vessel bioreactor, Uterine infection

1 Introduction

The lack of robust and physiologically relevant in vitro human endometrium models has been a limiting factor for studying host–microbe interactions at this site. Previous studies have utilized tissue explants or organotypic models in which epithelial cells from endometrium are cultivated in Matrigel matrix or fibrin–agarose scaffolds [1–4]. However, these approaches require the availability of fresh tissue for isolation of endometrial epithelial cells (EEC) and are not suited for high throughput screening. Alternatively, our novel human three-dimensional (3D) EEC model, developed using the HEC-1a cell line and the rotating wall vessel (RWV)

Myron Christodoulides (ed.), *Neisseria gonorrhoeae: Methods and Protocols*, Methods in Molecular Biology, vol. 1997, https://doi.org/10.1007/978-1-4939-9496-0_20,

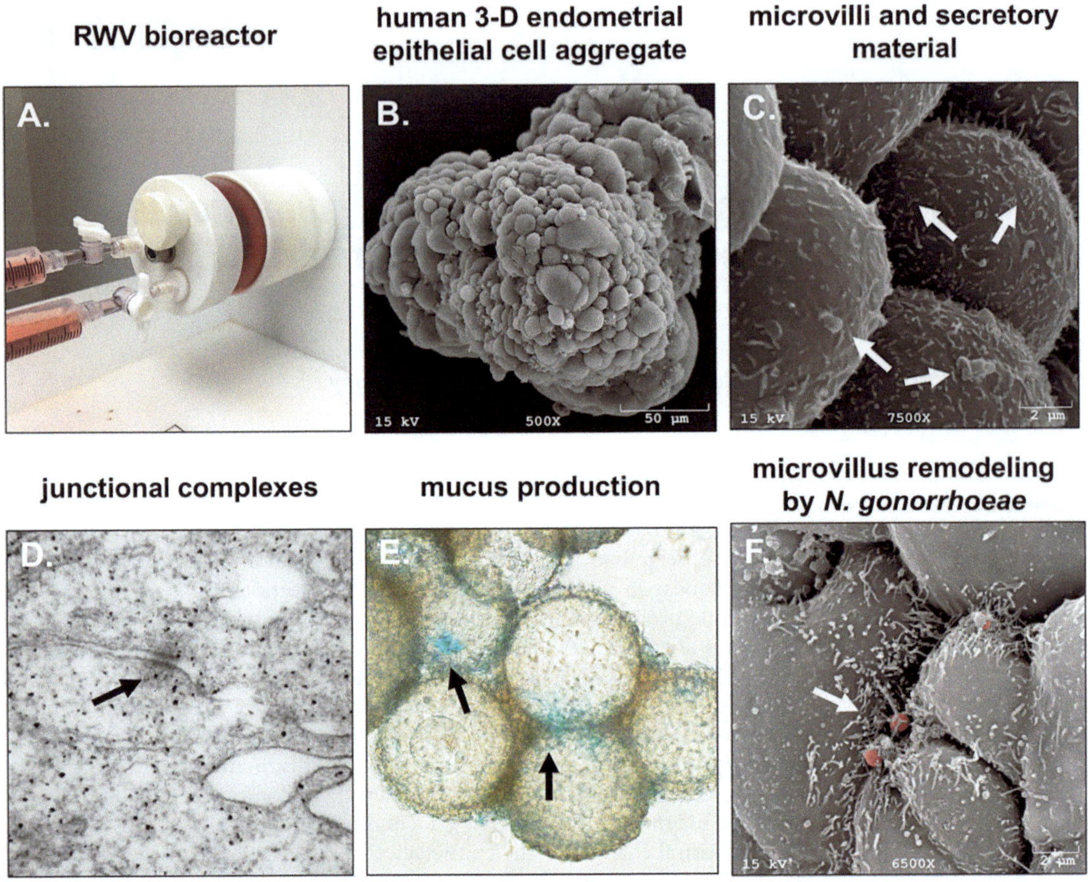

Fig. 1 The RWV bioreactor-derived 3D EEC model. (**a**) The bioreactor is used to generate a 3D human EEC model by combining endometrial cells, beads, and medium in a low-fluid shear environment, allowing the cells to self-assemble, grow, and differentiate. (**b–e**) Microscopy images of 3D EEC aggregates displaying hallmarks of differentiation including microvilli, mucus (stained with Alcian blue) and junctional complexes. (**f**) Microscopy images of 3D EEC aggregates infected with *N. gonorrhoeae*. Gonococci interact with several cells simultaneously and localize to folds and crevices of the 3D EEC model. Bacteria were pseudocolored red using Photoshop C5.1 software (Adobe). Arrows denote microvilli, secretory material, junctional complexes, mucus, and microvillus modeling by *N. gonorrhoeae*

bioreactor technology (Fig. 1a) can effectively be utilized to study host–microbe interactions [5].

The RWV bioreactor technology is designed to replicate the natural 3D low-fluid shear of the body that promotes tissue development. It is an optimized form of suspension culture that allows cells to gradually adhere to porous collagen coated beads [6, 7]. The beads' porosity enables the cells to respond to chemical and molecular gradients in three dimensions (i.e., at apical, basal, and lateral surfaces) in a manner akin to the response of the parental tissue in vivo [6]. RWV culture systems have been utilized for other tissue models [6, 8, 9]; however, our laboratory was the first to propose, construct, and validate RWV generated human 3D vaginal cell, endocervical cell, and EEC models for the evaluation of host interactions with commensal or pathogenic bacteria (i.e., endogenous vaginal bacteria, bacterial vaginosis-associated bacteria, or sexually transmitted pathogens) [5, 7, 10–15].

For our human 3D EEC model system, we selected the HEC-1A cell line, which was isolated in 1968 from a patient with endometrial adenocarcinoma and has been used widely to study the molecular mechanisms underlying the physiology and pathology of the human female reproductive tract (FRT) [16]. Our innovative human 3D EEC model replicates several key features of in vivo endometrial tissue, such as microvilli, junctional complexes, mucus and secretory material, and most notably, a 3D structure that allows for simultaneous microbe interactions between multiple adjacent cells, none of which are exhibited in traditional monolayer cell cultures or transwells (Fig. 1b–e) [5, 14]. The 3D EEC aggregates also express *Toll*-like receptor (TLR) isoforms and mucins in a pattern that largely mimics normal endometrial tissue [5]. We have also demonstrated the ability of 3D EEC aggregates to appropriately respond to a variety of microbial products as well as live bacteria, including *Neisseria gonorrhoeae* [5].

Gonococcal infections have been shown to be associated with chronic endometritis and pelvic inflammatory disease; therefore, understanding the mechanisms leading to those pathophysiological changes in the female upper reproductive tract is still imperative [17–19]. The human 3D EEC model can be utilized to study the functional impact of *N. gonorrhoeae* on the human endometrial epithelium. Infection of our 3D EEC model with *N. gonorrhoeae* demonstrated that the bacteria adhere to the surface of the 3D aggregates as microcolonies and significantly induced the expression and secretion of proinflammatory cytokines and chemokines in a dose-dependent manner. In contrast, endogenous vaginal bacteria (lactobacilli) did not elicit the production of proinflammatory immune mediators [5]. In addition, *N. gonorrhoeae* induced significant ultrastructural changes to the 3D endometrial epithelial cells, including microvillus remodeling as visualized by scanning electron microscopy (Fig. 1f). The 3D nature of our model allowed us to reveal intimate interactions of gonococci with multiple adjacent cells, mostly at their folds and crevices, which is not possible using epithelial cells grown in conventional monolayers [5, 20, 21]. Furthermore, using a well-characterized *N. gonorrhoeae pilT* deletion mutant, which produces nonretractable pili fibers, we showed differences in the host response to wild type and mutated *N. gonorrhoeae* strains [5, 22, 23].

Overall, our previous studies demonstrated that the human 3D EEC model is a robust experimental tool to study the host innate immune responses to microbial challenges in the upper FRT. In this chapter, we describe the methods to generate the human 3D EEC model using the RWV bioreactor technology and assess the host response (including ultrastructural/morphological changes to the epithelial cells; cytokine/chemokine secretion, and changes in the host gene expression profiles) following infection with *N. gonorrhoeae*.

2 Materials

2.1 Generation of the Human 3D EEC Model

2.1.1 Preparing the STLV Bioreactor

1. 95% (v/v) ethanol.
2. Sterile deionized H_2O.
3. Sterile 5 and 10 mL Luer-lock syringes.
4. Slow-turning lateral vessel (STLV) bioreactor system (55 mL autoclavable STLV bioreactor, rotator base and speed control power supply) (Synthecon).
5. Sterile one-way stopcocks (Medex Supply).
6. Autoclave sterilizer.
7. Class II biosafety cabinet.

2.1.2 Preparing Cytodex-3 Microcarrier Beads

1. Dulbecco's phosphate-buffered saline (DPBS), without calcium or magnesium, pH 7.4: commercial source, or prepare using a recipe of 2.7 mM KCl, 1.5 mM KH_2PO_4, 136.9 mM NaCl, and 8.9 mM $Na_2HPO_4 \cdot 7H_2O$ in distilled H_2O. Sterilize by autoclave at 121 °C and 2.68 kg/cm^2 for 15 min.
2. Cytodex-3 microcarrier beads.
3. Sterile polypropylene 50 mL conical tubes.
4. Autoclave sterilizer.
5. Complete McCoy's medium: McCoy's 5A medium (Iwata & Grace mod.) supplemented with 0.2 mM L-glutamine, 10% (v/v) heat-inactivated fetal bovine serum, penicillin (100 I. U.), and streptomycin (100 μg/mL). Stored at 4 °C, but used at 37 °C.
6. Class II biosafety cabinet.

2.1.3 Seeding Human EEC and Beads into the STLV Bioreactor

1. Human endometrial adenocarcinoma cell line HEC-1A (American Type Culture Collection HTB-112).
2. Complete McCoy's medium: McCoy's 5A medium (Iwata & Grace mod.) supplemented with 0.2 mM L-glutamine, 10% (v/v) heat-inactivated fetal bovine serum, penicillin (100 I. U.) and streptomycin (100 μg/mL). Stored at 4 °C, but used at 37 °C.
3. Slow-turning lateral vessel (STLV) bioreactor system (55 mL autoclavable STLV bioreactor, rotator base and speed control power supply) (Synthecon).

4. 0.25% (w/v) trypsin–0.1% (w/v) ethylenediaminetetraacetic acid (EDTA).
5. 0.4% (w/v) trypan blue stain.
6. Sterile polypropylene 50 mL conical tube containing prepared Cytodex-3 microcarrier beads.
7. Sterile one-way stopcocks (Medex Supply).
8. Sterile 5 and 10 mL Luer-lock syringes.
9. Sterile T25 cm^2, T75 cm^2, T150 cm^2 tissue culture flasks.
10. Disposable sterile serological pipettes.
11. Countess cell counting chamber slides (Invitrogen).
12. Countess automated cell counter (Invitrogen) or hemocytometer.
13. Humidified CO_2 incubator with 5% (v/v) CO_2, set to 37 °C.
14. Water bath.
15. Inverted microscope model Axiovert 40 CFL (Zeiss, or equivalent).
16. Class II safety cabinet.

2.1.4 Culturing the EEC in the STLV Bioreactor

1. Complete McCoy's medium: McCoy's 5A medium (Iwata & Grace mod.) supplemented with 0.2 mM L-glutamine, 10% (v/v) heat-inactivated fetal bovine serum, penicillin (100 I. U.) and streptomycin (100 μg/mL). Stored at 4 °C, but used at 37 °C.
2. Slow-turning lateral vessel (STLV) bioreactor system (55 mL autoclavable STLV bioreactor, rotator base and speed control power supply) (Synthecon).
3. Sterile one-way stopcocks (Medex Supply).
4. Sterile 5 and 10 mL Luer-lock syringes.
5. 10 mL serological pipette.
6. Petri dishes (35 mm).
7. Wide bore pipette tips.
8. Inverted microscope model Axiovert 40 CFL (Zeiss or equivalent).
9. Class II safety cabinet.

2.1.5 Harvesting Human 3D EEC Aggregates

1. Complete McCoy's medium without antibiotics: McCoy's 5A medium (Iwata & Grace mod.) supplemented with 0.2 mM L-glutamine and 10% (v/v) heat-inactivated fetal bovine serum. Stored at 4 °C, but used at 37 °C.
2. Slow-turning lateral vessel (STLV) bioreactor system (55 mL autoclavable STLV bioreactor, rotator base and speed control power supply) (Synthecon).

3. Sterile one-way stopcocks (Medex Supply).
4. Sterile 5 and 10 mL Luer-lock syringes.
5. 10 mL serological pipette.
6. Sterile polypropylene 50 mL conical tube.
7. Wide bore pipette tips.
8. Sterile tissue culture 24-well plates.
9. 0.25% (w/v) trypsin–0.1% (w/v) EDTA.
10. 0.4% (w/v) trypan blue stain.
11. Countess cell counting chamber slides (Invitrogen).
12. Countess automated cell counter (Invitrogen) or hemocytometer.
13. Inverted microscope model Axiovert 40 CFL (Zeiss, or equivalent).
14. Class II safety cabinet.

2.2 Infection of Human 3D EEC Model with *N. gonorrhoeae*

2.2.1 Preparing *N. gonorrhoeae* Inoculum

1. Cryopreserved stock of *N. gonorrhoeae* in GCP broth (15 g/L of proteose peptone no. 3; 1 g/L of starch; 4 g/L of K_2HPO_4; 1 g/L of KH_2PO_4; 5 g/L of NaCl) with 25% (v/v) glycerol. Sterilize by autoclave at 121 °C and 2.68 kg/cm^2 for 15 min. Store at −80 °C.
2. Solution A: 72 g/L of GC agar base (BD 212307) and 10 g/L of agar in deionized H_2O. Sterilize by autoclave at 121 °C and 2.68 kg/cm^2 for 15 min.
3. Solution B: 20 g/L of dried bovine hemoglobin (BD 211871) in ice-cold deionized H_2O. Sterilize by autoclave at 121 °C and 2.68 kg/cm^2 for 15 min.
4. Solution C: IsoVitalex Enrichment (BD 211876), rehydrated according to manufacturer's instructions.
5. GC agar supplemented with 1% (w/v) dried bovine hemoglobin and 1% (v/v) IsoVitalex Enrichment. Allow Solutions A and B to cool to 50 °C and add 500 mL of Solution B to Solution A under sterile conditions and mix thoroughly. Aseptically add 10 mL of rehydrated Solution C to the medium, mix and pour into sterile Petri dishes. Store at 4 °C.
6. Sterile Petri plates (90 mm plastic).
7. Sterile disposable inoculating loops.
8. Sterile Dulbecco's phosphate-buffered saline (DPBS).
9. Disposable 10 mm plastic cuvettes.
10. Spectrophotometer SP-830 (Turner, or equivalent).
11. Humidified CO_2 incubator with 5% (v/v) CO_2, set to 37 °C.
12. Vortex mixer.
13. Class II safety cabinet.

2.2.2 Infecting Human 3D EEC Aggregates with N. gonorrhoeae

1. 3D EEC aggregates.
2. Heat-inactivated *N. gonorrhoeae*.
3. Polypropylene 1.5 mL microfuge tubes.
4. GC agar supplemented with 1% (w/v) dried bovine hemoglobin and 1% (v/v) IsoVitalex Enrichment. Agar plates stored at 4 °C.
5. Water bath.
6. Humidified CO_2 incubator with 5% (v/v) CO_2, set to 37 °C.
7. Class II safety cabinet.

2.3 Methods to Assess Host Cell Response to N. gonorrhoeae Infection

2.3.1 Analysis of Host Cell Interactions with N. gonorrhoeae by Electron Microscopy

1. 3D EEC aggregates, infected with *N. gonorrhoeae* and uninfected cell aggregates.
2. Polypropylene 1.5 mL microfuge tubes.
3. Wide bore pipette tips.
4. Dulbecco's phosphate-buffered saline (DPBS).
5. 2.5% (v/v) glutaraldehyde in DPBS.
6. Reagents and materials for processing samples for scanning or transmission electron microscopy.
7. Scanning electron microscope model JSM-6300 (JEOL, or equivalent).
8. Transmission electron microscope model JEM-1200-EX (JEOL, or equivalent).
9. Class II safety cabinet.

2.3.2 Analysis of Cytokine/Chemokine Secretion

1. Cell culture supernatants from infected and uninfected 3D EEC aggregates.
2. Polypropylene 1.5 mL microfuge tubes.
3. Wide bore pipette tips.
4. Class II safety cabinet.
5. Commercial enzyme-linked immunosorbent assays (ELISA).
6. Customized Luminex-based cytometric bead array.

2.3.3 Gene Expression Analysis

1. 3D EEC aggregates, infected with *N. gonorrhoeae* and uninfected cell aggregates.
2. 20-G needles.
3. 1 mL tuberculin syringes.
4. Polypropylene 1.5 mL microfuge tubes.
5. Centrifuge for microfuge tubes.
6. Commercial RNA extraction kit.
7. Commercial cDNA synthesis kit for quantitative Real-Time Polymerase Chain Reaction (qRT-PCR).

8. Reagents and materials for qRT-PCR.
9. BioPlex 200 analyzer system (Bio-Rad, or equivalent).
10. qRT-PCR system.
11. Class II safety cabinet.

3 Methods

3.1 Generation of the Human 3D EEC Model

HEC-1A cells (ATCC HTB-112) are grown on collagen-coated dextran Cytodex-3 microcarrier beads in a STLV bioreactor (Synthecon) for 21 days to form differentiated human 3D EEC aggregates, as shown previously in a video demonstration (http://www.jove.com/video/3868/) [5] (*see* **Note 1**).

3.1.1 Preparing the STLV Bioreactor

1. Assemble the STLV bioreactor according to manufacturer's protocol.
2. To detoxify the bioreactor cover, open the ports with Luer caps and fill the bioreactor with 95% (v/v) ethanol and incubate overnight at room temperature. Remove the ethanol and fill the bioreactor with sterilized deionized H_2O and incubate overnight at room temperature. Repeat washing the bioreactor with sterilized deionized H_2O. Remove the deionized H_2O and loosen caps, center port and all screws using the tool supplied by the vendor.
3. Autoclave at 110 °C for 20 min in a sterilization pouch. After cooling, tighten the screws and screw on sterile one-way stopcocks to each side port.

3.1.2 Preparing Cytodex-3 Microcarrier Beads

1. Add 15 mL of DPBS to 250 mg of Cytodex-3 microcarrier beads in a 50 mL conical tube. Allow the beads to swell overnight at room temperature.
2. Autoclave the beads at 110 °C for 20 min and allow the beads to cool to room temperature.
3. Allow the beads to settle, remove DPBS, add complete McCoy's medium and swirl the tube to suspend and wash the beads. Repeat the wash steps with complete McCoy's medium twice. After the final wash, add 10 mL of complete McCoy's medium to the beads (*see* **Note 2**).

3.1.3 Seeding Human EEC and Beads into the STLV Bioreactor (See **Note 2**)

1. Thaw a vial with frozen HEC-1A cells rapidly in a 37 °C water bath and transfer the thawed cells dropwise into a T25 cm^2 cell culture flask containing 5 mL of prewarmed complete McCoy's medium. Incubate for 6 h to overnight in a humidified incubator with 5% (v/v) CO_2 at 37 °C to allow cells to attach to the flask. Change the medium and grow the cells to a confluency of 75–90%, observed visually with a light microscope.

2. Treat the cells with 0.5 mL of trypsin–EDTA solution for 5 min at 37 °C to detach the cells from the T25 cm^2 flask and suspend the cells in complete McCoy's medium. Transfer the cells to a T75 cm^2 flask. Grow the cells as described in **step 1** and transfer to a T150 cm^2 flask.
3. Grow HEC-1A cells as monolayers in complete McCoy's medium in a T150 cm^2 tissue culture flask, to a confluency of 75–90%. Trypsin-treat the cells as described in **step 2** and suspend the cells in 10 mL of complete McCoy's medium. Add 10 μL of the cell suspension to 10 μL of trypan blue stain, then load 10 μL of the mixture into cell counting chamber slides and quantify the cell concentration and viability by trypan blue exclusion using an automated cell counter. Alternatively, cells can be counted manually using a hemocytometer.
4. Add 5×10^6 of viable cells to the conical tube containing prepared Cytodex-3 microcarrier beads. Then, transfer the cell-bead suspension into the bioreactor through the center port using a serological pipette. Close the center port and place the bioreactor in a humidified incubator at 37 °C with 5% (v/v) CO_2 for 30 min without rotation to allow the cells to adhere to the beads.
5. Remove plungers from 5 and 10 mL syringes and resleeve plungers back into their wrappers to maintain sterility. Screw syringes into ports with open stopcocks. Fill the bioreactor with the complete McCoy's medium, replace plungers, remove bubbles using the plungers and close the stopcocks.
6. Fasten the seeded bioreactor onto the rotator base in a humidified incubator at 37 °C with 5% (v/v) CO_2 and set rotation on the speed control power supply at 20 rpm (*see* **Note 3**).

3.1.4 Culturing the EEC in the STLV Bioreactor (See Note 2)

1. Change the medium after the initial 4 days of continuous rotation and culture of cells in the bioreactor. To change the medium, remove the bioreactor from the rotator base. Tilt the bioreactor to allow cells/beads to settle. Remove plungers from syringes, open both stopcocks and remove ~75% of the medium from the side port using a 10 mL serological pipette. Exchange with new syringes and feed the bioreactor with fresh medium as described in **step 5** of Subheading 3.1.3.
2. Change the medium daily, after the initial 4-day culturing period.
3. Monitor the growth of 3D aggregates weekly by removing ~0.2 mL of aggregate suspension carefully from the bioreactor through the center port using wide bore pipette tips (*see* **Note 4**). Place the 3D aggregates into a small Petri dish, add 0.5–1 mL of medium and image using an inverted light microscope (Fig. 2).

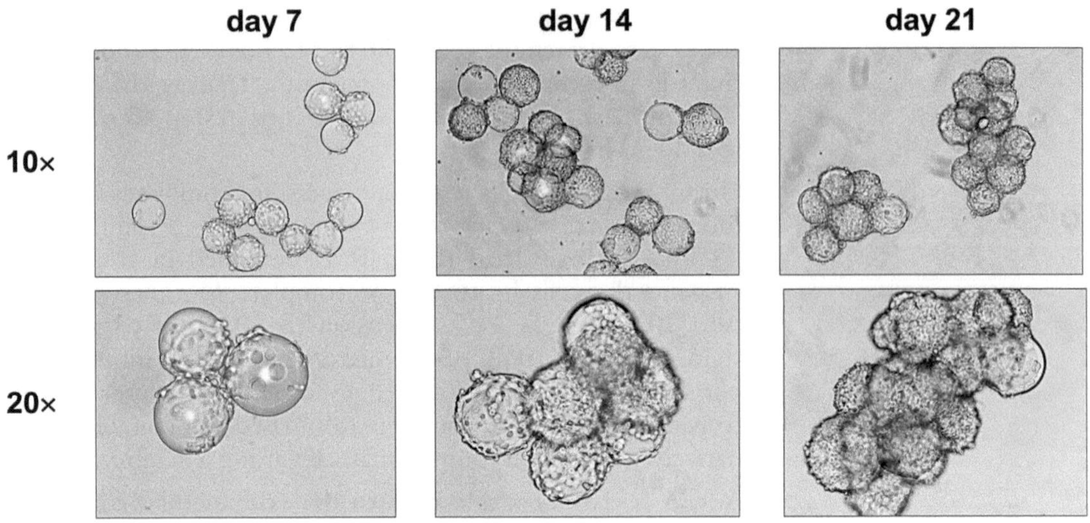

Fig. 2 Bright-field light micrographs of human 3D EEC aggregates after 7, 14, and 21 days of growth in the STLV bioreactor. The aggregates are sampled every 7 days and imaged by light microscopy to monitor the developmental progression. The aggregates begin to form after ~7 days. The complete formation of aggregates occurs by day 21

3.1.5 Harvesting Human 3D EEC Aggregates (See ***Note 2****)*

1. After 21 days in the bioreactor, human 3D endometrial epithelial cell aggregates are fully differentiated. Remove plungers from syringes, open both stopcocks and remove ~50% of the medium from the side port using a 10 mL serological pipette. Open the center port and carefully pour all contents into a 50 mL conical tube. Rinse the bioreactor twice with 5 mL of medium to remove any residual 3D aggregates and combine the rinse with the remainder of the 3D aggregates.

2. Allow the aggregates to settle, remove the medium and suspend the aggregates in 50 mL of fresh complete McCoy's medium without antibiotics (*see* **Note 5**). Swirl the tube and distribute 1 mL aliquots of the suspension (using wide bore pipette tips) into sterile 24-well plates for experimental manipulations (*see* **Note 4**).

3. Collect aggregates from a single well from the seeded sterile 24-plate and measure the cell count and viability with the trypsin-treatment and trypan blue exclusion method as described in **steps 2** and **3** of Subheading 3.1.3. Typically, when the bioreactor is seeded into two 24-well plates, the cell concentration is approximately 2×10^6 cells per well.

3.2 Infection of Human 3D EEC Model with *N. gonorrhoeae*

3.2.1 Preparing *N. gonorrhoeae* Inoculum (See Note 2)

1. Using a sterile loop, streak out *N. gonorrhoeae* from the frozen stock onto a GC agar plate. Incubate overnight at 37 °C in the humidified incubator with 5% (v/v) CO_2.
2. Pick several colonies with a sterile loop and streak out onto a fresh GC agar plate. Incubate for 16–18 h at 37 °C in the humidified incubator with 5% (v/v) CO_2 (*see* **Note 6**).
3. Harvest the gonococci from the plate using a sterile loop and place the bacteria into 1 mL of sterile DPBS (*see* **Note** 7). Suspend the bacteria thoroughly by pipetting or using a vortex mixer. Dilute the bacterial suspension appropriately and measure the optical density at the wavelength of $\lambda_{600\ \text{nm}}$ (OD_{600}) with a spectrophotometer and disposable cuvettes.
4. Adjust the bacterial suspension to an OD_{600} of 1.0, which is equivalent to ~1×10^9 bacteria/mL, determined by using a serial dilution and standard plating assay.

3.2.2 Infecting Human 3D EEC Aggregates with N. gonorrhoeae

1. Add the live bacterial suspension in DPBS at an appropriate multiplicity of infection (e.g., a MOI of 10 is 10 bacteria to 1 endometrial epithelial cell) to the wells containing 3D aggregates, in duplicate wells as a minimum. Typically, 3D aggregates are infected with 50–100 μL of bacterial suspension at MOIs ranging from 1 to 10 for 24 h (*see* **Note 8**).
2. Add the same volume of DPBS to other wells containing 3D aggregates to serve as a negative control.
3. Use heat-inactivated bacteria as an additional control. To inactivate gonococci, incubate the bacterial suspension for 20 min at 55 °C in a water bath. Cool the bacterial suspension to room temperature before adding to the wells containing the 3D aggregates. Confirm sterility of the heat-inactivated bacterial suspension using a standard plating assay.
4. Incubate infected and mock-treated 3D aggregates at 37 °C in the humidified incubator with 5% (v/v) CO_2 for up to 24 h.
5. Collect 3D aggregates and cell supernatants at various time points for further analyses as described in Subheading 3.3.

3.3 Methods to Assess Host Cell Response to *N. gonorrhoeae* Infection

3.3.1 Analysis of Host Cell Interactions with N. gonorrhoeae by Electron Microscopy (See Note 9)

1. Transfer infected or mock-treated 3D aggregates from a well to a microfuge tube using wide bore pipette tips (*see* **Note 4**). Allow aggregates to settle at the bottom of the tube for 2–3 min. Remove supernatant and add 0.5 mL of DPBS directly into the center of the tube to stir up aggregates. Allow the aggregates to settle and remove DPBS. Repeat washing two more times to a total of three washes.
2. Add 0.5 mL of 2.5% (v/v) glutaraldehyde in DPBS to fix the aggregates. Incubate for 30 min at room temperature and then place the tubes at 4 °C and store at this temperature for 24 h.

3. Remove the fixative and then wash the aggregates three times with DPBS as described in **step 1**. Store at 4 °C. Following the fixation, 3D aggregates can be processed for scanning and transmission electron microscopy as described previously [12, 13]. In our previous study, we have shown that *N. gonorrhoeae* adheres to the surfaces of cells in the human 3D EEC aggregates as microcolonies. Moreover, the microcolonies were often located at the crevices and folds between adjacent cells, and individual microcolonies interacted with the adjacent epithelial cells. Microvilli beneath and close to the bacteria are elongated, and the microcolonies often interact with multiple cells simultaneously [5].

3.3.2 Analysis of Cytokine/Chemokine Secretion

1. Transfer infected or mock-treated 3D aggregates from a well to a microfuge tube using wide bore pipette tips, at the desired time points (*see* **Note 4**). Allow aggregates to settle at the bottom of the tube for 2–3 min. Transfer the cell culture supernatants to a new microfuge tube. Save the aggregates for gene expression analysis (*see* Subheading 3.3.3). To avoid repeated freeze-thaw cycles, freeze the supernatant samples in small aliquots (≥120 μL). Store the cell culture supernatants at −80 °C until required.
2. The level of secreted cytokine/chemokine in the collected cell culture supernatants can be measured by commercial enzyme-linked immunosorbent assays (ELISA) or Luminex-based cytometric bead arrays following respective manufacturer's instructions. In our previous study, infection of human 3D EEC aggregates with *N. gonorrhoeae* strain MS11 for 24 h induced secretion of IL-8, TNFα, and MIP-3α (CCL20) [5]. A time course analysis can also be done to determine the kinetics of cytokine/chemokine secretion following different microbial stimuli (live bacteria, inactivated bacteria, microbial products, etc.) (*see* **Note 10**).

3.3.3 Gene Expression Analysis

1. Following the cell culture supernatant collection described in **step 1** of Subheading 3.3.2, add an appropriate volume of lysis buffer (typically 300 μL per well) from a commercially available RNA extraction kit. Aggregate lysates can be stored at −80 °C or directly processed for RNA extraction (*see* **Note 11**).
2. Homogenize the lysates using a 20-G needle and 1 mL tuberculin syringe. Following homogenization, centrifuge the lysates for 1 min at ≥10,000 × *g* to pellet the cell debris and beads. Carefully transfer the supernatant to a new microfuge tube. Do not transfer the beads as they clog the column membrane, resulting in low RNA yield. Afterward, follow respective manufacturer's instructions for completing RNA extraction. Store extracted total RNA samples at −80 °C for further

analyses. Typically, we obtain around 25 μg of total RNA from 3D aggregates collected from 2 wells.

3. Synthesize cDNA library from extracted RNA samples using a commercially available reverse transcription kit. Typically, we use 1 μg of total RNA for cDNA synthesis. cDNA libraries can be analyzed by real-time quantitative PCR assays using primers/probes specific for a gene of interest. In our previous study, we showed that infection of human 3D EEC aggregates with *N. gonorrhoeae* induced expression of *IL1B*, *IL8*, and *MIP3A* (*CCL20*) genes in a dose-dependent manner. Other targets, not limited to genes encoding cytokine/chemokines, can be also tested.

4 Notes

1. The detailed video protocol for generating a variety of RWV-derived 3D epithelial cell models has been created previously and can be used as a reference (http://www.jove.com/video/3868/) [7].
2. All steps should be done under BSL-2 conditions in a laminar flow hood.
3. Cultures must be rotated continuously, except for feeding or harvesting.
4. For all aggregate transfers, use special wide bore pipette tips to avoid aggregate shearing. Standard 1000 μL pipette tips can be cut ~2 cm from the point and sterilized.
5. For bacterial infection experiments, use McCoy's medium with all supplements except for penicillin and streptomycin (i.e., antibiotic-free medium).
6. The gonococcal pilin and colony opacity-associated (Opa) proteins frequently undergo antigenic and phase variation both in vivo and in vitro. Monitor pilation and Opa status of *N. gonorrhoeae* used for the infection experiments by microscopy observation of colony morphology and/or immunoblotting with monoclonal antibodies against pilin and Opa.
7. The highest viability of *N. gonorrhoeae* grown for 16–18 h on GC agar plates is found in colonies that are very closely spaced, but not confluent; thus these colonies are preferred for use in subsequent experiments. For the infection experiments, use a sterile loop to harvest *N. gonorrhoeae* colonies only from the high viability region of growth.
8. Confirm the MOI by serial dilution and standard plating assay. For each infection experiment, dilute the bacterial suspension used for the infection and plate on GC agar. Incubate plates for

3 days at 37 °C in the humidified incubator with 5% (v/v) CO_2 to determine the number of colony-forming units (CFU) in the suspension. Calculate the actual MOI by dividing number of CFU by number of cells in a well.

9. 3D aggregates can also be fixed with other application-specific fixatives (e.g., 4% (w/v) paraformaldehyde) for 30 min at room temperature for immunofluorescence or confocal microscopy analyses.

10. We routinely use human Milliplex magnetic bead panel arrays from Millipore, but other manufacturers' reagents are also suitable for cytokine/chemokine analysis by BioPlex.

11. We usually use RNeasy Mini kit from Qiagen or Quick-RNA MiniPrep from Zymo, but other manufacturers' reagents are also suitable for RNA extraction.

Acknowledgments

The authors would like to thank Dr. Magdalene So (Department of Immunobiology and BIO5 Institute, University of Arizona, Tucson, AZ) for providing *N. gonorrhoeae* strains, and Adriana Tonachio, and David Lowry (Electron Microscope Laboratory, Arizona State University, Tempe, AZ) for technical support with electron microscope analyses.

References

1. Illouz S, Dales JP, Sferlazzo K et al (2003) Effects of progestins of human proliferative endometrium: an in vitro model of potential clinical relevance. Int J Mol Med 12 (4):517–523
2. Sharma M, Shubert DE, Sharma M et al (2003) Antioxidant inhibits tamoxifen-DNA adducts in endometrial explant culture. Biochem Biophys Res Commun 307(1):157–164
3. Sharma M, Shubert DE, Sharma M et al (2003) Biotransformation of tamoxifen in a human endometrial explant culture model. Chem Biol Interact 146(3):237–249
4. Blauer M, Heinonen PK, Martikainen PM et al (2005) A novel organotypic culture model for normal human endometrium: regulation of epithelial cell proliferation by estradiol and medroxyprogesterone acetate. Hum Reprod 20(4):864–871. https://doi.org/10.1093/humrep/deh722
5. Laniewski P, Gomez A, Hire G et al (2017) Human three-dimensional endometrial epithelial cell model to study host interactions with vaginal bacteria and *Neisseria gonorrhoeae*. Infect Immun 85(3). https://doi.org/10.1128/IAI.01049-16
6. Barrila J, Radtke AL, Crabbe A et al (2010) Organotypic 3D cell culture models: using the rotating wall vessel to study host-pathogen interactions. Nat Rev Microbiol 8 (11):791–801. https://doi.org/10.1038/nrmicro2423
7. Radtke AL, Herbst-Kralovetz MM (2012) Culturing and applications of rotating wall vessel bioreactor derived 3D epithelial cell models. J Vis Exp (62):e3868. https://doi.org/10.3791/3868
8. Herbst-Kralovetz MM, Radtke AL, Lay MK et al (2013) Lack of norovirus replication and histo-blood group antigen expression in 3-dimensional intestinal epithelial cells. Emerg Infect Dis 19(3):431–438. https://doi.org/10.3201/eid1903.121029
9. LaMarca HL, Ott CM, Honer Zu Bentrup K et al (2005) Three-dimensional growth of extravillous cytotrophoblasts promotes

differentiation and invasion. Placenta 26 (10):709–720. https://doi.org/10.1016/j.placenta.2004.11.003
10. Winkle SM, Throop AL, Herbst-Kralovetz MM (2016) IL-36gamma augments host defense and immune responses in human female reproductive tract epithelial cells. Front Microbiol 7:955. https://doi.org/10.3389/fmicb.2016.00955
11. Radtke AL, Quayle AJ, Herbst-Kralovetz MM (2012) Microbial products alter the expression of membrane-associated mucin and antimicrobial peptides in a three-dimensional human endocervical epithelial cell model. Biol Reprod 87(6):132. https://doi.org/10.1095/biolreprod.112.103366
12. McGowin CL, Radtke AL, Abraham K et al (2013) *Mycoplasma genitalium* infection activates cellular host defense and inflammation pathways in a 3-dimensional human endocervical epithelial cell model. J Infect Dis 207 (12):1857–1868. https://doi.org/10.1093/infdis/jit101
13. Hjelm BE, Berta AN, Nickerson CA et al (2010) Development and characterization of a three-dimensional organotypic human vaginal epithelial cell model. Biol Reprod 82 (3):617–627. https://doi.org/10.1095/biolreprod.109.080408
14. Herbst-Kralovetz MM, Pyles RB, Ratner AJ et al (2016) New systems for studying intercellular interactions in bacterial vaginosis. J Infect Dis 214(Suppl 1):S6–S13. https://doi.org/10.1093/infdis/jiw130
15. Doerflinger SY, Throop AL, Herbst-Kralovetz MM (2014) Bacteria in the vaginal microbiome alter the innate immune response and barrier properties of the human vaginal epithelia in a species-specific manner. J Infect Dis 209 (12):1989–1999. https://doi.org/10.1093/infdis/jiu004
16. Kuramoto H, Tamura S, Notake Y (1972) Establishment of a cell line of human endometrial adenocarcinoma in vitro. Am J Obstet Gynecol 114(8):1012–1019
17. Reighard SD, Sweet RL, Vicetti Miguel C et al (2011) Endometrial leukocyte subpopulations associated with *Chlamydia trachomatis*, *Neisseria gonorrhoeae*, and *Trichomonas vaginalis* genital tract infection. Am J Obstet Gynecol 205(4):324.e1–324.e7. https://doi.org/10.1016/j.ajog.2011.05.031
18. Wiesenfeld HC, Hillier SL, Krohn MA et al (2002) Lower genital tract infection and endometritis: insight into subclinical pelvic inflammatory disease. Obstet Gynecol 100(3):456–463
19. Reekie J, Donovan B, Guy R et al (2018) Risk of pelvic inflammatory disease in relation to chlamydia and gonorrhea testing, repeat testing, and positivity: a population-based cohort study. Clin Infect Dis 66(3):437–443. https://doi.org/10.1093/cid/cix769
20. Griffiss JM, Lammel CJ, Wang J et al (1999) *Neisseria gonorrhoeae* coordinately uses Pili and Opa to activate HEC-1-B cell microvilli, which causes engulfment of the gonococci. Infect Immun 67(7):3469–3480
21. Higashi DL, Lee SW, Snyder A et al (2007) Dynamics of *Neisseria gonorrhoeae* attachment: microcolony development, cortical plaque formation, and cytoprotection. Infect Immun 75 (10):4743–4753. https://doi.org/10.1128/IAI.00687-07
22. Merz AJ, So M, Sheetz MP (2000) Pilus retraction powers bacterial twitching motility. Nature 407(6800):98–102. https://doi.org/10.1038/35024105
23. Wolfgang M, Lauer P, Park HS et al (1998) PilT mutations lead to simultaneous defects in competence for natural transformation and twitching motility in piliated *Neisseria gonorrhoeae*. Mol Microbiol 29(1):321–330

Chapter 21

In Vitro Models of Eye Infection with *Neisseria gonorrhoeae*

Colin P. Churchward and Lori A. S. Snyder

Abstract

Prophylaxis with silver nitrate and later antibiotics has significantly reduced the cases of infant blindness from gonococcal infection at birth to the point where it has all but been forgotten in the developed world as the devastating disease that it was in the pre-antibiotic era. As a result, while it is known that the bacteria are transmitted to the eyes during passage through the infected birth canal, little is known about *Neisseria gonorrhoeae* colonization of the eye and the establishment and progression of keratitis. Treatment failures due to rising antimicrobial resistance necessitate investigations into all aspects of gonococcal disease, including eye infections, so that new treatment strategies can be developed. Here we present models for *N. gonorrhoeae* eye infection using excised bovine corneas and coculture of gonococci with primary human corneal epithelial cells. These models can be used to explore the interactions of the bacteria with corneal tissues and cells and to investigate novel therapeutics against infection.

Key words Gonococcal keratitis, *Ophthalmia neonatorum*, Bovine corneal opacity and permeability test, Corneal cell coculture

1 Introduction

More than 50% of women visiting genitourinary medicine clinics, seeking treatment for a reproductive health concern, had no symptoms of their infection with *Neisseria gonorrhoeae* [1]. Undiagnosed gonorrhea infections in pregnancy are a risk for passing infection to the infant during childbirth. Up to 48% of infants born to mothers with gonorrhea will develop gonococcal *ophthalmia neonatorum*, an eye infection that can lead to blindness. Even with antibiotic prophylaxis, up to 10% of infants will still develop infection. *N. gonorrhoeae* infection of the eye causes ulceration of the cornea, perforation, and inflammation [2, 3], resulting in a characteristic swelling of the eye lids and a purulent yellowish discharge from the eyes.

Adults are also susceptible to gonococcal eye infections. This may be the origin of the myth that masturbation can cause blindness; an individual with genital gonorrhoea can inoculate the bacteria into their own eyes [4]. An increasing number of reports of

Myron Christodoulides (ed.), *Neisseria gonorrhoeae: Methods and Protocols*, Methods in Molecular Biology, vol. 1997, https://doi.org/10.1007/978-1-4939-9496-0_21,

adult eye infections are being reported due to multi-antibiotic resistant gonococci. Whereas previously a broad spectrum antibiotic would have successfully treated an adult eye infection before the gonococcus was identified as the causative agent by culture, such treatments now fail. In one report [5], a patient was left with significant reduction in visual acuity even after treatment with intravenous ceftriaxone and hourly ceftazidime drops. Another patient, who had received a 4-day course of chloramphenicol that failed to treat an unidentified eye infection, experienced swollen eye lids, purulent discharge, corneal melt and corneal perforation [2]. Following identification of the infecting organism as *N. gonorrhoeae*, treatment was changed to intravenous ciprofloxacin and ceftriaxone and hourly topical ofloxacin and erythromycin. However, there was still permanent damage to the cornea and the patient required multiple corneal grafts. A report of corneal perforation by *N. gonorrhoeae* resistant to ciprofloxacin and nalidixic acid, prompted the authors to call for specific guidance in the management of gonococcal keratitis in adults, especially in the face of growing antibiotic resistance [6]. With the rise in multi-antibiotic resistance, we are facing a future of untreatable gonorrhoea. Without effective treatment, infants and adults face a future of gonococcal blindness similar to that recorded in pre-antibiotic days [7].

A great deal of research focuses on sexually transmitted infection (STI) and its impact upon reproductive health. The molecular and genetic mechanisms that underlie the pathogenesis of gonococcal eye infections are poorly understood. Some aspects of gonococcal biology are common to gonococcal STI and eye infection, such as mechanisms for avoiding the immune system via antigen variation and survival within the host [8, 9].

Before the introduction of antimicrobial treatments, *ophthalmia neonatorum* was common and infections were a major cause of infant blindness [10]. In 1880, up to 79% of children in institutions for the blind had been blind since birth due to gonococcal *ophthalmia neonatorum* [11]. Treatment with silver nitrate began in the 1830s, but was not always successful. At the end of the nineteenth century, Carl Credé proposed the first prophylaxis, in which the eyes of newborns should be washed and treated with a single dose of 2% (w/v) silver nitrate shortly after birth [10]. After enforcement of this post-natal practice, the incidence rate in Dr. Credé's hospital dropped from 13.6% to 0.15%.

Although silver nitrate prophylaxis dramatically reduced the incidence rates of *ophthalmia neonatorum* [10], it was criticised for causing chemical conjunctivitis in 90% of infants within 6 h of application [12, 13]. The use of silver nitrate for prophylactic treatment of *ophthalmia neonatorum* was discontinued in UK hospitals in the mid-1950s [14] and replaced with a less toxic and more effective 1% (w/v) tetracycline ointment [15]. In 1982, the US

Centers for Disease Control and Prevention guidelines recommended 1% (w/v) tetracycline ointment or 0.5% (w/v) erythromycin ointment for prophylaxis, with the current recommendation being 0.5% (w/v) erythromycin [16]. The use of prophylactic treatment divides medical opinion between a substantial benefit and unnecessary use in regions of low gonorrhea prevalence.

Once ocular infection is established, it would normally be treated by antibiotics. Due to risk for the infant of development of disseminated gonococcal infections following *ophthalmia neonatorum*, it is recommended that patients are treated with both topical and intravenous antibiotics, as well as frequent saline irrigation of the eye to remove purulent discharge [2]. Concern remains about how to properly implement a prophylactic treatment or to choose the correct treatment for established *ophthalmia neonatorum*.

For humans, sight is probably the most important of the five senses and our heavy reliance on sight has driven evolution [17]. Defensive mechanisms have evolved to help protect the eyes from physical injury or infectious diseases. The eye is protected by the cornea, a completely transparent outer structure made of five layers: the epithelium; Bowman's membrane; stroma; Descemet's membrane; and endothelium. On top of the cornea sits the tear film, which has an aqueous layer coated with lipids to reduce evaporation. The eye is an immunologically privileged site and few, if any, leucocytes are normally found. Normally, the intact corneal epithelium and its ability to desquamate, provides a considerable physical barrier to bacterial colonization and penetration. Unlike other pathogens causing keratitis that can only enter the corneal stroma through physical breaches, gonococci can directly invade epithelial cells and cause damage. Defense against bacterial infection is also provided by the production of tears that contain antimicrobial peptides, 20–30% lysozyme, and 20–30% lactoferrin [18]. Lysozyme kills Gram-positive bacteria, while lactoferrin sequesters iron, limiting nutrients to the bacteria. Secretory immunoglobulin A (sIgA), which acts as a neutralizing antibody, is also present in tear fluid [19], although gonococci produce an IgA protease that can cleave sIgA [20].

In order to better understand the gonococcal factors that contribute to corneal melt and perforation and to identify new potential treatments for gonococcal keratitis [21–23], eye infection models are needed. Here we describe a whole tissue model system, originally modified from a model of herpes simplex virus infection [24], and a cell culture system for the investigation of gonococcal eye infections and for the testing of ophthalmic formulations that can be used to combat gonococcal infections.

2 Materials

2.1 Infection of Excised Bovine Corneas

2.1.1 Preparation of Corneas for Ex Vivo Infection

1. Bovine eyeballs collected as soon as possible after removal from slaughter house (*see* **Note 1**).
2. Sterile Phosphate Buffered Saline (PBS) (137 mM NaCl, 10 mM Na_2HPO_4, 1.8 mM KH_2PO_4, 2.7 mM KCl), pH 7.4.
3. Transport Medium: cold sterile PBS plus antibiotic–antimycotic solution (100 units/ml of penicillin, 100 μg/ml of streptomycin, and 0.25 μg/ml of Fungizone Antimycotic).
4. Sterile 90 mm petri dishes.
5. Tools for cornea removal: sharp scalpel, sharp scissors, two forceps, and sterile gauze.
6. 100% ethanol.
7. Sterile spot-plate.
8. Scaffold Medium: Minimum Essential Medium (MEM) with 1% (w/v) agarose. Add 0.2 g agarose to 20 ml of MEM and heat in a microwave just until it starts to boil.
9. Sterile 6-well cell culture plates.
10. Maintenance Medium: MEM plus nonessential amino acids (Gibco) and 50 μg/ml gentamicin.
11. Humidified CO_2 (5% v/v) incubator.

2.1.2 Preparation of Bacteria and Infecting Corneas

1. *N. gonorrhoeae* isolate of choice to infect corneas.
2. GC agar: add 18 g/l of GC base (Oxoid) to deionised H_2O and sterilise by autoclaving at 121 °C and 2.68 kg/cm^2 for 15 min. Once cooled to approximately 50–60 °C, add Kellogg supplements (final concentrations: 12.3 μM $Fe(NO_3)_3 \cdot 9H_2O$, 22.2 mM D-glucose, 0.68 mM L-glutamine, and 0.43 μM thiamine pyrophosphate) and pour into sterile petri dishes.
3. GC Broth: 15 g/l of protease peptone, 4 g/l of K_2HPO_4, 1 g/l of KH_2PO_4 and 5 g/l of NaCl. Sterilise by autoclaving at 121 °C and 2.68 kg/cm^2 for 15 min.
4. Stereo microscope for colony selection.
5. Spectrophotometer and cuvettes capable of reading 0.5 ml volume at $\lambda_{520\ \mathrm{nm}}$.
6. Humidified CO_2 (5% v/v) incubator.

2.1.3 Testing of Ophthalmic Formulations

1. Test formulations.
2. Sterile PBS.
3. Sterile 55 mm petri dishes.
4. Scaffold Medium.
5. Humidified CO_2 (5% v/v) incubator.

2.1.4 Method for Quantifying Colony Forming Units (CFU)

1. Detachment Buffer: 1% (w/v) saponin in PBS. Sterilize by passing through a 0.22 μm filter with a syringe.
2. Scalpel blade.
3. GC broth.
4. GC agar plates.
5. Humidified CO_2 (5% v/v) incubator.

2.1.5 Optional: Corneal Print Method

1. GC agar plates.
2. Humidified CO_2 (5% v/v) incubator.
3. Camera to record growth.

2.2 Infection of Human Corneal Epithelial Cells

2.2.1 Preparation of Primary Human Corneal Epithelial Cells

1. Primary human corneal epithelial cells (pHCEC).
2. Keratinocyte-SFM (serum-free) medium (Gibco), with L-glutamine, human recombinant epidermal growth factor 1-53 and bovine pituitary extract.
3. Sterile 75 cm^2 culture flask (T75).
4. Collagen (human type-1) Coating Matrix Kit (Gibco, or equivalent).
5. Cell Splitting Solution: 5.3 mM KCl, 0.44 mM KH_2PO_4, 4.2 mM $NaHCO_3$, 137.9 mM NaCl, 0.34 mM Na_2HPO_4, 5.6 mM D-glucose, 0.48 mM EDTA, 25 μM phenol red, and 21 μM (0.5 g/l) trypsin solution.
6. Inverted microscope (Motic AE31 or similar).
7. Cell Splitting Neutralizer Solution: Cell culture growth medium plus 10% (v/v) fetal bovine serum (decomplemented by incubation at 56 °C for 30 min).
8. Trypan blue and hemocytometer slide (optional).
9. Sterile 24-well cell culture plates.
10. 13 mm circular coverslips.
11. Centrifuge.
12. Humidified CO_2 (5% v/v) incubator.

2.2.2 Infecting pHCEC with Gonococci

1. Primary human corneal epithelial cells (pHCEC) grown to confluence in Keratinocyte-SFM medium (Gibco) in 24-well cell culture plates.
2. Trypan blue and hemocytometer slide, or automated cell counter.
3. Detachment Buffer: 1% (w/v) saponin (Sigma-Aldrich) in PBS.
4. GC broth.
5. Sterile 96-well microtiter plates.
6. Multichannel pipettes.

7. GC agar plates.
8. L-shaped spreaders.
9. Humidified CO_2 (5% v/v) incubator.

2.2.3 Log Reduction Experiments

All reagents and materials described in Subheading 2.2.2.

3 Methods

3.1 Infection of Excised Bovine Corneas

3.1.1 Preparation of Corneas for Ex Vivo Infection

1. Obtain fresh excised bovine eyeballs from a local slaughter house and transport in Transport Medium which contains antibiotics to help prevent bacterial contamination. The corneas should be prepared within 24 h of removal (*see* **Notes 2** and **3**).
2. Place eyeballs in sterile 90 mm petri dishes (Fig. 1a) and examine the surface of the cornea for any signs of damage. The corneas can often have abrasions or epithelial detachment due to damage during removal or transport. If the corneas have been allowed to dry for any reason they will appear slightly opaque. Discard any damaged corneas.
3. Make an incision just outside the edge of the cornea using a sharp scalpel: this is best done while holding the eyeball with sterile gauze (*see* **Notes 4** and **5**).
4. Insert a pair of sharp scissors into the incision and remove the cornea with a small amount of excess sclera with it (*see* **Notes 4** and **5**). Trim back the sclera to 2–4 mm from the edge of the cornea if required. Peel away the lens from the back of the cornea using two forceps (*see* Fig. 1b).
5. Rinse the corneas in fresh Transport Medium, then in sterile PBS, and place on sterile spot-plates with the epithelial layer facing down (*see* Fig. 1c).
6. Prepare the Scaffold Medium using a microwave. The Scaffold Medium will maintain the shape of the cornea and provide sufficient nutrients. Once the Scaffold Medium has cooled to approximately 55 °C, add between 1 and 2 ml until the corneas are full and allow to set at room temperature for a couple of minutes (*see* Fig. 1c).
7. Move the corneas, epithelial side down, to new sterile 6-well cell culture plates and add sufficient volume of Maintenance Medium to cover the whole of the cornea (approximately 5 ml). Place the plates in a humidified incubator at 37 °C with 5% (v/v) CO_2. The corneas are stable for up to 1 week but infections can be done after 24 h (*see* Fig. 1d) (*see* **Note 6**).

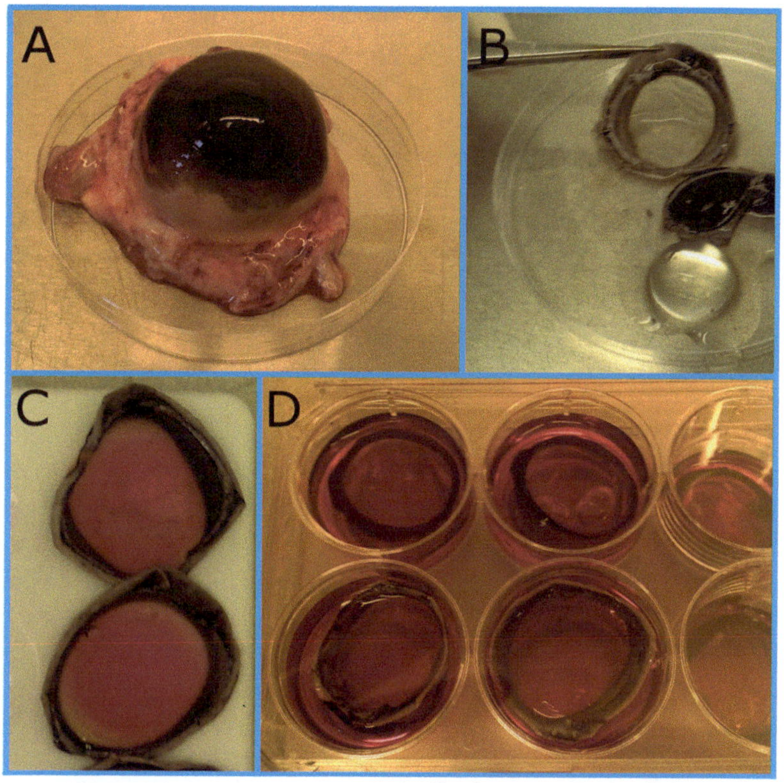

Fig. 1 Preparation of bovine corneas. (**a**) Excised bovine eyeball. (**b**) Cornea after removal of the lens and iris from the underside of the cornea. (**c**) The corneas in spot-plates after the addition of the Scaffold Medium. (**d**) Corneas being maintained, once removed. Corneas are placed epithelial layer-down in the culture medium and can be maintained for up to a week like this

3.1.2 Preparation of Bacteria and Infecting Corneas

1. Grow the desired *N. gonorrhoeae* isolate (for example, MS11, FA1090, FA19, or NCCP11945) on GC agar in a humidified incubator at 37 °C with 5% (v/v) CO_2 (*see* **Note** 7).
2. Select Piliated (P+) and Opa phenotypes (Opa+) by monitoring colony morphology [25] and/or western blot [26] (*see* **Notes 8** and **9**).
3. Remove selected colonies from the GC agar plate into GC broth at a volume more than 100 μl × the number of corneas to be tested. Adjust the optical density (OD) $\lambda_{520\ \text{nm}}$ to between 0.25 and 0.50 (equivalent to approximately 10^7 CFU/ml). Confirm this beforehand by doing CFU counts on dilutions measured at OD $\lambda_{520\ \text{nm}}$ (*see* **Note 10**).
4. Add 100 μl of the bacterial suspension to the center of the wells of a 6-well plate. Place the corneas epithelial layer-side down on the bacterial suspension. Make sure that the bacterial suspension does not come into contact with any other part of the cornea.

5. GC medium that has not been inoculated can be used as a negative control.
6. Incubate the samples for 1 h in a humidified incubator at 37 °C with 5% (v/v) CO_2 (*see* **Note 11**).

3.1.3 Testing of Ophthalmic Formulations

1. Remove the 6-well plates from the incubator and wash the corneas five times by pipetting 1 ml of sterile PBS on to the contact points of the cornea while holding the cornea vertically with sterile forceps.
2. Wash the whole cornea further by dipping into sterile PBS five times, taking care that the Scaffold Medium does not become dislodged from the back of the cornea.
3. Place the corneas onto a new sterile 55 mm petri dish so that the Scaffold Medium side of the cornea is facing down. Add 500 μl of ophthalmic formulations onto the dome of the cornea (*see* **Note 12**).
4. Leave the formulations on the cornea for 2 min, as was done previously [22], or for a treatment time suitable to test your ophthalmic formulations.
5. Wash the corneas by dipping them in sterile PBS ten times (*see* **Notes 13** and **14**).

3.1.4 Method for Quantifying Colony Forming Units (CFU)

1. To help recover the gonococci, incubate the corneas epithelial layer-down on a 100 μl spot of Detachment Buffer in 6-well plates in a humidified incubator at 37 °C with 5% (v/v) CO_2 for 10 min.
2. Pass a clean scalpel blade over the epithelial layer and then wash with the Detachment Buffer already in the well twice to remove any loose epithelial cells.
3. Recover 100 μl of Detachment Buffer and dilute into 900 μl of GC broth and then dilute 1/10 in GC broth (100 μl in to 900 μl).
4. To count CFU, plate duplicate samples of 100 μl over a whole GC plate for both dilutions. Incubate the plates in a humidified incubator at 37 °C with 5% (v/v) CO_2 for 48 h.
5. Use the following equations to calculate CFU:

$$\text{Total CFU in well, calculated from first dilution} = \text{CFU} \times \left(\frac{1000}{100}\right)$$

$$\text{Total CFU in well,calculated from second dilution} = \text{CFU} \times \left(\frac{1000}{100}\right) \times 10$$

$$\text{Log reduction} = \log_{10}(\text{Infection control mean total}) - \log_{10}(\text{Sample well mean total})$$

3.1.5 Optional: Corneal Print Method

1. Do this procedure after the corneas have been washed, following the formulation treatment (*see* **step 4** of Subheading 3.1.3) (*see* **Note 15**).
2. Hold the corneas horizontally using two sets of forceps and place them epithelial side-down onto a dried GC agar plate (*see* **Note 16**).
3. Wait 30 s. Do not move the cornea or apply additional pressure on it. It is best that this is done on a flat surface to stop the cornea from moving or sliding across the plate.
4. Two corneas can be done per plate: label each half of the plate and be careful to place the correct cornea on the correct side.
5. Incubate the GC plates in a humidified incubator at 37 °C with 5% (v/v) CO_2 for 48 h.
6. After incubation, remove the plates from the incubator and place on a white surface. A description of the amount of growth can be detailed in comparison to the controls.
7. Take photographs to record the amount of growth. This is better done in a bright area but not directly above artificial lighting.

3.2 Infection of Human Corneal Epithelial Cells

3.2.1 Preparation of Primary Human Corneal Epithelial Cells

1. Coat all cell culture flaks and plates with human Type I collagen, following the manufacturer's instructions (*see* **Note 17**).
2. Culture at least one T75 flask of low passage primary human corneal epithelial cells (pHCECs) to between 80 and 95% confluence using standard cell culture methods with Keratinocyte-SFM that does not contain any antibiotics. The coculture study will be done in sterile 24-well plates and one T75 flask will be enough to inoculate two 24-well plates.
3. Add 7 ml prewarmed of Cell Splitting Solution to the flask.
4. Incubate the flask at room temperature for up to 15 min until all cells have detached from the flask. Monitor detachment using an inverted microscope.
5. Add 8 ml of Trypsin Neutralizer solution and rock the flask for a few seconds.
6. Transfer the medium to a 15 ml centrifuge tube and remove a 20 μl sample for cell count and cell viability assay using trypan blue following standard cell culture techniques.
7. Centrifuge at 125 × *g* for 10 min.
8. Discarded the supernatant and suspend the cell pellet in 24 ml of pre-warmed cell culture medium.

9. Collagen-coat as many 24-well plates as will be required for the infection experiments. It is recommended that at least two wells should be used for each sample (*see* **Notes 17** and **18**).
10. Once the plates have been prepared, add 500 μl of the cell suspension into each well.
11. Incubate the 24-well plates in a humidified incubator at 37 °C with 5% (v/v) CO_2 and inspect daily. Cells can be infected once the cells have reached at least 80% confluence. This will take between 24–72 h depending on the cells used and culture conditions.

3.2.2 Infecting pHCEC with Gonococci

1. Count the pHCECs using standard cell culture techniques or an automated cell counter.
2. Prepare the bacterial inoculum as described in Subheading 3.1.2.
3. A Multiplicity Of Infection (MOI) of 10 is recommended (can be increased up to 100) (*see* **Note 19**). Therefore, calculate the number of bacteria required and prepare this amount in 50 μl of GC broth.
4. Add the 50 μl bacterial suspension to the appropriate control and sample wells.
5. Incubate the cells for 3 h in a humidified incubator at 37 °C with 5% (v/v) CO_2.

3.2.3 Log Reduction Experiments

1. Remove the cell culture medium from the wells.
2. Wash wells five times each with 1 ml sterile PBS per well to remove any unattached bacterial cells.
3. Add 100 μl of prewarmed (37 °C) Detachment Buffer.
4. Incubate in a humidified incubator at 37 °C with 5% (v/v) CO_2 for exactly 10 min.
5. Add 900 μl of GC broth to each well to stop the action of the Detachment Buffer.
6. Take 100 μl of this suspension and plate this over a whole 90 mm GC agar plate using an "L" shaped spreader. This will be designated as the neat sample.
7. Prepare a sterile 96-well plate containing four wells of 180 μl GC broth medium for every sample well in your experiment.
8. Take 20 μl of the cell suspension from the 24-well plate and put into the first well of each set of four.
9. Take a multichannel pipette capable of pipetting 20 μl and mix the first column of wells and mix this with the second column.
10. Repeat the last step with the newly inoculated wells until the fourth column of wells has been inoculated.

11. Divide the GC plates into quarters using a permanent marker: two quarters will be required per well as CFU counts should be done in a minimum of two replicates.
12. Using a pipette, drop 10 μl of each dilution of all samples onto a quarter of the GC agar plate and spread over a quarter of the plate using a loop.
13. Incubate the GC plates in a humidified incubator at 37 °C with 5% (v/v) CO_2 for 48 h.
14. Use at least two wells to represent each control or sample. Plate the contents of each well in duplicate for the CFU count and all of these estimates should be used to calculate the mean using the equations below:

    Log reduction calculation
    CFU = mean number of colonies at given dilution
    Dilution = dilution at which colony counts were taken

$$\text{Total CFU in well, calculated from neat well} = \text{CFU} \times 10$$

$$\text{Total CFU in well, calculated from } 10^{-1} \text{ to } 10^{-4}$$
$$= \text{CFU} \times \left(\frac{200}{10}\right) \times \left(\frac{1}{\text{Dilution}}\right) \times 5$$

$$\text{Log reduction} = \log_{10}(\text{Infection control mean total}) - \log_{10}(\text{Sample well mean total})$$

4 Notes

1. This model can contribute to the replacement, reduction, and refinement (3Rs) of use of live animals in scientific research. This model makes use of animal tissues that would otherwise be destroyed as a waste by-product of the food industry and uses them to understand pathogenesis mechanisms and for testing antimicrobials.
2. Plan to acquire additional excised bovine eyes, so that you can select the best ones for your experiments. Some will have been damaged during excision and transport; you want to use the best ones for your experiments.
3. Bovine corneas are used here as they have previously been shown to be a good model of infection. Porcine or ideally human eyes could potentially be used.
4. Keep cleaning your scissors and scalpel in ethanol during the excision of the cornea to reduce the introduction of contamination.

5. It is important to have sharp scissors that come to a point for cutting the corneas from the bovine eye balls. Make the first cut into the eyeball with a sharp scalpel, piercing through the eye, then put the point of the scissors into the hole and cut around the white of the eye. Using sharp scissors here is key.
6. Prepare more excised corneas than you will need, in case some of them develop a contamination issue. If the medium becomes cloudy or a mold develops, discard the tissue and move your unaffected corneas to a fresh 6-well plate containing pre-warmed medium.
7. *N. gonorrhoeae* viability decreases rapidly when the culture reaches stationary phase so the bacterial culture should not be incubated over 20 h (16–18 h is optimal).
8. Colonies are best picked from a GC plate using a sterile cotton/rayon swab while observing with a stereo microscope.
9. The proportion of P+ and Opa+ phenotype colonies can be increased by creating a freezer glycerol stock from P+ Opa+ colonies which can be used to produce the inoculum for each experiment. Glycerol stocks can be made by adding 15% (v/v) sterile glycerol to a heavy suspension of bacteria in GC broth.
10. It is good practice to also do a CFU count on the inoculum as well as an OD estimate. It will take 48 h to get this estimate but it helps confirm the original estimate and to measure variation in OD estimates across experiments.
11. Incubation times can be extended depending on the experimental aim as long as the experiments have been validated with appropriate controls.
12. Ophthalmic formulations may be new antibiotics or antimicrobials that are being tested for their killing activity against *N. gonorrhoeae*. For example, monocaprin has been investigated using this method [22].
13. The number of washes can be varied depending upon the viscosity of the formulation. Thicker formulations will require more washes; ointments should be avoided as they are not easily removed.
14. It is best to group the corneas into batches of four or five as the various steps involved in processing the samples are quite time consuming. It is good practice to do multiple controls and spread them across the batches.
15. The corneal prints are an optional step to give a qualitative result of the effectiveness of the treatment. Other possible experimental procedures that could be used are quantitative polymerase chain reaction and immunohistochemistry.

16. By this stage of the experiment the corneas have been though many experimental steps all of which would increase the risk of bacterial contamination. To help reduce the risk of contamination, the GC plates can be supplemented with 3 μg/ml of vancomycin. First check that this antibiotic does not suppress the growth of the gonococcal isolate, as some isolates of *N. gonorrhoeae* can be sensitive to vancomycin.
17. The plates can be precoated with collagen in bigger batches to save time. Check the manufacturer's instructions that come with the kit for details on how to precoat, appropriate storage of the plates, and the maximum storage length after coating.
18. If Live-Dead staining or any other microscopy techniques are required, then glass coverslips can be added to the wells before coating the plate. For standard sized 24-well plates, 13 mm diameter round coverslips are recommended. These should be autoclaved, stored in 70% (v/v) ethanol and dipped into sterile deionized H_2O before adding to the wells.
19. The multiplicity of infection (MOI) is the number of viable infecting bacteria divided by the number of host cells in a well. The best way to estimate the number of pHCECs in the well is count cells using standard cell culture techniques. Bacterial cell density should be estimated using optical density.

Acknowledgments

This work was supported by Sparks Children's Medical Research Charity grants 11KIN01 and 13KIN01.

References

1. Delpech V, Martin IM, Hughes G et al (2009) Epidemiology and clinical presentation of gonorrhoea in England and Wales: findings from the Gonococcal Resistance to Antimicrobials Surveillance Programme 2001–2006. Sex Transm Infect 85(5):317–321. https://doi.org/10.1136/sti.2008.034843
2. McElnea E, Stapleton P, Khan S et al (2015) Challenges in the management of *Neisseria gonorrhoeae* keratitis. Int Ophthalmol 35 (1):135–140. https://doi.org/10.1007/s10792-014-0032-8
3. Epling J (2012) Bacterial conjunctivitis. BMJ Clin Evid 2012. pii: 0704
4. Bodurtha Smith AJ, Holzman SB, Manesh RS et al (2017) Gonococcal conjunctivitis: a case report of an unusual mode of transmission. J Pediatr Adolesc Gynecol 30(4):501–502. https://doi.org/10.1016/j.jpag.2016.11.003
5. Bastion ML, Prakash K, Siow YC et al (2006) Bilateral corneal perforation in a sexually active adult male with gonococcal conjunctivitis. Med J Malaysia 61(3):366–368
6. Tipple C, Smith A, Bakowska E et al (2010) Corneal perforation requiring corneal grafting: a rare complication of gonococcal eye infection. Sex Transm Infect 86(6):447–448. https://doi.org/10.1136/sti.2010.044701
7. Johnson AP, Hughes G (2017) The prospect of untreatable gonorrhoea. BMJ 358:j3973. https://doi.org/10.1136/bmj.j3973
8. Hill SA, Masters TL, Wachter J (2016) Gonorrhea – an evolving disease of the new millennium. Microb Cell 3(9):371–389. https://doi.org/10.15698/mic2016.09.524
9. Cahoon LA, Seifert HS (2011) Focusing homologous recombination: pilin antigenic variation in the pathogenic *Neisseria*. Mol

Microbiol 81(5):1136–1143. https://doi.org/10.1111/j.1365-2958.2011.07773.x

10. Dunn PM (2000) Dr Carl Crede (1819–1892) and the prevention of *ophthalmia neonatorum*. Arch Dis Child Fetal Neonatal Ed 83(2): F158–F159
11. Laga M, Meheus A, Piot P (1989) Epidemiology and control of gonococcal *ophthalmia neonatorum*. Bull World Health Organ 67 (5):471–477
12. Lehrfeld L (1935) Limitations of use of silver nitrate in the prevention of *opthalmia neonatorum*. JAMA 104:1468–1469
13. Nishida H, Risemberg HM (1975) Silver nitrate ophthalmic solution and chemical conjunctivitis. Pediatrics 56(3):368–373
14. Schaller UC, Klauss V (2001) Is Crede's prophylaxis for *ophthalmia neonatorum* still valid? Bull World Health Organ 79(3):262–263
15. Rothenberg R (1979) *Ophthalmia neonatorum* due to *Neisseria gonorrhoeae*: prevention and treatment. Sex Transm Dis 6(2 Suppl): 187–191
16. Workowski KA, Bolan GA (2015) Sexually transmitted diseases treatment guidelines, 2015. MMWR Recomm Rep 64(Rr-03): 1–137
17. Fernald RD (2000) Evolution of eyes. Curr Opin Neurobiol 10(4):444–450
18. Sack RA, Nunes I, Beaton A et al (2001) Host-defense mechanism of the ocular surfaces. Biosci Rep 21(4):463–480
19. McClellan KA (1997) Mucosal defense of the outer eye. Surv Ophthalmol 42(3):233–246
20. Senior BW, Woof JM (2005) The influences of hinge length and composition on the susceptibility of human IgA to cleavage by diverse bacterial IgA1 proteases. J Immunol 174 (12):7792–7799
21. Butt U, ElShaer A, Snyder LA et al (2016) Fatty acid microemulsion for the treatment of neonatal conjunctivitis: quantification, characterisation and evaluation of antimicrobial activity. Drug Deliv Transl Res 6(6):722–734. https://doi.org/10.1007/s13346-016-0338-3
22. Churchward CP, Alany RG, Kirk RS et al (2017) Prevention of *ophthalmia neonatorum* caused by *Neisseria gonorrhoeae* using a fatty acid-based formulation. mBio 8(4). https://doi.org/10.1128/mBio.00534-17
23. Butt U, ElShaer A, Snyder LAS et al (2018) Fatty acid based microemulsions to combat *ophthalmia neonatorum* caused by *Neisseria gonorrhoeae* and *Staphylococcus aureus*. Nanomaterials (Basel) 8(1). https://doi.org/10.3390/nano8010051
24. Alekseev O, Tran AH, Azizkhan-Clifford J (2012) *Ex vivo* organotypic corneal model of acute epithelial herpes simplex virus type I infection. J Vis Exp (69):e3631. https://doi.org/10.3791/3631
25. Swanson J (1978) Studies on gonococcus infection. XII. Colony color and opacity variants of gonococci. Infect Immun 19:320–321
26. Black WJ, Schwalbe RS, Nachamkin I et al (1984) Characterization of *Neisseria gonorrhoeae* protein-II phase variation by use of monoclonal-antibodies. Infect Immun 45:453–457

Chapter 22

Use of Human Fallopian Tube Organ in Culture (FTOC) and Primary Fallopian Tube Epithelial Cells (FTEC) to Study the Biology of *Neisseria gonorrhoeae* Infection

A. Said Álamos-Musre, Alejandro Escobar, Cecilia V. Tapia, Myron Christodoulides, and Paula I. Rodas

Abstract

Epithelial cells represent one of the most important physical barriers to many bacterial pathogens. In the case of *Neisseria gonorrhoeae*, the epithelial cell response is critical because they are the main target of the tissue damage triggered by the pathogen, particularly when the organism reaches the Fallopian tube (FT). Although the irreversible damage triggered by *N. gonorrhoeae* in the FT has been previously reported (ectopic pregnancy, pelvic inflammatory disease and infertility), the mechanisms of gonococcal-induced tissue damage are not fully understood. In addition, the lack of animal models that efficiently mimic the human disease and the complexity of gonococcus–host interactions make studying gonococcal pathogenesis particularly difficult. The use of human immortalized cells is also limited, since a variety of commercial FT cell lines is not yet available. Finally, the phase and antigenic variation of many gonococcal surface molecules involved in attachment and invasion of epithelial tissues leads to a failure to reproduce results using different human cells lines used in previous studies. The FT organ in culture (FTOC) and primary human fallopian tube epithelial cell (FTEC) represent the closest ex vivo cell models to explore the biology of *Neisseria gonorrhoeae* during infection of the FT, since it is a natural host target of the gonococcus. In this chapter, we describe protocols to process human FT samples to obtain FTOC and FTEC and assess their response to infection with *Neisseria gonorrhoeae*.

Key words *Neisseria gonorrhoeae*, Human Fallopian tube, Explant, Primary epithelial cell, Infection

1 Introduction

The mucosal surface of the human female genital tract, in common with other mucosal surfaces (e.g., the respiratory and gastrointestinal tracts) represents a first line of defense against external injuries and microbial infection. It is composed of mechanical, chemical, and cellular components and mucosal epithelial cells act as a physical barrier and also as a stratified group of cells involved in ciliary function, motility, desquamation, and mucous secretion. They are involved also in releasing inflammatory modulators (cytokines and

Myron Christodoulides (ed.), *Neisseria gonorrhoeae: Methods and Protocols*, Methods in Molecular Biology, vol. 1997, https://doi.org/10.1007/978-1-4939-9496-0_22, © Springer Science+Business Media, LLC, part of Springer Nature 2019

chemokines) and other host markers, which trigger the recruitment of immune cells to the zone of injury [1]. Most of these functions are carried out by the mucosal barrier of the lower genital tract (vagina and ectocervix) [2], which acts synergistically with the genital microbiota [3] to prevent pathogens ascending to the upper regions of the genital tract (e.g., the uterus and Fallopian tubes (FTs)). FTs are arranged into longitudinal folds composed by both ciliated and secretory epithelial cells, which play a critical role in mammalian reproduction by eliciting spermatozoa transport and ovulation, fecundation and embryogenesis [4]. FT secretory epithelial cells have attracted particular interest, since they are involved in high-grade serous epithelial ovarian cancers, which is the most common ovarian cancer histological subtype [5–7]. In addition, FT secretory epithelial cells obtained from healthy human donors and subsequently immortalized have been used to develop in vitro cell models to characterize the phenotypic and genotypic differences between normal and malignant cells [6].

The FT is also one of the main targets of sexually transmitted pathogens such as *Chlamydia trachomatis* and *Neisseria gonorrhoeae*, which have developed successful strategies to overcome genital host barriers. These pathogens cause pelvic inflammatory disease (PID), an inflammatory disorder encompassing pathologies such as endometritis, salpingitis and tubo-ovarian abscess, which can lead to ectopic pregnancy and infertility if untreated or treated inappropriately [8, 9]. Both organisms display similar pathogenicity mechanisms, since they invade FT secretory cells directly and destroy mainly uninfected ciliated cells by inducing TNF-α-mediated loss of ciliary activity and cell sloughing [10]. In addition, other proinflammatory cytokines are produced by FTs in response to gonococcal infection, such as IL-1β, IL-6 and IL-8, which probably contribute to the massive tissue damage observed in gonococcal salpingitis [11]. Resolution of the inflammatory response and subsequent cell death involves infiltration of fibroblasts to the infection zone, which leads to permanent scarring, eventual functional impairment of FTs and irreversible infertility [12].

The first experimental evidence of tissue damage triggered by gonococcal infection in the FT was obtained from studies carried out with FT Organs in Culture (FTOCs), currently known as "FT explants" [13]. Most of our knowledge about gonococcal infection-induced FT tissue damage has come from experiments done with FTOCs and with primary cultures of FT epithelial cells (FTECs) obtained from patient donors [14–20]. However, the whole infective cycle of *N. gonorrhoeae* in the FT, particularly at the cellular and molecular level, has not been fully described. Gonococcus–host interactions are complex and depend not only on strain-specific differences prevalent among *N. gonorrhoeae* strains, but also on phase and antigenic variation occurring within many virulence factors [10, 21, 22]. The latter can be considered a mechanism of bacterial cell differentiation because they cause phenotypic divergence within a bacterial cell lineage, thus contributing to evasion of the immune response [23]. This explains the variety of results obtained in gonococcal infection of classical epithelial cell lines [24, 25]. In order to overcome this, natural cell targets of *N. gonorrhoeae* can be used to explore the cellular and molecular events involved in pathogenesis of this bacterium. In this chapter, we describe the FTOC and FTEC models of *N. gonorrhoeae* infection, from processing human FT samples through to infection and the protocols used to assess the bacteria-induced host cell responses.

2 Materials

All the reagent solutions for human sample processing for FTOC and FTEC models are prepared with cell culture grade consumables. A Class II biosafety (BL2) cabinet and a humidified cell culture incubator set to 37 °C with 5% (v/v) CO_2 are mandatory for handling both human and gonococcal samples as well as doing most of the protocols detailed in this chapter.

2.1 Fallopian Tube Organs in Culture (FTOC)

1. FT tissue samples. Samples should be provided by fertile donors undergoing hysterectomy or tubal ligation/resection for reasons unrelated to the experimental ex vivo studies. Exclusion criteria of human FT samples must include the occurrence of a sexually transmitted infection, tubal disease, pelvic inflammatory disease, or use of contraceptive methods within 3 months before surgery [26, 27] (*see* **Note 1**).
2. Sterile phosphate buffered saline (PBS), pH 7.4. Add 8 g/L of NaCl, 0.2 g/L of KCl, 1.15 g/L of Na_2HPO_4, and 0.2 g/L of KH_2PO_4 in distilled H_2O and autoclave at 121 °C and 2.68 kg/cm^2 for 15 min.
3. TC-199 cell culture base liquid medium containing Earle's salts, L-glutamine, phenol red, and sodium bicarbonate.
4. Complete TC-199 medium: add 10% (v/v) fetal bovine serum (FBS, not heat-inactivated), 1 mM sodium pyruvate, 10 μg/mL of human, recombinant insulin, 100 units/mL of penicillin and 100 μg/mL of streptomycin to the TC-199 cell culture base liquid medium.
5. 50 mL sterile conical centrifuge tubes.
6. 90 × 15 mm sterile Petri dishes.
7. Disposable scalpel, tweezers, and scissors for surgery.
8. Stereomicroscope.
9. 24-well cell culture plates.

2.2 Fallopian Tube Epithelial Cells (FTEC)

1. FT tissue samples.
2. Sterile phosphate buffered saline (PBS), pH 7.4.
3. TC-199 cell culture base liquid medium.
4. Complete TC-199 medium.
5. 0.25% (w/v) trypsin solution containing 2.21 mM ethylenediaminetetraacetic acid (EDTA), without sodium bicarbonate.
6. Hank's balanced salt solution (HBSS) with sodium bicarbonate, but with no phenol red, calcium chloride, or magnesium sulfate.
7. 0.4% (v/v) trypan blue solution.
8. 100% dimethyl sulfoxide (DMSO).
9. 100 μm cell strainers.
10. 50 mL sterile conical centrifuge tubes.
11. 1.5 mL sterile microcentrifuge tubes.
12. Sterile Pasteur pipettes.
13. 25 and 75 cm^2 tissue-culture (TC)-treated cell culture flasks (*see* **Note 2**).
14. 5 and 10 mL serological pipettes.
15. Sterile 90 mm × 15 mm Petri dishes.
16. 1.8 mL sterile cryogenic vials.
17. Hemocytometer.
18. Disposable scalpel, tweezers, and scissors for surgery.
19. Laboratory mini rocker.
20. Centrifuge with a swing bucket rotor for conical tubes (15 and 50 mL).
21. Stereomicroscope and inverted microscope.
22. −80 °C freezer or liquid nitrogen storage tank.

2.3 Expansion of FTEC Cultures

1. Cryovial containing frozen FTEC or cells seeded into T75 cm^2 flasks from directly prepared FT tissue samples.
2. Complete TC-199 medium.
3. 25 and 75 cm^2 tissue-culture (TC)-treated cell culture flasks.
4. HBSS.
5. 0.25% (w/v) Trypsin solution containing 2.21 mM EDTA, without sodium bicarbonate.
6. 0.4% (v/v) trypan blue solution.
7. Hemocytometer.
8. Sterile Pasteur pipettes.

9. 25 and 75 cm^2 tissue-culture (TC)-treated cell culture flasks (*see* **Note 2**).
10. 5 and 10 mL serological pipettes.
11. Centrifuge with a swing bucket rotor for conical tubes (15 and 50 mL).
12. 12- or 24-well TC-treated cell culture plates.
13. Water bath.
14. Inverted microscope.

2.4 Culture of *Neisseria gonorrhoeae*

1. *Neisseria gonorrhoeae* strain P9 variant 17 (Pil^+ Opa^+) [28] or other gonococcal strains (*see* **Note 3**).
2. GC agar base: add 10 g of proteose peptone, 1.01 g of corn starch, 4.03 g of K_2HPO_4, 1.01 g of KH_2PO_4, 5.03 g of NaCl, and 10.07 g of agar to 1 L of distilled H_2O and autoclave for 15 min at 121 °C and 2.68 kg/cm^2 [29].
3. BBL IsovitaleX™ Enrichment or Kellogg's Supplements. Supplement I (100×): add 40 g of glucose and 2 mg of thiamine pyrophosphate to 100 mL of distilled H_2O and sterilize by filtration using a 0.22 μm filter, then store at −20 °C. Supplement II (1000×): add 50 mg of $Fe(NO_3)_3 \cdot 9H_2O$ to 100 mL of distilled H_2O and sterilize by filtration using a 0.22 μm filter. Store at −20 °C [29].
4. Sterile 90 mm × 15 mm Petri dishes.
5. GC agar plates: add 10 mL of Supplement I and 1 mL of Supplement II per 1 L of GC agar base cooled to ~50 °C, prior to pouring into Petri dishes.
6. 10 μL sterile disposable inoculation loops.
7. Stereomicroscope.

2.5 Infection of FTOC with *Neisseria gonorrhoeae*

1. Infection Medium: TC-199 cell culture base medium supplemented with 1 mM sodium pyruvate and 10 μg/mL of human, recombinant insulin. Do not add antibiotics or FBS.
2. HBSS.
3. Lysis buffer: 1% (w/v) Sodium Dodecyl Sulfate (SDS) and 0.1% (w/v) sodium hydroxide (NaOH) prepared in distilled H_2O.
4. GC broth: add 1.5 g of proteose peptone, 0.4 g of K_2HPO_4, 0.1 g to KH_2PO_4, and 0.1 g of NaCl to distilled H_2O, adjust the pH to 7.2 with HCl and make up to a final volume of 100 mL with H_2O [29]. Autoclave for 15 min at 121 °C and 2.68 kg/cm^2 and store at room temperature.
5. GC agar plates.
6. 1.5 mL sterile microcentrifuge tubes.

7. 15 mL sterile conical centrifuge tubes.
8. 10 μL sterile disposable inoculation loops.
9. 0.22 μm syringe filters.
10. 96-well cell culture plate.
11. Spectrophotometer.
12. Ultracentrifuge.
13. Inverted microscope.

2.6 Infection of FTEC with *Neisseria gonorrhoeae*

All materials described in Subheading 2.5.

2.7 Gentamicin Protection Assay

1. Infection Medium.
2. 10 mg/mL stock gentamicin solution.
3. 1% (w/v) saponin in cell culture grade H_2O or PBS.
4. 15 mL sterile centrifuge tubes.
5. 1.5 mL sterile microcentrifuge tubes.
6. GC broth.
7. GC agar plates.
8. 1.5 mL sterile microcentrifuge tubes.
9. 96-well cell culture plate.
10. Multichannel pipette.

2.8 Cytotoxicity Assay

1. Supernatants of both uninfected and infected FTOC and FTEC cultures.
2. CytoTox 96® non–radioactive cytotoxicity assay.
3. 96-well plates.
4. ELISA microplate reader.

2.9 Immunofluorescence of FTEC

1. 12 mm poly-D-lysine coated coverslips.
2. 12-well cell culture plates.
3. PBS, pH 7.4.
4. 2% (w/v) paraformaldehyde (PFA) in cold PBS.
5. Alexa Fluor 555® phalloidin.
6. 0.2% (v/v) Triton X-100 in PBS.
7. 1% (w/v) bovine albumin serum (BSA) in PBS.
8. Fluoromount® aqueous mounting medium.
9. Microcentrifuge tubes.
10. Parafilm.
11. Tweezers.
12. Fluorescence microscope.

2.10 Gelatin Zymography

1. Bio-Rad minigel system and 1 mm thick gels with glass plates.
2. 70% (v/v) ethanol.
3. 1.5% (w/v) gelatin type A from porcine skin in double-distilled water (dd) H_2O. Store stock solutions in aliquots of 1 mL at −20 °C. Thaw the 1.5% (w/v) gelatin stock by incubation in a water bath at 35 °C. Check that the solution is completely liquid before use (*see* **Note 4**).
4. 12.5% polyacrylamide separating gel with 0.1% (w/v) gelatin: in the following order, mix 3.05 mL of ddH_2O, 2.655 mL of 1.5 M Tris–HCl buffer, pH 8.8, 4.185 mL of 30% (v/v) acrylamide–bisacrylamide solution (ratio of 37.5:1), 101.3 μL of 10% (w/v) SDS, and 49.5 μL of 10% (w/v) ammonium persulfate (APS). This recipe is to cast two gels. Cool the thawed 1.5% (w/v/) gelatin solution and add 675 μL to the 12.5% polyacrylamide separating gel solution to give 0.1% (w/v) gelatin and mix quickly (*see* **Note 5**). Add 4.95 μL of *N*,*N*,*N*′,*N*′-tetramethylethylenediamine (TEMED), mix again and immediately pour 5.5 mL of the separating gel solution between the glass plates leaving 1.0 cm approx. of space for the stacking gel. Overlay carefully with 1 mL of ddH_2O (*see* **Note 6**) and allow the gel to polymerize for 40 min approx.
5. 4% stacking gel solution: mix 3.71 mL of ddH_2O, 550 μL of 0.5 M Tris–HCl buffer, pH 6.8, 675 μL of 30% (v/v) acrylamide–bisacrylamide solution (37.5:1), 50 μL of 10% (w/v) SDS, and 30 μL of 10% (w/v) APS. Remove the water from the cast 12.5% separating gel and carefully remove any remaining water using filter paper, but avoid touching the polymerized gel with the paper. Add 7.5 μL of TEMED to the stacking gel solution and mix quickly. Pour 1.3 mL of the stacking gel solution above the polymerized separating gel and insert the comb. Let the gel to polymerize for ~20 min.
6. 4× sample loading buffer: 10 mL of 500 mM Tris–HCl buffer, pH 6.8, containing 0.1% (v/v) glycerol, 10% (w/v) SDS, and 0.1% (w/v) bromophenol blue. Distribute into aliquots of 1 mL and store at −20 °C. Thaw an aliquot of 4× loading buffer to room temperature during gel polymerization. Do not use a cooled aliquot when mixing with cell supernatants before gel loading.
7. Human recombinant matrix metalloproteinase (MMP)-2 and MMP-9 activity standard markers.
8. Supernatants from uninfected and infected FTOC or FTEC cultures.
9. 10× SDS-PAGE running buffer: 3% (w/v) Tris, 14.4% (w/v) glycine and 1% (w/v) SDS in ddH_2O. Store at room temperature. Dilute 100 mL of this buffer in 900 mL of ddH_2O to prepare 1 L of 1× SDS-PAGE running buffer.

10. 10× Zymography Renaturating (Washing) Buffer: 25% (v/v) Triton X-100 in ddH_2O. Store at room temperature [26]. Dilute 20 mL of this 10× stock with 180 mL of ddH_2O to make 200 mL of 1× Zymography Renaturating (Washing) Buffer.
11. 10× Zymography Development Buffer: 50 mM Tris–HCl buffer, pH 7.5, containing 200 mM NaCl, 5 mM $CaCl_2$, and 0.02% (w/v) sodium azide (NaN_3) [26]. Dilute 100 mL of this 10× Zymography Development Buffer with 900 mL of ddH_2O to make 1 L of 1× Zymography Development buffer.
12. Gel staining solution: 0.5% (w/v) Coomassie Blue R-250 in a solution of 50% (v/v) methanol, 10% (v/v) acetic acid, and 40% (v/v) distilled H_2O (5:1:4) [30] (*see* **Note 7**).
13. Gel destaining solution: gel staining solution without the Coomassie blue dye [30].
14. Optional: DryEase® Mini-Gel Drying System.

2.11 Total Protein Fractionation of FTEC

1. PBS, pH 7.4.
2. RIPA buffer: 50 mM Tris–HCl pH 8.0, containing 150 mM NaCl, 1% (v/v) Nonidet™ NP-40, 0.5% (w/v) sodium deoxycholate, 0.1% (w/v) SDS. Store at 4 °C.
3. Halt™ protease inhibitor cocktail, EDTA-free.
4. 1.5 mL sterile microcentrifuge tubes.
5. Ultracentrifuge.
6. Cell scrapers.
7. −20 °C freezer.

2.12 Cytoskeleton-Associated Protein Fractionation

1. Cytoskeleton extraction (CSK) buffer: 10 mM piperazine-*N*,*N*′-bis(2-ethanesulfonic acid) (Pipes) buffer, pH 6.8 containing 50 mM NaCl, 300 mM sucrose, 3 mM $MgCl_2$ and 0.5% (v/v) Triton X-100 [31].
2. SDS buffer: 15 mM Tris buffer, pH 7.5, containing 5 mM EDTA, 2.5 mM ethylene glycol-bis(β-aminoethyl ether)-*N*,*N*,*N*′,*N*′-tetraacetic acid (EGTA) and 1% (w/v) SDS [31].
3. ACK lysing buffer: 0.15 M NH_4Cl, 10 mM $KHCO_3$, 0.1 mM Na_2EDTA [31].
4. Halt™ protease inhibitor cocktail, EDTA-free.
5. 1.5 mL microcentrifuge tubes.
6. Ultracentrifuge.
7. Cell scrapers.

2.13 Tricine-SDS-PAGE and Western Blotting

1. Bio-Rad minigel and transfer tank system with 1 mm thick gels [32, 33].
2. AB-3 stock solution: dissolve 48 g of acrylamide and 1.5 g of bisacrylamide (each twice-crystallized) in 100 mL of ddH_2O. S-tore at 4 °C.
3. 3× Gel buffer: 3 M Tris–HCl buffer, pH 8.45, containing 0.3% (w/v) SDS. Store at 4 °C.
4. 16% polyacrylamide separating gel: mix 3.45 mL of ddH_2O, 5 mL of AB-3 stock solution, 5 mL of 3× gel buffer, 1.5 mL of 100% glycerol, 50 μL of 10% (w/v) APS, and 5 μL of TEMED. This recipe is to cast two gels. Mix and immediately pour 6 mL of the separating gel solution into the glass plates, leaving 1.0 cm of space for the stacking gel. Overlay carefully with 1 mL of ddH_2O (*see* **Note 6**). Let the gel polymerize for ~50 min.
5. 4% stacking gel solution: mix 3.95 mL of ddH_2O, 500 μL of AB-3 stock solution, 1.5 mL of 3× gel buffer, and 45 μL of 10% (w/v) APS. Remove the water from the cast 12.5% separating gel and carefully remove any remaining water using filter paper, but avoid touching the polymerized gel with the paper. Add 4.5 μL of TEMED to the stacking gel solution and mix quickly. Pour 1.5 mL of the stacking gel solution above the polymerized separating gel and insert the comb. Let the gel polymerize for ~20 min.
6. Buffer A: 150 mM Tris–HCl buffer, pH 7.0, containing 12% (w/v) SDS, 6% (v/v) mercaptoethanol, 30% (v/v) glycerol, and 0.05% Coomassie blue G-250 [33]. Store at 4 °C.
7. Buffer A/4: buffer A diluted with 3 volumes of ddH_2O.
8. 10–180 kDa PageRuler™ Prestained Protein Ladder.
9. 10× Anode buffer: 1 M Tris–HCl buffer, pH 8.9 [33]. Dilute 80 mL of this 10× buffer with 720 mL of ddH_2O to make 800 mL of 1× anode buffer.
10. 10× Cathode buffer: 1 M Tris–HCl buffer, pH 8.25, containing 1 M Tricine and 1% (w/v) SDS [33]. Dilute 20 mL of this 10× buffer with 180 mL of ddH_2O to make 200 mL of 1× cathode buffer.
11. Polyvinylidene difluoride (PVDF) membranes.
12. Transfer buffer for Western Blotting: 1× SDS-PAGE running buffer (*see* **step 9** of Subheading 2.10) with 20% (v/v) methanol. Mix 100 mL of 1× SDS-PAGE running buffer with 200 mL of methanol and 700 mL of ddH_2O to make 1 L of transfer buffer.
13. 100% methanol.
14. Glass trays or glass Petri dishes.

15. Clear plastic film.
16. TBS buffer: 25 mM Tris–HCl buffer, pH 8.0 containing 125 mM NaCl.
17. Membrane blocking buffer: TBS containing 0.1% (v/v) Tween® 20 and 5% (w/v) bovine serum albumin (BSA).
18. Membrane washing buffer: TBS containing 0.05% (v/v) Tween® 20 (TBST).
19. Antibody dilution buffer: TBST containing 1% (w/v) BSA.
20. Anti-FLAG®M2 antibody, diluted 1/5000 in antibody dilution buffer.
21. Human β-actin antibody, human E-cadherin antibody and human β-tubulin antibody. Dilute each antibody 1/5000 in antibody dilution buffer.
22. Anti-mouse IgG–horseradish peroxidase conjugate, diluted 1/10,000 in antibody dilution buffer.
23. SuperSignal West Pico Chemiluminescent substrate.
24. 15 mL conical tubes.
25. Clear plastic film.
26. Western blot digital imaging system.

3 Methods

3.1 Fallopian Tube Organs in Culture (FTOC)

We use the protocol described previously by McGee et al. [13] with some modifications.

1. Transport human FT samples in cold, complete TC-199 medium in a 50 mL conical tube. Process the samples within 2 h of collection.
2. Rinse the FT samples with cold PBS in order to wash away blood and then transfer it to a Petri dish using sterile tweezers.
3. Cut longitudinally and discard the connective tissue to expose the lumen. Wash strips of mucosal folds with TC-199 base medium.
4. Cut the FT samples into pieces of ~3–4 mm under a stereomicroscope and transfer them to a 24-well cell culture plate containing complete TC-199 medium. The FT explants must be completely covered by 1.0–1.5 mL of medium (Fig. 1).
5. Incubate at 37 °C with 5% (v/v) CO_2 for up to 24 h. FTOCs must then be used for infection (*see* Subheading 3.5), otherwise discard the sample.

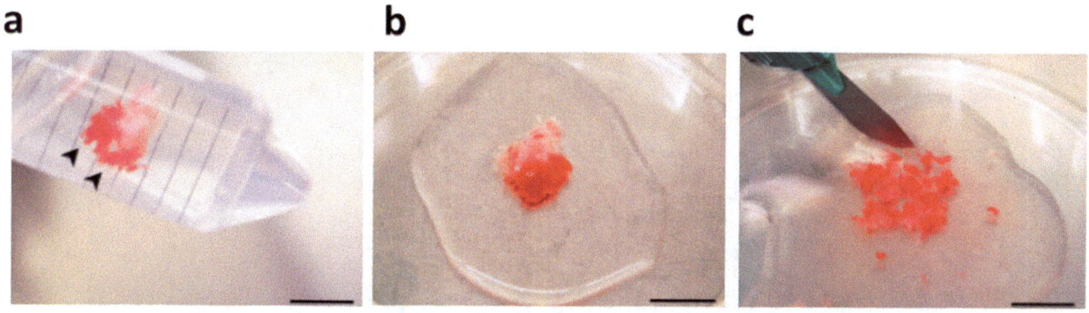

Fig. 1 Processing Fallopian tube samples. The FT sample must be transported in cold, sterile PBS (**a**). After rinsing with PBS, the FT sample is put into a sterile Petri dish (**b**) and minced into ~3–4 mm pieces under a stereomicroscope (**c**). Scale bars: 10 mm. Image taken from [6]

3.2 Fallopian Tube Epithelial Cells (FTEC)

We use a protocol described previously [6, 34] with some modifications [26, 32].

1. Transport human FT samples in cold, sterile PBS in a 50 mL conical tube. Store samples at 4 °C and process them within 24 h of collection.
2. Repeat **steps 2** and **3** of Subheading 3.1.
3. Cut the FT samples into smaller pieces (~1–3 mm) under a stereomicroscope and transfer them to a 50 mL conical tube containing 5 mL of 0.25% (v/v) Trypsin-EDTA solution. Incubate at 37 °C with 5% (v/v) CO_2 for 15 min.
4. Add 5 mL of complete TC-199 medium to inactivate trypsin, invert the tube twice to suspend the minced tissue and then leave for 10–15 min to sediment the bulky pieces.
5. Transfer to a cell strainer and a 50 mL conical tube. Wash the tissue pieces gently four times with HBSS.
6. Transfer the medium containing the cell suspension to 50 mL conical tubes and centrifuge at 1500 × *g* for 5 min at room temperature.
7. Wash the pellet once gently with complete TC-199 medium and remove and discard the supernatant using a Pasteur pipette. Suspend the cell pellet in 3 mL of complete TC-199 medium. Pipette the cell pellet gently to break up any cell clumps.
8. Transfer the cell suspension into three T75 cm^2 cell culture flasks each containing 14 mL of pre–warmed complete TC-199 medium.
9. Incubate for 2 h at 37 °C and 5% (v/v) CO_2 to enable adherence of fibroblasts and red blood cells to the plastic surface.
10. Carefully transfer the medium containing the unattached epithelial cells to 50 mL conical tubes and centrifuge at 1500 × *g* for 5 min at room temperature.

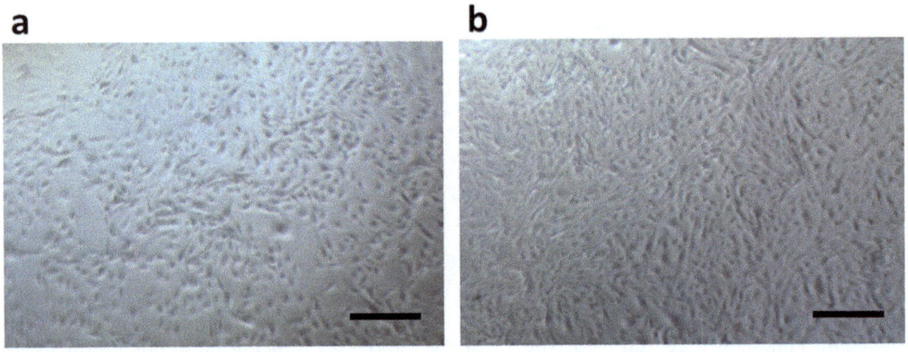

Fig. 2 Culture of FTECs. "Island" colonies of epithelial cells tend to expand independently until they reach each other, then a uniform layer of cells is observed in the cell culture flask. (**a**) Five days after seeding; (**b**) 7 days after seeding. Scale bars: 100 μm

11. The resulting cell pellet contains FTEC. To count the cells, suspend cells in 1 mL of complete TC-199 medium, then transfer 10 μL into a microcentrifuge tube and mix with 10 μL of trypan blue solution, incubate for 5 min at 37 °C and then place 10 μL of the mix onto a glass hemocytometer. View under an inverted microscope and count cells.
12. Transfer the cell suspension into a cryogenic vial and add DMSO to at final concentration of 10% (v/v) and store the vial in a −80 °C freezer or liquid nitrogen (*see* **Note 8**). Alternatively, seed the cells directly into T75 cm^2 cell culture flasks each containing 14 mL of pre–warmed complete TC-199 medium and incubate at 37 °C with 5% (v/v) CO_2 for ≥5 days or until morphologically round "island" colonies of epithelial cells are visible [6, 26, 32] (Fig. 2) (*see* **Notes 9–11**).
13. Remnant epithelial cells can be found in the supernatant discarded in **step 11**. Put into a T75 cm^2 cell culture flask and incubate for 24–48 h at 37 °C with 5% (v/v) CO_2, which allows these cells to adhere to the plastic surface of the flask.

3.3 Expansion of FTEC Cultures

1. Thaw a cryovial tube as described in **step 12** of Subheading 3.2 and containing FTEC, by transferring to a 37 °C water bath for 5 min. Immediately transfer the thawed contents to a sterile microcentrifuge tube using a Pasteur pipette and centrifuge for 5 min at 1500 × *g* and room temperature.
2. Repeat the centrifugation in **step 1** twice, suspending the cells with pre–warmed complete TC-199 medium.
3. Suspend the cells gently in 1 mL of pre–warmed complete TC-199 medium and transfer to a T75 cm^2 cell culture flask containing 14 mL of prewarmed complete TC-199 medium using a Pasteur pipette (*see* **Note 12**). If the cell density is low, use a T25 cm^2 cell culture flask containing 4 mL of pre–warmed complete TC-199 medium.

4. Incubate the cells at 37 °C with 5% (v/v) CO_2 for ≥3 days or until morphologically round "island" colonies of epithelial cells are visible [6, 26, 32] (Fig. 2) (*see* **Notes 10** and **11**). To remove unattached cells and debris, replace a half-flask volume with fresh complete TC-199 medium on days 2–3 using a serological pipette.
5. Observe the cells using an inverted microscope. Depending of the level of cell expansion and medium consumption, use serological pipettes to replace the medium with fresh pre-warmed complete TC-199 medium. Do this only when a monolayer of attached cells is observed at the bottom of the cell culture flask (Fig. 2).
6. If cells are growing well (with ≤70% confluence), expand them to two T75 cm^2 cell culture flasks by trypsin treatment. Remove the medium from the flask using serological pipettes, wash the attached cells twice and carefully with pre-warmed HBSS (each wash a volume of 2.5 mL for a T25 cm^2 flask, 5 mL for a T75 cm^2 flask) and then add pre-warmed trypsin-EDTA solution (0.5 mL for a T25 cm^2 flask, 1 mL for a T75 cm^2 flask). Incubate for 5–10 min at 37 °C with 5% (v/v) CO_2 until cell detachment, then add pre-warmed complete TC-199 medium to inactivate trypsin (2 mL for a T25 cm^2 flask, 4 mL for a T75 cm^2 flask). Transfer the cells to a 15 mL conical tube using a serological pipette and centrifuge 5 min at 1500 × *g* and room temperature. Wash the pellet once with pre-warmed complete TC-199 medium (3 mL for a T2 5cm^2 flask, 5 mL for a T75 cm^2 flask), suspend the cells gently and repeat the centrifugation. Suspend the cells into pre-warmed complete TC-199 medium and transfer to T75 cm^2 flasks containing 14 mL of the same medium. Incubate at 37 °C with 5% (v/v) CO_2 and check daily for cell attachment using an inverted microscope. Cells can continue to expand by repeating the trypsin digestion and seeding into new cell culture flasks (*see* **Note 13**).
7. Detach the FTEC grown in cell culture flasks by trypsin treatment as described in **step 6** and count cells as described in **step 11** of Subheading 3.2.
8. Seed the cells into 12 or 24-well cell culture plates. Use 1.0×10^5 cells/mL for 12-well plates or 1.0×10^4 cells/mL for 24-well plates. Incubate at 37 °C with 5% (v/v) CO_2 until ≥70% cell confluence in each well.

3.4 Culture of *Neisseria gonorrhoeae*

1. Plate bacteria from a stock of bacterial cell suspension (frozen in 10% (v/v) DMSO and GC broth supplemented with 1% Isovitalex™ Enrichment or Kellogg's Supplements I and II) onto GC plates (*see* **Note 14**). Grow bacteria for 18–24 h at 37 °C with 5% (v/v) CO_2.

2. Check the piliated (Pil^+) and opacity protein (Opa^+) colony phenotype by viewing the bacterial growth on the agar plates with a stereomicroscope (*see* **Note 3**). Select colonies with the Pil^+ Opa^+ phenotype and plate onto new GC agar plates using disposable inoculation loops to produce sufficient bacterial growth for infection of FTOC and FTEC. Incubate the GC agar plates for 18–24 h at 37 °C with 5% (v/v) CO_2 [29] (*see* **Notes 14** and **15**).

3.5 Infection of FTOC with *Neisseria gonorrhoeae*

1. Remove the cell culture medium from each well (*see* **steps 4** and **5** of Subheading 3.1). Wash the explants gently with 1 mL of pre-warmed PBS and add 1.0–1.5 mL of prewarmed Infection Medium. Incubate for 12 h at 37 °C with 5% (v/v) CO_2.
2. Collect gonococcal colonies previously streaked onto GC plates and check the phenotype as described in **step 2** of Subheading 3.4. Suspend the bacteria gently using a disposal inoculation loop into 1 mL of pre-warmed Infection Medium in a sterile microcentrifuge tube.
3. Mix 100 μL of the gonococcal suspension with 900 μL of Lysis Buffer. Transfer the volume into a UV cuvette and measure the absorbance at $\lambda_{260\ \mathrm{nm}}$ using a spectrophotometer. An absorbance at $\lambda_{260\ \mathrm{nm}}$ of 1 is equivalent to ~1.0×10^9 P9 gonococci/mL.
4. Calculate an inoculum of 2.0×10^5 P9 bacterium/mL per well. Prepare the suspension in 15 mL conical tubes containing pre-warmed Infection Medium. Remove the cell medium of each well, wash the FTOC once with pre-warmed PBS gently, and add 200–300 μL of bacterial suspension per well. Check that the FTOC is completely covered by the bacterial suspension. Do not discard the removed cell medium (*see* **step 6**).
5. Determine the number of viable gonococci in the organ cultures at the beginning of the experiment by plating out serial tenfold dilutions of the inoculum onto GC plates: add 180 μL of pre-warmed GC broth into 8 wells of a sterile 96-well plate and add 20 μL of the original inoculum (*see* **Note 16**). Dilute the suspension from the first well serially across the 8 wells and then add 10 μL of each dilution onto pre-warmed GC agar plates in duplicate. Incubate for 18–24 h at 37 °C with 5% (v/v) CO_2 and count the colony forming units (CFUs) [19].
6. At time 0 h and thereafter, recover the Infection Medium from each well (for bacteria-infected and for uninfected controls) and transfer to microcentrifuge tubes, centrifuge at 13,000 × *g* for 20 min to sediment cell debris and bacteria and transfer the supernatants to new microcentrifuge tubes. Sterilize the supernatant samples by filtration using 0.22 μm filters and syringes. Prepare aliquots of 100 μL and store at 4 °C

for subsequent cytotoxicity measurements (*see* **Note 17**). Store the remaining supernatants at −80 °C for measuring MMP activity, ELISA or other experiments (*see* **Note 18**).

7. Incubate at 37 °C and 5% (v/v) CO_2. Incubation time depends on the purpose of the experiment: times of 1–4, 6, 12 and 24 h have been reported for measurement of secreted cytokines by ELISA, gonococcal attachment and internalized bacteria by immunohistochemistry, Real-Time PCR and zymography [27, 35–37], and 24 h of infection has been previously reported for quantifying tissue damage by gonococcal lipooligosaccharide (LOS) [16].
8. The gonococcal attachment and internalization into FTOC can be studied by immunofluorescence using an anti–gonococcal serum and the corresponding FITC-conjugated antibody [37] or infecting explants with a GFP-expressing *N. gonorrhoeae* strain [38].

3.6 Infection of FTEC with Neisseria gonorrhoeae

1. Repeat **steps 1–3** of Subheading 3.5.
2. Calculate the inoculum required for infection of FTEC with a Multiplicity Of Infection (MOI) of 10 or 50, that is, 10 or 50 bacteria to 1 epithelial cell using the calculated number of cells seeded into the wells (*see* **step 8** of Subheading 3.3) [26, 39]. Prepare the MOI suspensions in 15 mL conical tubes containing pre–warmed Infection Medium, calculating 500 μL of bacterial suspension per well for 12-well plates or 250 μL per well for 24-well plates.
3. Remove the cell medium from each well, wash the FTEC once with pre–warmed HBSS gently, and then add the corresponding volume of bacterial suspension to each well. Maintain a sufficient number of control uninfected cell monolayers for each time point by adding Infection Medium alone, that is, no bacteria. Do not discard the removed cell medium (*see* **step 6** of Subheading 3.5).
4. Quantify the MOI as described in **step 5** of Subheading 3.5: this must be done to calculate the number of internalized bacteria.
5. Repeat **step 6** of Subheading 3.5.
6. Incubate the cells with bacteria for 4, 8, 12, and 24 h at 37 °C with 5% (v/v) CO_2 for measurement of cytotoxicity, MMP activity, ELISA, quantification of intracellular bacteria and other experiments [26].

3.7 Gentamicin Protection Assay

For FTEC cultures infected with *N. gonorrhoeae*, we use a protocol described previously [31] with some modifications [26].

1. Dilute the stock solution of 10 mg/mL gentamicin in Infection Medium to the working concentration of 100 μg/mL in a sterile 15 mL conical tube. Pre–warm this solution at 37 °C.
2. Wash the infected FTEC monolayers three times gently with prewarmed HBSS or PBS and then add 0.5 (for 24-well) or 1 mL (for 12-well) of gentamicin solution. Incubate for 2 h at 37 °C with 5% (v/v) CO_2 to kill extracellular bacteria.
3. Wash the FTEC monolayers three times gently with pre–warmed HBSS or PBS and then add 150 μL of 1% (v/v) saponin lysis solution to each well. Incubate for 15 min at room temperature to permeabilize the cells, which can also be scraped with a pipette tip to enrich the recovery of intracellular bacteria.
4. Transfer 50 μL of the bacteria–saponin solution into a well of a 96-well cell culture plate containing 50 μL of prewarmed GC broth (*see* **Note 16**). Dilute this suspension serially over an additional four wells containing 50 μL of pre–warmed GC broth. Plate 20 μL of each bacterial dilution, as well as the undiluted bacteria–saponin solution, in duplicate onto pre–warmed GC agar plates and incubate at 37 °C with 5% (v/v) CO_2 for 24 h.
5. Count CFU and determine the percentage of internalized bacteria by comparing with the initial count of gonococci made in **step 5** of Subheading 3.5. Consider the initial CFU counting as 100%.

3.8 Cytotoxicity Assay

1. Add supernatants from infected and control FTOC or FTEC monolayers into a 96-well microtiter plate.
2. Use the CytoTox® non-radioactive cytotoxicity assay to determine FTOC or FTEC damage during infection, by measuring the release of lactate dehydrogenase (LDH) into the cell supernatants, following the manufacturer's instructions [26].

3.9 Immunofluorescence of FTEC

1. Prepare FTEC as described in Subheadings 3.2 and 3.3, and seed the cells into 12- or 24-well cell culture plates containing 12 mm round poly-D-lysine-coated coverslips (*see* **Note 19**). Infect the cells with *N. gonorrhoeae* as described in Subheading 3.6.
2. After recovery and processing of the supernatants as described in **step 6** of Subheading 3.5 and after gentamicin treatment as described in Subheading 3.7, wash the cells gently with 1 mL of PBS, twice.

3. Fix the FTEC monolayers with cold 2% (w/v) PFA solution. Incubate for 20 min at room temperature.
4. Remove the 2% (w/v/) PFA solution gently with a Pasteur pipette and wash the fixed cells three times with 1 mL of PBS each (*see* **Note 20**).
5. Remove the PBS from each well and add 500 μL of 0.2% (v/v) Triton X-100 in PBS to permeabilize the cells (*see* **Note 21**). Incubate for 10 min at room temperature.
6. Remove the 0.2% (v/v) Triton X-100 in PBS solution and wash four times with PBS.
7. Add 500 μL of 1% (w/v) BSA in PBS to each coverslip-containing well. Incubate for 30 min at room temperature to reduce non–specific background staining.
8. Repeat the washing procedure of **step 4**. Dry the coverslips gently using paper tissue.
9. Dilute the Alexa Fluor 555® Phalloidin conjugate 1/200 in PBS in a microcentrifuge tube. Estimate the volume of conjugate required (25 μL per coverslip) (*see* **Note 22**).
10. Place 25 μL of Alexa Fluor 555® Phalloidin conjugate (1/200) onto a piece of Parafilm. Place the coverslips cell-side down onto the dye dilution and incubate for 20 min at room temperature. To avoid evaporation, keep the coverslips inside a covered container during the incubation.
11. Place the coverslips cell-side up into wells of a 12- or 24-well cell culture plate containing 500 μL of PBS, and wash four times with PBS.
12. Put 10 μL of bubble-free Fluoromount® mounting medium onto a glass slide, and then put the coverslip cell-side down onto the mounting medium. Incubate overnight at room temperature in a cool and dark place. When assembling coverslips over the mounting medium on a slide, leave ~3–5 mm of the slide free between them (*see* **Note 23**).
13. View the fixed samples slices using a fluorescence microscope, following the equipment manufacturer's instructions (Fig. 3).

3.10 Gelatin Zymography

The following instructions are for a Bio-Rad minigel system with 1 mm thick gels (*see* **Note 24**) [26, 27].

1. Clean the glass plates with 70% (v/v) ethanol and dry in air.
2. Assemble the minigel system according to the manufacturer's instructions.
3. Mix 1 volume of FTOC or FTEC supernatant(s) or MMP standard(s) with 3 volumes of 4× Sample Buffer (*see* **Note 25**). We usually use 8–12 μL of sample.
4. Load the MMP standard. If it is unavailable, use a typical protein ladder, but these contain β-mercaptoethanol, a

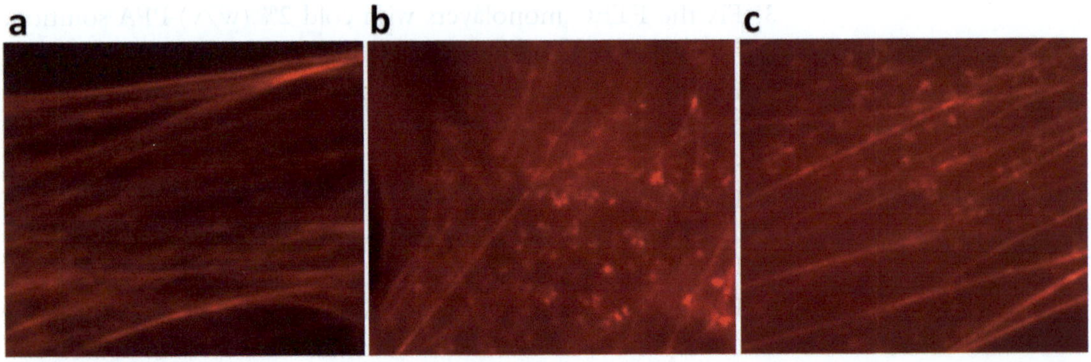

Fig. 3 Cytoskeleton rearrangement during infection of FTECs with *Neisseria gonorrhoeae*. Cells were infected with *N. gonorrhoeae* (Pil^+ Opa^+) at MOI of 50 during 8 h, then permeabilized cells were stained with Alexa Fluor 555® phalloidin (red). Note that gonococcal infection triggers an unusual cytoskeleton rearrangement with some point contact-like depositions shedding from stress fibers. (**a**) Uninfected cells. (**b**, **c**) FTECs infected with gonococci. Images were obtained by use of Olympus BX-61 Upright Microscope (100×)

reducing agent that could leak into the adjacent wells. Leave an empty well between the protein ladder-containing well and the supernatant sample-containing well to avoid mixing with the MMP-containing supernatant sample (*see* **Note 24**).

5. Load samples into the zymogram and assemble the gel unit. Start the electrophoresis with 50 V using a power supply until the samples reach the stacking gel bottom, and then run at 100 V through the separating gel until the blue dye front reaches the bottom of the gel.
6. Disconnect the electrophoresis system from the power supply and extract the gel using a spatula. Be careful not to touch the zymogram with your fingers. Quickly put the gel into a plastic container (*see* **Note 24**) and add 30–50 mL of 1× Zymography Renaturating (Washing) Buffer. Incubate at room temperature with gentle agitation for 15 min (*see* **Note 26**).
7. Rinse the zymogram three times, replacing with new 1× Zymography Renaturating Buffer each time.
8. Discard the 1× Zymography Renaturating Buffer and add 30 mL of 1× Zymography Development Buffer. Incubate for 30 min at room temperature with gentle agitation.
9. Replace with 50 mL of 1× Zymography Development Buffer and cover the plastic container with clear plastic film. Incubate for 12 h at 37 °C and gentle agitation (*see* **Notes 26** and **27**). Check that the zymogram is completely soaked, since the buffer tends to evaporate.
10. Rinse the zymogram three times for 5 min with ddH_2O and gentle agitation.

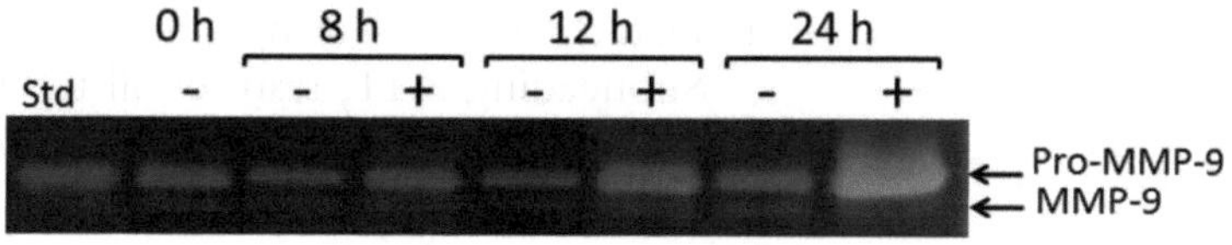

Fig. 4 Zymography of human MMP-9 secreted after infection of FTECs with *Neisseria gonorrhoeae*. Cells were infected with *N. gonorrhoeae* (Pil$^+$ Opa$^+$) at MOI of 50 as described in Methods (Subheading 3.6), then supernatants were recovered and handled as detailed in Subheadings 3.5 and 3.10. Image modified from [26]

11. Place the zymogram into a lid-covered plastic container and add 30–50 mL of gel staining solution. Incubate for 1 h at room temperature with gentle agitation.
12. Add 50 mL of gel staining solution and agitate gently at room temperature. Replace the solution when it becomes dye-saturated and repeat the procedure until clear bands of gelatin protease activity are visible against the dark dye background. Both zymogen (latent) and proteolytically active form of MMP are detected by zymography [26, 27] (Fig. 4).
13. Store the gels in ddH_2O and dry using the DryEase® Mini-Gel Drying System.

3.11 Total Protein Fractionation of FTEC

1. Infect FTEC monolayers with bacteria and treat with gentamicin as described in Subheading 3.6 and **steps 1** and **2** of Subheading 3.7, respectively.
2. Wash the cells three times gently with pre-warmed HBSS or PBS.
3. Add 500 μL of PBS to each well and scrape the well bottom vigorously with cell scrapers to detach and mechanically lyse the cells.
4. Recover all the PBS volume and transfer to microcentrifuge tubes. Centrifuge at 8000 × *g* for 5 min at 4 °C.
5. Discard the supernatant and add 50 μL of RIPA buffer, previously supplemented with 1× Halt™ Protease Inhibitor Single-Use Cocktail EDTA-Free.
6. Suspend the pellets gently and store at −20 °C [32].

3.12 Cytoskeleton-Associated Protein Fraction

1. Infect FTECs with bacteria and treat with gentamicin as described in Subheading 3.6 and **steps 1** and **2** of Subheading 3.7, respectively.
2. Wash cells as described in **step 2** of Subheading 3.11.
3. Add 300 μL of CSK buffer per well and store at 4 °C for 10 min.

4. Detach the cells using cell scrapers as described in **step 3** of Subheading 3.11, transfer all the well volume to microcentrifuge tubes and centrifuge at 18,000 × *g* for 30 min at 4 °C.
5. Remove the supernatant and suspend the pellet in 30 μL of SDS buffer (*see* **Note 28**).
6. Boil the samples for 5 min or until the samples are solubilized.
7. Cool the samples on ice and add 0.9 volume of ACK lysing buffer supplemented with 1× Halt™ Protease Inhibitor Single-Use Cocktail EDTA-Free.
8. Store at −20 °C until use [31].

3.13 Tricine-SDS-PAGE and Western Blotting

This protocol is suitable for detection of 3×FLAG-tagged proteins of *N. gonorrhoeae* in cell lysates of infected FTECs (unpublished data) using human cytoskeleton markers as a loading control. However, Tricine-SDS-PAGE can be also used to detect proteins of *N. gonorrhoeae* >30 kDa [32, 33]. The following instructions are valid for a Bio-Rad minigel and transfer tank system with 1 mm thick gels [32, 33].

1. Polymerize the separating and stacking gels. Remove the comb carefully and wash the wells with 1× cathode buffer using a syringe. Add 1× anode buffer gently to the upper chamber and the 1× cathode buffer to the lower chamber of the electrophoresis system.
2. Mix 4 μL of sample (FTEC cell lysate) with 12 μL of Buffer A (ratio 1:3) and incubate at 37 °C for 30 min.
3. Pre–run the gel using 15 V for 10 min.
4. Load 5 μL of protein ladder samples under the cathode buffer and assemble the gel unit. Start the electrophoresis with 30 V, until the samples reach the stacking gel bottom, then run at 100 V through the separating gel until the blue dye front reaches the gel bottom (~90 min) (*see* **Note 29**).
5. Pour 100% methanol and transfer buffer into two separate glass trays or glass Petri dishes.
6. Wet a sheet of PVDF membrane in 100% methanol until the color becomes slightly grey and then soak it in transfer buffer.
7. Assemble the transfer unit following the manufacturer's instructions. Place the PVDF membrane onto two blotting paper sheets and one sponge placed previously in the transfer cassette system.
8. Disconnect the electrophoresis system from the power supply and extract the gel using a spatula. Carefully place the gel onto the PVDF membrane and smooth the gel over the membrane to remove any air bubbles. Finish the transfer unit assembly by placing two blotting paper sheets onto the gel and ensuring no

air bubbles are trapped in the sandwich. Replace the second sponge and close the cassette.

9. Place the cassette into the transfer tank following the manufacturer's instructions. Fill the tank with transfer buffer, making sure all the unit is soaked by the buffer. Use a pipette to wet the upper side of the unit with buffer.
10. To avoid overheating of the transfer system, put a cooling unit provided by the manufacturer into the tank.
11. Close the system and run the transfer by applying a current of 200 mA for 2 h.
12. Disconnect the transfer system from the power supply and carefully disassemble the cassette. Check that the prestained protein ladder has completely transferred to the membrane.
13. Place the PVDF membrane into a container with 20–30 mL of membrane blocking buffer and incubate for 2 h at room temperature with gentle agitation.
14. Discard the blocking buffer and wash the membrane with membrane washing buffer four times, for 5 min each and gentle agitation.
15. Discard the membrane washing buffer and add 10 mL of the appropriate primary antibody (1/5000 dilution): Anti-FLAG®M2 antibody for gonococcal FLAG-tagged proteins (>30 kDa) [32], E-cadherin and β-tubulin antibody as cytoskeleton protein markers [31] or β-actin antibody for loading control of Western Blotting. Keep overnight at 4 °C with gentle agitation.
16. Remove the primary antibody and wash the membrane as described in **step 14**.
17. Add 10 mL of the horseradish peroxidase-labeled anti-mouse antibody conjugate (1/10,000 dilution). Incubate for 2 h at room temperature with gentle agitation.
18. Remove the secondary antibody and wash the membrane as described in **step 14**.
19. Prewarm the chemiluminescent substrate to room temperature and prepare 2 mL in a conical tube protected from light, following the manufacturer's instructions. Place the PVDF membrane, face up, on a piece of clear plastic film, then cover with the substrate solution and cover with the rest of clear plastic film to form a wrap. Remove air bubbles using a glass pipette rolling over the wrap. Avoid exposing to light.
20. Place the wrapped membrane into a digital imaging system and image follow the manufacturer's instructions.

4 Notes

1. Patient consent is mandatory and must be previously approved by the local Institutional Review Board. Additionally, human samples are considered as biohazards, thus handling and disposal must be done under biohazard safety regulations.
2. Cell culture plastic flasks and plates with tissue culture (TC)-treated surface are suitable to grow and expand any type of epithelial cell because their hydrophilic, negatively charged surface favors cell attachment. We usually use Falcon® or Corning® consumables.
3. The use of *N. gonorrhoeae* variants, that is, pilus (Pil) or/and phase-variable colony opacity (Opa) phenotype, depends on the purpose of the experiment. Pilus is involved in initial attachment to FTOC but not in invasion directly; however, it can inhibit release of proinflammatory cytokines such as IL-1β and TNF-α [35]. Opa proteins are mainly required in subsequent events (intimate adherence and bacterial internalization) alongside type-IV pilus [14].
4. Thawed gelatin solutions cannot be frozen. Discard the unused solution after the aliquot is warmed to 35 °C.
5. Be careful not to cool more than to hand temperature, otherwise gelatin becomes solid. The solution must be completely clean: if "opalescent waves" appear in the solution it means that the gelatin has solidified and not correctly homogenized. Discard the solution and repeat.
6. Add drops of dH_2O carefully over the cast separating gel until the drops join to create an aqueous layer. Use a micropipette.
7. Gel staining solution can be used up to five times.
8. FTECs can be stored at −80 °C, but it is strongly recommended to use liquid nitrogen for long-term storage (>6 months).
9. FTECs can also be characterized by immunohistochemical staining with antibodies to ICAM-1 [34], cytokeratin [5, 34] and HLA.DR (MHC class II antigen) [40].
10. FTECs incubation time is sample-dependent. Check FTECs every day using an inverted microscope in order to track the formation of "islands" of cells (Fig. 2). Moreover, cell proliferation rates vary considerably with each sample and are sometimes insufficient for in vitro culture [6]. If cells are not proliferating, discard them and repeat the procedure with a new FT sample.
11. Contamination with fibroblasts is a serious problem during processing of FT samples. Usually these cells grow and expand

quickly into the flask (<12 h post seeding) with a typically enlarged, spindle-shaped morphology covering the flask surface. Unfortunately, no extra steps can be done once fibroblasts start to grow into the cell culture and it is preferable to discard these cultures.

12. FTEC attachment can be improved by using collagen-coated cell culture flasks.
13. As a primary cell culture, cell passage is only possible for 3–4 times depending on the FT sample, otherwise cells become senescent and die.
14. Streak gonococcal colonies onto GC plates gently, otherwise no bacterial growth will occur.
15. The Pil^+ and Opa^+ phenotype of *N. gonorrhoeae* can be confirmed also by Western Blot using bacterial lysates and the specific monoclonal anti-pilus antibody SM1 [41] and the anti-Opa protein antibody B33 [42].
16. Use GC broth without Isovitalex™ Enrichment or Kellogg's supplements for serial dilution and plating of *N. gonorrhoeae* after gentamicin protection assay [29]. Do not use PBS as a buffer for bacterial suspension.
17. LDH measurement must be done up to 8 h after collection of cell supernatants, since LDH has a half-life of approximately 9 h when released into cell culture media [43].
18. Supernatants of infected cells can be stored at −80 °C up to 3 months for MMP and cytokine quantification and for up to 6 months for MMP activity. Avoid repeated freezing and thawing cycles, since MMPs lose gelatinase activity.
19. Poly-D-lysine-coated coverslips can be prepared using the following recipe: coat coverslip with 1 mg/mL of poly-D-Lysine in cell culture grade H_2O for 2 h in a Class II cabinet. Wash the coverslips gently with H_2O twice and a single wash with absolute ethanol, air dry on a piece of filter paper and keep them in a dry container until use. Expose to UV-light for 5 min before cell seeding [44].
20. PFA-fixed FTECs can be stored overnight at 4 °C in 2 mL of PBS in each well of the cell culture plate. The immunofluorescence staining protocol then can be continued.
21. Do not use a cold solution of 0.2% (v/v) Triton X-100–PBS, but prewarm at room temperature before use.
22. Be careful with direct light exposure of dye stock and working solutions, or use opaque microcentrifuge tubes.
23. Avoid any air bubbles in the mounting medium, since they can scatter light and interfere with the fluorescence observation.
24. Since MMP activity is inhibited by many reducing agents commonly found in SDS-PAGE solutions (e.g., β-mercaptoethanol

in both sample loading buffer and protein ladder solutions), prepare separate reagents and solutions for zymography as well as using separate plastic containers for subsequent incubations [45].

25. Do not boil cell supernatants for zymography, as this will irreversibly inactivate MMPs [45].
26. Do not incubate two or more zymograms in a plastic container in the subsequent steps of renaturation/staining/destaining. Use an individual container per gel.
27. Although picograms of MMP-2 and -9 activity are detectable in zymograms [45], zymography incubation can be extended to 24–72 h depending of the FTOC/FTEC sample and volume loaded into the gel. In this case, check that the gel is completely soaked up by the development buffer.
28. The supernatant contains the cytoplasm fraction and can be used also in Western blot.
29. Run tricine gels in a cool room because they may warm up [33].

Acknowledgments

This work was supported by PCI-CONICYT "*Apoyo a la Formación de Redes Internacionales Para Investigadores en Etapa Inicial*" REDI170370 (P.I.R., C.V.T., A.E., M.C.), UNAB DI-10-17/RG (P.I.R.), and FONDECYT regular 1180666 (A.E., P.I.R.).

References

1. Nasu K, Narahara H (2010) Pattern recognition via the toll-like receptor system in the human female genital tract. Mediators Inflamm 2010:1–12. https://doi.org/10.1155/2010/976024
2. Wira CR, Patel MV, Ghosh M et al (2011) Innate immunity in the human female reproductive tract: endocrine regulation of endogenous antimicrobial protection against HIV and other sexually transmitted infections. Am J Reprod Immunol 65:196–211. https://doi.org/10.1111/j.1600-0897.2011.00970.x
3. Rampersaud R, Randis TM, Ratner AJ (2012) Microbiota of the upper and lower genital tract. Semin Fetal Neonatal Med 17:51–57. https://doi.org/10.1016/j.siny.2011.08.006
4. Lyons RA, Saridogan E, Djahanbakhch O (2006) The reproductive significance of human Fallopian tube cilia. Hum Reprod Update 12:363–372. https://doi.org/10.1093/humupd/dml012
5. Lawrenson K, Notaridou M, Lee N et al (2013) In vitro three-dimensional modeling of fallopian tube secretory epithelial cells. BMC Cell Biol 14:43. https://doi.org/10.1186/1471-2121-14-43
6. Karst AM, Drapkin R (2012) Primary culture and immortalization of human fallopian tube secretory epithelial cells. Nat Protoc 7:1755–1764. https://doi.org/10.1038/nprot.2012.097
7. Takeuchi K, Maruyama I, Yamamoto S et al (1991) Isolation and monolayer culture of human fallopian tube epithelial cells. In Vitro Cell Dev Biol 27A:720–724
8. Brunham RC, Gottlieb SL, Paavonen J (2015) Pelvic inflammatory disease. N Engl J Med 372:2039–2048. https://doi.org/10.1056/NEJMra1411426
9. Duarte R, Fuhrich D, Ross JDC (2015) A review of antibiotic therapy for pelvic

inflammatory disease. Int J Antimicrob Agents 46:272–277. https://doi.org/10.1016/j.ijantimicag.2015.05.004

10. Edwards JL, Butler EK (2011) The pathobiology of *Neisseria gonorrhoeae* lower female genital tract infection. Front Microbiol 2:102. https://doi.org/10.3389/fmicb.2011.00102
11. Singer M, Ouburg S (2016) Effect of cytokine level variations in individuals on the progression and outcome of bacterial urogenital infections—a meta-analysis. Pathog Dis 74:ftv126. https://doi.org/10.1093/femspd/ftv126
12. Skerlev M, Čulav-Košćak I (2014) Gonorrhea: new challenges. Clin Dermatol 32:275–281. https://doi.org/10.1016/j.clindermatol.2013.08.010
13. McGee ZA, Johnson AP, Taylor-Robinson D (1976) Human fallopian tubes in organ culture: preparation, maintenance, and quantitation of damage by pathogenic microorganisms. Infect Immun 13:608–618
14. McGee ZA, Johnson AP, Taylor-Robinson D (1981) Pathogenic mechanisms of *Neisseria gonorrhoeae*: observations on damage to human fallopian tubes in organ culture by gonococci of colony type 1 or type 4. J Infect Dis 143:413–422
15. Gregg CR, Melly MA, Hellerqvist CG et al (1981) Toxic activity of purified lipopolysaccharide of *Neisseria gonorrhoeae* for human fallopian tube mucosa. J Infect Dis 143:432–439
16. Cooper MD, McGraw PA, Melly MA (1986) Localization of gonococcal lipopolysaccharide and its relationship to toxic damage in human fallopian tube mucosa. Infect Immun 51:425–430
17. Melly MA, McGee ZA, Rosenthal RS (1984) Ability of monomeric peptidoglycan fragments from *Neisseria gonorrhoeae* to damage human fallopian-tube mucosa. J Infect Dis 149:378–386
18. Stephens DS, McGee ZA, Cooper MD (1987) Cytopathic effects of the pathogenic *Neisseria*. Studies using human fallopian tube organ cultures and human nasopharyngeal organ cultures. Antonie Van Leeuwenhoek 53:575–584
19. Johnson AP, Clark JB, Osborn MF et al (1980) A comparison of the association of *Neisseria gonorrhoeae* with human and guinea-pig genital mucosa maintained in organ culture. Br J Exp Pathol 61:521–527
20. McGee ZA, Jensen RL, Clemens CM et al (1999) Gonococcal infection of human fallopian tube mucosa in organ culture: relationship of mucosal tissue TNF-alpha concentration to sloughing of ciliated cells. Sex Transm Dis 26:160–165
21. Virji M (2009) Pathogenic neisseriae: surface modulation, pathogenesis and infection control. Nat Rev Microbiol 7:274–286. https://doi.org/10.1038/nrmicro2097
22. Merz AJ, So M (1997) Attachment of piliated, Opa− and Opc− gonococci and meningococci to epithelial cells elicits cortical actin rearrangements and clustering of tyrosine-phosphorylated proteins. Infect Immun 65:4341–4349
23. Vink C, Rudenko G, Seifert HS (2012) Microbial antigenic variation mediated by homologous DNA recombination. FEMS Microbiol Rev 36:917–948. https://doi.org/10.1111/j.1574-6976.2011.00321.x
24. Swanson KV, Jarvis GA, Brooks GF et al (2001) CEACAM is not necessary for *Neisseria gonorrhoeae* to adhere to and invade female genital epithelial cells. Cell Microbiol 3:681–691
25. Dehio C, Gray-Owen SD, Meyer TF (1998) The role of neisserial Opa proteins in interactions with host cells. Trends Microbiol 6:489–495
26. Rodas PI, Pérez D, Jauffret C et al (2017) Modified profile of matrix metalloproteinase-2 and -9 production by human Fallopian tube epithelial cells following infection in vitro with *Neisseria gonorrhoeae*. J Infect Dis 215:452–455. https://doi.org/10.1093/infdis/jiw568
27. Juica NE, Rodas PI, Solar P et al (2017) *Neisseria gonorrhoeae* challenge increases matrix metalloproteinase-8 expression in fallopian tube explants. Front Cell Infect Microbiol 7. https://doi.org/10.3389/fcimb.2017.00399
28. Lambden PR, Heckels JE, James LT et al (1979) Variations in surface protein composition associated with virulence properties in opacity types of *Neisseria gonorrhoeae*. J Gen Microbiol 114:305–312
29. Dillard JP (2011) Genetic manipulation of *Neisseria gonorrhoeae*. Curr Protoc Microbiol. Chapter 4:Unit4A.2. https://doi.org/10.1002/9780471729259.mc04a02s23
30. D'Ascenzo S, Giusti I, Millimaggi D et al (2004) Intrafollicular expression of matrix metalloproteinases and their inhibitors in normally ovulating women compared with patients undergoing in vitro fertilization treatment. Eur J Endocrinol 151:87–91
31. Rodríguez-Tirado C, Maisey K, Rodríguez FE et al (2012) *Neisseria gonorrhoeae* induced disruption of cell junction complexes in epithelial cells of the human genital tract. Microbes Infect 14:290–300. https://doi.org/10.1016/j.micinf.2011.11.002

32. Rodas PI, Álamos-Musre AS, Álvarez FP et al (2016) The NarE protein of *Neisseria gonorrhoeae* catalyzes ADP-ribosylation of several ADP-ribose acceptors despite an N-terminal deletion. FEMS Microbiol Lett 363:fnw181. https://doi.org/10.1093/femsle/fnw181
33. Schägger H (2006) Tricine-SDS-PAGE. Nat Protoc 1:16–22. https://doi.org/10.1038/nprot.2006.4
34. Utreras E, Ossandon P, Acuña-Castillo C et al (2000) Expression of intercellular adhesion molecule 1 (ICAM-1) on the human oviductal epithelium and mediation of lymphoid cell adherence. J Reprod Fertil 120:115–123
35. Velasquez L, García K, Morales F et al (2012) *Neisseria gonorrhoeae* pilus attenuates cytokine response of human fallopian tube explants. J Biomed Biotechnol 2012:491298. https://doi.org/10.1155/2012/491298
36. Maisey K, Nardocci G, Imarai M et al (2003) Expression of proinflammatory cytokines and receptors by human fallopian tubes in organ culture following challenge with *Neisseria gonorrhoeae*. Infect Immun 71:527–532
37. Fernandez R, Nelson P, Delgado J et al (2001) Increased adhesiveness and internalization of *Neisseria gonorrhoeae* and changes in the expression of epithelial gonococcal receptors in the Fallopian tube of copper T and Norplant users. Hum Reprod 16:463–468
38. Christodoulides M, Everson JS, Liu BL et al (2000) Interaction of primary human endometrial cells with *Neisseria gonorrhoeae* expressing green fluorescent protein. Mol Microbiol 35:32–43
39. Ortiz MC, Lefimil C, Rodas PI et al (2015) *Neisseria gonorrhoeae* modulates immunity by polarizing human macrophages to a M2 profile. PLoS One 10:e0130713. https://doi.org/10.1371/journal.pone.0130713
40. Imarai CM, Rocha A, Acuña C et al (1998) Endocytosis and MHC class II expression by human oviductal epithelium according to stage of the menstrual cycle. Hum Reprod 13:1163–1168
41. Virji M, Heckels JE, Potts WJ et al (1989) Identification of epitopes recognized by monoclonal antibodies SM1 and SM2 which react with all pili of *Neisseria gonorrhoeae* but which differentiate between two structural classes of pili expressed by *Neisseria meningitidis* and the distribution of their encoding sequences in the genomes of *Neisseria* spp. Microbiology 135:3239–3251. https://doi.org/10.1099/00221287-135-12-3239
42. Griffiss JM, Lammel CJ, Wang J et al (1999) *Neisseria gonorrhoeae* coordinately uses pili and opa to activate HEC-1-B cell microvilli, which causes engulfment of the gonococci. Infect Immun 67:3469–3480
43. Riss TL, Moravec RA (2004) Use of multiple assay endpoints to investigate the effects of incubation time, dose of toxin and plating density in cell-based cytotoxicity assays. Assay Drug Dev Technol 2:51–62
44. Sawasdichai A, Chen H-T, Abdul Hamid N et al (2010) In situ subcellular fractionation of adherent and non-adherent mammalian cells. J Vis Exp (41). pii: 1958. https://doi.org/10.3791/1958
45. Troeberg L, Nagase H (2003) Zymography of metalloproteinases. Curr Protoc Prot Sci Chapter 21:Unit 21.15. https://doi.org/10.1002/0471140864.ps2115s33

Chapter 23

A Natural Mouse Model for *Neisseria* Persistent Colonization

Katherine Rhodes, Mancheong Ma, and Magdalene So

Abstract

We have developed a natural mouse model to study persistent colonization by commensal *Neisseria*. The system couples the ordinary lab mouse with *Neisseria musculi* (Nmus), a commensal in the oral cavity and gut of the wild mouse, *Mus musculus*. The pairing of Nmus with its natural reservoir circumvents host restriction barriers that have impeded previous studies of *Neisseria* in vivo behavior. The model allows, for the first time, for the dissection of host and neisserial determinants of asymptomatic colonization. Inoculation procedures are noninvasive and susceptibility to Nmus colonization varies with host genetic background. In colonized mice, bacterial burdens are detectable up to 1-year post inoculation, making it an ideal model for the study of persistence. As Nmus encodes several *Neisseria gonorrhoeae* (and *Neisseria meningitidis*) host interaction factors, the system can be used to query the in vivo functions of these commonly held genes and factors. Nmus also encodes many pathogenic *Neisseria* vaccine targets including a polysaccharide capsule, making the model potentially useful for vaccine development. The ease of genetic manipulation of Nmus enhances the feasibility of such studies.

Key words Mouse model, Colonization and persistence, Commensal *Neisseria*, Pathogenic *Neisseria*

1 Introduction

1.1 Why Is It Important to Understand Commensalism?

Commensal organisms (microbiota, normal flora) play a critical role in physiology [1–7]. They are required for gut and immune system development, and prevent pathogen colonization. Perturbation of the composition of these microbial communities is strongly linked to chronic, non-infectious diseases such as obesity, inflammatory bowel disease, diabetes (Type I and II) and autoimmunity.

How does a commensal organism colonize and persist in the body in the presence of a robust immune system? Little is known about the mechanisms of commensal–host coadaptation. Understanding the molecular basis of these interactions will improve disease prevention and contribute to our fundamental knowledge of immunology and microbiology. Except for two species, all members of the *Neisseria* genus are commensals that inhabit a wide

Myron Christodoulides (ed.), *Neisseria gonorrhoeae: Methods and Protocols*, Methods in Molecular Biology, vol. 1997, https://doi.org/10.1007/978-1-4939-9496-0_23,

range of hosts, from rodents, felines, and bovines to monkeys and man [8, 9]. Animal-adapted commensal *Neisseria* have been detected or isolated from the oral cavity and feces of their hosts, suggesting they establish niches along the alimentary tract. The niche of human-adapted commensals is generally assumed to be the upper respiratory tract, but little has been done to examine the extent of their colonization range. The two pathogenic species in the genus are *Neisseria gonorrhoeae* (Ngo) and *Neisseria meningitidis* (Nme), which only infect humans [8]. Ngo and Nme inhabit the oral cavity and urogenital tract and are generally assumed to be restricted to these sites. Whether they also colonize the gut is an open question.

Ngo and Nme exhibit a commensal-like behavior; at fairly high frequency, they colonize the oral cavity and urogenital tract without eliciting overt symptoms of disease [10, 11] (http://citeseerx.ist.psu.edu/viewdoc/summary?doi=10.1.1.177.8777). Asymptomatic carriage is an important facet of infection, as it is key to person-to-person transmission. The mechanisms that promote asymptomatic carriage of Ngo and Nme are little understood. As commensal *Neisseria* are closely related to Ngo and Nme, they may provide a means to understanding asymptomatic infections by these pathogens.

1.2 A Natural Mouse Model for Commensal Neisseria Persistent Colonization

We have developed a natural small animal model for dissecting the mechanisms of *Neisseria* colonization and persistence, from the standpoint of both host and microbe [12]. Our model pairs the lab mouse, which does not normally harbor *Neisseria*, with *Neisseria musculi* (Nmus) a commensal of wild mice [13]. Pairing the mouse with a member of its normal flora circumvents barriers imposed by host restriction, a problem that plagues current models for studying the human-specific pathogenic *Neisseria* species [14–17] (*see* **Note 1**).

Nmus encodes many host interaction genes of the pathogenic *Neisseria*, including a complete set of genes for Type IV pilus (Tfp) biogenesis and function [12]. The Tfp fiber is generally assumed to mediate colonization [18, 19], based on in vitro studies, but in vivo function has never been tested. As final proof of principle for the validity of our model for investigating *Neisseria*-host interactions, the role of Tfp in colonization was tested using the mouse model. Nmus Δ*pilE*, which cannot produce the Tfp fiber subunit, PilE, fails to colonize CAST and B6 mice [12]. The Nmus-mouse model is therefore useful for querying the in-host function of other proteins that modulate Tfp activity, and the interaction genes/factors held in common by commensal and pathogenic *Neisseria* (*see* **Note 2**).

Finally, it should be noted that commensal *Neisseria* naturally colonize rhesus macaques (RM) [20] and persist in them for at least 6 years [21]. RM *Neisseria* are closely related to human commensal species [22] and encode many host interaction factors of the pathogenic species [21] (Weyand and So unpublished). Thus, pairing RM *Neisseria* with the macaque holds promise for understanding *Neisseria*-host interactions. However, the logistics and expense of the model, and the limited number of non-human primate facilities make the model inaccessible to most investigators. In this chapter, we provide a detailed protocol for Nmus colonization of mice that is readily accessible to investigators.

2 Materials

2.1 Screen for Autochthonous Neisseria Species

2.1.1 Sample Collection

1. Mouse (*see* **Note 3**).
2. Ear tag system or ear punch.
3. 100 μl aliquots of sterile Dulbecco's phosphate buffered saline without magnesium and calcium (DPBS).
4. 1.5 ml Sterile Eppendorf tubes.
5. BBL CultureSwab Plus Transport system, Amies gel with charcoal.
6. Laminar flow hood: sterilize by swabbing with 70% (v/v) ethanol in distilled H_2O.
7. Sterile catch pan.

2.1.2 Oral Sample Preparation

1. Gonococcal Medium Base (GCB) agar plate (BD Difco): add 15 g/L of Proteose Peptone No. 3, 1 g/L of Cornstarch, 4 g/L of K_2HPO_4, 1 g/L of KH_2PO_4, 5 g/L of NaCl, 10 g/L of agar to 1 L of distilled H_2O. Autoclave at 121 °C and 1.05 kg/cm^2 for 40 min. Cool the medium to 55 °C and add 1% (v/v) Kellogg's Supplement I and 0.1% (v/v) Kellogg's Supplement II and appropriate antibiotics (*see* **Note 4**).
2. Kellogg's Supplement I: 400 mg/ml of glucose, 10 mg/ml of L-glutamine, 20 μg/ml of thiamine pyrophosphate dissolved in distilled H_2O and sterilized by filtration (0.22 μm filter). Store in aliquots at −20 °C.
3. Kellogg's Supplement II: 500 μg/ml of Iron (III) Nitrate in distilled H_2O and sterilized by filtration (0.22 μm filter). Store in aliquots at −20 °C [23].
4. Humidified incubator set at 37 °C with 5% (v/v) CO_2.
5. Sterile plating beads or glass spreader rod.

6. GC broth (GCBL): add 15 g/L of Proteose peptone #3, 1 g/L of KH_2PO_4, 4 g/L of K_2HPO_4, 1 g/L of NaCl, and 20% (v/v) glycerol to 1 L of distilled H_2O, and adjust pH to 7.2 [23]. Autoclave at 121 °C and 1.05 kg/cm^2 for 40 min.
7. Wire cutters.
8. Plate spinner.
9. Bunsen burner.
10. Vortex mixer.
11. 70% (v/v) ethanol.
12. Glass spreader rod or disposable inoculating plastic loops.
13. 1.5 ml sterile Eppendorf tubes.

2.1.3 Fecal Sample Preparation

1. Same materials as described in Subheading 2.1.2.
2. Sterile P1000 pipette tips.
3. Tared balance.

2.2 Preparing the Neisseria musculi Inoculum

1. *Neisseria musculi* strains/isolates, stored at −80 °C in GCBL with 20% (v/v) glycerol.
2. GCB agar plates containing 40 μg/ml of rifampicin, and any additional selective antibiotics appropriate to the strain being tested.
3. Sterile CLASSIQSwabs (Copan, or equivalent).
4. Sterile plastic cell scrapers.
5. DPBS without calcium and magnesium.
6. GCBL broth.
7. Humidified incubator set at 37 °C with 5% (v/v) CO_2.
8. Dissecting microscope.
9. Spectrophotometer.
10. Disposable plastic cuvettes.
11. 1.5 ml sterile Eppendorf tubes.
12. Vortex mixer.

2.3 Inoculation of Neisseria musculi into the Oral Cavity

1. Prescreened mice with Identification (ID) numbers.
2. *Neisseria musculi* inoculum.
3. P200 pipet.
4. Aerosol-resistant tips.
5. Vortex mixer.
6. Sterile catch pans.

3 Methods

3.1 Screen for Autochthonous Neisseria Species (See Note 5)

3.1.1 Sample Collection

1. On day 1, tag or band each mouse to give it an identification (ID) number.
2. Sterilize the laminar flow hood. Collect the necessary equipment and place it in the hood, along with a sterile catch pan and the cage of naïve mice.
3. Gently restrain the mouse so as to prevent the forelimbs from touching the mouth or nose.
4. Submerge a BBL culture swab into an aliquot of DPBS; remove excess liquid by gently pressing the swab along the side of the tube. Collect an oral sample by gently inserting the wetted swab into the oral cavity of the mouse, taking care that it does not reach the esophagus. Using a swirling motion, pass the swab over the palate, cheek pouches, sublingual tissue, and tongue. Return the swab to the receptacle, making sure it does not touch the side of the tube. Record the mouse ID on the side of the transport tube.
5. Use a sterile, labeled Eppendorf tube to collect a fecal sample from each mouse (*see* **Note 6**). Use the inside edge of the tube to scoop the sample into the tube, taking care not to allow the fecal pellet to contact any unsterilized surface or gloved fingers.
6. Proceed to sample preparation and plating as described in Subheadings 3.1.2 and 3.1.3.

3.1.2 Oral Sample Preparation

1. Dispense 1 ml of GCBL containing 20% (v/v) glycerol into prelabeled sterile Eppendorf tubes.
2. Dip the metal cutting edge of a pair of wire cutters in 70% (v/v) ethanol and flame to sterilize. Wait for the pliers to cool.
3. Remove the oral swab from the media receptacle in **step 4** of Subheading 3.1.1 without touching the metal shaft.
4. Insert the cotton bud of the swab into the appropriate tube with GCBL containing 20% (v/v) glycerol. Use the sterile wire cutters to clip the cotton bud from the shaft, dropping the bud into the storage media. Cap the tube firmly and vortex at high speed for 1 min.
5. Repeat **step 2** between each swab.
6. Pulse each sample on the vortex mixer for 2 s to suspend any settled material.
7. Dispense 100 μl of each sample onto a labeled, prewarmed GCB plate with appropriate antibiotics (*see* **Note 4**).
8. Spread the liquid evenly over the agar surface using a sterilized glass rod or disposable loop. Incubate the plate for 48 h at 37 °C in an atmosphere with 5% (v/v) CO_2.

9. Check plates for growth of bacteria. When screening for endogenous *Neisseria*, exclude any mouse that is positive for either oral or fecal CFUs (*see* **Note** 7).

3.1.3 Fecal Sample Preparation

1. Weigh each fecal sample collected from **step 5** in Subheading 3.1.1 in its tube on a tared balance scale and record the weight.
2. Dispense 1 ml of GCBL containing 20% (v/v) glycerol into each tube, taking care to prevent cross contamination.
3. Crush each fecal pellet using a sterile filtered P1000 pipette tip. Vortex each sample for ~1 min at high speed until homogenous.
4. Repeat **steps 6–9** of Subheading 3.1.2.

3.2 Preparing the Neisseria musculi Inoculum (See Note 8)

1. Use a sterile swab to inoculate a GCB plate containing 40 μg/ml of rifampicin with Nmus from a pure freezer stock, without allowing the stock to thaw, and streak for single colony isolation (*see* **Note 9**).
2. Incubate the plate at 37 °C with an atmosphere of 5% (v/v) CO_2 for 48 h to obtain colonies large enough to visualize colony morphological type.
3. Use a dissecting microscope to view and collect ten rough colonies onto a single sterile swab. Streak a lawn onto a GCB plate containing 40 μg/ml of rifampicin with the collected colonies. Repeat five times, for a total of 50 colonies streaked to lawns on five separate plates. Repeat for each experimental strain.
4. Incubate these plates for an additional 16–18 h at 37 °C with an atmosphere of 5% (v/v) CO_2.
5. Prepare a sterile Eppendorf tube with 1 ml of sterile DPBS.
6. Harvest the plates that had been streaked to lawns in **step 3**. Collect the bacterial lawn from each plate using a sterile cell scraper, taking care not to gouge the agar. Inadvertent collection of agar with the lawn will skew absorbance readings.
7. Deposit the collected lawns into the Eppendorf tube containing 1 ml of DPBS, to create a stock suspension.
8. Vortex the suspension at maximum speed for at least 1 min to disperse any aggregates.
9. Prepare a second Eppendorf tube with 900 μl of DPBS. Dilute the stock suspension 1/10 and vortex at high speed for an additional minute.
10. Transfer the diluted bacterial suspension to a clean cuvette. Measure the optical density (OD) at $\lambda_{600\ nm}$ using a spectrophotometer blanked against DPBS.

11. Calculate the OD $\lambda_{600\ nm}$ of the undiluted stock suspension. Calculate the total volume needed to dose all mice with 50 μl of inoculum per mouse, leaving a minimum of 100 μl excess for long-term storage. Using these values, calculate the volume of stock suspension necessary to create an inoculum where OD $\lambda_{600\ nm} = 2.0$.
12. Prepare the inoculum as calculated in **step 11**, using a new sterile Eppendorf tube and the appropriate volumes of DPBS diluent and stock suspension. Vortex the inoculum for 1 min at high speed. Move on to Subheading 3.3 immediately. Samples remain viable in DPBS for approximately 4–6 h at room temperature.

3.3 Inoculation of *Neisseria musculi* into the Oral Cavity (See Note 10)

1. Prepare the biosafety cabinet as described in **step 2** of Subheading 3.1.
2. Manually restrain each mouse as described in **step 3** of Subheading 3.1, ensuring that the animal cannot move its head or forelimbs.
3. Briefly vortex the inoculum to suspend any bacteria that may have settled to the bottom of the tube.
4. Withdraw 50 μl of inoculum from the tube using a sterile filtered P200 pipette. Insert the pipette tip into the oral cavity of the restrained mouse, taking care not to abrade the tissues. Slowly dispense the inoculum. Depending on the size of the mouse, and how quickly it swallows the inoculum, this may have to be done in two stages.
5. Continue to restrain the mouse until the majority of the inoculum has been swallowed. This will take up to 2–3 min. It is vital that the mouse does not move from an immobile, supine position before this time, to prevent inoculum loss and deviation between delivered burdens within the experimental group.
6. Release the mouse into the sterile catch pan and repeat for the remaining mice (*see* **Note 11**).
7. Sample mice at designated time points post inoculation following **steps 3–6** of Subheading 3.1. Prepare samples for plating as indicated in Subheadings 3.1.2 and 3.1.3 or storage (*see* **Note 4**).

4 Notes

1. One oral dose of Nmus (6×10^6 CFUs), pipetted gently into the oral cavity of the mouse, will achieve lasting colonization of the oral cavity and gut. Nmus is recovered in large quantities from these sites every week, for as long as 1 year. The animals remain healthy throughout the studies. No administration of

antibiotics or hormones, or invasive procedures or surgery is necessary. Since the system involves a commensal in its native host, mice do not need to be derived to introduce human markers, for example, receptor proteins for gonococcal virulence factors, to mimic human infection. Horizontal transmission between cage mates has not been detected, allowing the cohousing of naïve and inoculated mice.

2. Nmus is easy to grow and manipulate genetically; many reagents, genetic tools, and genome databases are available for both *Neisseria* and the lab mouse [23–25] (http://www.sanger.ac.uk/resources/downloads/bacteria/neisseria.html), allowing rapid dissection of microbial and host determinants of colonization, and of the elusive process of persistence.
3. Host genetics is a major determinant of colonization. CAST/EiJ (CAST) and A/J strains of lab mice are colonized by Nmus at high frequency, and C57BL/6J (B6) at intermediate frequency. All studies to date have been done with mice from 4–6 weeks of age. No sex bias has been detected. NOD/LtJ, NZO/HILtJ, PWK/PhJ, WSB/EiJ, and 129S2/SvlmJ mice are resistant to colonization. A second major colonization determinant is innate immunity. B6 MyD88$^{-/-}$ mice are exquisitely sensitive to Nmus colonization, unlike their parental wt B6. By contrast, B6 Rag1$^{-/-}$ mice are no more susceptible than B6. Many genetic tools are available for manipulating the lab mouse, such as the Collaborative Cross [24], which allows for identification of mouse alleles and immune pathways that determine susceptibility/resistance to Nmus colonization.
4. The GCB agar formulation can be used for both prescreening and experimental sample propagation. The antibiotics used in each assay must be adapted as appropriate. To screen for autochthonous *Neisseria* prior to inoculation, samples must be plated on GCB agar containing 3 μg/ml of Trimethoprim and 2 μg/ml of Vancomycin. This medium selects for growth of *Neisseria* species and discourages growth of other commensals. The experiments described in ref. 12 were done using Nmus AP2365, a spontaneously arising rifampicin-resistant strain. Samples from AP2365-inoculated mice should be selected on GCB agar containing 40 μg/ml of rifampicin, and/or the antibiotic specific to the selective marker introduced to the Nmus strain of interest. Samples can be stored at −80 °C indefinitely in GCBL with 20% (v/v) glycerol.
5. At least 3 days prior to inoculation, prescreen all mice for the presence of endogenous *Neisseria*. This is done by plating oral cavity and fecal samples on agar that selects for *Neisseria* species (*see* **Note 10**). This will avoid complications that interfere with interpretation of colonization data. We note that we have not

detected *Neisseria* in any of the lab mice we have used [12]. The sampling collection and preparation method is also used for assaying experimental animals post inoculation.

6. If working in pairs, one person should restrain the mouse while the other collects the fecal samples and oral swabs. If working alone, the experimenter is advised to place each mouse in a sterilized container or catch basin and collect the fecal sample (s) from the animal prior to swabbing their oral cavity. This increases the efficiency of fecal sample collection, especially when working with CAST/EiJ strains.
7. Thorough genetic and biochemical protocols for identifying native *Neisseria* species can be found in ref. 13.
8. Preparation of the Nmus inoculum is straightforward, but good aseptic technique is crucial to prevent contamination that could confound study results. All steps should be performed in a sterile biosafety cabinet, or near a flame if working on the bench top.
9. Do not freeze and thaw a frozen stock of *Neisseria*. To avoid this problem, multiple freezer stocks should be prepared, and Nmus should be taken from frozen stock by scraping the surface of the frozen culture.
10. Inoculating the mouse oral cavity is a slow process; insufficient time in the oral cavity can result in artificially poor colonization efficiency and variability within experimental groups. To minimize potential variability arising from the inoculation process, all mice in the experiment should be inoculated by the same person. Mice should be inoculated and sampled at approximately the same time of day to further reduce variability between time points and experimental groups.
11. If inoculating multiple strains of Nmus, it is vital to sterilize the hood and instruments between groups. All materials that had contact with each group of mice should be decontaminated or replaced, before moving to the next bacterial strain to prevent cross contamination.

Acknowledgments

Funding for this project was provided by NIH 1R56A124665-01 awarded to M. So.

References

1. Sommer F, Backhed F (2013) The gut microbiota-masters of host development and physiology. Nat Rev Microbiol 11 (4):227–238. https://doi.org/10.1038/nrmicro2974
2. Campisi L, Barbet G, Ding Y et al (2016) Apoptosis in response to microbial infection induces autoreactive TH17 cells. Nat Immunol 17(9):1084–1092. https://doi.org/10.1038/ni.3512

3. Kugelberg E (2016) Autoimmunity: infection stimulates self-antigen presentation. Nat Rev Immunol 16(9):534–535. https://doi.org/10.1038/nri.2016.91
4. Liu X, Zeng B, Zhang J et al (2016) Role of the gut microbiome in modulating arthritis progression in mice. Sci Rep 6:30594. https://doi.org/10.1038/srep30594
5. Linehan JL, Harrison OJ, Han SJ et al (2018) Non-classical immunity controls microbiota impact on skin immunity and tissue repair. Cell 172(4):784–796 e718. https://doi.org/10.1016/j.cell.2017.12.033
6. Deriu E, Liu JZ, Pezeshki M et al (2013) Probiotic bacteria reduce *Salmonella typhimurium* intestinal colonization by competing for iron. Cell Host Microbe 14(1):26–37. https://doi.org/10.1016/j.chom.2013.06.007
7. Evans CM, Pratt CB, Matheson M et al (2011) Nasopharyngeal colonization by *Neisseria lactamica* and induction of protective immunity against *Neisseria meningitidis*. Clin Infect Dis 52(1):70–77. https://doi.org/10.1093/cid/ciq065. ciq065 [pii]
8. Bennett JS, Bratcher HB, Brehony C et al (2014) The genus *Neisseria*. The prokaryotes – alphaproteobacteria and betaproteobacteria, 4th edn. Springer-Verlag, Berlin, Heidelberg
9. Liu G, Tang CM, Exley RM (2015) Non-pathogenic *Neisseria*: members of an abundant, multi-habitat, diverse genus. Microbiology 161(7):1297–1312. https://doi.org/10.1099/mic.0.000086
10. Janda W, Bohnhoff M, Morello J et al (1980) Prevalence and site-pathogen studies of *Neisseria meningitidis* and *N. gonorrhoeae* in homosexual men. JAMA 244:2060–2064
11. Caugant DA, Maiden MC (2009) Meningococcal carriage and disease—population biology and evolution. Vaccine 27(Suppl 2): B64–B70. https://doi.org/10.1016/j.vaccine.2009.04.061. S0264-410X(09)00615-X [pii]
12. Ma M, Powell DA, Weyand NJ et al (2018) A mouse model for *Neisseria* colonization. Infect Immun. https://doi.org/10.1128/IAI.00839-17
13. Weyand NJ, Ma M, Phifer-Rixey M et al (2016) Isolation and characterization of *Neisseria musculi* sp. nov., from the wild house mouse. Int J Syst Evol Microbiol 66 (9):3585–3593. https://doi.org/10.1099/ijsem.0.001237
14. Rice PA, Shafer WM, Ram S et al (2017) *Neisseria gonorrhoeae*: drug resistance, mouse models, and vaccine development. Annu Rev Microbiol 71:665–686. https://doi.org/10.1146/annurev-micro-090816-093530
15. Weyand NJ (2017) *Neisseria* models of infection and persistence in the upper respiratory tract. Pathog Dis 75(3). https://doi.org/10.1093/femspd/ftx031
16. Johswich KO, McCaw SE, Islam E et al (2013) In vivo adaptation and persistence of *Neisseria meningitidis* within the nasopharyngeal mucosa. PLoS Pathog 9(7):e1003509. https://doi.org/10.1371/journal.ppat.1003509
17. Join-Lambert O, Lecuyer H, Miller F et al (2013) Meningococcal interaction to microvasculature triggers the tissular lesions of purpura fulminans. J Infect Dis 208 (10):1590–1597. https://doi.org/10.1093/infdis/jit301
18. Merz AJ, So M (1997) Attachment of piliated, Opa– and Opc– gonococci and meningococci to epithelial cells elicits cortical actin rearrangements and clustering of tyrosine-phosphorylated proteins. Infect Immun 65 (10):4341–4349
19. Join-lambert O, Morand PC, Carbonelle E et al (2010) Mechanisms of meningeal invasion by a bacterial extracellular pathogen, the example of *Neisseria meningitidis*. Prog Neurobiol 91(2):130–139. https://doi.org/10.1016/j.pneurobio.2009.12.004
20. Vedros NA, Hoke C, Chun P (1983) *Neisseria macacae* sp. nov., a new *Neisseria* species isolated from the oropharynges of rhesus monkeys (*Macaca mulatta*). Int J Syst Bacteriol 33(3):515–520
21. Weyand NJ, Wertheimer AM, Hobbs TR et al (2013) *Neisseria* infection of rhesus macaques as a model to study colonization, transmission, persistence, and horizontal gene transfer. Proc Natl Acad Sci U S A 110(8):3059–3064. https://doi.org/10.1073/pnas.1217420110. 1217420110 [pii]
22. Bennett JS, Jolley KA, Earle SG et al (2012) A genomic approach to bacterial taxonomy: an examination and proposed reclassification of species within the genus *Neisseria*. Microbiology 158(Pt 6):1570–1580. https://doi.org/10.1099/mic.0.056077-0
23. Dillard JP (2011) Genetic manipulation of *Neisseria gonorrhoeae*. Curr Protoc Microbiol. Chapter 4:Unit4A.2. https://doi.org/10.1002/9780471729259.mc04a02s00
24. Aylor DL, Valdar W, Foulds-Mathes W et al (2011) Genetic analysis of complex traits in the emerging Collaborative Cross. Genome Res 21(8):1213–1222. https://doi.org/10.1101/gr.111310.110. gr.111310.110 [pii]
25. Jolley KA, Maiden MC (2010) BIGSdb: scalable analysis of bacterial genome variation at the population level. BMC Bioinformatics 11:595. https://doi.org/10.1186/1471-2105-11-595

Chapter 24

Female Mouse Model of *Neisseria gonorrhoeae* Infection

Erica L. Raterman and Ann E. Jerse

Abstract

Mouse models of infection are important tools in the study of infectious disease or host the development of products to prevent or treat infections. The estradiol-treated mouse model of *Neisseria gonorrhoeae* genital tract infection has proved to be a valuable system for determining the importance of gonococcal factors that mediate evasion of host innate effectors in vivo or host gonococcal adaptation to hormonally driven host factors in females. Examination of mechanisms that *Neisseria gonorrhoeae* uses to subvert the host immune response also has been greatly aided by this whole model system, as have studies on the consequence of antibiotic resistance mutations on gonococcal fitness in vivo and the search for new antibiotics to treat antibiotic-resistant infections. The strict human specificity of *N. gonorrhoeae* limits the ability of experimental murine infection to mimic human infection. However, in recent years, the development of transgenic mice and protocols for supplementing mice with human factors has improved animal modeling of gonorrhea. To date, however, because the mouse estrous cycle is much shorter than the human reproductive cycle, all reported gonorrhea mouse models require treatment with estradiol and antibiotics to maintain an estrus-like state and suppress the overgrowth of inhibitory commensal flora that occurs under the influence of estrogen to allow sustained *N. gonorrhoeae* infection. In this chapter, we detail the methods used to (1) prepare the mice for experimental infection with *N. gonorrhoeae*, (2) inoculate mice and quantitatively culture vaginal swabs for noncompetitive and competitive infection experiments, and (3) monitor the host innate immune response to infection.

Key words *Neisseria gonorrhoeae*, Female mouse model of gonorrhea infection, Estrous cycle, Estradiol, Polymorphonuclear leukocytes, Cytokines, Chemokines

1 Introduction

Neisseria gonorrhoeae (*Ng*), the causative agent of gonorrhea, is a common sexually transmitted pathogen that maintains a reservoir solely in the mucosae of humans. *Ng* establishes infection of naïve hosts by commandeering host factors and expressing a variety of colonization factors, and can cause repeated infections through phase and antigenic variation of surface molecules and suppression of the adaptive immune response [1–9]. While limited by host restrictions, the female mouse model of *Ng* genital tract infection has proved to be a valuable system for determining the importance

Myron Christodoulides (ed.), *Neisseria gonorrhoeae: Methods and Protocols*, Methods in Molecular Biology, vol. 1997, https://doi.org/10.1007/978-1-4939-9496-0_24,

of gonococcal factors that mediate evasion of host innate effectors in vivo and bacterial adaptation to hormonally driven host factors in females (reviewed in [10]). This model has also provided insight into the genetic basis of the spread of antibiotic resistance [11–13], and allowed delineation of mechanisms by which *N. gonorrhoeae* suppresses the host adaptive response [14, 15]. The development of products to control this increasingly antibiotic resistant pathogen has also been greatly accelerated by having a small animal model for preclinical in vivo efficacy testing of antibiotics [16, 17], vaginal microbicides [18, 19], vaccines, and immunotherapies [20–25].

Genital tract pathogens are often highly adapted to their host, which includes the evolution of mechanisms to evade or capitalize on hormonally driven changes in histology and innate defenses that influence susceptibility to infection in the female genital tract. Consistent with the influence of reproductive hormones on susceptibility to genital tract infections is the need to utilize progesterone to increase susceptibility of laboratory mice to herpes simplex virus [26] and *Chlamydia trachomatis* [27] and estradiol for *Mycoplasma genitalium* [28], *Ng* [10] and *Candida albicans* [29]. The estrous cycle of female BALB/c mice lasts for 4–5 days and consists of the following four stages: proestrus, estrus, metestrus, and diestrus. A fifth stage termed anestrus is also observed in mice that have stopped cycling, which occurs when female mice are not exposed to male mice [30]. Each stage is defined by the types of immune cells and epithelial cells present in the vaginal tract (Fig. 1).

Work to develop a mouse model of *Ng* genital tract infection began in the 1970s and 1980s when it was discovered that female mice could be colonized with *Ng* following vaginal inoculation, as long as they were inoculated during the proestrus stage of the estrous cycle [31, 32]. Recovery of *Ng* was transient, however, with clearance occurring upon transition into the metestrus phase, which limited the feasibility of using female mice as a surrogate host for *N. gonorrhoeae*. This problem was solved by Taylor-Robinson and colleagues in 1990 who demonstrated that treatment of mice with 17β-estradiol prolongs the duration of infection with *Ng* [33]. Germ-free mice were used to eliminate the potential of inhibitory commensal flora interfering with infection. This model was reestablished 10 years later by our laboratory using slow-release 17β-estradiol pellets and antibiotics to promote long-term *Ng* infection [34]. Protocols have now been developed that support short-term (7–10 days) or longer-term (12–14 days) infection [35] that differ in cost and may be chosen based on the needs of the experiment being conducted. The need to use germ-free mice or antibiotics stems from the increased proliferation and turnover of vaginal cells that occurs under the influence of estrogen, which results in higher concentrations of free glycogen and thus overgrowth of the native vaginal microbiota. The choice of

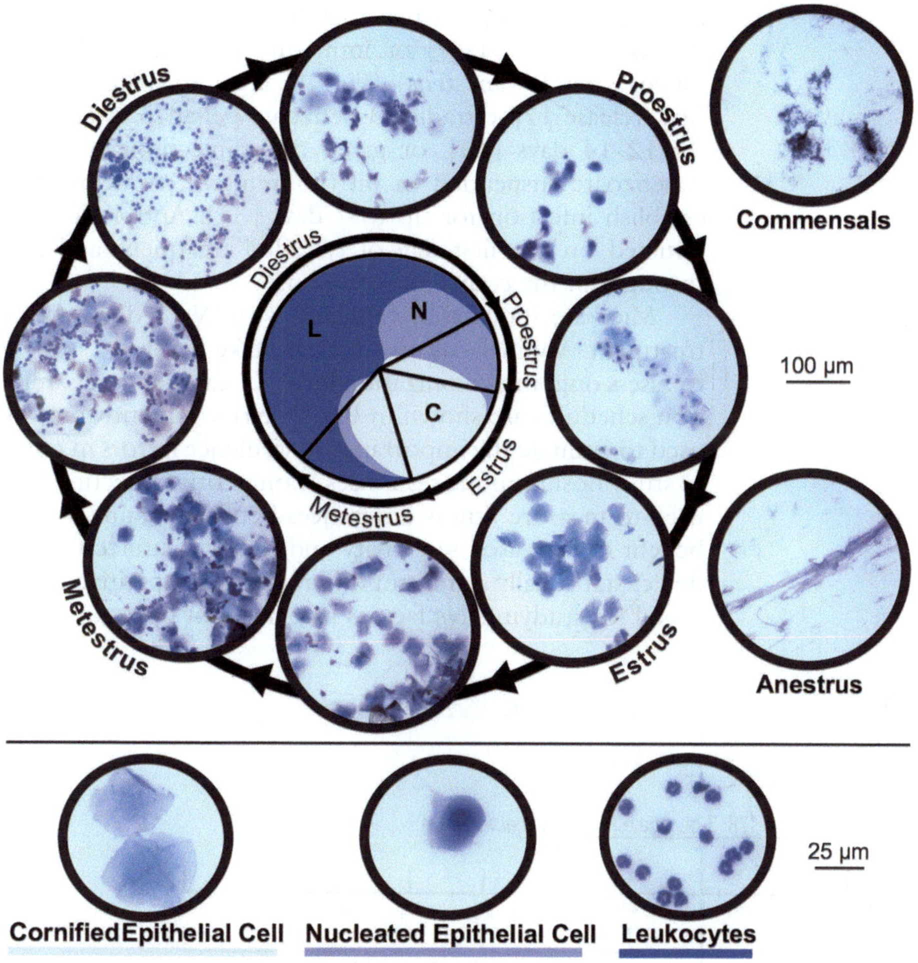

Fig. 1 Mouse estrous cycle. The estrous cycle consists of four stages (estrus, metestrus, diestrus, and proestrus), each with its own distinctive mix of cell types. Stained vaginal smears representing the four stages of the mouse estrous cycle, including intermediate stages, are depicted in the outer circle. Relative proportions of nucleated epithelial cells (N), cornified epithelial cells (C), and leukocytes (L) are depicted in the center of the images. Isolated images of each cell type are shown in the bottom panel. Smears that show the presence of vaginal commensal bacteria and the anestrus stage when cycling ceases are shown in the panels of the right

antibiotic treatment may vary with husbandry practices or when using adaptations of the *Ng* infection model for polymicrobial studies such as examination of the effect of H_2O_2-producing *Lactobacillus* sp. on *Ng* in vivo [36] and comparison of *Ng*/*Chlamydia* coinfection with infection by either pathogen alone [37].

Several inbred and outbred strains of mice have been used in this model, and to establish experimental *Ng* infection, mice in the diestrus or anestrus stages are identified by cytological differentiation of vaginal cells 2 days before bacterial inoculation to better synchronize the cycles and take advantage of the low commensal load in these stages. Mice that have a predominance of vaginal PMNs (diestrus stage) or immature epithelial cells, PMNs and mucus strands (anestrus) (Fig. 1) are implanted with a 21-day slow-release 17β-estradiol 3-benzoate pellet to establish infection for 12–14 days [10], or given three injections of 17β-estradiol 3-benzoate suspended in oil or Premarin, as detailed below to establish infection for up to 7 days [35]. Antibiotic treatment is initiated on the first day of estradiol treatment and maintained throughout the course of infection.

Mice are inoculated vaginally with *Ng* 2 days after estradiol treatment. Vaginal cultures may be taken daily or less frequently to assess ongoing colonization levels. Treatment and sample collection schedules are shown in Fig. 2. Noncompetitive infections are used to evaluate the importance of virulence factors in infection and to study host immune responses. Competitive infections can also be used to measure fitness differences between strains, which has the benefit of increased sensitivity and thus can utilize fewer mice for detect statistically signficant differences. Competitive infections are useful for studying *Ng* factors that may not be essential, but aid in

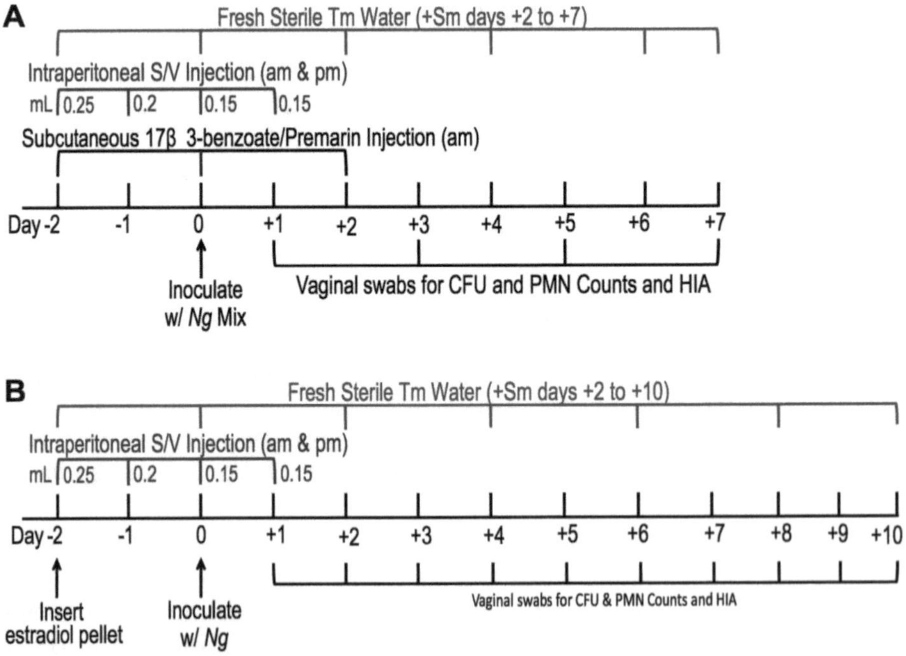

Fig. 2 Competitive and noncompetitive infection experimental schedules. The treatment and sample collection schedules for (**a**) competitive infections and (**b**) noncompetitive infections are shown. Concentrations of 17 β-estradiol 3-benzoate, Premarin, and streptomycin/vancomycin (S/V) solutions used for the injections, as well as the streptomycin (Sm) and trimethoprim™ concentrations used in the drinking water, are described in Subheading 3.2. The time points for collection of vaginal swabs to determine bacterial colonization levels, PMN influx, and the presence of commensals following inoculation with *Ng* on Day 0 are shown. HIA, heart-infusion agar for isolation of commensal bacteria

the colonization of, and persistence in, the lower reproductive tract. Only noncompetitive infections, however, can discern which gonococcal genes are essential for infection. An inflammatory response can be measured by determining the relative number of PMNs in stained vaginal smears and chemokine and cytokine concentrations in vaginal washes. During noncompetitive infections, a cyclical pattern of bacterial recovery is often observed, which reflects the influence of progesterone on colonization and/or host inflammatory responses [38, 39]. This pattern is not observed in mice that have had their ovaries surgically removed [38]. Typical results for single and competitive infections are shown in Figs. 3 and 4, respectively.

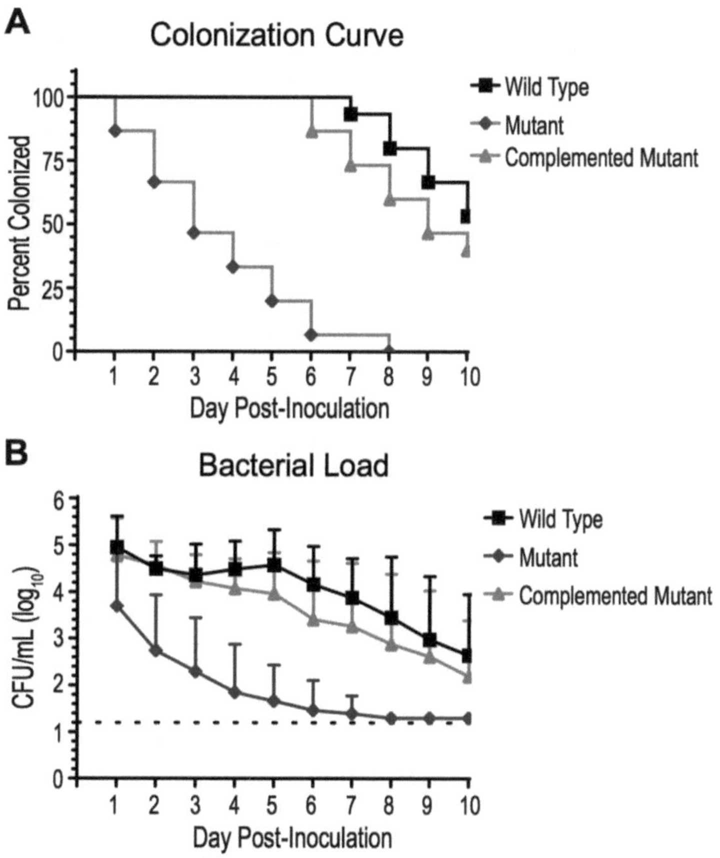

Fig. 3 Example results from a noncompetitive infection experiment with a wild-type strain, an attenuated isogenic mutant and complemented mutant. Typical (**a**) percentage of mice infected over time (colonization curve) and (**b**) number of gonococci recovered (bacterial load) (average CFU/mL ± SD) are shown. The graphs were created using GraphPad Prism software. Data for the colonization curves are analyzed using the log-rank Mantel–Cox test, and the bacterial load data are analyzed using the repeated measures two-way ANOVA with Bonferroni post hoc test

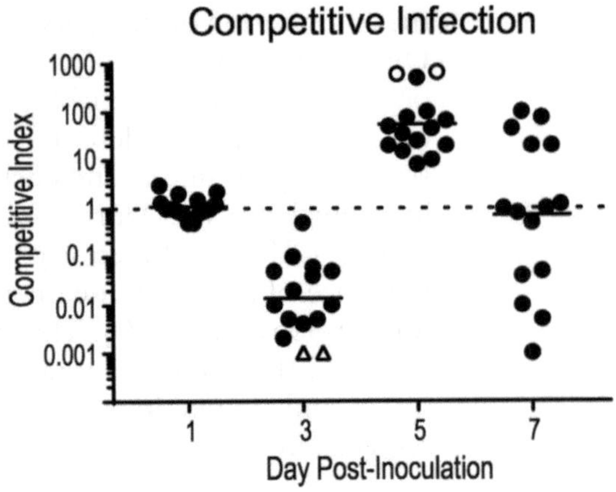

Fig. 4 Typical results from a competitive infection experiment. The attenuation or hyper-fitness of a mutant bacterial strain is determined by calculating the competitive index (CI) of mutant to wild-type [(mutant CFUs/wild-type CFUs)^{Output}/(mutant CFUs/wild-type CFUs)^{Input}] for each mouse at each time-point sampled. The bar represents the geometric mean of the cumulative CIs at each time point. A CI equal to 1 (dashed line) indicates that both strains competed similarly in vivo (Day 1), a CI of less than 1 indicates that the mutant was attenuated in vivo (Day 3), and a CI of greater than 1 indicates that the mutant strain outcompeted the wild-type strain in vivo (Day 5). An open triangle indicates that only wild-type bacteria were recovered from that mouse at that time point, and an open circle indicates that only mutant bacteria were recovered from that mouse at that time point

In this chapter, we detail the methods used to (1) prepare the mice for experimental infection with *N. gonorrhoeae*, (2) inoculate and quantitatively culture mice used in noncompetitive and competitive infection experiments, and (3) monitor the host innate immune response to infection.

2 Materials

2.1 Staging of Mice

1. BALB/C mice, bred in house animal facility or available from a commercial breeder (*see* **Note 1**).
2. Sterile 1× Phosphate Buffered Saline, pH 7.4 (*see* **Note 2**).
3. Polyester-tipped sterile applicators.
4. Glass microscope slides.
5. Colored marker pens.
6. Hemacolor® stain kit (EMD Millipore, or equivalent).
7. Light microscope.

2.2 Antibiotic and Hormone Treatment

1. 5 mg/mL of 17β-estradiol 3-benzoate dissolved in sesame oil by sonication for 5–10 s using an ethanol-sterilized probe (*see* **Note 3**), and sterilized by filtration using a 0.2 μm disc filter and syringe.
2. 2.5 mg/mL Premarin (Pfizer) dissolved in endotoxin-free distilled H_2O.
3. 5 mg 17β-estradiol slow-release pellets (Innovative Research of America).
4. Sterile scalpel, medical gauze, and blunt-end tweezers.
5. 0.4 mg of vancomycin hydrochloride and 2.4 mg streptomycin sulfate dissolved in 100 mL of endotoxin-free distilled H_2O and filter-sterilized through a 0.2 μm disc filter with syringe.
6. 0.4 g/L of trimethoprim dissolved in tap water, which has been previously autoclaved at 121 °C and 2.68 kg/cm^2 for 15 min.
7. 0.4 g/L trimethoprim and 5 g/L streptomycin sulfate dissolved in autoclaved tap water.
8. Sterile 1 mL Monoject™ tuberculin disposable syringes.
9. 27 gauge × ½ in. PrecisionGlide™ single-use beveled needles.

2.3 N. gonorrhoeae Inoculum Preparation and Non-Competitive Infection of Mice

1. *Neisseria gonorrhoeae* strains and isolates, stored at −80 °C.
2. GC agar plates: 36 g of Difco™ Gonococcal (GC) Base, 5 g of agar in 1 L of distilled H_2O. To prepare, rinse 2 L Erlenmeyer flasks with 70% (v/v) nitric acid in H_2O to remove any residual detergent (*see* **Note 4**). Rinse well 3–4 times with H_2O. Add GC base and agar to H_2O and boil for 20 min (*see* **Note 5**). Autoclave on liquid cycle for 20 min at 121 °C and 2.68 kg/cm^2 and cool to approximately 55 °C. Then add 10 mL of Kellogg's Supplement I (400 g of glucose, 20 mg of cocarboxylase, 10 g of L-glutamine dissolved in 1 L of distilled H_2O) and 1 mL of Kellogg's Supplement II (12 mM $Fe(NO_3)_3$ per 1 L of distilled H_2O). Pour into petri dishes and keep at room temperature for 2 days, flipping the plates upside down after the first day (*see* **Note 6**). Store plates at 4 °C until use.
3. Sterile 100 × 15 mm petri dishes.
4. 1× Phosphate Buffered Saline, pH 7.4.
5. Sterile 5 mL disposable syringes.
6. 25 mm 1.2 μm Sterile Puradisc™ syringe filters.
7. Cotton-tipped wooden-shaft sterile applicators.
8. 10 mL glass test tubes.
9. Spectrophotometer.
10. Humidified incubator set at 35 °C with 7% (v/v) CO_2.

2.4 *N. gonorrhoeae* Recovery from Vaginal Swabs

1. 1× Phosphate Buffered Saline, pH 7.4.
2. Polyester-tipped sterile applicators.
3. GC broth: 15 g of protease peptone #3, 4 g of dibasic potassium phosphate, 1 g of monobasic potassium phosphate, 5 g of NaCl dissolved in 1 L of distilled H_2O. Autoclave at 121 °C and 2.68 kg/cm^2 for 20 min to sterilize and store at 4 °C until used.
4. 0.5% (w/v) saponin in GC broth: dissolve 0.5 g of saponin in 100 mL of H_2O and add 1 mL of this solution per 10 mL of GC broth.
5. GC agar plates supplemented with vancomycin, colistin, nystatin, and trimethoprim sulfate (VCNT) antibiotics (*see* **Note 7**). Add VCNT supplement to cooled GC agar medium at the same time as the Kellogg's Supplements, as per manufacturer's instructions.
6. Humidified incubator set at 35 °C with 7% (v/v) CO_2.

2.5 Alterations in Protocol for Competitive Infections

1. All materials in Subheadings 2.3 and 2.4.
2. GC agar plates with 100 μg/ml of streptomycin sulfate (for total number of CFU).
3. GC agar plates containing the appropriate antibiotic for gonococci carrying the appropriate resistance cassette (e.g., 100 μg/mL of streptomycin sulfate, or 50 μg/mL of kanamycin sulfate, or 1–10 μg/mL of chloramphenicol, or 50 g/mL of spectinomycin, or 1–2 μg/mL of erythromycin) (mutant CFU).

2.6 Monitoring of the Immune Response

1. 1× phosphate buffered saline, pH 7.4.
2. Polyester-tipped sterile applicators.
3. Glass microscope slides.
4. 1.5 mL Eppendorf tubes.
5. Hemacolor® stain kit (EMD Millipore, or equivalent).
6. Light microscope.
7. Benchtop microcentrifuge.
8. ELISA assays for cytokines and chemokines.

2.7 Monitoring of the Vaginal Microbiota

1. 1× phosphate buffered saline, pH 7.4.
2. Polyester-tipped sterile applicators.
3. Heart Infusion Agar (HIA) plates: add 40 g of Difco™ HIA per 1 L of distilled H_2O. Autoclave on a liquid cycle at 121 °C and 2.68 kg/cm^{2}for 20 min. Cool the agar to 55 °C and pour into petri plates. Let agar set at room temperature, and store at 4 °C.
4. Humidified incubator set at 35 °C with 7% (v/v) CO_2.

5. Glass microscope slides.
6. BD BBL™ Gram stain kit, or equivalent.
7. Light microscope.

3 Methods

3.1 Staging of Mice

1. House 5–6-week-old female BALB/c mice for at least 10 days in a female only room of the animal facility (*see* **Note 8**).
2. On Day −2, code the mice in each cage using indelible ink to make colored stripes on their tails (e.g. Mouse 1 = Blue, Mouse 2 = Double Blue, Mouse 3 = Red, Mouse 4 = Double Red, Mouse 5 = Unmarked) (*see* **Note 9**). Label a corresponding grid on the glass slides for each cage of mice.
3. Moisten an applicator in sterile PBS. Hold a mouse by the base of the tail and gently insert the swab into the vagina and rotate it 2–3 times. Rub the applicator on the corresponding grid section on the glass slides to transfer the cells for staining.
4. Once the slides are dry, stain the cells using the Hemacolor stain and allow to dry. Use the microscope to examine the slides and determine which stage of the estrous cycle each mouse is in using the following criteria (*see* **Note 10**) (Fig. 1):

   Proestrus: Mix of nucleated and cornified epithelial cells.

   Estrus: Mostly or entirely cornified epithelial cells—commensal bacteria are often present.

   Metestrus: Largely cornified epithelial cells with a few nucleated cells and a few PMNs—commensal bacteria are often present.

   Diestrus: Mostly PMNs with a few cornified and nucleated epithelial cells.

   Anestrus: Few cells; those cells seen are mostly parabasal epithelial cells (small, condensed nucleus) that may be in large clusters and some PMNs and increased mucus, often seen as long strands.
5. Transfer the mice into new cages to consolidate the mice in diestrus and anestrus for the experiment. If necessary, color the tails of the mice again so that each cage does not contain mice with the same tail colors (this will be important later for tracking *N. gonorrhoeae* infection in individual mice for the duration of the experiment).

3.2 Antibiotic and Hormone Treatment

Estradiol treatments need to be given on Days −2, 0, and +2, unless slow-release estradiol pellets are used. Administer the estradiol treatment as follows, depending on the estradiol formulation used:

1. For 17β-estradiol 3-benzoate, inject the mice subcutaneously with 100 μL of a 5 mg/mL solution using a 27 gauge needle on a disposable 1 mL tuberculin syringe (*see* **Note 11**).
2. For Premarin, inject the mice subcutaneously with 200 μL of a 2.5 mg/mL solution.
3. For 17β-estradiol slow-release pellets, hold the mice by the scruff and make a small 3–4 mm incision in the skin using a sterile scalpel. Insert the estradiol pellet using blunt-ended tweezers to push the pellet through the incision. Gently guide the pellet under the skin using your fingers to ensure the mouse cannot remove the pellet by scratching. This procedure causes minimal pain, and no anesthetic or analgesic is required.
4. Administer injections of 4 mg/L of vancomycin, 24.0 mg/L of streptomycin (*see* **Note 12**) on Days −2 through +1 via intraperitoneal site according to the following schedule:

   Day −2: inject 0.25 mL of the antibiotic solution in the afternoon.

   Day −1: inject 0.2 mL of the antibiotic solution in the morning and afternoon, at least 5 h apart.

   Day 0: inject 0.15 mL of the antibiotic solution in the morning and afternoon, at least 5 h apart.

   Day +1: inject 0.15 mL of the antibiotic solution in the morning and afternoon, at least 5 h apart.
5. Replace H_2O on Day −2 with the 0.4 mg/L trimethoprim H_2O. Replace with fresh trimethoprim H_2O every other day for the duration of the experiment, adding 2.4 mg/L of streptomycin sulfate to the water after Day +1.

3.3 N. gonorrhoeae Inoculum Preparation and noncompetitive infection of Mice

1. Streak the *N. gonorrhoeae* strains from the −80 °C freezer onto GC agar plates without antibiotics on Day −2 and incubate for 24 h in a humidified incubator at 35 °C with 7% (v/v) CO_2 (*see* **Note 13**). Subculture each strain onto 5–10 fresh GC agar plates on Day −1 and incubate for 18–21 h in a humidified incubator at 35 °C with 7% (v/v) CO_2.
2. Collect colonies using the cotton-tipped applicator and suspend colonies in 4 mL of sterile PBS in a glass test tube. Collect only closely spaced discreet colonies from agar plates (*see* **Note 14**).
3. Filter the bacterial suspension through a 1.2 μm disc filter using a 5 mL syringe into a new glass test tube to remove bacterial aggregates (*see* **Note 15**).
4. Measure the optical density (OD) of the suspension with a spectrophotometer at $\lambda_{600\ nm}$ and adjust the OD to 0.08 using PBS. For inocula containing ~1–2 × 10^6 CFU/20 μL, dilute the suspension twofold in PBS, which should give a

concentration of ~5×10^7–1×10^8 CFU/mL. This suspension must be used to inoculate the mice within 20 min or the bacteria will begin to lyse.

5. Inoculate the lower genital tracts of the female mice with the prepared bacterial suspension at least 3 h after the Day 0 estradiol injection. Hold each mouse by the base of the tail and expel 20 μL of the suspension into the vagina using a micropipette. The inoculum for each mouse depends on the *N. gonorrhoeae* strain (*see* **Note 16**). Suspend each mouse upside down by the tail, allowing its paws to grasp the wire lid of the cage, for 1 min before releasing it into the cage (*see* **Note 17**).
6. Perform serial dilutions of the bacterial suspension in PBS and plate the dilutions onto GC agar plates. Incubate at 35 °C with 7% (v/v) CO_2 for 24 h. Count the colonies on each plate, and calculate the total inoculum as Colony Forming Units (CFU)/ mouse.

3.4 *N. gonorrhoeae* Recovery from Vaginal Swabs

For noncompetitive infections, vaginal samples are typically taken daily up to Day 10 post-inoculation.

1. Hold each mouse by the base of its tail and gently insert a PBS-moistened polyester-tipped swab into the vagina, and rotate it 2–3 times. Suspend the swab material in 100 μL of GC broth +0.05% (w/v) saponin (*see* **Note 18**).
2. Serially dilute the samples out to the 10^{-4} dilution and plate onto GC-VCNTS plates. Incubate in a humidified atmosphere at 35 °C with 7% (v/v) CO_2 for 24 h.
3. Count the plates and calculate the total bacterial load as CFU/mL.

3.5 Alterations in Protocol for Competitive Infections

1. Follow the same inoculation preparation protocol as detailed in Subheading 3.3 for each of the two *N. gonorrhoeae* strains to be competed. Mix the strains 1:1 before inoculating the mice (*see* **Note 19**). Plate equal volumes of the mixed inoculum onto GC-streptomycin agar plates and GC-streptomycin agar plates containing the antibiotic appropriate for the selectable marker present in one of the tested *N. gonorrhoeae* strains. Calculate the starting ratio of the two strains using the equation:

   [CFUs Strain#2 (selectable media counts)/CFUs Strain#1 (GC-strep counts – Selectable media counts)].

2. Follow the same protocol for *N. gonorrhoeae* recovery as detailed in Subheading 3.4. For competitive infections, vaginal samples are typically cultured every other day starting on Day +1 up to Day +7. Plate equal volumes of the diluted vaginal

swab suspensions onto GC-streptomycin agar plates and GC-streptomycin agar plates containing the antibiotic appropriate for the selectable marker present in one of the tested *N. gonorrhoeae* strains. Use the following equation to determine the competitive index (CI):

$$\mathrm{CI} = \frac{(\mathrm{CFUs\ Strain\ \#2/CFUs\ Strain\ \#1})^{\mathrm{Output}}}{(\mathrm{CFUs\ Strain\ \#2/CFUs\ Strain\ \#1})^{\mathrm{Input}}}.$$

3.6 Monitoring of the Inflammatory Response

1. To monitor vaginal PMN influx, streak a glass slide with the same applicator used to swab for the CFU counts, and use the same grid pattern on the glass slide as used when staging the mice, as described in **steps 2** and **3** of Subheading 3.1. Repeat **step 4** of Subheading 3.1. Under the microscope, count the number of PMNs per every 100 cells to obtain the estimated percentage of PMN influx.
2. For the vaginal washes to monitor cytokine/chemokine levels, wash the vaginal tract with 33 μL of sterile PBS by pipetting the liquid in and out of the vagina several times. Place the suspension in a labeled 1.5 mL Eppendorf tube. Repeat the procedure twice more for a total collection volume of 99 μL. Centrifuge the samples at 8000 × *g* for 3 min and transfer the supernatants to new tubes. Store the samples at −80 °C until ready to analyze by ELISA (*see* **Notes 20** and **21**).

3.7 Monitoring of the Vaginal Microbiota

1. Divide an HIA plate into quadrants (one for each mouse), and touch the plate with the same applicator used to swab for the CFU counts, i.e. the applicator described in **step 1** of Subheading 3.4. Streak for isolates in each quadrant, and incubate the plates in a humidified incubator set to 35 °C with 7% (v/v) CO_2 for 24 h.
2. Make note of any growth and do a Gram stain to further identify any commensals. The most common commensals observed are listed in Table 1.

4 Notes

1. Experimental infection of BALB/c and C57/BL6 mice is the best-characterized. *Ng* infection of BALB/c mice results in a proinflammatory cytokine and chemokine response and an influx of vaginal PMNs; in contrast, *Ng* infection of C57/BL5 mice does not induce an inflammatory response [10, 40].
2. Sterile PBS can be procured from commercial sources, or prepared as follows: add 8 g/L of NaCl, 0.2 g/L of KCl,

Table 1
Possible commensal organisms isolated from vaginal swabs

Commensal	Visible description and action
Gram-negative rods	Large, shiny colonies (oxidase negative unless *Pseudomonas* sp.). Remove mouse from study and move cage mates to fresh cage
Gram-positive cocci	White colonies (usually *Staphylococcus* spp.) and yellow colonies (usually *Micrococcus* spp.) generally do not inhibit *Ng* infection; may need to isolate mice to keep from spreading, but continue culturing
Gram-positive rods	Pinpoint colonies (usually *Lactobacillus* spp.) do not inhibit *Ng* infection
Gram-positive large cells in either single cells or filaments	Large creamy colonies or fuzzy, sporulating colonies (usually yeast or fungi) generally do not clear infection. Mice can be treated with an over-the-counter antifungal vaginal cream if growth on GC-streptomycin agar occludes *N. gonorrhoeae* colonies

1.15 g/L of Na_2HPO_4, 0.2 g/L of KH_2PO_4 in distilled H_2O and autoclave at 121 °C and 2.68 kg/cm^2 for 15 min.

3. 17β-Estradiol 3-benzoate is difficult to dissolve and will not form a solution unless sonicated. The solution should be used right away, as the 17β-estradiol 3-benzoate will precipitate if left at room temperature for too long.
4. *N. gonorrhoeae* is very sensitive to detergent left in the flasks after washing. Rinsing with diluted nitric acid ensures that any residual detergent is removed.
5. The boiling step helps to more fully dissolve the starch in the GC agar. *N. gonorrhoeae* will still grow on the plates if this step is skipped, but the plates will not be usable with a spiral plater and an automated plate counter, as the flecks of undissolved starch will be read as colonies by the counter software.
6. *N. gonorrhoeae* is sensitive to the moisture content of the plates. The 2 day drying protocol consistently produces plates that are ideal for *N. gonorrhoeae* growth.
7. VCNT antibiotics can be added to isolate *N. gonorrhoeae* from murine vaginal mucus. We use VCNT from Becton Dickinson, which contains 300 μg of vancomycin, 750 μg of colistin, 1250 units of nystatin and 500 μg of trimethoprim lactate per 1 mL reconstituted solution. Vancomycin inhibits gram-positive contaminants, colistin inhibits gram-negative bacteria, including *Pseudomonas* species, nystatin suppresses the growth of yeasts and trimethoprim inhibits *Proteus* species. Other suppliers can be used.
8. Male pheromones arrest the female estrous cycle in the estrus phase, resulting in lowered numbers of mice in the proper

phase to begin the experiment. Husbandry requirements should also include the housing of mice in cages with autoclaved bedding, food and water to lower the commensal load.

9. Approximately 70% of the mice will be in the diestrus or anestrus stage at any given time, requiring two separate stagings spaced 3 days apart to catch most of the mice in the proper state to begin the experiment. Therefore, the total number of starting mice should be greater (we use 20–25% more) than the number intended for the experiment, as some mice will not be in the correct phase at either of the stagings. Tail markings have to be darkened every 5–6 days due to fading. We recommend using blue, red, purple, brown, and/or black markers as yellow, orange and green markers tend to fade more quickly.

10. Rarely are commensal bacteria isolated from diestrus-stage mice or mice in anestrus; however, if bacteria are seen in vaginal smears, it is best not to use these mice to minimize the chance of these commensal bacteria inhibiting infection.

11. Due to the viscosity of the oil, it is easier to load the syringe before the 27 gauge needle is attached.

12. The problem of inhibitory commensal flora can be specific to the breeder or to husbandry practices. In the protocols described here, only streptomycin treatment is required for *N. gonorrhoeae* infection; if commensal overgrowth is a problem, trimethoprim can be added to the drinking water to prevent overgrowth of *Pseudomonas* spp. or members of the *Enterobacteriaceae*, which can affect the survival of *Ng*. Vancomycin can be used to suppress Gram-positive flora, although unless present in high numbers, Gram-positive flora, including murine *Lactobacillus* sp., do not typically inhibit experimental *Ng* murine infection. When giving all three antibiotics, vaginal yeast infections may sometimes result, which do not inhibit *Ng* but make it difficult to isolate *Ng* on the selective agar used. In this case, over-the-counter yeast creams can be used to hold back yeast overgrowth. It is important that *Ng* strains used are resistant to the antibiotics chosen. Many *Ng* strains are naturally streptomycin resistant and most are resistant to trimethoprim and vancomycin.

13. The strains must be streptomycin-resistant (i.e. the strains have a mutation in *rpsL* that confers resistance) to be used successfully in the mouse model. Construction of streptomycin-resistant strains can be achieved by introducing the *rpsL* gene from a streptomycin resistant strain by natural transformation [12].

14. It is important to use isolated colonies, so that fewer dead bacteria are included in the inoculum.

15. Filtration through 1.2 μm disc filter pores removes bacterial clumps, which are formed by piliated gonococci or gonococci that express Opa proteins that confer a highly opaque colony morphology due to bacterial aggregation.
16. Passage of *Ng* strains through mice is not required to establish infection. For most *Ng* strains tested, 10^5–10^6 colony-forming units (CFU) establishes infection in >80% of mice [10, 34]. For some more virulent strains, such as strain H041, an inoculum as low as ~1×10^4 CFU per mouse is sufficient to infect >80% of mice.
17. Inverting the mice allows gravity to carry the inoculum deeper into the vaginal tract.
18. *N. gonorrhoeae* tends to form aggregates. Addition of saponin to the dilution medium helps to break apart clumping, thus producing a more accurate final CFU count.
19. Not all *N. gonorrhoeae* strains and mutants have similar CFU to OD ratios. It is therefore important to pretest the inoculum before the day of inoculation to be sure that the correct number of bacteria is inoculated and that a 1:1 ratio is obtained for competitive infections.
20. A PMN influx is usually observed on Days 3–5 post-inoculation in BALB/c mice and inflammatory cytokine and chemokine levels (e.g. TNF-α, IL-1α, IL-1β, IL-2, MIP-1α, and KC) usually spike around the same time [10]. C57/BL6 mice are colonized at similar levels but do not exhibit a PMN influx or proinflammatory response to infection [40]. It is critical that uninfected mice be tested in parallel to provide a base-line for these immune parameters, which may be altered as the effect of the estradiol diminishes over time.
21. Cytokine and chemokine levels can be quantified using individual ELISA kits and/or multiplex platforms.

References

1. Cornelissen CN (2008) Identification and characterization of gonococcal iron transport systems as potential vaccine antigens. Future Microbiol 3(3):287–298. https://doi.org/10.2217/17460913.3.3.287
2. Criss AK, Seifert HS (2012) A bacterial siren song: intimate interactions between *Neisseria* and neutrophils. Nat Rev Microbiol 10 (3):178–190. https://doi.org/10.1038/nrmicro2713
3. Edwards JL, Apicella MA (2004) The molecular mechanisms used by *Neisseria gonorrhoeae* to initiate infection differ between men and women. Clin Microbiol Rev 17(4):965–981
4. Liu Y, Feinen B, Russell MW (2011) New concepts in immunity to *Neisseria gonorrhoeae*: innate responses and suppression of adaptive immunity favor the pathogen, not the host. Front Microbiol 2:52. https://doi.org/10.3389/fmicb.2011.00052
5. Ngampasutadol J, Ram S, Gulati S et al (2008) Human factor H interacts selectively with *Neisseria gonorrhoeae* and results in species-specific complement evasion. J Immunol 180 (5):3426–3435
6. Quillin SJ, Seifert HS (2018) *Neisseria gonorrhoeae* host adaptation and pathogenesis. Nat

Rev Microbiol 16(4):226–240. https://doi.org/10.1038/nrmicro.2017.169

7. Ram S, Cullinane M, Blom AM et al (2001) Binding of C4b-binding protein to porin: a molecular mechanism of serum resistance of *Neisseria gonorrhoeae*. J Exp Med 193(3):281–295

8. Zhu W, Tomberg J, Knilans KJ et al (2018) Properly folded and functional PorB from *Neisseria gonorrhoeae* inhibits dendritic cell stimulation of CD4(+) T cell proliferation. J Biol Chem 293(28):11218–11229. https://doi.org/10.1074/jbc.RA117.001209

9. Zhu W, Ventevogel MS, Knilans KJ et al (2012) *Neisseria gonorrhoeae* suppresses dendritic cell-induced, antigen-dependent CD4 T cell proliferation. PLoS One 7(7):e41260. https://doi.org/10.1371/journal.pone.0041260

10. Jerse AE, Wu H, Packiam M et al (2011) Estradiol-treated female mice as surrogate hosts for *Neisseria gonorrhoeae* genital tract infections. Front Microbiol 2:107. https://doi.org/10.3389/fmicb.2011.00107

11. Kunz AN, Begum AA, Wu H et al (2012) Impact of fluoroquinolone resistance mutations on gonococcal fitness and *in vivo* selection for compensatory mutations. J Infect Dis 205(12):1821–1829. https://doi.org/10.1093/infdis/jis277

12. Vincent LR, Kerr SR, Tan Y et al (2018) In vivo-selected compensatory mutations restore the fitness cost of mosaic penA alleles that confer ceftriaxone resistance in *Neisseria gonorrhoeae*. mBio 9(2). https://doi.org/10.1128/mBio.01905-17

13. Warner DM, Shafer WM, Jerse AE (2008) Clinically relevant mutations that cause derepression of the *Neisseria gonorrhoeae* MtrC-MtrD-MtrE efflux pump system confer different levels of antimicrobial resistance and *in vivo* fitness. Mol Microbiol 70(2):462–478. https://doi.org/10.1111/j.1365-2958.2008.06424.x

14. Liu Y, Liu W, Russell MW (2014) Suppression of host adaptive immune responses by *Neisseria gonorrhoeae*: role of interleukin 10 and type 1 regulatory T cells. Mucosal Immunol 7(1):165–176. https://doi.org/10.1038/mi.2013.36

15. Song W, Condron S, Mocca BT et al (2008) Local and humoral immune responses against primary and repeat *Neisseria gonorrhoeae* genital tract infections of 17beta-estradiol-treated mice. Vaccine 26(45):5741–5751. https://doi.org/10.1016/j.vaccine.2008.08.020

16. Butler MM, Waidyarachchi SL, Connolly KL et al (2018) Aminomethyl spectinomycins as therapeutics for drug-resistant gonorrhea and chlamydia coinfections. Antimicrob Agents Chemother 62(5). https://doi.org/10.1128/AAC.00325-18

17. Schmitt DM, Connolly KL, Jerse AE et al (2016) Antibacterial activity of resazurin-based compounds against *Neisseria gonorrhoeae in vitro* and *in vivo*. Int J Antimicrob Agents 48(4):367–372. https://doi.org/10.1016/j.ijantimicag.2016.06.009

18. Spencer SE, Valentin-Bon IE, Whaley K et al (2004) Inhibition of *Neisseria gonorrhoeae* genital tract infection by leading-candidate topical microbicides in a mouse model. J Infect Dis 189(3):410–419. https://doi.org/10.1086/381125

19. Zeitlin L, Hoen TE, Achilles SL et al (2001) Tests of Buffergel for contraception and prevention of sexually transmitted diseases in animal models. Sex Transm Dis 28(7):417–423

20. Gulati S, Zheng B, Reed GW et al (2013) Immunization against a saccharide epitope accelerates clearance of experimental gonococcal infection. PLoS Pathog 9(8):e1003559. https://doi.org/10.1371/journal.ppat.1003559

21. Cole JG, Jerse AE (2009) Functional characterization of antibodies against *Neisseria gonorrhoeae* opacity protein loops. PLoS One 4(12):e8108. https://doi.org/10.1371/journal.pone.0008108

22. Liu Y, Hammer LA, Liu W et al (2017) Experimental vaccine induces Th1-driven immune responses and resistance to *Neisseria gonorrhoeae* infection in a murine model. Mucosal Immunol 10(6):1594–1608. https://doi.org/10.1038/mi.2017.11

23. Plante M, Jerse A, Hamel J et al (2000) Intranasal immunization with gonococcal outer membrane preparations reduces the duration of vaginal colonization of mice by *Neisseria gonorrhoeae*. J Infect Dis 182(3):848–855. https://doi.org/10.1086/315801

24. Zhu W, Chen CJ, Thomas CE et al (2011) Vaccines for gonorrhea: can we rise to the challenge? Front Microbiol 2:124. https://doi.org/10.3389/fmicb.2011.00124

25. Shaughnessy J, Gulati S, Agarwal S et al (2016) A novel factor H-Fc chimeric immunotherapeutic molecule against *Neisseria gonorrhoeae*. J Immunol 196(4):1732–1740. https://doi.org/10.4049/jimmunol.1500292

26. Parr MB, Kepple L, McDermott MR et al (1994) A mouse model for studies of mucosal immunity to vaginal infection by herpes simplex virus type 2. Lab Investig 70(3):369–380

27. Pal S, Tifrea DF, Zhong G et al (2018) Transcervical inoculation with *Chlamydia trachomatis* induces infertility in HLA-DR4 transgenic and wild-type mice. Infect Immun 86(1). https://doi.org/10.1128/IAI.00722-17
28. McGowin CL, Spagnuolo RA, Pyles RB (2010) *Mycoplasma genitalium* rapidly disseminates to the upper reproductive tracts and knees of female mice following vaginal inoculation. Infect Immun 78(2):726–736. https://doi.org/10.1128/IAI.00840-09
29. Yano J, Kolls JK, Happel KI et al (2012) The acute neutrophil response mediated by S100 alarmins during vaginal *Candida infections* is independent of the Th17-pathway. PLoS One 7(9):e46311. https://doi.org/10.1371/journal.pone.0046311
30. Dalal SJ, Estep JS, Valentin-Bon IE et al (2001) Standardization of the Whitten Effect to induce susceptibility to *Neisseria gonorrhoeae* in female mice. Contemp Top Lab Anim Sci 40(2):13–17
31. Braude AI (1982) Maxwell Finland Lecture. Resistance to infection with the gonococcus. J Infect Dis 145(5):623–624
32. Streeter PR, Corbeil LB (1981) Gonococcal infection in endotoxin-resistant and endotoxin-susceptible mice. Infect Immun 32(1):105–110
33. Taylor-Robinson D, Furr PM, Hetherington CM (1990) *Neisseria gonorrhoeae* colonises the genital tract of oestradiol-treated germ-free female mice. Microb Pathog 9(5):369–373
34. Jerse AE (1999) Experimental gonococcal genital tract infection and opacity protein expression in estradiol-treated mice. Infect Immun 67(11):5699–5708
35. Raterman EL, Macintyre A, Jerse A (2018) Comparison of use of 17β-estradiol 3-benzoate and premarin for promoting experimental *Neisseria gonorrhoeae* genital tract infection in female mice. J Microbiol Methods, submitted for publication
36. Muench DF, Kuch DJ, Wu H et al (2009) Hydrogen peroxide-producing lactobacilli inhibit gonococci in vitro but not during experimental genital tract infection. J Infect Dis 199(9):1369–1378. https://doi.org/10.1086/597390
37. Vonck RA, Darville T, O'Connell CM et al (2011) Chlamydial infection increases gonococcal colonization in a novel murine coinfection model. Infect Immun 79(4):1566–1577. https://doi.org/10.1128/IAI.01155-10
38. Cole JG, Fulcher NB, Jerse AE (2010) Opacity proteins increase *Neisseria gonorrhoeae* fitness in the female genital tract due to a factor under ovarian control. Infect Immun 78(4):1629–1641. https://doi.org/10.1128/IAI.00996-09
39. Simms AN, Jerse AE (2006) *In vivo* selection for *Neisseria gonorrhoeae* opacity protein expression in the absence of human carcinoembryonic antigen cell adhesion molecules. Infect Immun 74(5):2965–2974. https://doi.org/10.1128/IAI.74.5.2965-2974.2006
40. Packiam M, Veit SJ, Anderson DJ et al (2010) Mouse strain-dependent differences in susceptibility to *Neisseria gonorrhoeae* infection and induction of innate immune responses. Infect Immun 78(1):433–440. https://doi.org/10.1128/IAI.00711-09

Chapter 25

Experimental Human Infection with *Neisseria gonorrhoeae*

Marcia M. Hobbs and Joseph A. Duncan

Abstract

Experimental infection of male volunteers with *Neisseria gonorrhoeae* is safe and reproduces the clinical features of naturally acquired gonococcal urethritis. The human model is useful for testing the importance of putative gonococcal virulence factors for urethral infection in men and the model presents opportunities to examine host immune responses that may be exploited or improved in development and testing of gonococcal vaccines. In this chapter, we describe methods for production, characterization, and storage of *N. gonorrhoeae* stocks for experimental human challenge, preparation and delivery of inoculum suspensions, monitoring experimental infection, and statistical considerations for data analysis.

Key words *Neisseria gonorrhoeae*, Gonococcal urethritis, Experimental human infection

1 Introduction

The exclusive adaptation of *Neisseria gonorrhoeae* to its human host has long challenged the use of animal models to study gonorrhea and has hampered vaccine development. A murine model, in which estradiol-treated female mice are colonized in the lower reproductive tract with ascending infection in approximately 20% of infected mice, has greatly facilitated systematic vaccine testing [1] (*see* also Chapter 24 by Raterman and Jerse). The relatively recent development of transgenic mice that relieve some of the host restrictions has the potential to enhance the relevance of mouse studies to human disease caused by *N. gonorrhoeae* [2–4]. Nevertheless, experimental human urethral infection in male volunteers, established decades ago as a safe and highly relevant model of natural infection in men [5], remains an important tool in gonorrhea vaccine and microbial pathogenesis research.

Experimental gonorrhea is restricted to male subjects; women are excluded due to potential reproductive complications from ascendant gonococcal infection. Inoculation is accomplished by the instillation of a bacterial suspension through a catheter inserted into the distal end of the anterior urethra. After inoculation,

Myron Christodoulides (ed.), *Neisseria gonorrhoeae: Methods and Protocols*, Methods in Molecular Biology, vol. 1997,
https://doi.org/10.1007/978-1-4939-9496-0_25,

infection can be monitored by bacterial recovery in urogenital specimens; nucleic acid amplification-based testing (NAAT) may also prove to be a useful method to monitor infection in the model. Host responses can be assessed by enumeration of white blood cells in urine and urethral exudates and measurement of soluble immune mediators including antibodies and cytokines or mRNA in urogenital specimens and serum. Clinical manifestations of gonococcal infection are monitored by daily interview and physical examination. Under currently approved protocols, subjects receive prompt effective antibiotic treatment when (1) they request treatment regardless of signs, symptoms or positive cultures, (2) purulent urethral discharge is observed by the examining clinician or reported by the subject, or (3) at the end of the experimental period, usually 5 days after inoculation, regardless of whether or not the subject is judged to have been infected. A negative *N. gonorrhoeae* NAAT from urine obtained 3–7 days after antibiotic treatment is required to document clearance of experimental infection. A final follow-up interview by telephone is conducted approximately 7 days after the test of cure to assess study-related adverse events.

Two *N. gonorrhoeae* strains, FA1090 and MS11mkC, have been used in the majority of experimental human infection studies [5]. In this chapter, we describe current methods for production, characterization, and storage of *N. gonorrhoeae* stocks for experimental human challenge, preparation and delivery of inoculum suspensions, and monitoring experimental infection. We note that rigorous safety and ethical review are essential for the protection of human subjects in experimental gonococcal infection studies; however, the details of these and other regulatory aspects of study design are beyond the scope of the methods described herein.

2 Materials

2.1 *Neisseria gonorrhoeae* Working Banks

Biosafety level 2 facilities are required in a research laboratory and in a location near (adjacent to, or within a 1–2 min walk, in the same building) the clinical site(s) where subjects will be inoculated and specimens to monitor experimental infection will be obtained. In addition, ready access to an accredited clinical microbiology laboratory is needed for microbial characterization of *N. gonorrhoeae* strains to be used for experimental human infection.

2.1.1 Bacterial Strains and Media

1. *Neisseria gonorrhoeae* strains FA10910, MS11mkC, or other wild-type or isogenic mutant strains, stored at −70 °C in Freezer Storage Medium (FSM).
2. *Staphylococcus aureus* and *Escherichia coli* bacteria stored as frozen stocks at −70 °C.

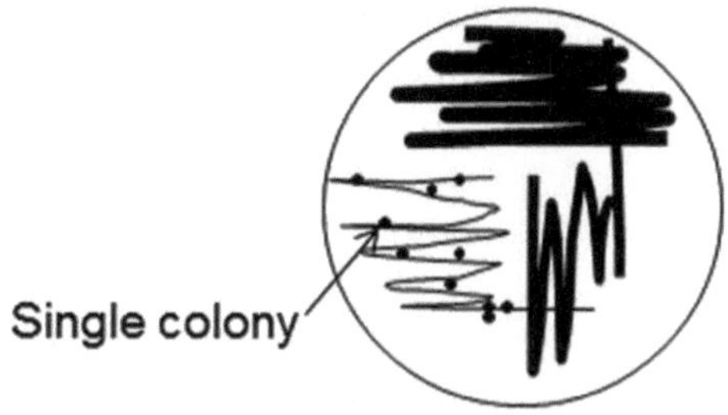

Fig. 1 Streaking pattern for areas of heavy growth and isolated bacterial colonies on agar plates

3. Freezer Storage Medium (FSM): mix 3 g of Trypticase soy broth powder (*see* **Note 1**) and 25 mL of glycerin in a 100 mL graduated cylinder. Add distilled, deionized (dd) H_2O to 100 mL final volume. Mix with a magnetic stirrer until all the ingredients are dissolved. Dispense 0.5 mL into sterile 2 mL freezer storage vials and cap loosely. Autoclave for 15 min at 121 °C and 2.68 kg/cm^2 with slow exhaust. Remove from the autoclave and cool to room temperature. Tighten the caps and store vials at 4 °C. Label racks with "FSM" and the "date of preparation," leaving individual tubes to be labeled with strain information as they are used.

   Each batch is tested for sterility and ability to support gonococcal viability. To test for sterility, choose several vials chosen at random from the batch, and plate 100 μL onto GCB agar plates and incubate overnight in a humidified incubator set at 37 °C with 5% (v/v) CO_2. The plates must be free of microbial growth after 24 h.

   To test gonococcal viability, inoculate several vials chosen at random from the batch with gonococci grown for 18–24 h on GCB agar plates. Mix well and freeze vials at −70 °C for at least 24 h. Without allowing the contents of the vials to thaw, scrape a small amount of frozen bacterial suspension from each test vial with a sterile microbiological loop and streak a fresh GCB agar plate as shown in Fig. 1. Incubate the plates at 37 °C with 5% (v/v) CO_2 for 24 h. Plates should contain areas of heavy growth and isolated gonococcal colonies as shown.

4. GCB agar plates (*see* **Notes 1** and **2**): add 36 g of Difco™ GC medium base per liter of ddH_2O to a flask with a magnetic stir bar. Autoclave for 15 min at 121 °C and 2.68 kg/cm^2 on a slow exhaust cycle. After autoclaving, place the flask in a 56 °C water bath to cool. When cool, add 10.0 mL of Supplement I and 1.0 mL of Supplement II per liter of medium. Mix thoroughly on a magnetic stir plate to incorporate the supplements. Working on a clean, disinfected benchtop free from drafts, pour ~10 mL of medium per plastic petri plate (100 mm × 15 mm). Avoid bubbles in the medium. Invert plates when cooled and

solidified. Select two plates at random from each batch of medium; streak one plate with *N. gonorrhoeae* and leave the remaining plate uninoculated. Incubate test plates overnight for GCB. Batches that support the growth of *N. gonorrhoeae* and yield no contaminant growth on the uninoculated plate are suitable for use in experimental human infection studies.

5. GCB-VCNT agar plates: aseptically reconstitute the inhibitor mixture (Remel) with sterile H_2O, mix and add 10 mL of reconstituted solution per litre of cooled GCG agar medium. The concentrations of inhibitors in the reconstituted solution are: 400 μg/mL of vancomycin; 750 μg/mL of colistin; 1250 units/mL of nystatin; 625 μg/mL of trimethoprim lactate. Mix thoroughly on a magnetic stir plate to incorporate the supplements and inhibitors. Working on a clean, disinfected benchtop free from drafts, pour ~10 mL of medium per plastic petri plate (100 mm × 15 mm). Avoid bubbles in the medium. Invert plates when cooled and solidified. Store the inverted plates at 4 °C in plastic sleeves, dated and labeled. Prepare the plates within 2 weeks of an experimental infection start date. Test for sterility and ability to support gonococcal growth as described above for GCB agar plates, except incubate the plates for 48 h for GCB-VCNT.
6. Supplement I: add 400 g of glucose (dextrose) to a large glass beaker and add ddH_2O to approximately 900 mL volume. Carefully add a magnetic stir bar and mix on a heated stir plate until sugar is dissolved. Cool the solution completely to room temperature and add 10 g of L-glutamine and 20 mg of cocarboxylase, and mix. When all chemicals are in solution, transfer to a graduated cylinder (reserving the stir bar) and adjust the volume to 1 L with ddH_2O. Mix thoroughly and sterilize by filtration using a 0.22 μm filter, and aseptically dispense 100 mL aliquots into sterile, labeled foil-wrapped reagent bottles. Store Supplement I at 4 °C.
7. Supplement II: add 1.25 g $Fe(NO_3)_3{\cdot}9H_2O$ into a 250 mL graduated cylinder, and add ddH_2O to the 250 mL mark. Mix with a magnetic stirrer and when $Fe(NO_3)_3{\cdot}9H_2O$ is completely dissolved, filter-sterilize the solution using a 0.22 μm filter. Aseptically dispense 25 mL aliquots into sterile labeled reagent bottles and store Supplement II at 4 °C.
8. 100 mm × 15 mm sterile plastic petri plates.
9. Humidified incubator set to 37 °C with 5% (v/v) CO_2.
10. 2 mL sterile freezer storage vials.
11. Sterile microbiological loops.
12. LB agar plates (*see* **Note 1**): add 20 g of LB agar powder per litre of ddH_2O to a flask with a magnetic stir bar. Autoclave for

15 min at 121 °C and 2.68 kg/cm^2 on slow exhaust cycle. After autoclaving, place the flask in a 56 °C water bath to cool. Working on a clean, disinfected benchtop free from drafts, pour ~10 mL medium per plastic petri plate (100 mm × 15 mm). Avoid bubbles in medium. Invert plates when cooled and solidified. Store inverted plates at 4 °C in plastic sleeves dated and labeled LB agar. Select four plates at random from each batch of medium; streak one plate each with *N. gonorrhoeae*, *S. aureus* and *E. coli*, and leave the remaining plate uninoculated. Incubate the test plates overnight in a humidified incubator set at 37 °C with 5% (v/v) CO_2. Batches that support the growth of *S. aureus* and *E. coli* but not *N. gonorrhoeae* and yield no contaminant growth on the uninoculated plate are suitable for use in experimental human infection studies.

2.1.2 Preparation of Working Banks with a Single Strain

1. *Neisseria gonorrhoeae* strains FA10910, MS11mkC, or other wild-type or isogenic mutant strains, stored at −70 °C in Freezer Storage Medium (FSM).
2. Freezer Storage Medium (FSM).
3. GCB agar plates.
4. GCB-VCNT agar plates.
5. Supplement I.
6. Supplement II.
7. 100 mm × 15 mm sterile plastic petri plates.
8. Stereomicroscope with a sub-stage light reflector.
9. Humidified incubator set to 37 °C with 5% (v/v) CO_2.
10. Sterile cotton-tipped swabs.
11. Sterile microbiological loops.
12. Sterile glass tubes that fit into chamber of spectrophotometer, with lids.
13. 2 mL sterile freezer storage vials.
14. LB agar plates.
15. 1× sterile Dulbecco's phosphate-buffered saline (PBS), without calcium and magnesium. Commercially prepared buffer contains 0.20 g/L of KCl, 0.2 g/L of KH_2PO_4, 8.0 g/L of NaCl, and 1.15 g/L of Na_2HPO_4 (anhydrous) at a final pH of 7.4 ± 0.1 and osmolality of 276 ± 10 mOsm. It contains ≤0.25 endotoxin units/mL and is negative for *Mycoplasma species* as determined by culture. Store the product at 2–30 °C for up to 36 months.
16. 10 cc sterile syringe fitted with 1.2 μm filter.

17. Monoclonal antibodies and reagents for confirming piliation status (Pil), opacity protein (Opa) and lipooligosaccharide (LOS) expression.
18. Commercial whole-genome sequencing service.
19. Benchtop vortex mixer.
20. Genesys 30 spectrophotometer, or equivalent.

2.1.3 Preparation of Working Banks with Two Strains for Competitive Infections

1. All materials described in Subheadings 2.1.1 and 2.1.2.
2. GCB broth (*see* **Note 1**): add 15 g of proteose peptone, 5 g of NaCl, 4 g of K_2HPO_4, 1 g of KH_2PO_4 per liter of ddH_2O. Mix well and dispense into reagent bottles, autoclave for 15 min at 121 °C and 2.68 kg/cm^2 and store at room temperature.

2.2 Catheter Preparation

1. Sterile No. 8 paediatric French feeding catheter.
2. Scissors or scalpel.
3. Sterilization indicator strip.
4. Self-sealing pouches.
5. Low temperature ethylene oxide sterilization facility.

2.3 Preparation of Inocula for Experimental Human Infection

1. *Neisseria gonorrhoeae* characterized working banks.
2. 10 μL sterile microbiological loops.
3. GCB agar plates.
4. Humidified incubator set to 37 °C with 5% (v/v) CO_2.
5. Stereomicroscope with a sub-stage light reflector.
6. Oxidase reagent (BBL oxidase reagent droppers or equivalent).
7. Dead air box with a clean, decontaminated surface.
8. Sterile PBS, pH 7.4.
9. Sterile cotton-tipped swabs.
10. Sterile toothpick or wooden applicator stick.
11. Filter paper.
12. Genesys 30 spectrophotometer, or equivalent.
13. 50 mL sterile conical centrifuge tubes.
14. 2 mL sterile tubes.
15. FSM.
16. 2 mL sterile freezer storage vials.
17. LB agar plates.

2.4 Inoculation Procedure

1. Consented human male subjects (*see* **Note 3**).
2. Sterile gloves and waterproof plastic drapes.
3. 70% (v/v) isopropyl alcohol wipes, individually packaged.

4. Gas-sterilized No. 8 pediatric French catheter and hemostat clamp.
5. 0.9% (v/v) NaCl, sterile saline.
6. *Neisseria gonorrhoeae* suspension for inoculation.
7. Filtered pipette tip and P-1000 micropipettor (Pipetman or equivalent).
8. 50 mL sterile centrifuge tube.
9. Benchtop centrifuge.
10. GCB agar plates.

2.5 Monitoring Experimental Infection

1. Antibiotics.
2. Sterile urine specimen cups encircled with a marker at the 50 mL line.
3. 15 and 50 mL sterile centrifuge tubes for urine processing.
4. Sterile swabs with calcium alginate fiber tip for urethral swab collection.
5. GCB agar plates for bacterial recovery.
6. Assays for bacterial and/or host gene expression (e.g., transcriptomics, WGS).
7. Assays for host responses (e.g., for soluble immune mediators).

2.6 Statistical Considerations

1. One-sided Fisher's exact test.
2. Kaplan–Meier survival curves.
3. Competitive Indices (CI).

3 Methods

3.1 Neisseria gonorrhoeae Working Banks

3.1.1 Bacterial Strains and Media

1. Prepare lots of as many identical single-use vials for each *N. gonorrhoeae* strain as is practical for the intended study design. For example, working banks of wild-type gonococci for use in studies to compare experimental prevention or treatment methods, or as controls in isogenic mutant studies of gonococcal pathogenesis should contain ≥100 vials, to maximize comparability over time. Stocks of individual isogenic mutant strains for more limited use in individual pathogenesis studies may contain 20 or fewer vials. In addition, take into account the number of vials needed for characterization and stability testing, when determining the size of the working bank. Periodically take account of working bank inventory, and prepare a new working bank from previously characterized stock before it is depleted. All steps below should be done in a Biosafety 2 level cabinet to minimize the likelihood of contamination.

3.1.2 Preparation of Working Banks with a Single Strain

1. Use a sterile microbiological loop to transfer *N. gonorrhoeae* bacteria from a frozen stock vial of the desired strain and spread onto GCB agar plates in a pattern to yield isolated colonies. Incubate for ~24 h in a humidified atmosphere containing 5% (v/v) CO_2 at 37 °C.
2. Examine gonococcal growth through a stereo microscope with a sub-stage light reflector for the desired piliation and opacity characteristics. Historically, selected colonies are piliated (shiny, tightly domed colonies) and primarily opacity protein (Opa)-negative (transparent, as opposed to the sparkling, opaque quality of colonies formed by Opa+ gonococci).
3. Expand gonococcal cultures in a single passage from individual colonies. Pick 1–5 colonies, depending on the desired size of the working bank stock, and streak one single colony per GCB agar plate to maximize bacterial growth, of the previously characterized isolate.
4. Incubate for ~24 h in a humidified atmosphere containing 5% (v/v) CO_2 at 37 °C and visually inspect the plates carefully. Discard any plates with visible contaminating colonies.
5. Harvest 24 h bacterial growth from GCB plates containing only *N. gonorrhoeae* growth using sterile cotton-tipped swabs. Suspend material from all plates into FSM in a single, sterile tube (*see* **Note 4**). Use approximately 2 mL of FSM for each plate of *N. gonorrhoeae* to be harvested. Cap the tube and vortex-mix to create a uniformly dispersed suspension.
6. Aseptically transfer 0.1 mL of the gonococcal suspension to sterile freezer storage vials labeled clearly with the strain designation, date of preparation and a notation that the contents are for investigational use only. Store the vials at ≤−70 °C (*see* **Note 5**).
7. Confirm the purity of the frozen product by plating 20 μL from one vial onto GCB agar and streaking to obtain areas of heavy growth and areas of isolated colonies. Plate the remainder of the contents of the vial onto one LB agar plate. Incubate all agar plates for ~24 h in a humidified atmosphere containing 5% (v/v) CO_2 at 37 °C. If any bacterial growth is observed on LB agar, or if no growth is observed on GCB agar after incubation, discard the entire lot of vials. Submit bacterial growth on GCB agar to the clinical microbiology laboratory for biochemical identification as *N. gonorrhoeae* and antimicrobial susceptibility testing to confirm susceptibility to the antibiotics specified for treatment of experimental infection (*see* **Note 6**).
8. From another vial of a lot that passes the purity and identification tests, confirm the key bacterial characteristics of the frozen bacteria. Plate the contents of a test vial onto GCB to obtain isolated colonies, and repeat **step 3** to test single passage

growth as needed to meet the goals of the study. Typically variable characteristics such as piliation, Opa protein and LOS expression are confirmed for each lot (*see* **Note** 7). For *N. gonorrhoeae* strain FA1090, monoclonal antibodies (mAbs) that recognize the Opa proteins are available [6]. Use a cocktail of these mAbs in standard immunoblots of ~100 colonies to estimate the proportion of the population that is Opa-negative. Confirm piliation by quantitative transformation efficiencies of 10^{-4} or higher, and identify the specific *pilE* sequence by DNA sequencing. Confirm the expression of the lacto-N-neotetraose LOS epitope that can be sialylated by western blot with mAb 3F11 of proteinase K-treated bacterial lysate separated by SDS-PAGE [7] (*see* **Note 8**).

9. Purify DNA from gonococcal strains for Whole Genome Sequencing (WGS, *see* **Note 9**), which is strongly recommended for each lot of *N. gonorrhoeae* used in experimental human infection studies. Identical sequences for strains prepared after even minimal passage from previously characterized lots are unlikely, due to the plasticity of the gonococcal genome. However, WGS data allow comparisons between strains that may help explain unexpected experimental infection outcomes that may result from unanticipated differences not recognized by strain characterization as described above.
10. The concentration of *N. gonorrhoeae* in working stock vials stored at <−70 °C is not specified, but bacterial suspensions are quite dense. Bacteria grown on GCB agar from 20 μL of suspension should yield areas of heavy growth as well as areas of isolated colony growth consistent with typical microbiological streaking technique. Such gonococcal suspensions are stable for many years at −70 °C. To document the stability of stocks for clinical use in experimental human infection studies, enumerate the cfu/mL from a single vial every 18 months after initial preparation by streaking the contents on GCB agar plates and viable counting. If the concentration of *N. gonorrhoeae* in the vial is $<10^5$ cfu/mL, a new stock should be generated. Any remaining vials from a stock that fails stability testing should be removed from product storage location (*see* **Note 5**) and labeled not for human use, but may be used to generate a new working stock.
11. Determine the relationship between optical density of filtered bacterial suspension in PBS and cfu/mL for each working stock of *N. gonorrhoeae* prior to use in experimental human infection (*see* **Note 10**). From 18–24 h bacterial growth on GCB agar, harvest gonococci with a sterile cotton-tipped swab and prepare a suspension of by swirling the bacteria into 6 mL of sterile PBS in a sterile spectrophotometer tube (*see* **Note 4**). The amount of bacteria collected in three swift strokes is

usually sufficient. We measure absorbance at $\lambda_{550 \text{ nm}}$ (A_{550}) in a Genesys 30 spectrophotometer. Read and record absorbance of the initial, unfiltered suspension. Attach a sterile 1.2 μM filter to the barrel of a sterile 10 cc syringe from which the plunger has been removed (*see* **Note 11**). Carefully pour the gonococcal suspension from the tube into the syringe, insert the plunger and gently force the fluid gently through the filter, collecting the filtered suspension into a sterile tube. Read and record the absorbance of the filtered suspension. Prepare a series of dilutions in PBS for quantitative culture on GCB agar. Incubate the GCB agar plates in a humidified atmosphere with 5% (v/v) CO_2 at 37 °C and determine the cfu/mL in the filtered suspension by viable counting. Repeat with multiple solutions across a range of optical densities.

3.1.3 Preparation of Working Banks with Two Strains for Competitive Infections

Competitive infections with mixed gonococcal inocula allow identification of fitness differences between strains that may not be evident in noncompetitive infections with individual strains. Typically, mixed inocula contain approximately equal numbers of the two strains.

1. Prepare individual working stocks for each strain and characterize them as described in Subheading 3.1.1.
2. Plate material from a single vial from each frozen gonococcal stock onto GCB agar plates and incubate for ~24 h in a humidified atmosphere containing 5% (v/v) CO_2 at 37 °C.
3. Prepare a suspension of each individual strain with bacteria harvested from an agar plate with a sterile cotton swab, swirled into GCB broth (*see* **Note 4**). Estimate the bacterial concentration from the optical density of the suspension as described in **step 11** in Subheading 3.1.2.
4. To prepare the desired mixture, combine approximately equal concentrations of the two strains, add an equal volume of 2× FSM, mix thoroughly, and dispense aliquots into clearly labeled vials and freeze as described in **step 6** of Subheading 3.1.2 (*see* **Note 5**). Subculture the contents of one vial of the mixture onto GCB agar, plating a dilution to yield 100–300 cfu per plate, and enumerate the cfu of each strain. Individual strains may be identified by PCR with specific primers or by subculture onto differential, selective media to determine the proportion of each strain in the mixture. Aim for a 50:50 ratio of the strains and discard any mixtures exceeding a 45:55 ratio. Characterize the mixed stocks and do periodic stability testing as described above in **steps 8** and **10** of Subheading 3.1.2.

3.2 Catheter Preparation

1. Remove sterile No. 8 pediatric French feeding catheters from their original packaging, and cut to 8 cm in length using clean scissors or scalpel (sterility is not important at this stage as

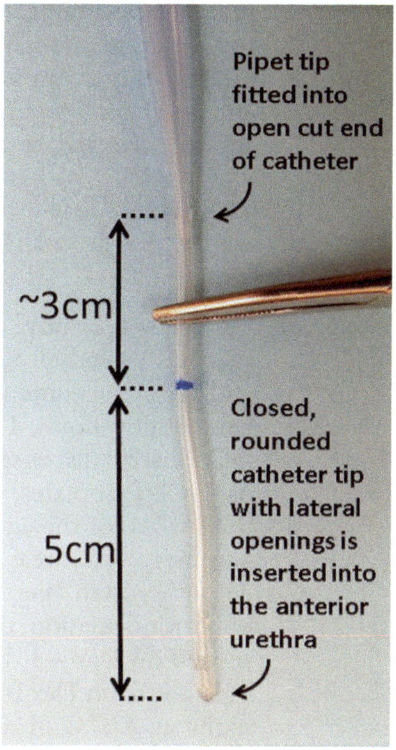

Fig. 2 Prepared catheter fitted with pipette tip shown with hemostat in the position as described after delivery of the inoculum suspension to prevent backflow

catheters will be resterilized before use). Mark the catheter with a permanent laboratory marker at the 5 cm mark; this will guide insertion to the proper depth into the anterior urethra. Encircle the outside of the catheter completely (Fig. 2).

2. Insert one catheter and a sterilization indicator strip into a self-sealing pouch. Ensure that the open end of the catheter is orientated toward the end of the pouch to be opened for use. Peel off the protective strip, fold along the perforation, and press firmly to seal the pouch. Prepare batches of 10–50 individually packaged catheters and transport them to a local facility for sterilization.
3. Low temperature ethylene oxide sterilization is important to minimize shrinkage of the plastic catheters, which can occur with steam sterilization and compromise the accuracy of the marked guide for insertion into the urethra. After sterilization, label catheters with the sterilization date and store at ambient temperature, protected from light. Catheters remain sterile indefinitely, unless the package is opened or damaged. For urethral inoculation, do not use catheters more than 1 year after the sterilization date. At least 48 h before use, inspect

pouches carefully. Use only catheters in undamaged pouches. If the outer pouch is opened, torn or otherwise damaged, the intact catheter can be transferred to a new pouch, resterilized and used for inoculation. Discard any catheter that is bent, twisted, crushed, or otherwise damaged.

3.3 Preparation of Inocula for Experimental Human Infection

In this protocol, Day 0 refers to the day of inoculation, Day −1 is the day before inoculation, and Days 1, 2, 3 etc. are days after inoculation.

1. On Day −1, remove a single vial of a gonococcal working stock that has passed all required lot-release tests from the freezer, and allow the contents to thaw at room temperature in a clean laminar flow hood. Thawing takes no more than a few minutes: immediately dispense the material aseptically onto five individual GCB agar plates, with approximately 20 μL per plate. Streak the surface of the agar with a sterile microbiological loop in a pattern to yield areas of dense growth as well as areas of isolated colony growth (Fig. 1). Bacterial growth should be 18–24 h old for inoculation; determine the time that cultures should be started on Day −1, based on the expected time of inoculation of subjects on Day 0. Incubate inverted GCB agar plates overnight at 37 °C in an humidified atmosphere containing 5% (v/v) CO_2.
2. On Day 0, 1–2 h before the expected inoculation time, thoroughly examine GCB agar plates by eye and with a microscope for purity and correct colony phenotype. Discard any plates containing contaminating bacteria. Mark areas of the plates containing gonococcal growth with undesirable phenotypes (for example Opa+ or nonpiliated morphology) by circling the area with a marker on the outside of the plate. Avoid these areas when preparing the inoculum suspension.
3. From an area with isolated colony growth, remove several colonies from each plate and test with oxidase reagent. First, saturate a piece of filter paper with oxidase reagent (BBL oxidase reagent dropper or equivalent). Using a sterile toothpick or wooden applicator stick (*see* **Note 12**), lift each isolated colony and streak it across the wet area of the filter paper. Bacterial material must test positive, that is, turn dark purple within 30 s, in order to use growth from that plate for the inoculum suspension. Return all plates passing preinoculation testing to the incubator until transport to laboratory at facility where subjects will be inoculated (*see* **Note 13**).
4. When all subjects in the cohort to be inoculated with a given gonococcal suspension have been prepared, as described in Subheading 3.4, remove the GCB agar plates containing 18–24 h bacterial growth from the incubator. It is not

necessary to work in a laminar flow ventilated hood at this stage, but it is helpful to work in a dead air box on a clean, decontaminated surface to minimize air flow and reduce the likelihood of contamination (*see* **Note 14**). Harvest the bacterial growth from the agar plates with a sterile cotton swab and prepare a suspension in 6 mL of sterile PBS in a sterile glass tube that will fit into the spectrophotometer chamber. Determine the optical density, adjust with more bacteria or more PBS as needed, and filter as described in **step 11** of Subheading 3.1.2.

5. Make a final dilution from the filtered suspension in approximately 10 mL of PBS in a sterile 50 mL conical centrifuge tube to achieve the desired inoculum concentration, based on previous work with the specific bacterial product lot. Immediately remove 1 mL from this tube to a sterile 2 mL tube. Securely cap the 50 mL tube containing the final inoculum suspension and assign a study team member to deliver it promptly to the physician who will inoculate the subjects as described in Subheading 3.4. Meanwhile, another team member should use the 1 mL of suspension that was transferred to the sterile 2 mL tube to prepare three labeled freezer vials with 0.1 mL of bacterial suspension added to 0.5 mL of FSM in each tube, and freeze immediately at ≤ -70 °C. This provides an archive of the inoculum suspension for future reference if needed.
6. Also from the 1 mL of inoculum suspension, plate 0.05–0.1 mL, undiluted onto LB agar to confirm purity. Furthermore, make serial dilutions of the suspension and plate onto GCB agar to determine the actual cfu/mL contained in the suspension by viable counting. Quantitative cultures should be plated before the first subject is inoculated and repeated with any remaining inoculum suspension that is returned to the lab after the last subject is inoculated, in order to document any potential loss of viability during the time it takes to complete the procedure.

3.4 Inoculation Procedure

1. Recline the subjects on a hospital bed, and ensure that they don sterile gloves immediately before inoculation. Cover the pelvic area with a waterproof plastic drape, and extend the penis through a hole in the drape and cleanse with 70% (v/v) isopropanol wipe. Remove a No. 8 pediatric French catheter, cut to 8 cm in length, from its sterile pouch, holding it between the cut opening and the 5 cm mark. Dip the closed end of the catheter into a container with sterile saline to moisten. This is inserted gently by the physician into the urethra until the 5 cm mark on the catheter is just visible at the opening of the urethra (*see* **Note 15** and Fig. 2).

2. Load 0.4 mL of gonococcal suspension prepared as described above in Subheading 3.3 into a filtered pipette tip on a Pipetman P-1000 micropipettor and immediately replace the cap on the tube. Fit the tip of the micropipettor into the open end of the end of the catheter that has already been inserted into the urethra; take care not to insert the catheter further than the 5 cm mark. Deliver the inoculum slowly into the catheter by depressing the plunger of the micropipettor. When the first stop of the micropipettor has been reached, clamp the catheter with a hemostat (Fig. 2), and then gently remove the entire assembly of pipette tip with catheter and hemostat still attached from the urethra in one smooth motion (*see* **Note 15**).
3. Instruct the subject to apply gentle pressure to the end of the penis and hold it upright for 30 min after inoculation to prevent leakage of the bacterial suspension. Collect the catheter, pipette tip, any liquid remaining in the catheter and any spilled drops of fluid that can be collected from the drape and deposit into an empty, sterile 50 mL centrifuge tube labeled with the subject's ID, and return these tubes with undelivered inoculum suspensions to the lab after all subjects have been inoculated.
4. Repeat **steps 1–3** for each subject in the cohort. After the last subject has been inoculated, return remaining inoculum suspension to the laboratory for postinoculation counts to be determined as described in **step 6** of Subheading 3.3.
5. Load the 50 mL tubes containing pipette tips, catheters and undelivered fluid into a benchtop centrifuge and spin briefly to collect all the fluid in the bottom of the tube. Remove the pipette tip and catheter, and measure the volume in the tube. Subtract this volume from the original 0.4 mL volume loaded into the pipette to determine the actual volume delivered to each subject.
6. Inoculate GCB agar plates with serial dilutions of preinoculation and postinoculation cultures made in sterile PBS and incubate for 24–48 h at 37 °C in a humidified atmosphere with 5% (v/v) CO_2. Count colonies and calculate the average of the preinoculation and postinoculation counts, and adjust for the volume delivered to determine the bacterial dose delivered to each subject in the cohort, that is, cfu delivered = volume delivered (mL) × cfu/mL in inoculum suspension.

3.5 Monitoring Experimental Infection

1. Physicians will perform targeted genital exams daily after inoculation and query subjects to determine signs and symptoms of infection. For pathogenesis studies, in which the timing and features of experimental infection with different strains of *N. gonorrhoeae* are being compared, subjects are housed as

inpatients in a clinical research unit to facilitate consistent examination and specimen collection first thing every morning for up to 5 days after inoculation (*see* **Note 16**). For studies designed to test preventive treatments, in which infectivity or time to clinical disease is the main outcome, subjects may be managed as out-patients, with daily return visits to the clinical research unit scheduled as early in the day as is practical for examinations and specimen collection.

2. Administer prompt and effective antibiotic treatment to subjects when one of the following occurs: (1) subject requests treatment, (2) urethral discharge is observed by clinician or reported by subject, or (3) 5 days after inoculation, regardless of whether or not the subject was judged to have been infected. For experimental infection with *N. gonorrhoeae* FA1090, treatment is a single observed 400 mg oral dose of cefixime. If a subject cannot swallow the oral pill, treatment is a single 250 mg intramuscular dose of ceftriaxone.
3. Monitor experimental gonococcal infection by bacterial recovery from urogenital specimens including urine and urethral swabs; bacteria may also be recovered from semen if the study design dictates. After the genital exam and before antibiotic treatment, subjects in inpatient studies provide up to 50 mL of first-void urine each morning. For outpatient studies, subjects provide up to 50 mL of first-catch urine at daily clinic visits. On the day on which a subject is treated, a urethral swab is obtained. Urine and urethral swabs can be processed for gonococcal culture on GCB agar and bacterial and/or host gene expression as dictated by the study design (*see* **Notes 17–19**).
4. Host responses can be monitored by enumeration of white blood cells in urine and measurement of soluble immune mediators including antibodies and cytokines in urine and serum (*see* **Notes 17–19**).

3.6 Statistical Considerations

1. Noncompetitive infections. For experimental infections initiated with inocula containing an individual strain, the proportion of subjects that become infected is the measure of infectivity. In pathogenesis studies, the hypothesis is frequently whether the wild-type strain is more infectious than an isogenic mutant that lacks a predicted essential virulence factor. Sample size considerations for such noncompetitive infections are summarized in Table 1. Typically, we enroll subjects in groups of up to 4, which is the maximum number of inpatient rooms in our clinical research facility. Assuming the true probabilities of infection are 80% with wild-type and 4% with mutant inocula, eight men per group provides 89% power to detect a difference using a one-sided Fisher exact test and significance level of 0.025.

Table 1
Sample size considerations for noncompetitive infection studies

Infection status					
Wild-type		Mutant			
Yes	No	Yes	No	*P*-value[a]	Number of subjects needed for analysis
8	0	0	4	0.002	12
		1	3	0.018	
		2	2	0.091	
		2	6	0.003	16
		3	5	0.013	
		4	4	0.038	
7	1	0	4	0.010	12
		1	3	0.066	
		1	7	0.005	16
		2	6	0.020	
		3	5	0.059	
6	2	0	4	0.030	
		0	8	0.003	16
		1	7	0.020	
		2	6	0.066	
5	3	0	4	0.071	
		0	8	0.012	16
		1	7	0.059	

[a]*P*-value is from the one-sided Fisher's exact test with $\alpha = 0.025$

Compare infectivity of wild-type *N. gonorrhoeae* and an isogenic mutant using Fisher's exact test. An interim analysis may include results from up to eight subjects inoculated with the wild-type *N. gonorrhoeae* strain only and at least four subjects inoculated with the isogenic mutant only. With a clearly directional hypothesis that a mutant is less infectious than the wild-type strain, one-sided statistics are appropriate. Final analysis of wild-type vs. isogenic mutant noncompetitive infections may include comparisons of Kaplan–Meier survival curves for the two strains.

2. Competitive infections. For isogenic mutants that do not demonstrate statistically significant differences in infectivity, competitive infection studies are useful to determine whether the mutant exhibits a fitness defect compared to the wild-type strain. Competitive infections, initiated with equivalent numbers of wild-type bacteria and isogenic mutant bacteria, allow demonstration of a difference in fitness between two strains using fewer subjects. For competitive infections with mixed inocula, sample sizes are determined by the expected standard deviation of the mean proportion of strains among gonococci

recovered from infected subjects. Because calculations are based on the numbers of bacteria in each cultured specimen, constituting larger sample sizes than infectious dose calculations based on the number of infected subjects, mixed infections can provide a more accurate indication of attenuation using fewer subjects than traditional comparisons of infectivity using separate, inocula with individual strains [8].

For competitive infections initiated with inocula containing a mixture of wild-type and isogenic mutant strains, the competitive index (CI) of the mutant compared to wild-type is the measure of relative fitness. The precise proportion of each strain in the inoculum is determined for each experimental cohort as described in **step 6** of Subheading 3.4. The CI is defined as the ratio of colony-forming units (cfu) of the two strains recovered from urine cultures on the day of treatment (output) compared to the ratio of strains in the inoculum (input) using the equation: CI = mutant cfu(output)/wild-type cfu(output) ÷ mutant cfu(input)/wild-type cfu(input). If the two strains have equal fitness in experimental human urethral infection, the CI will be close to 1. If the wild-type strain has a fitness advantage, the CI < 1, and if the mutant has a fitness advantage, the CI > 1.

4 Notes

1. Maintain certificates of origin and analysis to document use of BSE-free herds for all ingredients made from animal sources that are used to prepare and store bacteria for use in experimental human infection.
2. *N. gonorrhoeae* strains should be grown on GCB agar without added inhibitors for working stock preparation to allow visualization of potential contaminants.
3. Eligible subjects meeting all inclusion criteria and none of the exclusion criteria provide written informed consent after all study procedures are explained in detail and they have had the opportunity to ask questions. To date, we have included only subjects with no reported history of *N. gonorrhoeae* (and other sexually transmitted infections) to minimize the impact of immune responses to past pathogen exposure on experimental outcomes. Other study designs may require different inclusion/exclusion criteria (Table 2).
4. Take care not to gouge the surface of the agar when harvesting bacteria from the plates. Vigorously swirl swabs in liquid for several seconds to dislodge bacteria, and press swabs firmly against the side of the tube to wring out as much liquid as possible before discarding swabs.

Table 2
Inclusion and exclusion criteria for human challenge experiment

Inclusion criteria	• Healthy man ≥18 and <36 years old • Able and willing to be located easily by providing street address and telephone number (land line and/or cell phone number) • Willingness to provide written informed consent • Able and willing to attend all study visits • Able and willing to abstain from all sexual activity until completion of study and follow-up test for gonorrhea is negative • Acceptable medical history by screening evaluation • No clinically significant abnormalities on physical exam • Urinalysis: leukocyte esterase value and white blood cell count within normal limits • Total complement (CH50) value within normal limits • Urine negative for chlamydia, gonorrhea, and Trichomonas • Negative HIV, syphilis, and HCV serologic test results • Negative HBV core and surface antibodies or results consistent with immunization (negative HBV core antibody/positive HBV surface antibody) • Denies history of sexually transmitted infections including gonorrhea, chlamydia, syphilis, HIV, HBV, and HCV • Denies history of bleeding diathesis • Denies history of seizures (due to reports of seizures with ciprofloxacin) • Denies history of cancer, except basal cell carcinoma of the skin >5 years ago • Denies history of drug abuse • Denies history of genitourinary surgery
Exclusion criteria	• Student or employee under the direct supervision of any of the study investigators • Any known immunodeficiencies including complement deficiency, antibody deficiency, chronic granulomatous disease, or HIV infection • Psychiatric disorders that could interfere with the integrity of the data or subject safety • Unstable depression or depression that could compromise the subject's ability to comply with protocol requirements • Heart murmur or heart disease • Anatomic abnormality of the urinary tract • Any antibiotic treatment in the past 30 days, or azithromycin in the past 60 days • Self-reported chemotherapy within the past year • Current steroid use, except for topical application • Allergy to penicillin, cephalosporins or ciprofloxacin or to lidocaine • Treatment with medications in the previous month that are contraindicated with cefixime, ceftriaxone or ciprofloxacin and that cannot be withheld for the doses given in the study • Serum creatinine level <0.7 or >1.75 mg/dL • Serum alanine aminotransferase (ALT) level <lower limit of normal or >105 U/L • White blood cell count <2.5 or $>15.0 \times 10^9$/L • Absolute neutrophil count (ANC) <1.5 or $>7.5 \times 10^9$/L and deemed clinically significant by the study physician. For African Americans, ANC values as low as 1.3×10^9/L are allowed • Hemoglobin level <12.0 g/dL or above upper limit of normal • Urinalysis: Qualitative protein level >1+ or red blood cell count >10/high powered field

5. Several lots may be created at one time to maximize the likelihood that at least one will pass lot release testing. Clearly identify each lot prepared from an independent bacterial suspension, and promptly discard any lot that fails quality control testing. After quality control testing, store selected lot(s) in a separate location in the freezer where only product that has passed all lot-release tests is kept, to avoid inadvertent use of untested material for experimental infections.
6. *N. gonorrhoeae* strain FA1090 is naturally resistant to streptomycin and susceptible to chloramphenicol, cefixime, ceftriaxone and ciprofloxacin. During isogenic mutant constructions, intermediate strains that are susceptible to streptomycin are created [9]. To document the antimicrobial susceptibility pattern in final constructs, we confirm resistance to streptomycin and susceptibility to the other four antibiotics listed for each lot of bacteria prepared for experimental human use. Other *N. gonorrhoeae* strains may require different antibiotic susceptibility testing.
7. All FA1090 stocks characterized to date have been predominantly Opa-negative, piliated, that is, expressing the PilE sequence reported by Seifert et al. [10] and expressing the 3F11 LOS epitope to maintain comparability with our previous experimental infection studies. Other phenotypic characteristics may be desired and selected based on specific goals of the research.
8. These characterization methods are not produced in detail here but can be found in the accompanying references.
9. Genomic DNA for WGS has successfully been prepared with the ArchivePure DNA purification kit Protocol 18 (5-PRIME, USA) as instructed by the manufacturer with the following modifications: 2.2 μL of RNase A solution; 2.5 μL of proteinase K mixed thoroughly by inverting the tube at least 25 times and incubated for 40 min at 37 °C; centrifugation time increased to 6 min for all steps, as previously described [11].
10. Do not assume that the relationship between optical density of filtered bacterial suspension in PBS and cfu/mL is the same for different strains or for different lots of a given strain of *N. gonorrhoeae*; this must be determined empirically for each lot of each strain. There may not be a strictly linear relationship between optical density and viable counts over a broad range of absorbance values; the goal is to establish an absorbance range for filtered suspensions that corresponds to a known range of concentrations, which will enable adjustment of the inoculum for the dose required just before inoculation of human subjects. Dose response data from experimental human infection studies over many years indicate that the ID_{50} is approximately

10^5 cfu for *N. gonorrhoeae* strain FA1090 and approximately 10^3 cfu for strain MS11mkC [5]. At present, there are insufficient data from experimental human infections to estimate infectious doses of other *N. gonorrhoeae* strains.

11. Filtration through 1.2 μM pores breaks up bacterial clumps typical of piliated gonococci without shearing pili and provides a more uniform suspension. Take care not to touch anything with the rubber tip of the plunger after it has been removed from the syringe. Work quickly with gonococcal suspensions in PBS, as viability begins to decrease within 30 min in the buffer.
12. Use only sterile wooden implements to pick colonies for oxidase testing. Do not use a sterilized metal microbiological loop, as this will inhibit the oxidase reaction.
13. Transport agar plates containing *N. gonorrhoeae* (either working bank cultures for inoculation or clinical specimens plated for gonococcal recovery) between laboratories in candle jars. We use metal canisters with tight-fitting lids. Load plates into the canister, place a votive candle on top of the stack of plates, light the candle and fit the lid onto the canister. The flame will consume oxygen until it is extinguished, approximating a 5% (v/v) CO_2 atmosphere inside the canister. Typically, a 37 °C incubator without CO_2 is used in the laboratory located at or near the clinical research unit; store plates in candle jar inside incubator until transport to the research laboratory. Transport canister(s) upright at ambient temperature. Remove plates from the candle jar and transfer to 37 °C incubator with humidified atmosphere containing 5% (v/v) CO_2 in the research laboratory as soon as possible.
14. A dead air box provides a circulation-free workspace in the laboratory near the clinical research unit. Select a box that will accommodate the Genesys 30 Spectrophotometer to be used to prepare gonococcal inoculum suspensions with sufficient working space. Before and after use, wipe down work surfaces with 70% (v/v) ethanol and allow them to dry.
15. The inoculation process runs smoothly with a team of three people: the physician, who will place the catheter and deliver the inoculum, should don sterile gloves and avoid touching nonsterile items until the loaded pipettor is presented by an assistant wearing nonsterile laboratory gloves. Before inoculation, the assistant should open a packet containing an isopropanol wipe to be used to cleanse the penis and present it to the physician, open the package containing the sterile catheter so that the physician can remove it, open a container of sterile saline so that the physician can moisten the catheter, load the pipettor with the inoculum suspension and hand it to the physician, who will have just inserted the catheter into the

anterior urethra. Once the physician has dispensed the inoculum, and before the catheter is removed from the urethra, the assistant should clamp the catheter between the urethra and the pipette tip to prevent backflow (Fig. 2). Then, in a unified movement, the physician and assistant should remove the pipettor–catheter–hemostat assembly, collecting everything in a labeled empty tube. The physician and assistant should practice all the steps before actual inoculation. The third team member should help the subject don his sterile gloves, reassure him during the procedure, note the time of inoculation and instruct the subject to apply pressure to the penis for 30 min.

16. For inpatient studies, subjects may leave the clinical research unit during the day to attend classes or go to work; they must return to the clinical research unit by 9 pm and sleep in the unit. Between 6:30 and 7:00 am, subjects are awakened, examined, and interviewed for symptoms before specimens are obtained. If none of the outcomes triggering antibiotic treatment has occurred, normally only urine is obtained. If a subject reports symptoms or the clinician observes urethral discharge and treatment is warranted, a urethral swab is obtained before the urine specimen. It is critical that specimen processing and plating for gonococcal cultures occur with minimal delay; lab space near the inpatient clinical research unit is ideal. Once plates have been inoculated, they can be transported in candle jars from the processing lab to a more distant research lab for incubation at 37 °C with 5% (v/v) CO_2. If prolonged storage in the processing lab is needed, incubate candle jars at 37 °C if a CO_2 incubator is not available.

17. Do not refrigerate urine before processing. From 50 mL of urine, decant 8–10 mL into a 15 mL centrifuge tube for white cell counts. Spin the remaining urine promptly at approximately 1500 × *g* at room temperature for 10 min. Save urine supernatant as needed, and process urine pellet for quantitative culture, plating dilutions on GCB agar with appropriate antibiotics (e.g., vancomycin, colistin, nystatin, and trimethoprim) to reduce growth of potential contaminants. Spin 15 mL tubes at approximately 400 × *g* at room temperature for 10 min, decant or save supernatant as needed, gently suspend the pellet in residual urine and count white blood cells (predominantly PMNs) in a hemacytometer. Take care to label all tubes clearly and record data for bacteriuria and pyuria.

18. To process urethral swabs for gonococcal culture, place the swab into a small volume (0.2–0.5 mL) of FSM in a sterile vial; cut the wire shaft, cap the tube, and vortex vigorously to suspend material in the medium. This suspension may be plated undiluted on GCB agar to determine whether the

specimen is positive or negative for *N. gonorrhoeae*, or serial dilutions may be plated for enumeration of bacterial burden and recovery of individual colonies. We typically plate undiluted swab suspensions to determine positive culture status. Occasionally, urethral swab cultures are positive when urine cultures are negative. A subject is considered infected if urine or urethral cultures are positive on the day of antibiotic treatment. Urethral swab suspensions remaining after aliquots are removed for culture can be stored in FSM at −70 °C for future recovery and characterization of viable *N. gonorrhoeae* or further processed for analysis of bacterial or host gene expression.

19. Detailed procedures for processing urogenital specimens or blood for RNAseq analyses of bacterial and host gene expression during experimental infection are not well established and are beyond the scope of this chapter.

References

1. Jerse AE, Deal CD (2013) Vaccine research for gonococcal infections: where are we? Sex Transm Infect 89(Suppl 4):iv63–iv68
2. Jerse AE, Wu H, Packiam M et al (2011) Estradiol-treated female mice as surrogate hosts for *Neisseria gonorrhoeae* genital tract infections. Front Microbiol 2:107
3. Li G, Jiao H, Yan H et al (2011) Establishment of a human CEACAM1 transgenic mouse model for the study of gonococcal infections. J Microbiol Methods 87:350–354
4. Zarantonelli ML, Szatanik M, Giorgini D et al (2007) Transgenic mice expressing human transferrin as a model for meningococcal infection. Infect Immun 75:5609–5614
5. Hobbs MM, Sparling PF, Cohen MS et al (2011) Experimental gonococcal infection in male volunteers: cumulative experience with *Neisseria gonorrhoeae* strains FA1090 and MS11mkC. Front Microbiol 2:123
6. Barritt D, Schwalbe R, Klapper D et al (1987) Antigenic and structural differences among six proteins II expressed by a single strain of *Neisseria gonorrhoeae*. Infect Immun 55:2026–2031
7. Schneider H, Griffiss JM, Boslego JW et al (1991) Expression of paragloboside-like lipooligosaccharides may be a necessary component of gonococcal pathogenesis in men. J Exp Med 174:1601–1605
8. Beuzon CR, Holden DW (2001) Use of mixed infection with *Salmonella* strains to study virulence genes and their interactions *in vivo*. Microbes Infect 3:1345–1352
9. Johnston DM, Cannon JG (1999) Construction of mutant strains of *Neisseria gonorrhoeae* lacking new antibiotic resistance markers using a two gene cassette with positive and negative selection. Gene 236:179–184
10. Seifert HS, Wright CJ, Jerse AE et al (1994) Multiple gonococcal pilin antigenic variants are produced during experimental human infections. J Clin Invest 93:2744–2749
11. Costa-Lourenco A, Abrams AJ, Dos Santos KTB et al (2018) Phylogeny and antimicrobial resistance in *Neisseria gonorrhoeae* isolates from Rio de Janeiro, Brazil. Infect Genet Evol 58:157–163

Index

A

B

C

Myron Christodoulides (ed.), *Neisseria gonorrhoeae: Methods and Protocols*, Methods in Molecular Biology, vol. 1997, https://doi.org/10.1007/978-1-4939-9496-0, © Springer Science+Business Media, LLC, part of Springer Nature 2019

V

W

MIX
Papier aus verantwortungsvollen Quellen
Paper from responsible sources
FSC® C105338
FSC
www.fsc.org

If you have any concerns about our products,
you can contact us on
ProductSafety@springernature.com

In case Publisher is established outside the EU,
the EU authorized representative is:
Springer Nature Customer Service Center GmbH
Europaplatz 3, 69115 Heidelberg, Germany

Printed by Libri Plureos GmbH
in Hamburg, Germany